AF239589

FARM, HUNT, FEAST, CELEBRATE

FARM, HUNT, FEAST, CELEBRATE

ANIMALS AND SOCIETY IN
NEOLITHIC, BRONZE AND IRON AGE
NORTHERN FRANCE

GINETTE AUXIETTE & LAMYS HACHEM

© 2021 Individual authors

Published by Sidestone Press, Leiden
www.sidestone.com

Lay-out & cover design: Sidestone Press

Cover design: Auxiette G. and Hachem L.

ISBN 978-94-6426-021-2 (softcover)
ISBN 978-94-6426-022-9 (hardcover)
ISBN 978-94-6426-023-6 (PDF e-book)

Contents

ACKNOWLEDGEMENTS

We extend our sincerest thanks to all of the members of our research team, Trajectoires (UMR 8215 CNRS formerly URA 12), who have placed their confidence in us for over thirty years; to some we owe our training on the archaeological excavations in the Aisne valley, to others, invaluable exchanges of ideas and a flow of fruitful discussions on various topics that have enriched this publication. We also wish to thank the archaeologists working for the INRAP (*Institut National de Recherches Archéologiques Préventives*, formerly the AFAN), the University of Paris I, Panthéon-Sorbonne, the CNRS (*Centre National de la Recherche Scientifique*) and the *Ministère de la Culture* who provided us with faunal collections, images and site plans. This publication would not have been possible without the financial support of the INRAP, the CNRS the University Paris I Panthéon-Sorbonne, the ASAVA (*Association pour le Sauvetage Archéologique de la Vallée de l'Aisne*) and the Conseil Général de l'Aisne. Finally, we wish to thank Rhoda Cronin-Allanic who agreed to translate this work, and Michael Ilett who reread the text in its entirety. This book represents the sum total of our combined knowledge gained over the course of our zooarchaeological research and from our field experience, on the one hand regarding the Neolithic period, and on the other regarding The Bronze and Iron Ages.

PREFACE

In a few short years our view of animals has changed. Previously, they were divided up into three categories: domestic animals, there to be eaten; wild animals, rather dangerous and confined to areas that are themselves wild if not distant; and decorative pet animals. Today, we are dealing with the sixth mass extinction of species, to the extent that there are hardly any large wild mammals left, apart from the ones penned, counted, cared for and even "culled" in the so-called "natural" reserves. Furthermore, we have come to realise that the consumption of meat by an exponentially growing human population poses serious ecological problems in terms of land and access to water. Lastly, concern for what is termed "animal welfare", which has long been anecdotal, is now a growing issue, at least in the western world.

So it is in this context that this major work by two experienced zooarchaeologists, Ginette Auxiette and Lamys Hachem, must be seen. Over and above the archaeological scholarship, five millennia of tangible relationships between animals and human societies in the Paris Basin have been rigorously observed and measured. These five millennia span from the arrival of the first sedentary farmers towards the end of the sixth millennium, to the emergence of the first Celtic states at the end of the first millennium BC. It is quite exceptional to have such continuous and detailed records for a single extensive geographical area; this is largely the fruit of half a century of rescue excavations. In fact, the whole Paris Basin region played a pioneering role in the admittedly late development of preventive archaeology in France. The systematic monitoring of sand and gravel quarries in the Aisne valley, as well as the valleys of the Oise, Marne, Yonne and Seine, enabled archaeologists to save evidence of long-term human occupation from irreversible destruction. Since then preventive archaeology has been extended to other forms of development, although even today not all are exhaustively investigated.

As is shown here by zooarchaeology, domestication clearly caused transformations not only in the physiology and behaviour of animals, but also in human societies, in their diet as well as in their symbolic behaviour, because since the dawn of time societies have viewed themselves through animals, from the walls of Palaeolithic caves, the Easter lamb, the four evangelists, heraldry with its lions and eagles, to the Fables of La Fontaine. Thus, far from being a straightforward process involving the identification and counting of animal bones, zooarchaeology, which for a long time was even neglected on many archaeological excavations, including the most prestigious, has become a comprehensive discipline capable of informing us about seasonality, slaughter ages, the effects of work on the joints of draft or pack animals, the practice of castration, the consumption of dairy products, herd size, the extent of grazing land, the ways of butchering meat, the exploitation of wool, skins and fur, etc. To this we can add ethnographic comparisons which can often be very relevant and enlightening.

The principal interest of a regional study with such a long time-range is that it highlights broad trends and developments. One of the effects of domestication, which removes animals from their natural surroundings, is that the large mammals become smaller over time, to the extent that Iron Age cattle are about half the size of their early Neolithic counterparts. Likewise, in the relative frequencies of animals eaten, we see a gradual decrease in cattle and an increase in pigs, which are more closely linked to sedentism and the production of large quantities of refuse and which, in terms of actuel numbers, make up about two thirds of the livestock herd at the end of the Iron Age. This period also sees standardized and centralized butchery of cattle in the *oppida* sites. Sheep and goats, which predominate in the Near East and the Balkans, occupy a relatively marginal position and, despite a slight rise in the Late Bronze/Early Iron

Age, are only represented by a single animal per household at the end of the period. Also, Bronze Age societies appear to have eaten much less meat. There is however evidence for dairy products, although their appearance has sometimes been assigned to a late date because of the absence of lactase in adults, necessary for digestion of milk. Lastly, hunting, still important earlier on in the Neolithic, gradually declined as natural spaces became more limited, to end up almost exclusively as a leisure and prestige activity, although the symbolism of the opposition between wild and domestic remained significant.

Beyond food in the strict sense, this study provides innovative insights into the wider functioning of society. Hence, analysis of meat consumption in the houses of an LBK hamlet reveals clear dietary differences and preferences linked to both house size and location within the settlement. Later, the existence of large ceremonial banquets is evident from the Middle Neolithic onwards and continues into the Bronze Age and Iron Age.

Finally, the symbolic aspect of animals is a constant and takes multiple forms; the role of the aurochs and the domestic bull, the largest animals in this westernmost region of Europe, continues a tradition that first emerged in the Near East. The deposition of sacrificed animals, sometimes mixed with human remains, is another constant and evokes the totemism that exists in many traditional societies. In numerous instances, particularly in funerary contexts, wild boar or pig are a symbol of masculinity, while red deer, roe deer and even sheep are feminine symbols.

Without revealing more of the content of this fascinating book, it is worth emphasising that in recent decades archaeology in general, and zooarchaeology in this particular case, have made spectacular contributions to our understanding of pre-literate societies, touching on realms that range from the most material to the most intangible.

Jean-Paul Demoule

"Qui ne sait l'histoire que par les imprimés du temps en conçoit à peine le squelette."
Charles Pinot Duclos, Morceaux choisis (1810).

INTRODUCTION

Multiple studies have been conducted on the importance of food within the lives of populations who, at a very early date, integrated animals into their diet and also into the cults and rituals that punctuate the lives of human beings. From hunter-gatherers to confirmed agro-pastoralists, who invested in increasingly complex modes of production and control over resources, faunal remains act as witnesses that allow us to find answers to fundamental questions regarding the place of domestic and collective meat consumption and cultural manifestations in the daily lives of our ancestors.

While the historian explores written archives, archaeologists, or more precisely zooarchaeologists in our case, decipher remains that are mainly found by sifting through the rubbish tips of our ancestors. In this way, zooarchaeologists try to discover what prompted human beings to rear animals while at the same time exploring, interactions between social entities, and the place of each people and communities within societies that were becoming more and more hierarchized and within which they played a role and occupied a well-defined place.

It is a difficult task to try and ascertain the original intention through the lens of hundreds of thousands of bone fragments: why and how were animals reared, consumed, transformed and revered?

The corpus at our disposal, which is continually growing thanks primarily to rescue excavations, allows us to propose orientations, choices and prevailing aspects of the societies in question.

Taken individually, these pieces of evidence are certainly interesting but what they can tell us is limited. Taken together, however, the large number of studies covering territories, or parts thereof, and various cultural entities allow us to identify specificities, at various scales, and to examine in detail the evolution of practices over the course of centuries.

An approach limited to describing the general trends observable in the available data can mask the regional and local particularities, as well as the chronological and status-based variations that we describe throughout this work. This is why we have decided to adopt a rather different approach, focussed on interpreting the evidence in social and cultural terms. Our approach is based on the processing of a very large amount of data, from sources as varied as settlements, assembly places, cemeteries and other distinctive sites. We have organised our presentation according to the nature of the sites and the various cultures that existed in the northern half of France over the last five millennia BC. Intra-site and inter-site spatial analysis of a portion of the data has also been one of the keys to gaining certain levels of understanding and interpretation of the societies in question. All of these analyses have allowed us to propose a social reading of the populations concerned.

By virtue of the fact that most of this data derives from consumption refuse, it provides essential information to help us understand the social relationships established and maintained by the members of a given community, as well as their relationships with animals.

The application of archaeozoological analyses to later prehistoric societies in Europe provides us with various degrees of insight into these relationships, be they economic, social, political, status-related, or spiritual. The practices identified vary greatly over time and space. This has already been demonstrated in articles and other publications, providing examples from different countries and societies (deFrance, 2009; Gumerman, 1997; Twiss, 2008).

In order to understand the relationships that unite food and culture, it is essential to consider the nature of the meals (domestic, feasting), their structure and the array of artefacts that are associated with them. In the words of Powers and Powers, cited in Gumerman,

"Feasts may in fact satisfy hunger, but they are seen as having some intrinsic social value which transcends the nutritive function of eating. Feasts have social goals achieved by cultural means." (Gumerman, 1997, 83).

This is what we seek to demonstrate in this book, by examining the societies of Northern France from the first sedentary groups to the more evolved societies. In the earliest periods, societies were probably organized in clans and lineages (Godelier, 2013), before evolving into increasingly complex chiefdoms, and eventually into regional political entities (Brun and Ruby, 2008). It has enabled us to seize the complexity of their actions and beliefs (from the material to the ideal) over a period of 5000 years (fig.1).

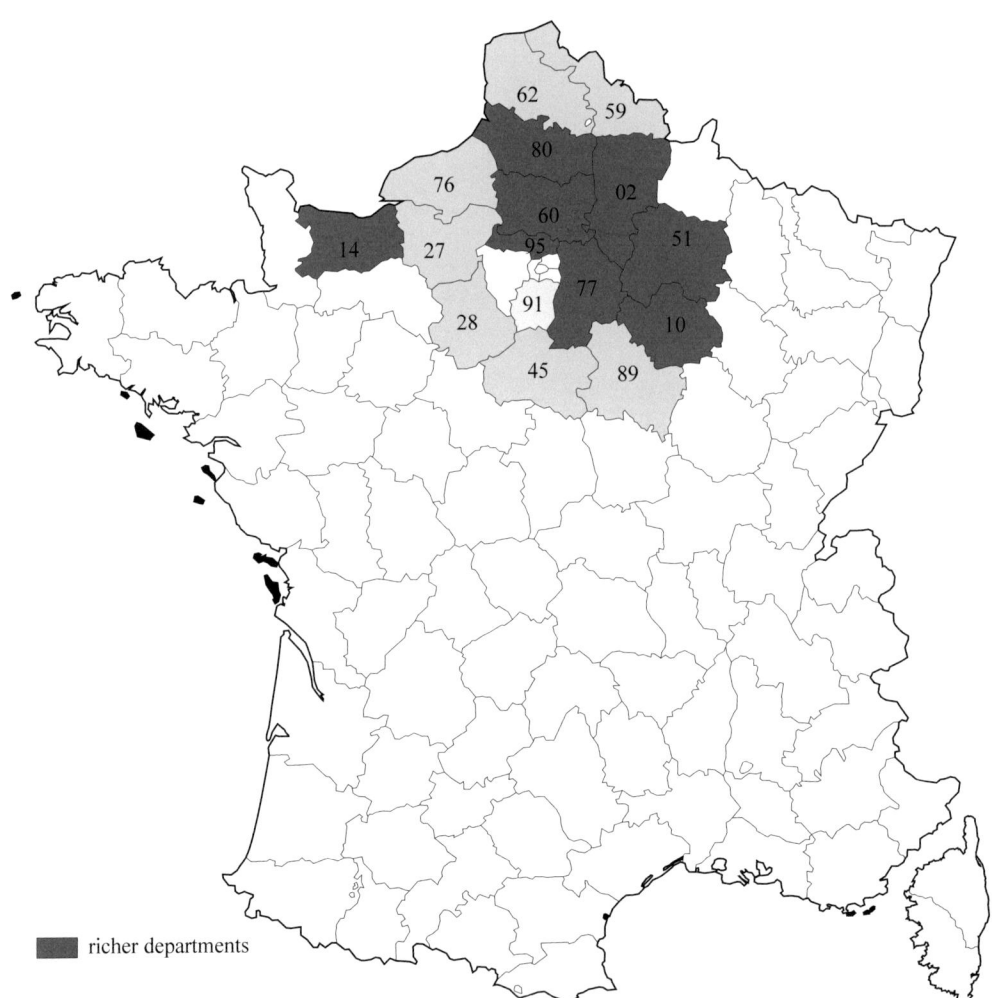

Figure 1: France, main geographical areas covered by the study.

richer departments

1. CHRONO-CULTURAL BACKGROUND

The Neolithic

The first appearance of the Linear Pottery Culture (LBK) in the Paris basin, and the pace of its subsequent development, can be reliably dated by comparing the pottery styles with the longer sequences established for Alsace and Lorraine (Lefranc, 2007; Blouet *et al.*, 2013) On these grounds, settlement of the Paris basin most likely starts in the middle LBK, around 5200 BC (fig. 2). The middle LBK in the Paris basin is nevertheless difficult to evaluate, due to limited evidence for house-plans and associated finds. This is also the case with the next stage, the Late LBK. Reflecting the spread of groups westwards from the Rhine and the Moselle river basins, settlement in the middle and Late LBK is restricted to the south-eastern part of the Paris basin, as attested by a small number of sites on the upper Marne and Seine rivers in the Champagne region (Tappret and Villes, 1996; Laurelut, 2010).

The latest or final stage of the LBK period is better documented and covers a more extensive area. Formerly refered to as the Rubané Récent du Bassin Parisien (RRBP), it has been re-named Rubané Final du Bassin de la Seine (RFBS) (Ilett and Meunier, 2013). This stage dates approximately to 5100-5000 BC (Dubouloz, 2003a). Out of a total of about 50 LBK sites discovered in the Paris basin, a large majority date to the Final LBK. The ceramic sequence for the Final LBK can be divided into three main phases (Constantin and Ilett, 1997; Meunier, 2012).

The Final LBK sees a major shift of settlement out of the Champagne region, to the north and west (Ilett, 2010). First the Aisne and Yonne valleys are settled, together with the Seine valley a short distance downstream from the confluence with the river Aube (Dubouloz *et al.*, 2005; Ilett, 2012; Meunier, 2012). By the end of the Final LBK, settlement has spread further westwards, into the Seine-Yonne confluence zone, as well as the middle Oise and lower Marne valleys. Two sites have been discovered in Normandy near Caen, although these remain outliers to the main distribution area (Billard *et al.*, 2014).

The Final LBK is replaced in the Paris basin by the Blicquy/ Villeneuve-Saint-Germain (BVSG) (Constantin, 1985; Meunier, 2012). This new cultural entity, which is in many respects closely related to the Final LBK, starts around 5000 BC and lasts until about 4700 BC), (Dubouloz, 2003a). The BVSG ceramic sequence has been divided into four principal stages (Constantin *et al.*, 1995; Lanchon, 2008; Meunier, 2012). Many excavated BVSG sites are attributed to the two final stages of the sequence (Lanchon, 2012). To date, well over 150 BVSG sites have been discovered in the Paris basin. They are not only found in most of the areas previously occupied by the LBK, but also occur widely in the western part of the Paris basin, with outliers as far as Brittany. The BVSG thus represents a significant increase in the geographical extent of early farming settlement at this time (fig. 3).

Faunal studies for the Early Neolithic are most numerous for two regions – the Aisne valley for the LBK and the Marne valley for the BVSG – both of which have a long history of archaeological research (fig. 3). The archaeozoological studies that we have undertaken are based on an approach that associates bone remains with their archaeo-logical context, in this case mainly houses; it also takes account of other studies (ceramic, lithic...).

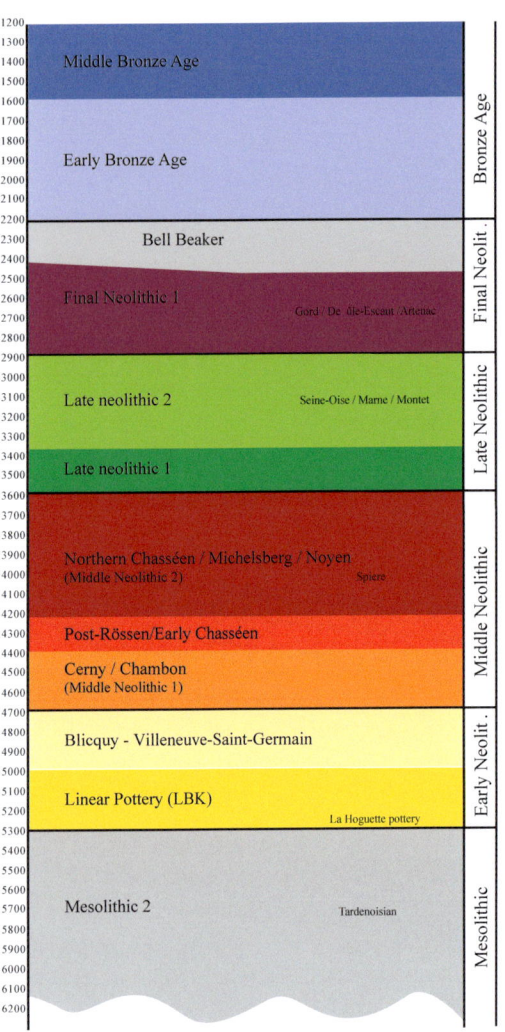

Figure 2: Chrono-cultural framework of the Neolithic period in the Paris Basin (chart: M. Ilett).

The main river valleys, which acted as vital routes for communication and exchange towards the west, were the sites of choice for LBK settlements. It was within this geographical context that the first village communities were organised and that these communities would go on to establish and expand their territories (Dubouloz, 2017; Bostyn *et al.*, 2018a).

Most LBK sites were established on the alluvial valley floor and were situated less than a kilometre from the nearest water source. Over the course of subsequent millennia, Neolithic populations tended to diversify and extend the areas occupied to include adjacent valleys and plateaus (Le Bolloch *et al.*, 1986; Dubouloz *et al.*, 2005; Chartier, 2010). This change in settlement patterns probably originated in exploration of the landscape during the LBK, through hunting expeditions, the search for wild plants and wood and prospection for mineral raw materials.

Settlements in the Aisne valley can be divided into three categories depending on the duration of their occupation (Ilett, 2012): small sites like

Pontavert "le Port aux Marbre" or Missy-sur-Aisne "le Culot" (fig. 4) with a single phase of occupation (estimated at ± 20 years); sites with several occupation phases with a hiatus between each; large villages with continuous occupation (estimated at ± 100 years). Long duration sites benefitted from territories principally made up of open plains with easily accessible natural resources. This was not the case for short duration sites which tended to be located in areas that were not easily exploitable.

Long duration sites are not numerous but Cuiry-lès-Chaudardes "les Fontinettes" which, with an area of 6 ha, is one of the largest settlement sites known in the North of France, is a good example, along with Menneville "Derrière-le-Village", Bucy-le-Long "la Fosselle" (fig. 5) and Chassemy "le Grand Horle".

At a larger scale of analysis, that of socio-political integration, the spatial distribution of the different site types in the Aisne valley indicates that three of the long duration settlements were located at intervals of 18km from each other (*i.e.* a return journey of one day's walk). This suggests that these large sites acted as centres for territories measuring approximately 20km in length (Dubouloz, 2012a, fig. 6), this hypothesis is supported by similar findings for LBK sites in Germany, on the Aldenhoven Plateau (Lüning, 1988, 1998) and in north Hesse. Due to the heterogeneous nature of the available documentation, it is difficult to analyse the processes that led to the construction of these territories.

Figure 3: Distribution of LBK and Blicquy-Villeneuve-Saint-Germain sites mentioned in the text (map: L. Bedault).
a. LBK settlements in Europe with published faunal remains.
b. LBK settlements in the Paris Basin : 1. Pont-Saint-Maxence "le Jonquoire"; 2. Osly-Courtil "la Terre Saint-Mard"; 3-4. Bucy-le-Long "la Fosselle"- Bucy-le-Long "la Héronnière/la Fosse Tounise"; 5. Missy-sur-Aisne "le Culot"; 6. Chassemy "le Grand Horle"; 7. Presles-et-Boves "les Bois Plantés"; 8. Concevreux "les Jombras "; 9. Cuiry-lès-Chaudardes "les Fontinettes"; 10. Pontavert "le Port-aux-Marbres"; 11-13. Berry-au-Bac "le Chemin de la Pêcherie"- Berry-au-Bac "la Croix Maigret"- Berry-au-Bac "le Vieux Tordoir"; 14-16 Menneville (habitat) "Derrière le Village"- Menneville (enclosure) "Derrière le Village"- Menneville "la Bourgignotte"; 17. Juvigny "les Grands Traquiers"; 18-19.

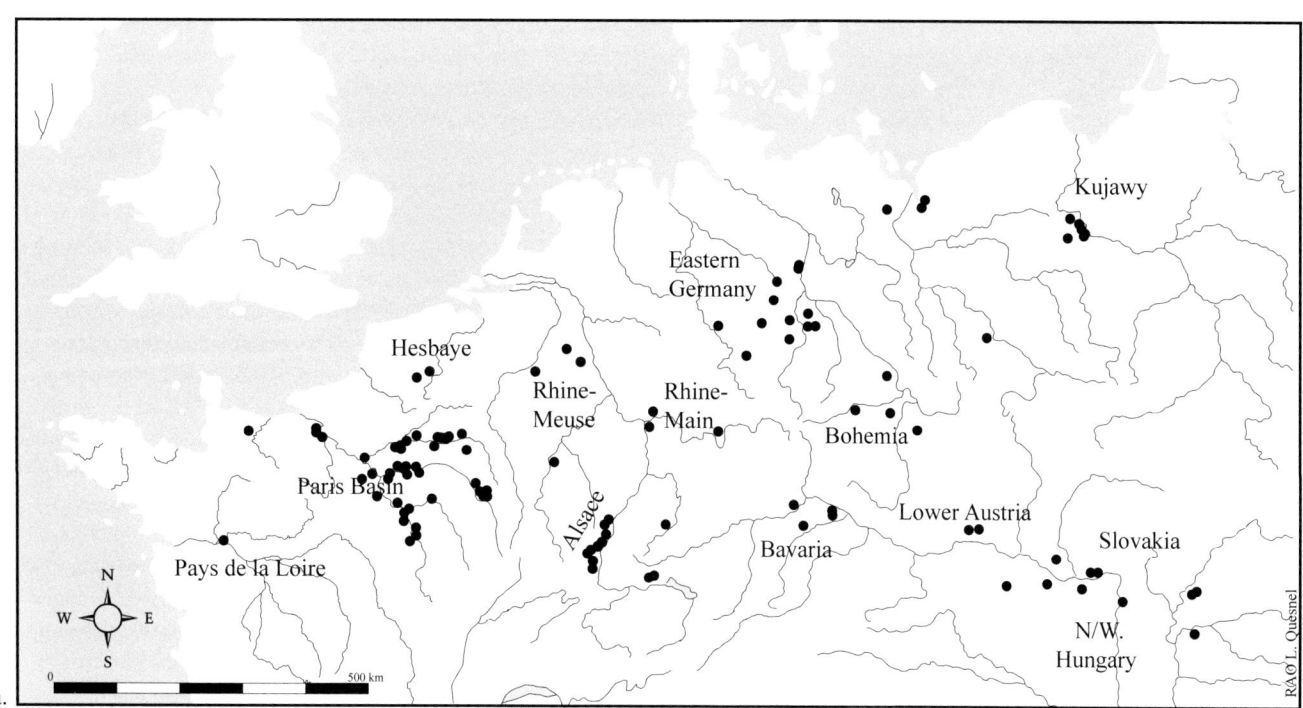

a.

Larzicourt "Champ Buchotte"- Larzicourt "Ribeaupré";
20. Norrois "la Raie des Lignes"; 21. Orconte "les Noues";
22. Armeau ; 23. Balloy "les Reaudins"; 24. Chaumont "les
Grahuches"; 25. Champlay "les Carpes".
c. BVSG settlements in the Paris basin : 26. Colombelles
"le Lazarro"; 27. Mondeville "Haut-Saint-Martin"; 28. Jort ;
29. Léry "le Chemin du Port"; 30. Poses "Sur la Mare"; 31.
Incarville-Val de Reuil ; 32. Aubevoye "la Chartreuse"; 33.
Maurecourt "la Croix de Choisy"; 34-35. Longueil-Sainte-
Marie "la Butte de Rhuis II"- Longueil-Sainte-Marie "la
Butte de Rhuis III"; 36. Pontpoint "le Fond du Rambourg";
37. La Croix-Saint-Ouen "le Pré des Iles"; 38. Trosly-Breuil
"les Obeaux"; 39. Villeneuve-Saint-Germain "les Grandes
Grèves"; 40-42. Bucy-le-Long "la Fosse Tounise"- Bucy-
le-Long "le Fond du Petit Marais"- Bucy-le-Long "le Fond
du Grand Marais"; 43. Missy-sur-Marne "le Culot"; 44.
Tinqueux "la Haubette"; 45. Luzancy "le Pré aux Bateaux";
46. Changis-sur-Marne "les Pétreaux"; 47. Mareuil-les-
Meaux "les Vignolles"; 48. Vignely "la Porte aux Bergers";
49. Fresnes-sur-Marne "les Sablons"; 50-51. Jablines "la
Pente de Croupeton"- Jablines "les Longues Raies"; 52.
Rungis "les Antes"; 53. Chelles "ZAC les Tuileries"; 54.
Neuilly-sur-Marne "la Haute Ile"; 55. Reuil-Malmaison "Rue
Marollet"; 56. Neauphle-le-Vieux "le Moulin de Lettrée";
57. Maisse "l'Ouche de Beauce"; 58. Buthiers-Boulancourt
"le Chemin de Malesherbes"; 59-60. Marolles-sur-Seine
"le Chemin de Sens"- Marolles-sur-Seine "les Prés-Hauts";
61-62. Barbey "le Chemin de Montereau"- Barbey "le
Buisson Rond"; 63. La Saulsotte "les Grèves de Frécul";
64. Villeneuve-la-Guyard "Prépoux"; 65-66. Passy "la
Sablonnière"- Passy "les Graviers"; 67. Gurgy "les Grands
Champs"; 68. Beaufort-en-Vallée "le Boule Rot".

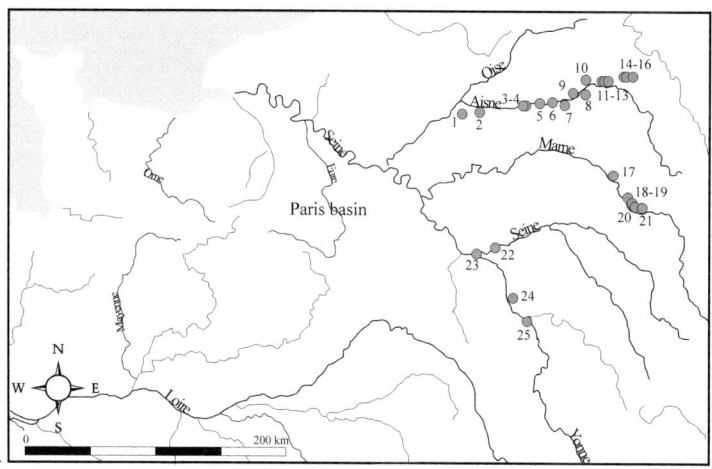

b.

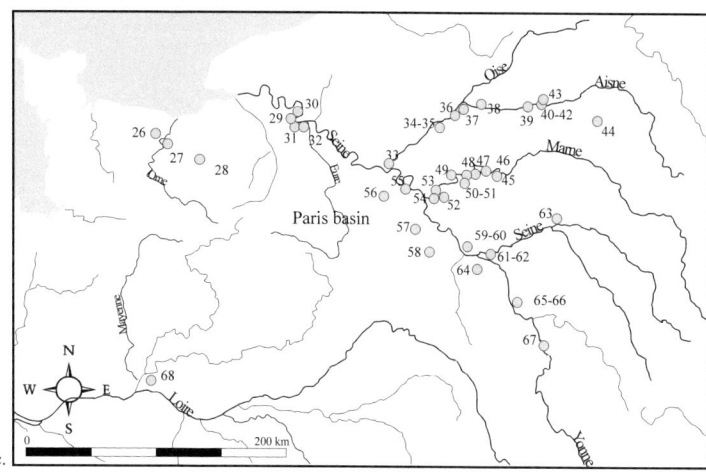

c.

However, certain socio-economic mechanisms can be more easily identified at a smaller scale, that of the "micro-area". In fact this is an intermediate scale, between the immediate territory of a settlement and the larger communal territory (Plateaux, 1990). The Bucy-le-Long micro-area, defined by a river meander, is one of the best documented examples in the Aisne valley. Over the course of a relatively long chronological sequence, we observe the shifting and relocation of two or three settlements within an area of about 10 km², which allowed the establishment and long-term development of a small agro-pastoral community.

Over the course of the 150 years or so of the LBK in Picardy, the number of settlements in the valley grows steadily (strictly in terms of new sites): from 0 (?)-2 sites per territory at the beginning of the sequence to 5-6 at the end. However, the sites apparently vary in their density of occupation (Ilett and Hachem, 2001; Ilett, 2012), particularly with regard to the distinction between major and secondary sites. Hamlets made up of 5 to 7 contemporary houses can most often be seen as major sites, while minor sites rarely have more than 2-3 houses. Furthermore, the population grows between the beginning and end of the sequence and house sizes increase.

During the Bliquy/Villeneuve-Saint-Germain, the situation becomes more complex compared to the LBK. The territory exploited is more extensive and, while there is still a preference for sites on the valley floors, settlements begin to be established on the edges of the plateaus. We also observe increasing diversity in the kinds of site, with settlements featuring houses (fig. 6). Occupations consisting solely of isolated silo-type pits, and tertiary flint acquisition/extraction sites (Lanchon, 2012; Bostyn and Denis, 2016).

Of the 150 or so sites recorded in the Paris Basin, about thirty of these are located in the Marne valley (fig. 3). The relationship between blade-producing sites (*e.g.* Jablines), which are spaced at intervals of c.15 km, and other villages is still under investigation as are various other aspects of the territorial and regional organisation (Bostyn and Lanchon, 2003; Lanchon, 2006, 2012; Bostyn *et al.*, 2018a).

These changes in house layouts and territorial occupation reflect underlying upheavals in Neolithic society.

Significant economic, social and ideological shifts become evident in the Middle Neolithic which can be divided into two periods; these are referred to in French as *Néolithique moyen I* (Middle Neolithic I) and *Néolithique moyen II* (Middle Neolithic II). These periods correspond

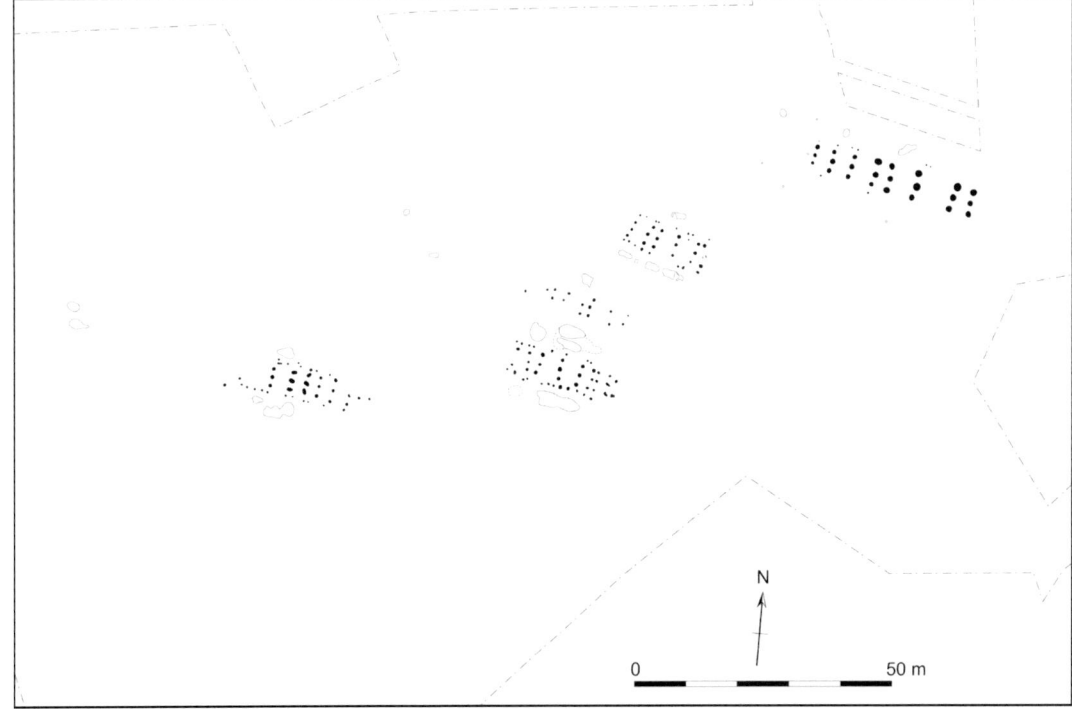

Figure 4: Missy-sur-Aisne "le Culot" (Aisne), a small LBK site (after Ilett 2012, fig. 4, CAD: C. Monchablon, Inrap).

N

0   50 m

respectively to the German *Späte Mittelneolithikum* and *Jungneolithikum* (fig. 2).

The first period begins with what is called the "Cerny" culture (4600-4200 BC.). It is divided into two successive phases, Cerny-Videlles and Cerny-Barbuise (Louboutin and Simonin, 1997), it can also be found in the more western part of France (Marcigny *et al.*, 2010); the relationship between Cerny and Rössen in the Paris basin is still an open question (Dubouloz and Lanchon, 1997). Houses are no longer built in the Danubian tradition and new forms of village settlement arise but are sometimes difficult to identify archeologically. Nevertheless, a study of settlement in the Paris basinindicates that circular buildings were being erected from the early phase of the Cerny in the valleys of the Seine and Yonne as well as in the North of France; in contrast, regions to the west witnessed the construction of rectangular buildings in a continuation of the Danubian tradition (Bostyn *et al.*, 2016).

More easily observable is a desire to mark out "territory" for both the living and the dead through the construction of ditch and palisade enclosures which can enclose areas covering several hectares. These enclosures were assembly sites where ceremonial practices took place including the deposition of pottery and bones. They are primarily found in the Seine and Yonne valleys.

In parallel, we see the construction of imposing funerary monuments of earth and wood which appear to have served as the burial places of a small emerging elite (Demoule *et al.*, 2007). First discovered at Passy in the Yonne (hence the term *"Structures de type Passy"* or STPs), then in the Caen region in Normandy, these structures are composed of a ditch or palisade which generally defines an oval area, sometimes over 200m in length (fig. 7). Burials were placed in pits within these enclosures and the graves were possibly covered by an earth mound. Some of these graves have yielded exceptional finds including the remains of animals or objects made of bone and teeth which were placed close to the body of the deceased.

Figure 5: Bucy-le-Long "la Fosselle" (Aisne), a long duration LBK site (after UMR 8215 Trajectoires, CAD: C. Monchablon, Inrap).

Subsequently, in the Middle Neolithic II, the northern Chasséen Culture developed in the western part of the Paris Basin. At the same time, north-eastern France saw the rise of another cultural entity, the Michelsberg. This culture appeared around 4200 BC and replaced the Rössen Culture.

The remains of buildings are rare and often poorly identified but generally consist of large rectangular structures.

In contrast, numerous enclosures are known from the period (Liétar, 2017). The term "enclosure" in fact masks a multitude of forms although they share a common principle: they are defined by ditches that tend to enclose a circular area and they generally lack internal structures apart from occasional large buildings.

Analysis of the general characteristics of the 132 recorded Neolithic enclosures belonging to the Cerny, Chasséen, northern Chasséen, Michelsberg (fig. 8) and Spiere Group cultural periods (*i.e.* 4600-3800 BC) in northern France suggests that there was significant intensification of economic, political and social activity over a period of several centuries (Dubouloz, 2018). From 4400 BC, in the Rössen III (Late Rössen) and Post-Rössen (4400-4250 BC) periods, networks of enclosures developed; these probably served as a basis for enclosures of the Middle Neolithic II (4250-3950 BC), with evidence for continuity of occupation and architectural style between these two periods on certain sites. The Middle Neolithic II is marked by a densification of occupation and a diversification of enclosure morphology (Liétar and Giligny, 2016).

A shared cultural substrate unites various Middle Neolithic II groups, which nonetheless display clear regional differences: within the Paris

Figure 6: Jablines "la Pente de Croupetons" (Seine-et-Marne), plan of the BVSG site (after Lanchon *et al.* 1997, fig. 1, modified, CAD: Y. Lanchon).

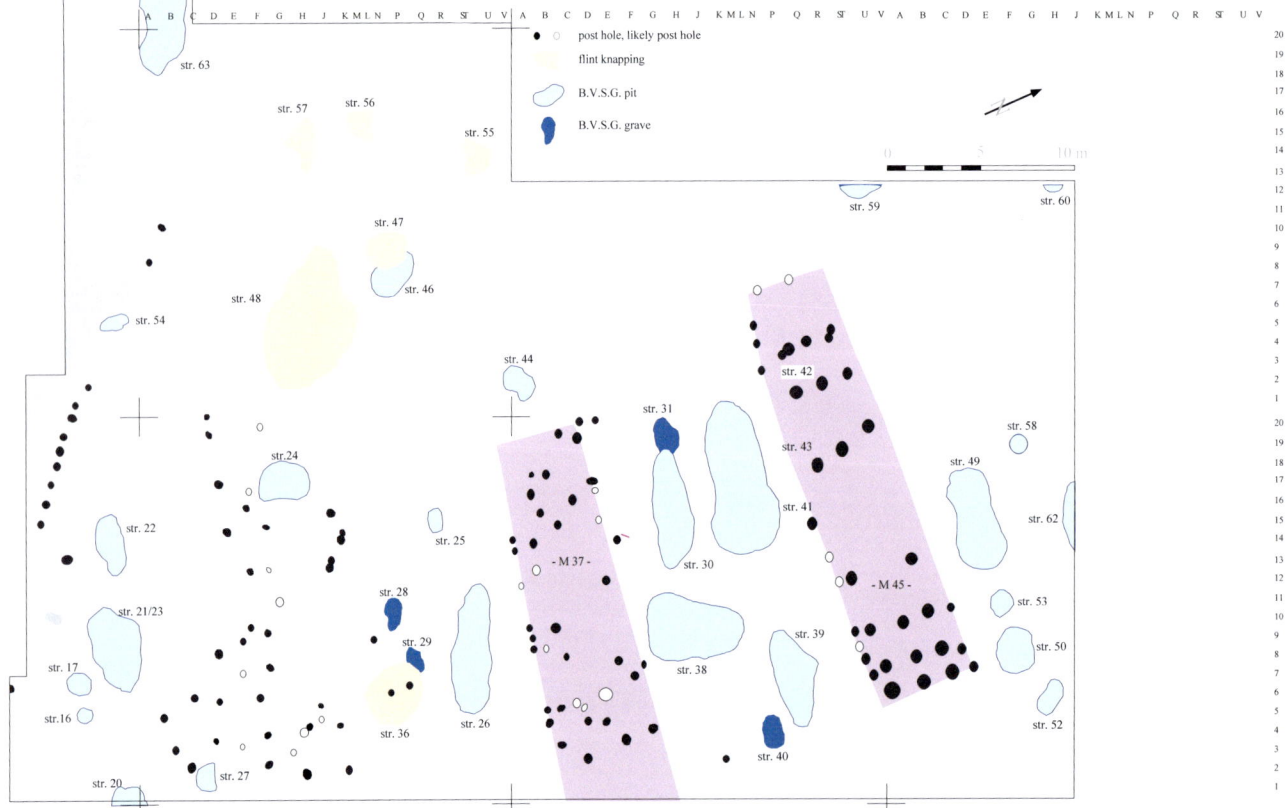

JABLINES, "la Pente de Croupeton" (Seine-et-Marne)

Basin (fig. 9) the Aisne valley becomes the epicentre of the Michelsberg Culture, the Marne valley falls within the same cultural complex, the Oise valley becomes the centre of the northern Chasséen, and part of the Seine valley is occupied by the Noyen Group (Dubouloz *et al.*, 1991; Dubouloz, 1998; Bostyn *et al.*, 2011a). Further north, in the Pas-de-Calais, the Spiere Group emerges (Bostyn *et al.*, 2011b; Colas *et al.*, 2016).

These enclosures, which usually encompass an area of between 1 and 10 ha, are interpreted as communal assembly sites which were principally established on valley bottoms, but which also occur on plateau edges and on plateau slopes (Dubouloz, 2018). In terms of morphology, "very simple" (defined by either a single palisade or ditch) and "simple" (defined by a single palisade and ditch) enclosures predominate. "Complex" (defined by multiple ditches and palisades) and "very complex" (defined by complex, deep enclosing elements with several phases of repair or rebuilding) are

much less numerous. Further study is required to determine the durations of occupation of these sites but they appear to have been relatively short-lived, perhaps less than a century.

Along with the enclosures of the Middle Neolithic II in northern France, we observe other types of settlement sites that produce various kinds of refuse and deposits. Nonetheless, in our opinion, two main categories of context can be distinguished: domestic pits and occupation layers.

Thus, we observe sites composed essentially of groups of pits. Such sites are rare and take the form of groups of pits which are sometimes spread out over several hectares.

We also encounter layers which are the remains of Neolithic occupation levels whose preservation is due to particular topographical situations. For example, these layers are often found in valley bottoms, reflecting occupation close to wet zones alongside water courses (*e.g.* waste material from habitation situated close to a waterway).

Figure 7: Fleury-sur-Orne "les Hauts de l'Orne" (Calvados), simplified monument plan (after Ghesquière *et al.* 2019, fig. 1, modified, CAD: E. Ghesquière, Inrap).

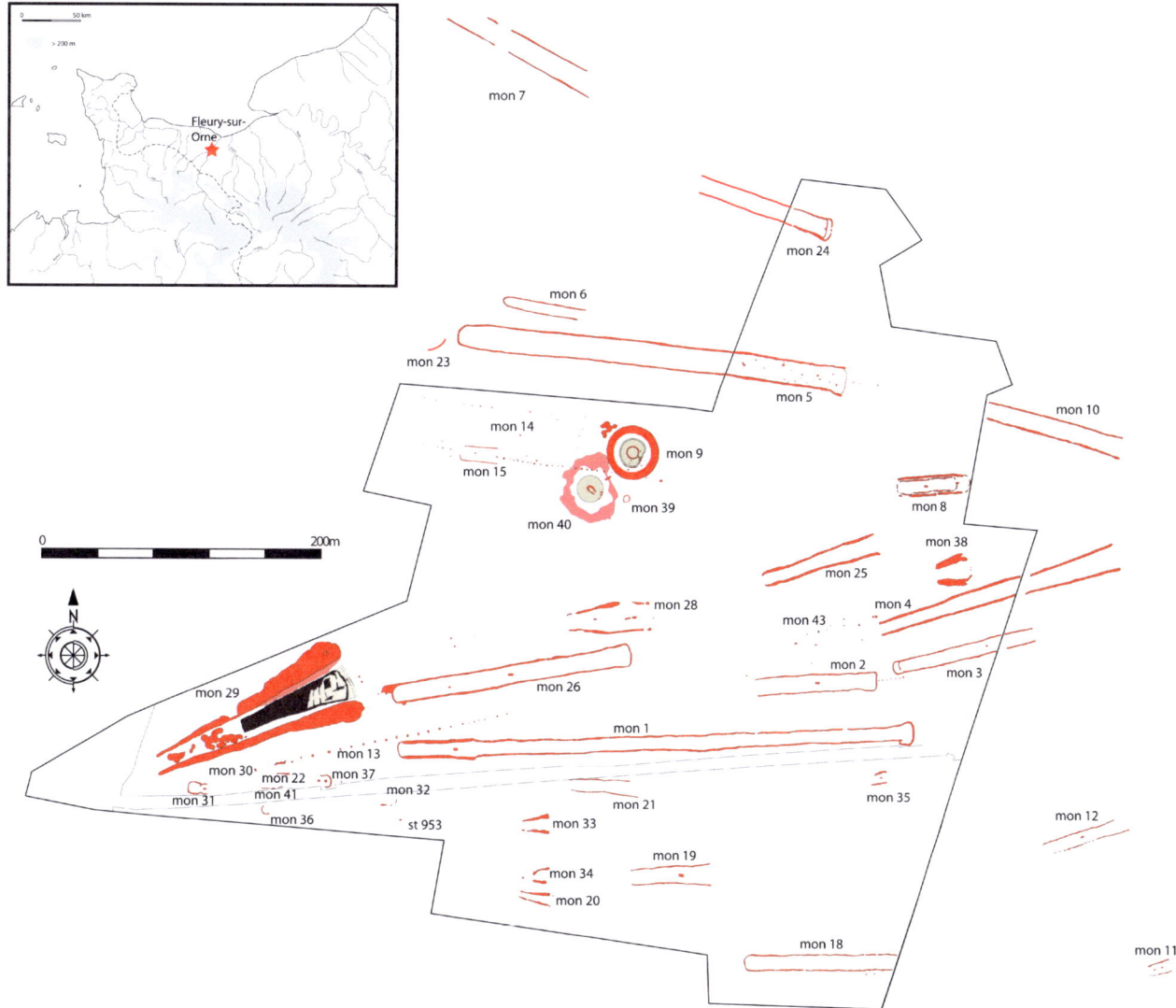

Figure 8: Bazoches-sur-Vesle "le Bois de Muisemont" (Aisne), plan of the Middle Neolithic 2 (Michelsberg) enclosure (CAD: C. Monchablon, B. Van den Bosche, J. Dubouloz; after Dubouloz 2018, fig. 7, modified).

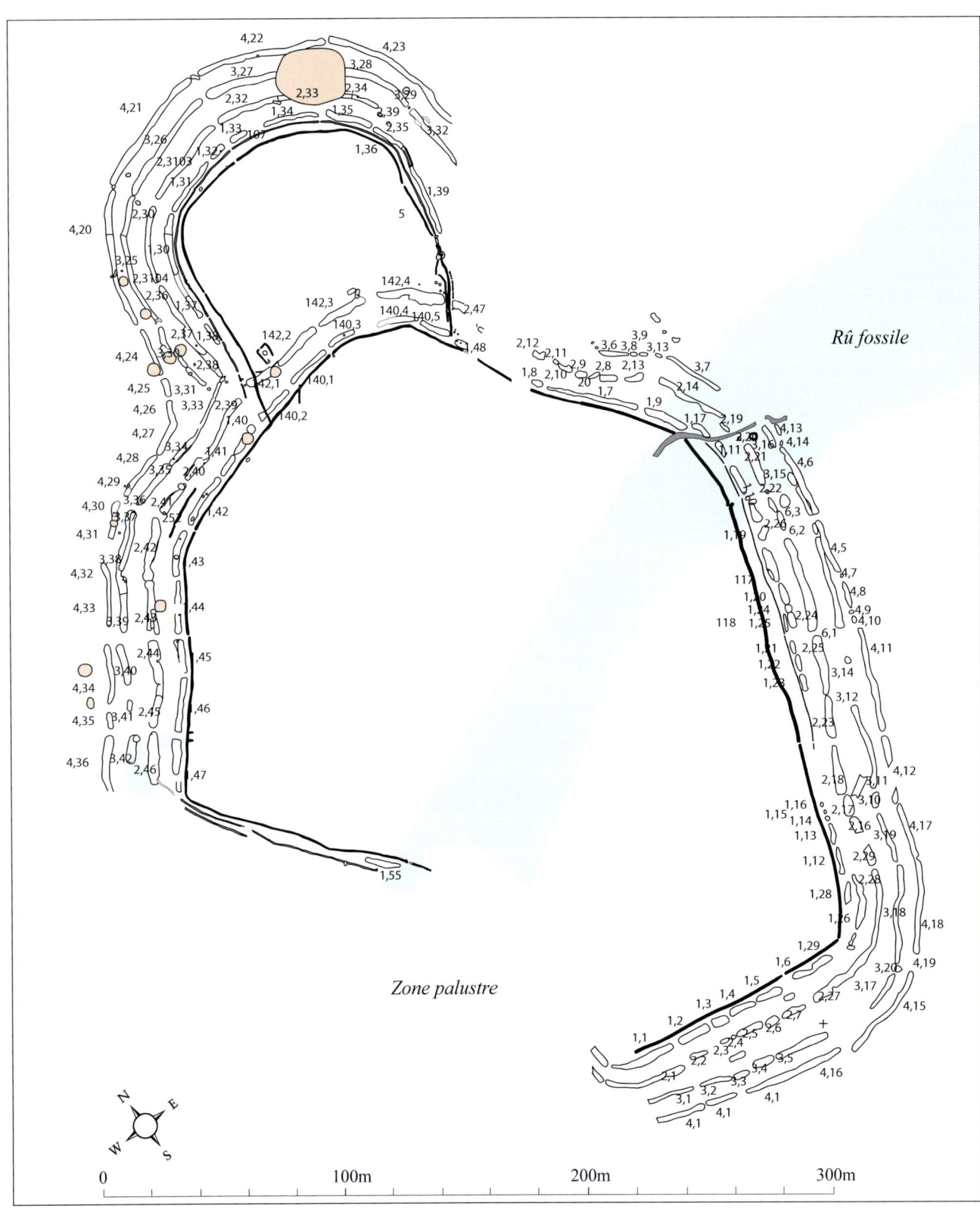

The Late Neolithic covers the period from 3,600 to 2,900 BC; it is divided into three phases on the basis of technological and radiometric criteria. As the data stands at the moment, Phase 2 (3,350-3,300 BC) the best documented phase, formerly known as the Seine-Oise-Marne Culture, is contemporary with the Horgen (Cottiaux *et al.*, 2014a). Data is scarce for the Early Phase (3,600-3,350 BC) and the Final Phase (after 3000 BC).

Until recently, the identification and the definition of the Late Neolithic in the Paris basin essentially depended on archaeological data from burial contexts; specifically collective burials, and particularly what are known as *allées couvertes*, which tend to be poorly preserved as a result of millennia of agricultural and construction activity (Chambon and Salanova, 1996). Settlement sites were largely unknown.

A collectively produced overview of archaeological research on the Neolithic in Picardy (focusing on characterization of sites, chronology, finds, etc.) underlines this disparity and highlights the scarcity of documentation regarding the Late/Final Neolithic (Dubouloz *et al.*, 2005). It was principally from the 1980s onwards that settlement

sites began to be recognized in the region, and particularly in the valleys of the Oise and Aisne where they have been found in various locations: on the banks of a water channel, on alluvial terraces and at the base of hillslopes. For the most part, these sites take the form of simple isolated pits or sedimentary layers containing waste material, the remnants of ancient floor levels or dumping areas (fig. 10). As yet no Neolithic village plans have been revealed in the region and no major house sites have been identified.

An overview of settlement has been attempted on a micro-regional scale in the Lower Marne valley; the study focuses on a roughly 70 km long portion of the valley that was judged to be particularly suitable for such an investigation (Cottiaux *et al.*, 2014a; Pastre *et al.*, 2000). The twenty sites

Figure 9: Distribution of Middle Neolithic II sites in northern France (map after Dubouloz 2018, modified).

identified, which were often discovered fortuitously, included eight collective burials and about ten occupation sites of various forms. Thus, we observe a flint mining site, an axe production site, three occupation levels, and eleven sites composed of dug features. The latter are represented by a few pits with low quantities of finds, suggesting a small number of occupants.

The digging of pits was a common practice, but it is not a principal characteristic of sites dating to this period, a fact that hinders the gathering of data. Moreover, for the Marne valley, the scarcity of sites dating from the Late Neolithic to the Middle Bronze Age is accentuated due to erosion of the archaeological levels; often all that remains are sites that were trapped beneath deep colluvium at the foot of the valley sides, between the alluvial plain and the edge of the plateau (Brunet *et al.*, 2004).

Distribution maps of sites in Picardy show a greater variability in geographical distribution than is evident for earlier periods of the Neolithic (Dubouloz *et al.*, 2005). In fact, instead of the initial concentration in the principal valleys (almost 90% of sites), we find a wider distribution on the plateaus and in minor valleys (more than 50% of sites are located outside the main valleys at the end of the sequence). This development indicates the desire and growing capacity of Neolithic communities to move away from the banks of the main rivers, and then from the plateau edges, in order to occupy the wider available landscape. While the number of collective tombs recorded in the valleys and on the plateaus constitutes a statistical bias, it is conceivable that the burial locations are in some way related to the locations of the corresponding settlements. Thus we indirectly see a widening of the occupied area to encompass various components of the landscape, a trend that was perceptible during the periods preceding the Late Neolithic and which reflects processes of demographic expansion and economic change.

Because of the disparate and incomplete nature of the archaeological record, there is a dearth of archaeozoological studies for the Late Neolithic, all the more so because, to date, very few (untransformed) animal offerings have been discovered in collective burials. Personal ornaments of bone, antler and teeth are present, and testify to a bone-working industry (Sohn, 2008), but apart from occasional references, these artefacts will not be dealt with in this book.

Figure 10: Mareuil-les-Meaux "les Lignières" (Seine-et-Marne), occupation layer dating to the Late Neolithic: the building dates to the Late Neolithic or Bell Beaker/Early Bronze Age period (after Brunet *et al.* 2018, fig. 4).

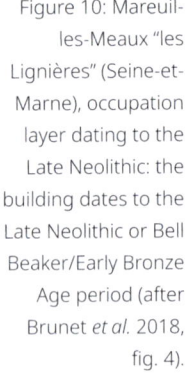

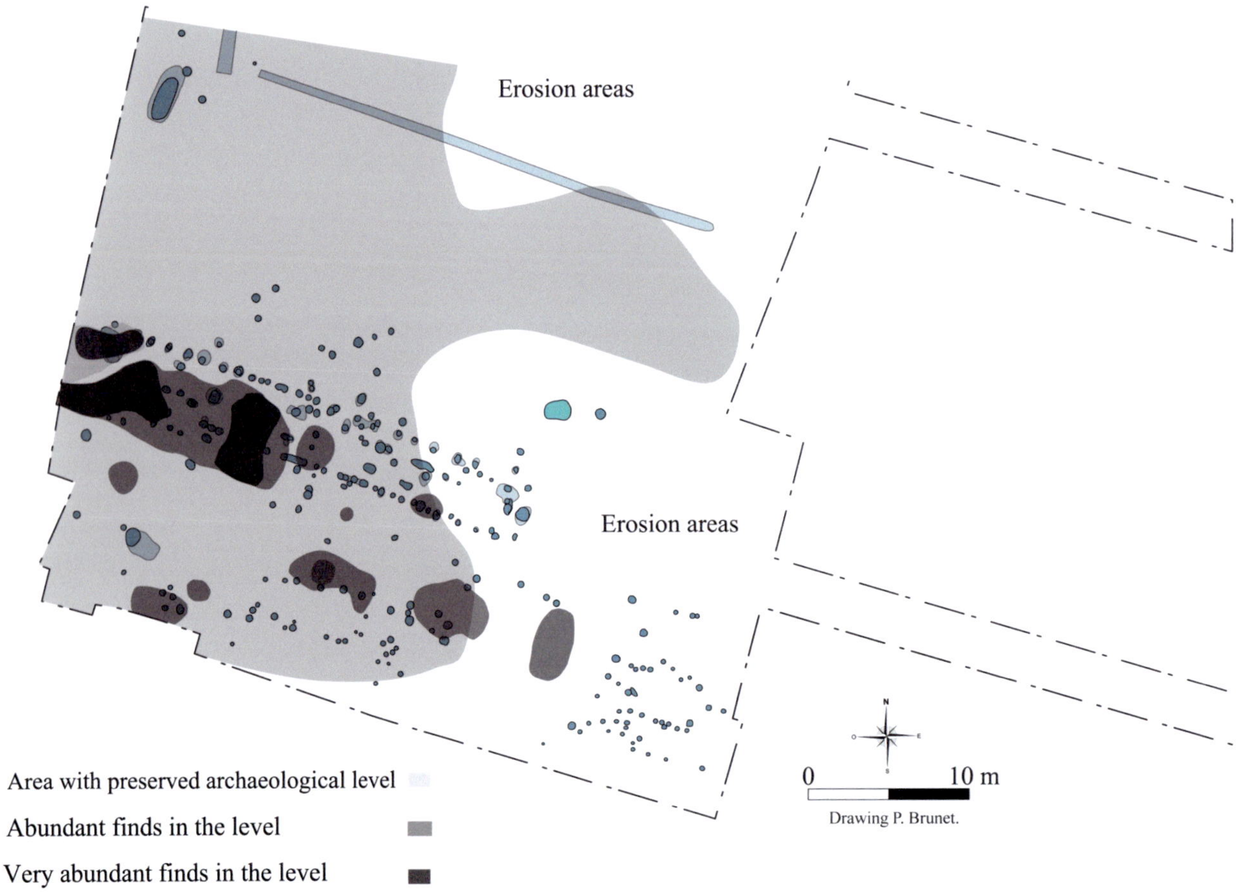

Erosion areas

Erosion areas

N

0   10 m

Drawing P. Brunet.

Area with preserved archaeological level

Abundant finds in the level

Very abundant finds in the level

The study of Late Neolithic occupation, spanning the period between 2,900 and 2,300 BC has allowed this period to be broken down into three phases. The first phase encompasses several groups including the "Deûle-Escaut" in the North and the "Gord" in the Paris Basin. The second phase, (from 2,400 BC), which is only known through a small number of isolated graves, corresponds to the Bell Beaker culture. The third, the Epi-Bell Beaker culture, is only attested to by a handful of inconspicuous occupation sites.

The definition of Final Neolithic cultural groups predating the Bell Beaker culture is one of the notable developments of recent years and, in contrast to the situation for the Late Neolithic, most of the evidence is based on domestic habitation sites. By this period, there are far fewer collective burials, even though it is likely that their numbers are under-estimated due to changes in funeral practices, such as the absence of deposited ceramic vessels at the entrances to the monuments (Sohn, 2007).

Over the five centuries between 2,900 and 2,400 BC, we observe an evolution in the construction of post-built structures, with a shift from the traditional rectangular buildings to apsidal buildings, although both forms did co-exist (Joseph *et al.*, 2011).

A more in-depth study, based on twenty sites associated with the principal expansion of the Deûle-Escaut group during the first half of the 3rd millennium BC, indicates that two types of spatial organisation existed. Together with settlements enclosed by a post-built palisade, there are also more numerous "open" settlements. In both cases, the sites contain buildings of various sizes. Certain buildings are monumental in scale and undoubtedly had an ostentatious character (fig. 11). More numerous are buildings of average or even small size. The multiplicity of these settlements reflects differences in their status, which is expressed mainly in the social effort invested in the construction of palisades and houses; in most cases a substantial population would have been required.

One possible hypothesis as to the territorial organisation is as follows:

"Villages or hamlets in unenclosed contexts, consisting of a large building (between 100 and 200 m²) associated with smaller structures, were dominated by settlements characterized by larger buildings (greater than 200 m²) and surrounded by a palisade. Finally, above this level, was a site whose ostentatious architecture was intended to reinforce the political power of a community over a territory." (Praud, 2012, 112).

This spatial organisation is echoed in the archaeological finds associated with these sites and which is found in the building foundations and associated pits. There is palaeobotanical evidence for the preparation and storage of grain; preparation and consumption; the presence of ceramic objects such as loom weights and spindle whorls highlights activities such as the spinning and weaving of plant fibres (*e.g.* linen, nettle, bark, etc.); technological and functional studies of the lithic and bone industries and of the grinding equipment show us what types of activity were practiced – hunting, agricultural tasks, wood working, leather working, grinding, crushing, etc. Judging from the sites that have been investigated, the transformation of plant fibres into textiles seems to have been one of the most widespread activities.

The geographical contexts of settlement sites dating to the Final Neolithic are varied, just as they are for the Late Neolithic. All areas from the valley floor to the loamy plateaus are intensively occupied, and particularly zones adjacent to wetland, as well as slopes

For the Bell Beaker period in Picardy, the number of recorded sites decreases significantly. Several sites are known in the Somme department, in the Oise and in the Aisne (Dubouloz *et al.*, 2005). These are mainly located on the alluvial terraces but also occur on the plateaus. The evidence for funerary practices includes both individual graves and collective tombs. The rest of the evidence takes the form of scattered finds without clear contexts.

Sites on the cusp between the Neolithic and the Bronze Age are extremely rare and mainly take the form of burials and scattered finds.

The Bronze Age

For the Early and Middle Bronze Age (2300-1350 BC; Bronze A1 to C2), traces of settlement are rare and characterised by isolated structures and occupation layers (fig. 12). Most of the settlement sites are simple isolated farms which have left few visible traces in the ground.

In the Late Bronze Age (1150-930 BC; Hallstatt A2/B1), sites are more numerous and tend to be easier to detect; they generally take the form of small units, which can extend over 100m or so. Virtually no house plans are known in Picardy although they are well documented in neighbouring regions. In Normandy, the houses are circular in plan and in the North of France we find rectangular houses that also doubled up as byres. Non-funerary, ditched enclosures have been identified in the departments of Calvados, Nord and Pas-de-Calais (Mondeville,

Abbeville, Etaples, Blainville, and Guînes for example). A number of fortified sites, specialized in metallurgy, are characteristic of the middle Oise valley at this period but the tradition did not last.

In the Late Bronze Age IIIb – Early Hallstatt (930-630 BC; Hallstatt B2/B3-Hallstatt C) we observe a marked increase in the size of settlement sites. Small farms of a few hundred square metres co-exist with villages that can be over 1 ha in extent (fig.13) and (fig. 14). Enclosed sites occur occasionally (Brun, 2015; Chancerel A., Marcigny C., 2006; Le Goff, 2009; Marcigny and Talon, 2009; Peake *et al.*, 2011; Riquier *et al.*, 2012).

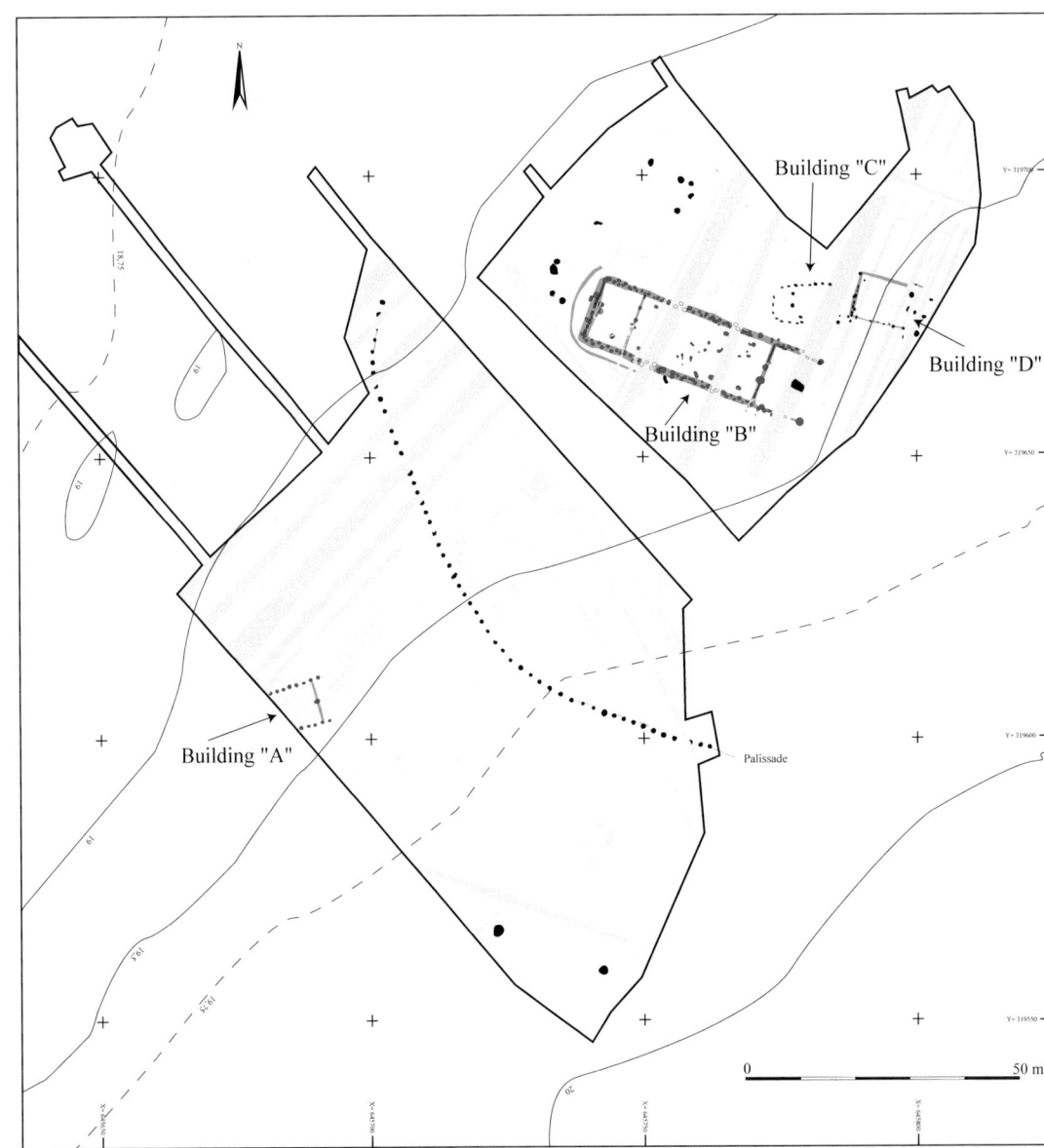

Figure 11: Houplin-Ancoisne "le Marais de Santes" (Nord), plan of the Final Neolithic site (after Praud *et al.* 2007, fig. 2; CAD: C. Benoit, L. Michel).

	Phase	Date
Early Bronze Age	Bronze A1	2300-1800 BC
	Bronze A2	1800-1600 BC
	Bronze B1	1600-1500 BC
Middle Bronze Age	Bronze B2	
	Bronze C1	1500-1350 BC
	Bronze C2	
Late Bronze Age	Bronze D	1350-1250 BC
	Hallstatt A1	1250-1150 BC
	Hallstatt A2	1150-1050/1020 BC
	Hallstatt B1	1050/1020-930 BC
	Hallstatt B2-B3	930-800 BC

	Phase	Date
Early Iron Age	Hallstatt C	800-640/630 BC
	Hallstatt D1	640/630-530 BC
	Hallstatt D2-D3	530/475-450 BC
	La Tène A1A2	475-450-400 BC
	La Tène B1	400-325 BC
Middle Iron Age	La Tène B2	325-270/250 BC
	La Tène C1	270/250-200/180 BC
Late Iron Age	La Tène C2	200/180-150/130 BC
	La Tène D1a	150/130-120/110 BC
	La Tène D1b	120/110-90/85 BC
	La Tène D2a	90/85-60/50 BC
	La Tène D2b	60/50-30/20 BC

Figure 12: Chrono-cultural framework, Bronze and Iron Ages.

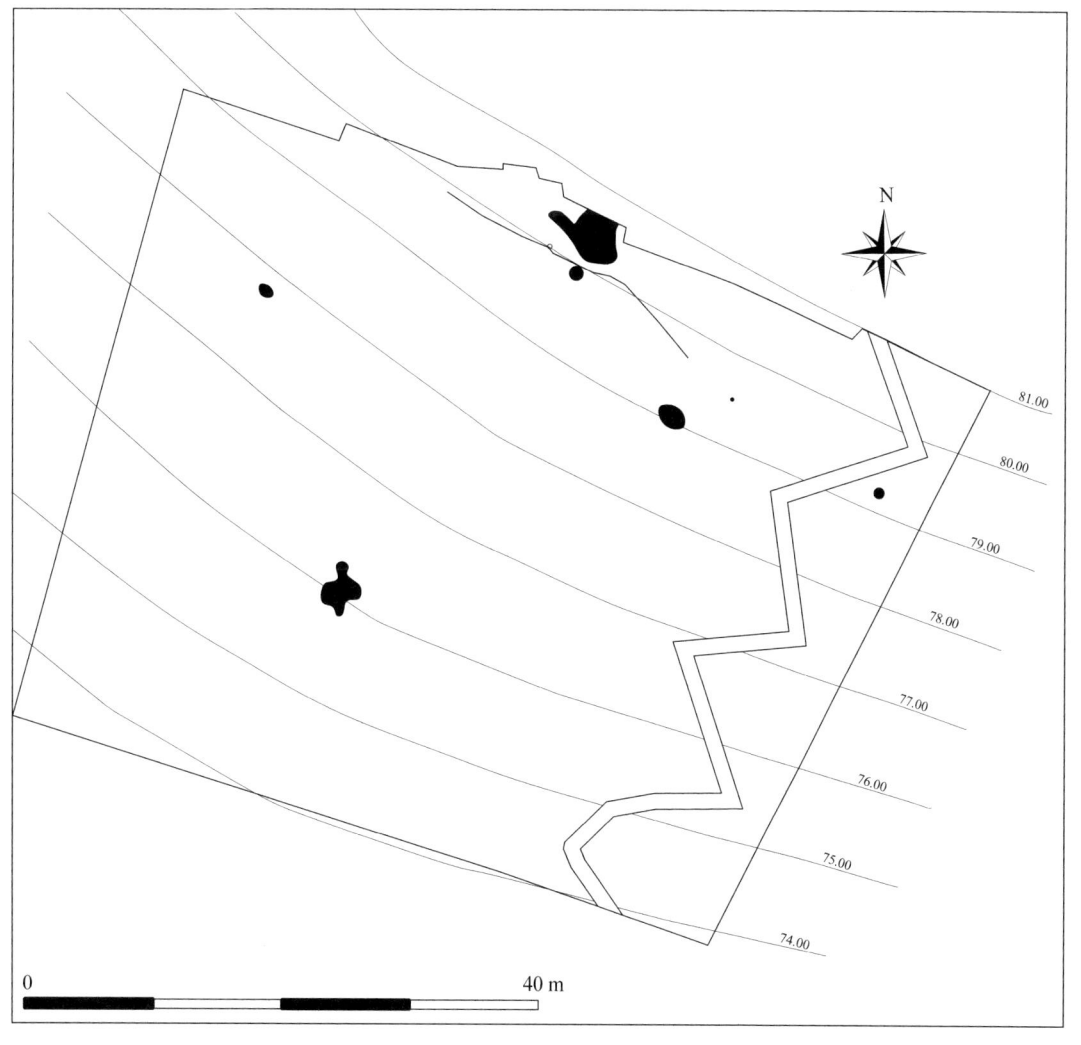

Figure 13: Pasly "les Coteaux" (Aisne), Late Bronze Age, open settlement (CAD: INRAP, S. Desenne).

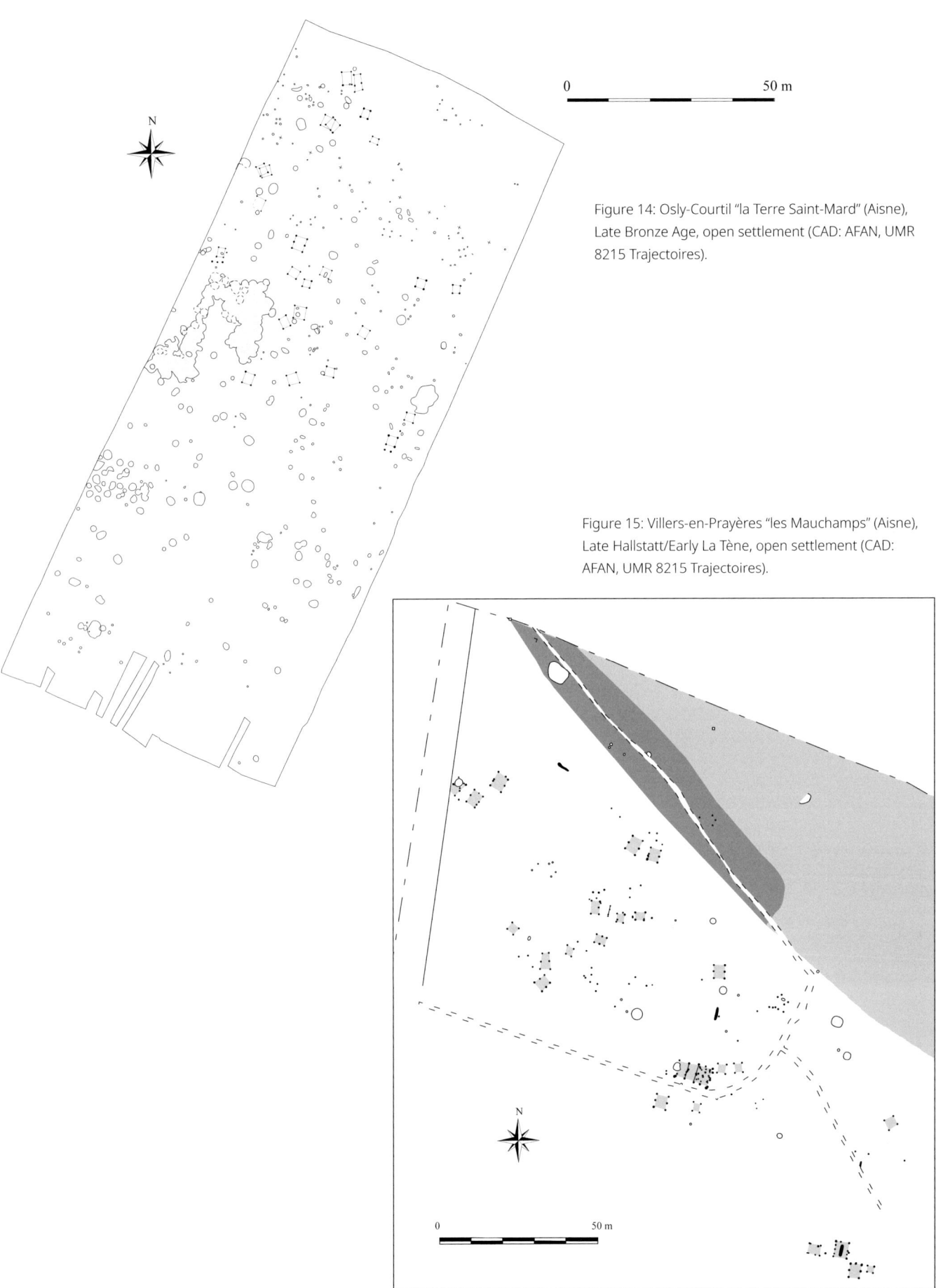

Figure 14: Osly-Courtil "la Terre Saint-Mard" (Aisne), Late Bronze Age, open settlement (CAD: AFAN, UMR 8215 Trajectoires).

Figure 15: Villers-en-Prayères "les Mauchamps" (Aisne), Late Hallstatt/Early La Tène, open settlement (CAD: AFAN, UMR 8215 Trajectoires).

The Iron Age

The last five hundred years BC are generally divided into four chrono-cultural units according to changes observed in house forms and material culture (Brun *et al.*, 2005b; Demoule, 1999; Gransar *et al.*, 1999; Issenmann, 2009; Le Goff, 2009; Malrain, 2000; Malrain and Pinard, 2006; Menez, 2008; Pion *et al.*, 1990, 1996; Riquier *et al.*, 2012). The first chrono-cultural division corresponds to Late Hallstatt and the Early La Tène (530-250 BC; Hallstatt D3/La Tène A1/A2/B2/B3), the second to the Middle La Tène, roughly the 3rd century BC

(270-180 BC), the third coincides with the 2nd century BC or the beginning of Late La Tène finale (200-85 BC; La Tène C2 et D1), and the fourth corresponds to the end of Late La Tène, or the 1st century BC (90-30/20 BC, La Tène D2, fig. 12).

In the Late Hallstatt final/Early La Tène (450-250 BC), we observe both unenclosed and enclosed settlements (fig. 15), (fig. 16), (fig. 17).

In the first category we find farmsteads made up of clusters of buildings and pits of various shapes and sizes. The buildings can be loosely or densely arranged, depending on the site. The

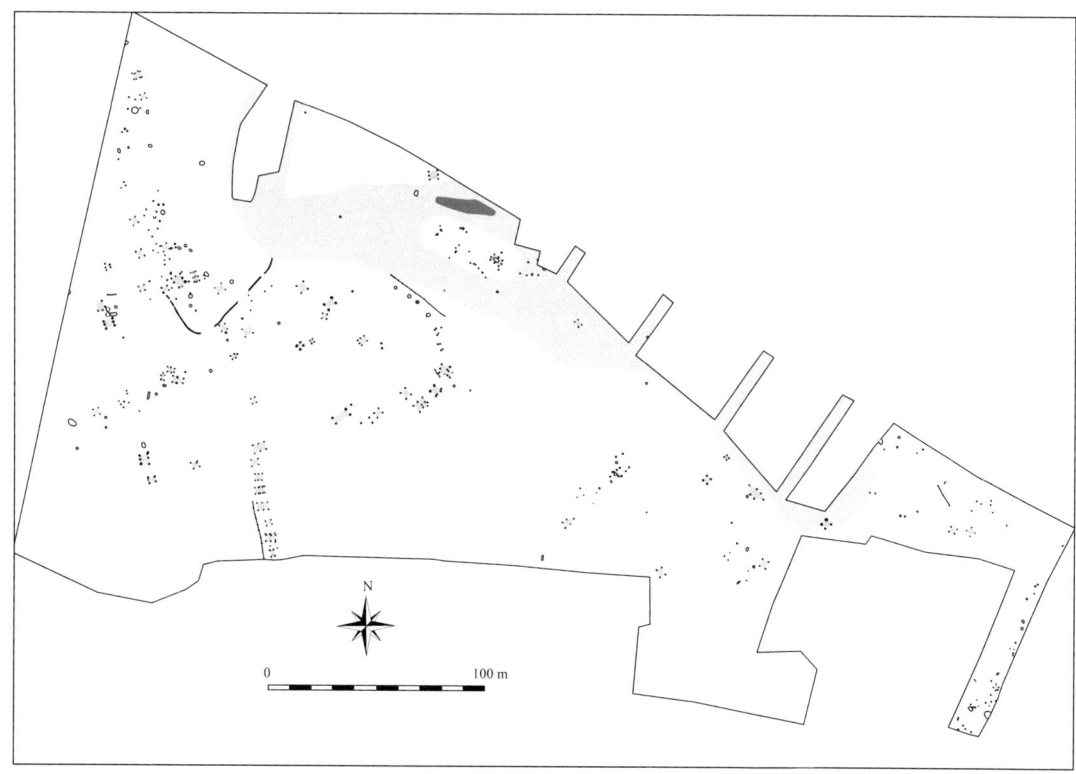

Figure 16: Ciry-Salsogne "le Bruit" (Aisne), Late Hallstatt/ Early La Tène, open settlement (CAD: AFAN, UMR 8215 Trajectoires).

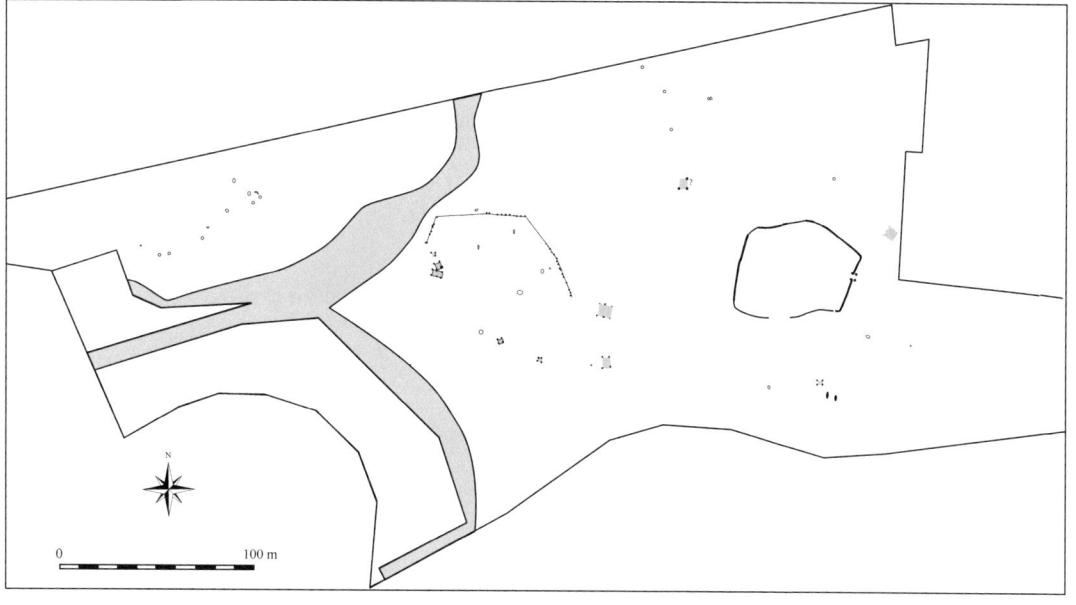

Figure 17: Bazoches-sur-Vesle "les Chantraines" (Aisne), Late Hallstatt/Early La Tène, enclosed and open settlement (CAD: AFAN, UMR 8215 Trajectoires).

simplest farmsteads are composed of a few pits and houses, loosely arranged, and yield little in the line of refuse, and, in particular, very little bone. It is possible that these sites were only occupied for part of the year. Certain farms were organised as small nuclei of pits and buildings, with the buildings sometimes being several tens of metres apart. The spaces between these nuclei were generally devoid of domestic features apart from silos, some of which contained unusual faunal assemblages (see chapter 2). A study of the finds from several sites, most notably in the Aisne, has allowed the total duration of occupation of each nucleus to be estimated at two or three generations (*i.e.* 25 to 30 years). Agriculture was orientated towards cereal and vegetable production. The importance of these crops is demonstrated by the large capacity and number of storage structures (silos) and by the occurrence of querns.

Occasionally human inhumations are found in these silos, sometimes accompanied by animals, and sometimes animal remains occur on their own. This practice may represent a form of ritual appropriation of the space, either as a foundation ritual or an abandonment ritual. Likewise, the presence of articulated animal carcass parts, lacking typical butchery marks, in certain pits indicates that symbolic or religious rituals were practiced on a number of sites (Auxiette, 2000a; Gransar *et al.*, 2007).

We also observe the establishment of open settlements composed of pits and houses that are more tightly arranged and which were occupied for long durations. Palisades were constructed to subdivide the spaces within the settlements but they do not completely enclose the sites. This type of settlement yields much more remains than the types already described.

As regards enclosed settlements, certain examples are enclosed by a palisade and others by a ditch. In the first case, the settlement surrounded by the palisade sometimes features a monumental entrance (fig. 18) and several storage structures supported by four or six posts (granaries).

This centralisation of storage facilities, correlated with the presence of unusual faunal remains (*e.g.* pit 484 at Bucy-le-Long "le Grand Marais" (Auxiette, 2000a), lends this category of site a special status, higher than that of the simple farms described above. Settlements enclosed by a ditch are quite rare in this period but certain examples have yielded very large quantities of remains. The structure of the settlement, the quality of the

Figure 18: Bucy-le-Long "le Grand Marais" (Aisne), Late Hallstatt/Early La Tène, enclosed and open settlement (CAD: AFAN, UMR 8215 Trajectoires).

N

0   20 m

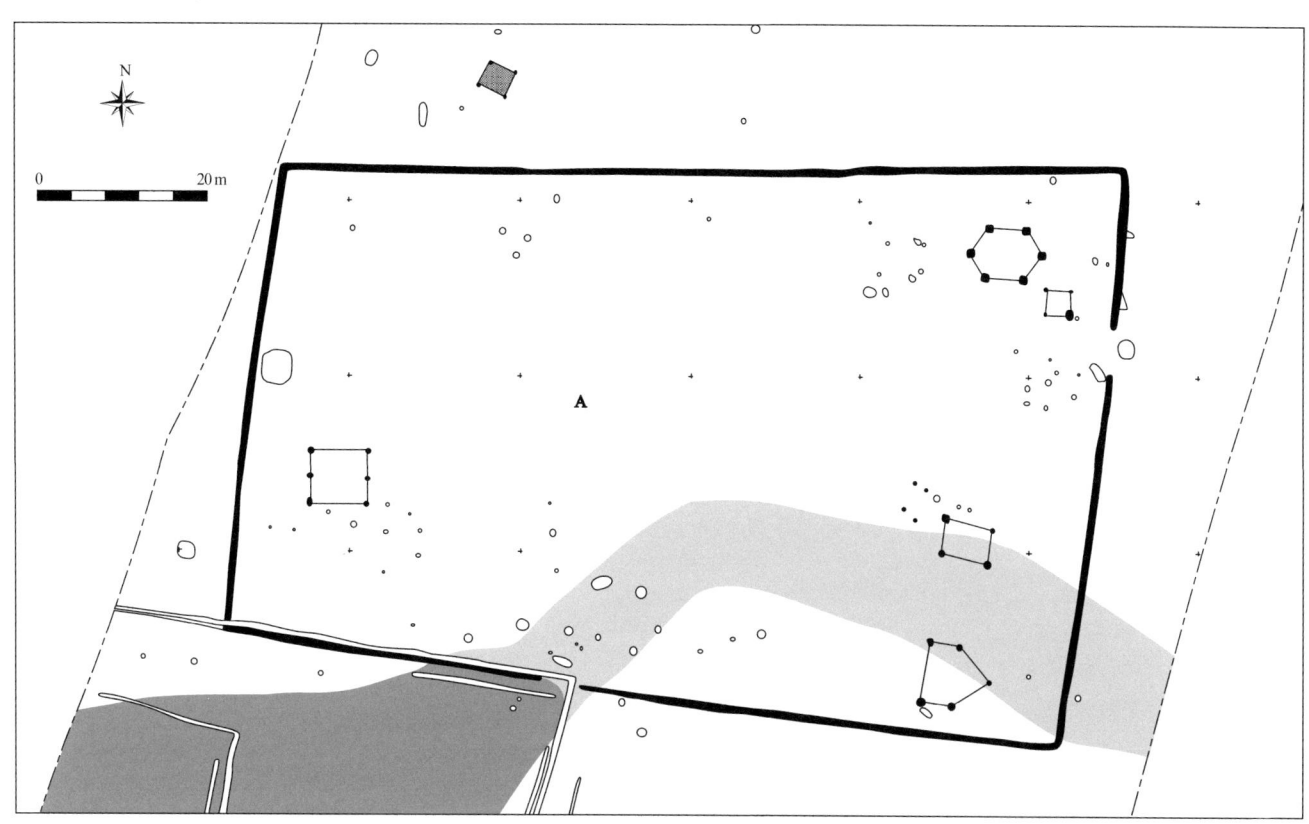

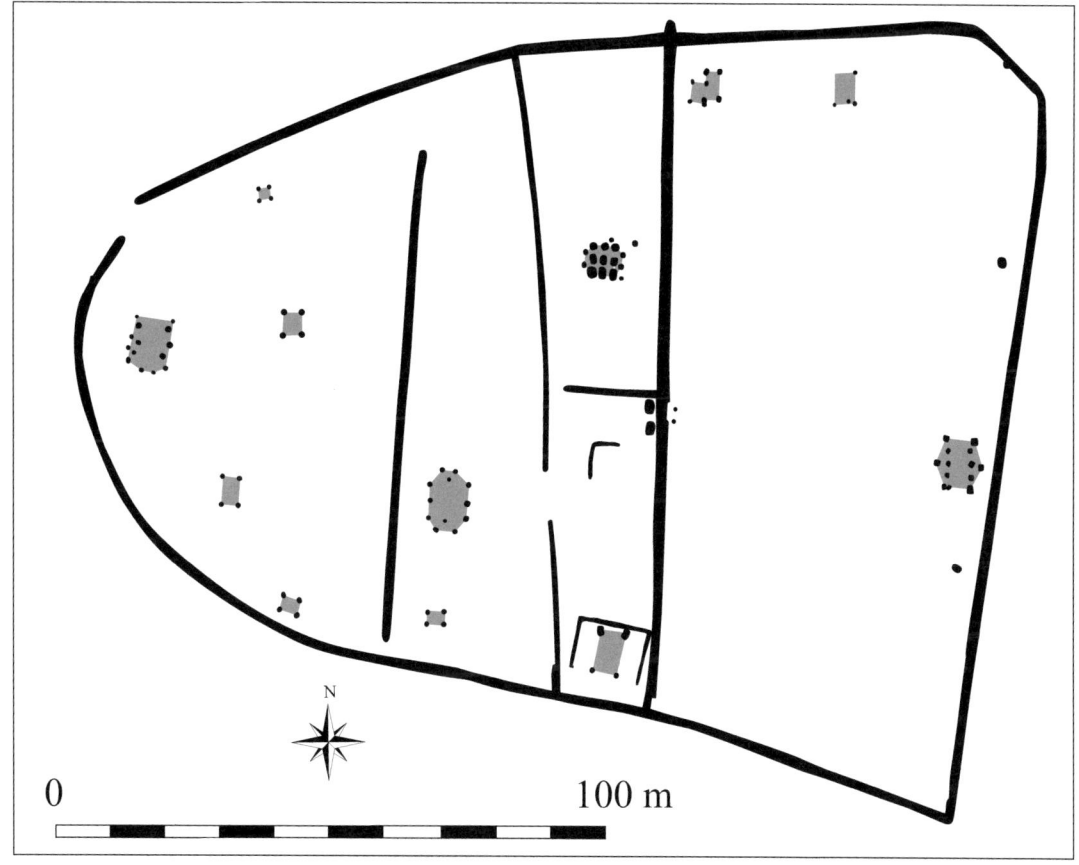

Figure 19: Bazoches-sur-Vesle "la Foulerie" (Aisne), Late La Tène (LTD1), enclosed settlement (CAD: AFAN, UMR 8215 Trajectoires).

Figure 20: Bazoches-sur-Vesle "les Chantraines" (Aisne), Late La Tène (LTD1), enclose settlement (CAD: AFAN, UMR 8215 Trajectoires).

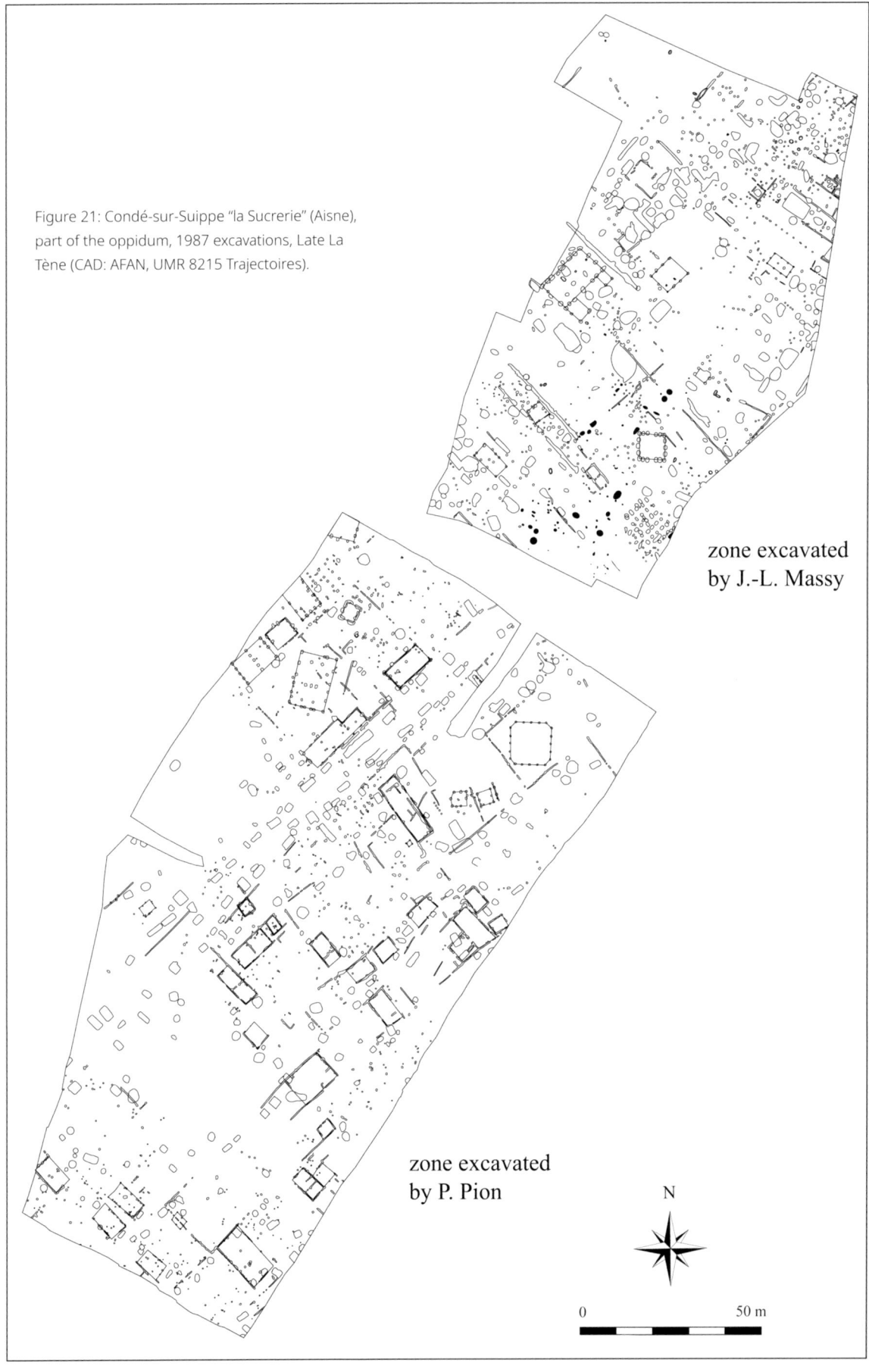

Figure 21: Condé-sur-Suippe "la Sucrerie" (Aisne), part of the oppidum, 1987 excavations, Late La Tène (CAD: AFAN, UMR 8215 Trajectoires).

zone excavated by J.-L. Massy

zone excavated by P. Pion

N

0   50 m

finds and the activities attested to are the principal indicators of high status on the hierarchical scale of sites. Such sites were home to the elites who controlled production and exchange networks (Gransar, 2001; Issenmann, 2009; Le Goff, 2009).

In the Caen Plain, an investigation extending over several hundreds of hectares revealed a large network of pathways which had developed around enclosed farmsteads.

Known Middle La Tène (325-180 BC) settlements are unequally distributed within the regions. The recorded sites range from a few pits and ditch segments to complete enclosures. In the Aisne

valley, where the occupation of the territory is well-documented, very few sites are known for this period. Sometimes it seems as if we are witnessing a desertion of the landscape, or perhaps the absence of evidence is the result of a profound change in the way structures were anchored in the ground. Alternatively, there may have been a change in site preference, with a shift towards the plateaus, for example (Malrain *et al.*, 2013).

In the Late La Tène (200-30/20 BC; La Tène C2 to La Tène D2), forms of settlement become more diversified. In the La Tène C2 (200-130 BC), enclosures tend to be curvilinear and the duration of

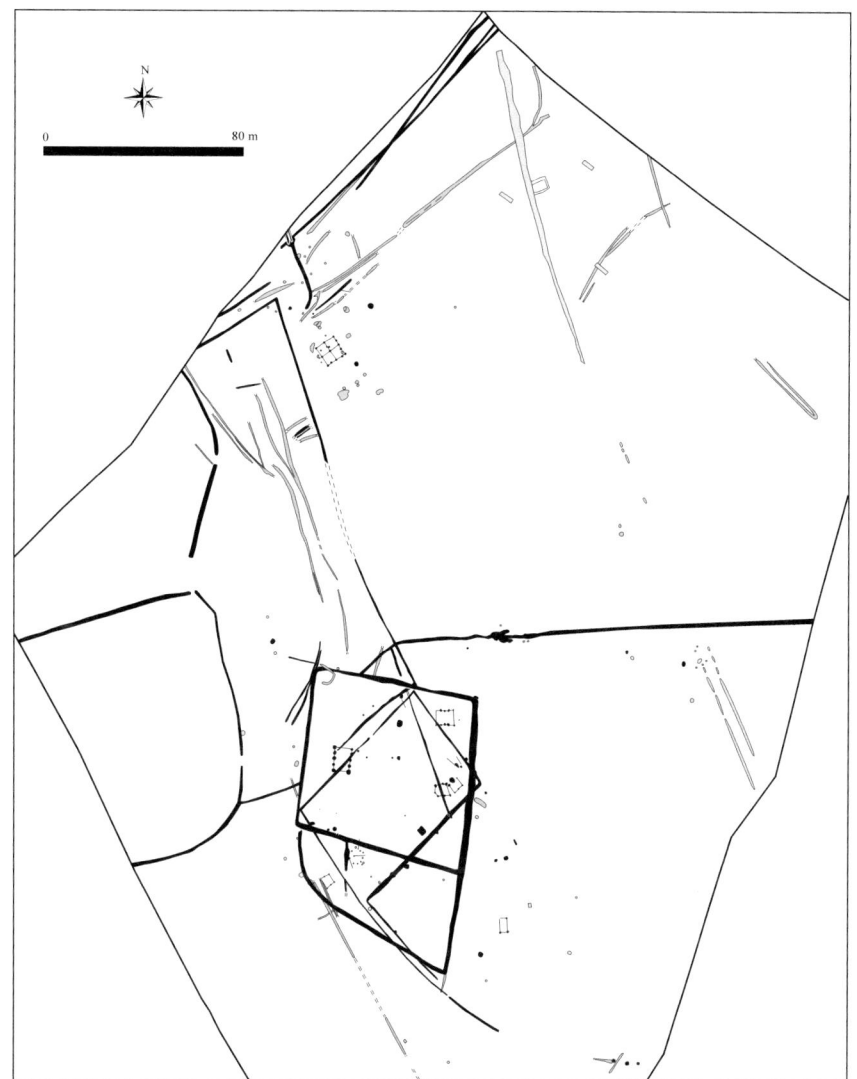

Figure 22: Braine "la Grange des Moines" (Aisne), Late La Tène (LTD2), enclose settlement (CAD: AFAN, UMR 8215 Trajectoires).

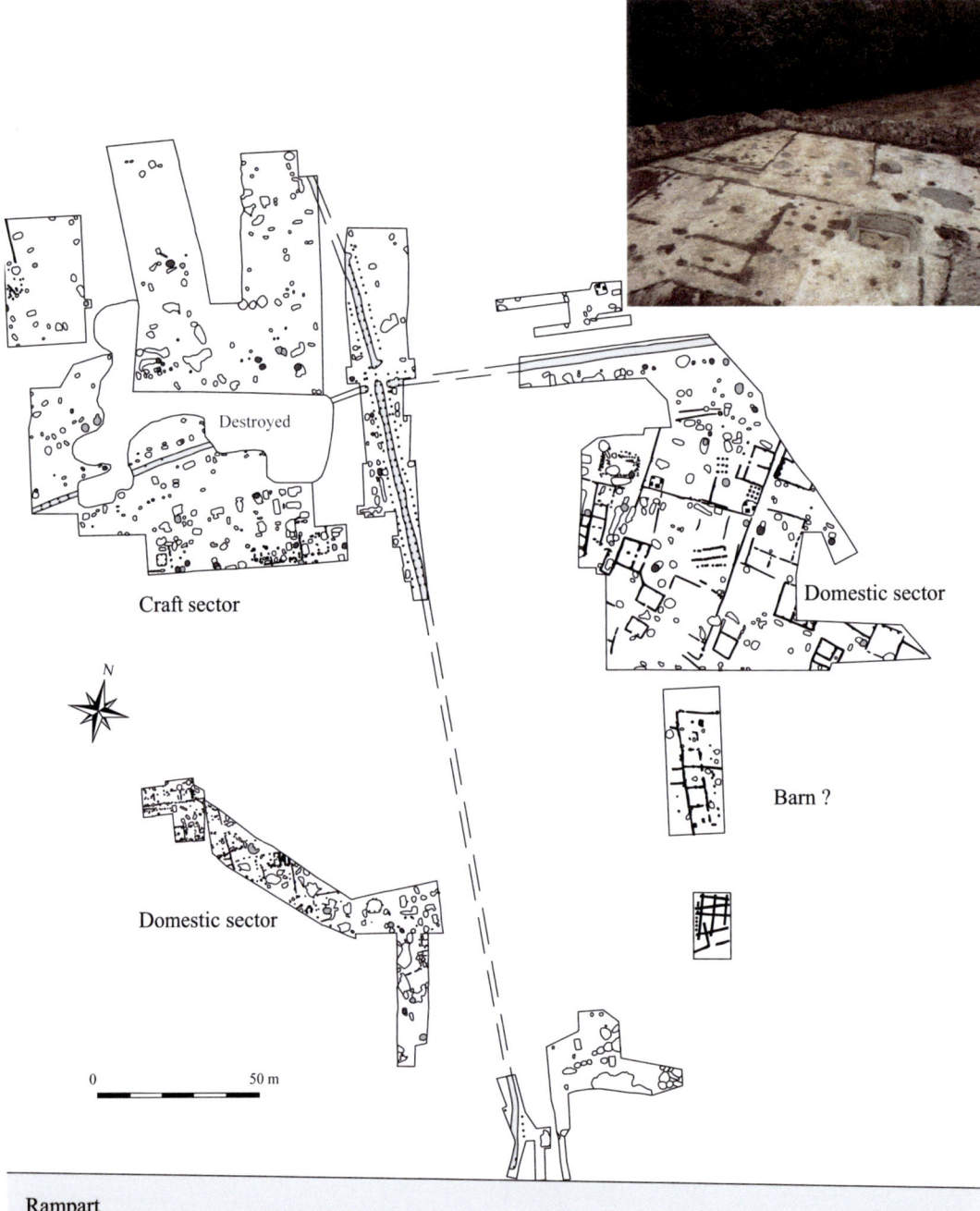

Destroyed

Craft sector

N

Domestic sector

Domestic sector

Barn ?

0   50 m

Rampart

Ditch

Ditch

Figure 23: Villeneuve-Saint-Germain "les Grandes Grèves" (Aisne), part of the oppidum, Late La Tène (CAD and photo: UMR 8215 Trajectoires).

occupation appears not to exceed two generations. In the La Tène D1a (150-110 BC.), we witness the appearance of farmsteads surrounded by rectilinear ditches during the second half of the 2nd century BC (fig.19 and fig.20).

The interior space is divided into areas with specific functions. The houses are imposing in size and are located strategically relative to the monumental entrance to the site. These systems of enclosing ditches, with impressive banks, delimit private space but are essentially ostentatious and in no way defensive. The presence of large numbers of amphorae is one of many indicators of the high status of the residents. Such sites are described as high ranking farms.

In the La Tène D1b (120-85 BC) and La Tène D2a and b (90-30/20 BC), the enclosed farm endures in different forms that reflect hierarchical ranking (Malrain, 2000) (fig. 21), (fig. 22), (fig. 23).

The first oppida appear around 120 BC; these proto-urban centres mark a shift towards a very hierarchized and structured state-like society. They also acted as centres for the production and diffusion of struck or cast coinage.

The farms can be small in size, without ostentatious connotations, featuring very few buildings and a relatively short duration of occupation. Others are more imposing but have more or less the same structure as the modest examples. Finally, we observe high-ranking farms characterised by rich equipment of all types and large constructions that were clearly designed to impress; they are also characterised by the presence of rarer categories of finds (Menez, 2008).

Among the oppida, in the territory of the Suessiones we note the sites of Villeneuve-Saint-Germain (Constantin *et al.*, 1982; Constantin and Debord, 1982; Debord, 1995, 1993, 1990, 1982; Ruby and Auxiette, 2010) and Pommiers (Brun and Robert, 1988) and in the territory of the Remi, the sites of Condé-sur-Suippe and Reims (Neiss *et al.*, 2015; Pion *et al.*, 1997), all of which have only been partially excavated. In the best documented examples, we observe a very clear organisation of space with streets and open areas serving domestic plots which include houses, storage structures (cellars and granaries), wells and pits for various uses. Two oppida have revealed sectors devoted to specialised activities such a metal working, coin production, butchery, the preparation of pelts, etc. The fact that they functioned as residences for elites is now widely accepted. The establishment of these urban centres, which centralised economic, political and apparently religious functions, brought about significant change in the areas of production and consumption. In fact, the Gaulish populations created an urban model that differed from the canonical Mediterranean model, with domestic units that functioned like those of farmsteads where storage, milling and the rearing of small animals were carried out on a household scale.

As throughout the Paris Basin, and in the North of France generally, the sites investigated are not stratified and, apart from a few notable exceptions (Damary and Bucy-le-Long in the Aisne, for example), the ground levels are not preserved. It is estimated that erosion has been responsible for the removal of 20-60 cm of soil, depending on the area. The preservation of architectural features and sub-surface storage features is poor and a non-negligible (but difficult to quantify) quantity of finds has been lost. It is also likely that a portion of the domestic waste was dumped outside the settlements.

2. DOMESTIC CONSUMPTION ON HABITATION SITES

HOUSES, VILLAGES, ENCLOSURES, UN-ENCLOSED SETTLEMENTS

The Neolithic

The Early Neolithic

About 200 Early Neolithic settlements in Temperate Europe have produced faunal assemblages that have been studied or are undergoing study (fig. 3). Of these, about a hundred have been dated to the LBK (from the earliest LBK through to the Final LBK). With very few exceptions, all LBK settlements in the Paris basin are located to the east of the Seine, with a significant concentration in the Aisne valley.

Several overviews of the fauna of the Danubian early Neolithic have already been published, referred to as the Middle and Late Neolithic in central European terminology (Lichardus and Lichardus-Itten, 1985), including works by the following: (Müller, 1964; Bökönyi, 1974; Bogucki, 1988; Döhle, 1993; Arbogast, 1994; Hachem, 1999, 2011a; Tresset, 1996; Marciniak, 2005; Bedault, 2009, 2012). For the Blicquy in Belgium, only two sites have yielded preserved skeletal material, a fact that for the moment prevents a comprehensive overview of the fauna associated with this cultural group. We can only note that the results agree with those already obtained for the Blicquy/Villeneuve-Saint-Germain [BVSG] in the Paris Basin. In recent decades, excavations have greatly increased our knowledge by revealing significant faunal corpora (apart from Blicquy), particularly in the Paris Basin. This is primarily due to the differential preservation of bone; unlike many of the loess regions in temperate Europe, the Paris basin has calcareous subsoils which favour the preservation of bone. However, the situation is undoubtedly also a function of the evolution of research, and of its resourcing and focus, which has contributed to the amassing of new information over the past forty years.

In the context of the LBK, each housing unit consists of the house itself as well as associated lateral pits: the latter were probably originally created for the extraction of clay required in the building of the house. The pits were subsequently used for dumping waste material; this waste material, at least in part, reflects the activities that took place within the house and in its immediate surroundings (Allard et al., 2013; Coudart, 1998; Lanchon et al., 1997; Soudský, 1969). Several models have been proposed for the organisation of LBK domestic space and for over 40 years have provided the basis for reflection and discussion regarding LBK settlement, both in terms of domestic architecture and activity areas (Coudart, 1998; Modderman, 1970; Soudský, 1962; Soudský and Pavlů, 1972) and the organisation of space within the villages (Claßen, 2005; Květina, 2010; Květina and Končelová, 2013; Lüning, 2005, 1982; Pavlů, 2016, 2010; Rück, 2013; Stehli, 1989; Zimmerman, 2012).

The combination of distinct individual building plans, well-preserved bone remains, and an archaeological approach focused on understanding the functioning of the household within the village context, has allowed us to construct elaborate socio-economic models.

Another area that has provided faunal data for the Early Neolithic, and which has benefitted from the presence of a long-established research team, is the Bassée (the alluvial plain of the Seine between Montereau-Fault-Yonne and Nogent-sur-Seine, including the confluence with the Yonne). Unfortunately the recording of faunal data here has not been conducted on a house-by-house basis which means that there is limited potential for comparative studies. The Troyes Plain and its environs, located in the Grand Est administrative region, has been the subject of intensive archaeological prospection over the past decade. Several Early Neolithic sites have been identified, but the bone remains and building plans are poorly preserved, which again limits the potential for comparisons with the models developed for the Aisne and Marne valleys.

The key elements of this corpus have been dealt with within the framework of a number of collective projects: Collective Research Actions ("ACR") in the Aisne valley (Liétar, 2017) and in the Lower Marne valley (Lanchon, 2006; Bostyn and Lanchon, forthcoming), Collective Research Projects ("PCR") in the Aisne valley (Hamon and Allard, 2010), the Bassée (Mordant, 2006), (Giligny, 2006) and in the Troyes Plain (Riquier, 2017).

Lastly, a number of faunal studies were carried out in the context of doctoral dissertations, focussing on the LBK (Hachem, 1995)1995 as well as the BVSG (Bedault, 2004, 2005, 2012), and many archaeozoological studies conducted by L. Hachem were carried out for Inrap's excavation reports.

Villages: rules and variations

In what follows, the results of the archaeozoological studies are presented by region, according to parameters which reflect the geographical characteristics, but also the history of archaeological research carried out in the particular territory.

We will focus principally on archaeozoological data for the valleys of the Aisne and Marne and for the Troyes Plain in our attempts to define village structure in the Early Neolithic because, in all three cases, the bone waste has been fully recorded in databases and has been analysed on the basis of individual household units. Data from the other

regions of the North of France, such as the Seine-Yonne confluence, the lower Seine valley (Tresset, 1996), Oise and Alsace (Arbogast, 1994) will complement this information but will not be used to draw comparisons at the scale of the domestic unit since the publications do not provide spatial references for the fauna at this level of detail.

Faunal remains have been analysed from about ten Early Neolithic sites in Champagne. As is the case for LBK sites in other regions, the sites in the Troyes Plain are located close to watercourses (Riquier and Meunier, 2014) and they take the form of low-density villages made up of large and small houses, similar to villages observed in the Paris basin (Bostyn et al., 2018a). However, as yet, no long-duration settlement site has been fully excavated due to recent or past damage.

The middle LBK, known as the "Rubané moyen champenois" [RMC], is the earliest phase in the Champagne region.

Four sites have provided RMC faunal assemblages[1] ; these include three houses each featuring a single pit. The sites were all excavated several decades ago. In fact no sites from this period have been excavated since the 1980s.

The total number of bone remains is quite low, i.e. 2,826 bones. The small number of structures and the small size of the sample mean that there is a danger that the data does not reflect the true diversity of the consumption.

The second chronological phase in the Champagne region is the Late LBK, followed by the Final LBK and faunal assemblages for this period have been recovered from a larger number of sites[2]. The samples vary significantly in size.

Faunal remains have also been retrieved from sites dated to the Blicquy/Villeneuve-Saint-Germain (BVSG) period in Champagne [3]. The material came from isolated pits rather than from lateral pits.

These faunal assemblages are very small, consisting of about 100 bones.

The Aisne valley has yielded the highest density of Final LBK (RFBS) settlements in the Paris Basin. In the context of a pioneering rescue excavation programme developed since the mid-1970s, 80 km of the valley have been studied by a team composed of members of the Paris 1 University and the former ERA 12 of the CNRS, now known as the

UMR 8215 Trajectoires. The long-term project in the Aisne valley – representing 40 years of archaeological fieldwork – has produced very large quantities of data; 20 sites, 90 houses and 80 burials have been discovered and excavated. Such exceptional archaeological evidence has provided invaluable insights into the relative chronology of these different occupations, the organisation of the territory and the economy of the LBK inhabitants (Ilett *et al.*, 2006; Hamon and Allard, 2010; Dubouloz, 2012b; Ilett, 2012).

The radiocarbon dates for the RFBS settlements in the Aisne valley fall between 5100 and 4900 BC (Dubouloz, 2003a). Three ceramic stages have been defined after a seriation of the ceramic assemblages (Ilett, 2012; Blouet *et al.*, 2013). Not all of these stages are represented on every site, apart from at Cuiry-lès-Chaudardes which had the longest duration of occupation.

Of the ten sites studied, the following six were selected for thorough analysis because of the size of their faunal assemblages, their well-defined chronological sequences and the clarity of their house plans[4].

Two archaeozoological analyses have been fully published in monographs dealing with the LBK sites of Berry-au-Bac "Le Chemin de la Pêcherie" and Cuiry-lès-Chaudardes "Les Fontinettes". For the others, we refer to a forthcoming article, which details the fauna on a house-by-house basis for each site (Hachem, 2018a) and to the appendices.

Several Blicquy/Villeneuve-Saint-Germain sites in the valleys of the Aisne and its tributary, the Vesle, have produced faunal assemblages that have already undergone detailed study (Bedault, 2012)[5].

In contrast to the other valleys of the Paris basin- the Aisne, Oise and Bassée- the Marne valley has not benefitted from a programme of systematic monitoring of gravel quarrying and so our knowledge of the number and density of sites in this region is less detailed. Certain areas, such as the Jablines meander, are relatively well-documented, while others, such as Isles-les-Meldeuses (to the east of Meaux), have not been surveyed, and yet others are extremely built-up, as is the case for the entire area downstream of the study area. Therefore, plans of the Neolithic villages, apart from those in the hamlets of Changis-sur-Marne and Mareuil-lès-Meaux, are incomplete, which may cause a bias in our conclusions.

Having said that, the data collected for the Early Neolithic is of good quality and the sources are numerous: the corpus includes about forty sites, thirty-three individual houses, sixteen burials, about ten sites identified as secondary sites, and several 100s of kilograms of finds. This territory was colonised as part of a westward expansion which resulted in the establishment of a greater number of BVSG sites than in the Aisne valley, for example, while the numbers of sites attributed to the LBK are very low.

The Early Neolithic period spans about four centuries but the fauna, which was studied in its entirety (Bedault, 2012), was analysed using the same protocol as was used for the LBK in the Aisne valley, *i.e.* data provided by the bone waste was considered in the context of individual house units.

The studied period, the Early Neolithic, lasted approximately four centuries, and a regional chronology comprised of four phases has been developed on the basis of ceramic decoration: the Early-, Middle-, Late- and Final BVSG (Lanchon, 2008). Radiocarbon dates for the BVSG occupation lie between 4950 and 4650 BC (Lanchon, 2012).

The sites take the form of villages composed of large houses with associated lateral pits, but can also take other singular or complementary forms such as groups of isolated silo-type pits or sites that were dedicated to the acquisition/extraction of tertiary flint (Lanchon, 2012).

For the Marne valley, the faunal composition per house and per site, and analysis of this data, has been detailed in a work which constitutes a fundamental reference (Bedault, 2012)[6].

The proportions of animals consumed in the villages vary as a function of at least four factors that have been previously identified for the Aisne valley (Hachem, 1999) and which have proved appropriate for LBK sites outside this area: chronology, house size, site environment, and house location within the village.

Let us now look in more detail at the first determining factor, chronology.

In the Champagne Middle Linear Pottery Culture, domestic fauna predominate, accounting for over 90% of the NISP while the percentage of wild animals varies between 2 and 3%, an exception being the site of "Champs Buchotte" in Larzicourt where the percentage of wild fauna reaches 10%. Among the domesticated species, cattle are predominant (62%), followed by caprines (22.7%), and finally by pigs (10.3%).

The wild species are comprised of red deer, wild boar, aurochs and roe deer, with a more significant assemblage of wild boar at Larzicourt "Champs Buchotte" and a preference for aurochs at Larzicourt "Ribeaupé".

In the succeeding period, the Late and Final LBK, domestic animals remain predominant: this is particularly the case for the houses at "les Bordes" and "Clos II" in Buchères, and those at Pont-sur-Seine "Marnay", Lesmont "Les Graveries" and Bréviandes "Zac St Martin" (94 to 96% of NISP).

Cattle are in the majority and represent more than half (about 56%) of the fauna consumed at Bréviandes and at Buchères "les Bordes", while they account for a little less than half (44-46%) at Pont-sur-Seine and Lesmont. However, they account for a much lower proportion at Buchères "Clos II" (10 %) where sheep and goats constitute the primary source of meat (57%). Caprines are also important on the sites of "les Bordes" and Buchères "Parc Logistique de l'Aube", and in Lesmont and Bréviandes (de 27 %) they occupy second position. The third domestic species in order of importance in the diet is pig, which trails far behind cattle and caprines at Buchères "les Bordes" (4.5%), but is a little more important on other sites (10 to 21%). Dog is barely represented which leads us to doubt that it was in fact a source of meat.

The four common large wild game species of the Neolithic (red deer, wild boar, aurochs and roe deer) are present but in variable proportions. All four are present on the sites of Buchères "Parc Logistique de l'Aube" and Bréviandes "Pont-sur-Seine", but only roe deer is present at Buchères "Clos II" and only aurochs at Buchères "les Bordes".

Small game is represented by fur-bearing animals such as badger, beaver, fox, marten, weasel, squirrel, and also by hedgehog. Birds, amphibians and fish are recorded from three sites: Pont-sur-Seine, Bréviandes and Buchères "les Bordes".

Cyprinid vertebrae have been recovered from the site of Buchères "les Bordes", which is evidence that fishing took place at the end of winter or the beginning of spring. Both cyprinid and pike vertebrae were found at Bréviandes, which indicates fishing during the spring and at the end of the summer (Frontin, 2014, 2017).

There is also evidence for the hunting of birds.

Finally, frog remains are systematically found on LBK sites in the Paris basin which leads us to suggest that they were deliberately captured and probably eaten.

Examination of the discarded anatomical parts for the domestic species does not reveal any obvious particularities. The bones appear to have been discarded in an *ad hoc* manner, either in concentrations or in pits. Limb bones, which provided meat, are the most numerous remains. Particular attention is drawn to the presence of new-born animals of all three domestic species within the waste material: this indicates that animals were reared within the settlements themselves. The presence of young animals, in greater numbers than old animals, strongly suggests that livestock rearing was geared towards the production of meat.

As regards the wild animal carcass parts, we observe a puzzling correspondence in the incidence of aurochs remains between the site of Pont-sur-Seine "Marnay" and the site of Buchères "les Bordes" (D39).

At Pont-sur-Seine, a group of several aurochs bones was retrieved from the southern pit of House n°1. The remains included the distal humerus of a male aurochs (an anatomical part that is extremely rare in waste material for this species), which was found lying flat on the bottom of the pit. It was accompanied by three right tali (ankle bones) and one left talus from three female aurochs and two or three cows (or perhaps young aurochs[7]) respectively. The break in the humerus was abnormally blunt, which suggests that the bone had been used for some purpose, and its end appears to bear traces of ochre.

Similarly, at Buchères, a cluster of twelve aurochs bones were discovered in the southern lateral pit of an LBK house. Two individual animals have been identified: a male and a female. Several long bones can be refitted and belong to the female. A pair of mandibles present in the remains indicates the presence of an individual aged between 5 and 6 years.

Since aurochs had a particular symbolic importance in the Early Neolithic, this similarity in the remains from two houses, located in two different settlements, can hardly be accidental, even less so when we consider that both sets of remains were found in the southern lateral pits of two large houses; in these cases it seems that we are, in fact, looking at deliberate deposition.

A number of common traits are therefore shared by all of the LBK sites in the Champagne region: livestock rearing predominates, caprines represent an important proportion of the herd and pigs are rather poorly represented. Dog does not appear to have been eaten. The incidence of game varies depending on the site and the most common large species are present, *i.e.* red deer, wild boar, aurochs and roe deer. Smaller, fur-bearing wild species are also present. In addition, there is evidence for fishing and the hunting of birds.

It is in the succeeding period, the Final LBK of the Seine Basin (*i.e.* RFBS: *Rubané Final du Bassin de la Seine*), that the first farmers arrived in the Aisne valley.

The fauna provide reliable chronological indicators via the proportions of domestic and wild animals consumed. The study is based on data from fifty-three RFBS houses, half of which are located on the sites of Berry-au-Bac "le Vieux Tordoir", Menneville "Derrière le Village" , Missy-sur-Aisne "le Culot", Bucy-le-Long "la Fosselle" and Berry-au-Bac "le Chemin de la Pêcherie" (see appendices); the other half are located on the site of Cuiry-lès-Chaudardes (Hachem, 1995, 2011a). In total a considerable number of bones were examined [8].

If we characterise LBK consumption on the basis of the data from all of these sites, then it emerges that about 80% of the animals consumed were domestic. If we look at the different species involved, it is evident that cattle were predominant (accounting for 60% of the remains), caprines come next in order of importance (+20%) and pigs are the least numerous (c. 15%) (fig. 24). Dog is extremely poorly represented.

It is probable that when a certain demographic threshold was reached, the Neolithic community established a new village and took part of the livestock herd with them. Wild animals were not domesticated locally, as this would have been an overly complex and time-consuming process. In fact, as has been demonstrated in a metric study carried out on bovine bones in the Aisne valley (Hachem, 1995, 2001, 2011a), domestic cattle are readily distinguishable from wild aurochs. A few hybrid animals, the results of mating between wild and domesticated individuals, undoubtedly existed but they were the exception. The cattle were, therefore, domestic very early on and followed the colonisation process.

Metric analyses of pigs and wild boars show a clear distinction between the domestic and wild forms. Thus, the local domestication of pigs can

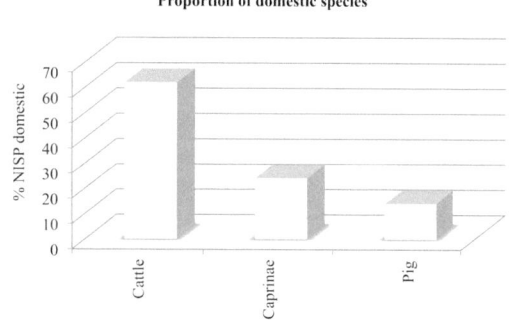

Proportion of domestic species

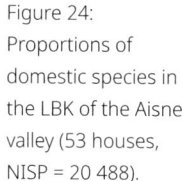

Figure 24: Proportions of domestic species in the LBK of the Aisne valley (53 houses, NISP = 20 488).

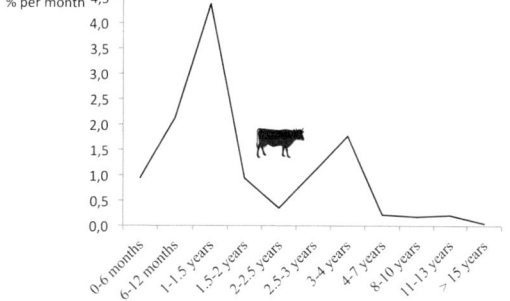

Figure 25: Slaughter pattern for cattle based on teeth, Cuiry-lès-Chaudardes, LBK (NR = 141).

probably also be ruled out (Hachem, 1995, 2011a). As regards sheep and goats, it is generally accepted that they originated in the Near East.

The exploitation of cattle for the production of meat was a priority for the first farmers. In the RFBS, animals were slaughtered at a young age and very few adult animals were allowed to live beyond the age of 6.5 years: this pattern indicates that cattle rearing was geared towards meat production. While age data has been collected for all of the sites studied, the slaughter curves for cattle at Cuiry-lès-Chaudardes (fig. 25) are taken to be representative because of the large number of mandibles present and the homogenous nature of the sample (Hachem, 2011a). Therefore, this site is taken as an example, especially since comparisons with the other sites do not reveal any fundamental differences. Two slaughter peaks have been identified, one for calves aged less than 2 years and a second for animals aged between 2 and 4 years of age. Very young animals are rare. Similar age classes have been identified for sites in Alsace (Arbogast, 1994) and eastern Germany (Müller, 1964).

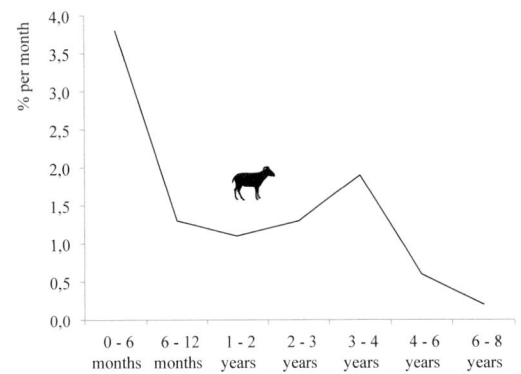

Figure 26: Slaughter pattern for caprines based on teeth, Cuiry-lès-Chaudardes LBK, (MNI = 56).

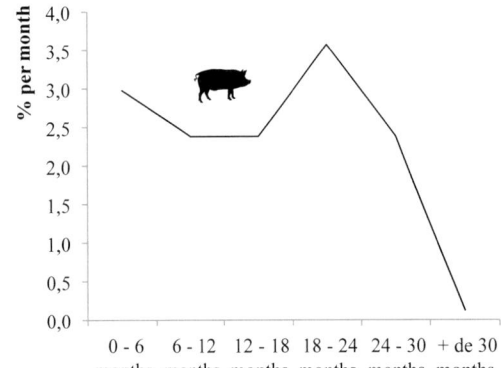

Figure 27: Slaughter pattern for *suinae* based on teeth, Cuiry-lès-Chaudardes, LBK (MNI = 19).

of slaughter of animals aged between 2 and 3 years, a surge in killing between the ages of 4 and 5 years, and the retention of a handful of older animals within the herd (Méniel, 1984; Arbogast, 1994).Data from Cuiry-lès-Chaudardes, and from other sites in the Aisne valley, tends to follow the first scenario.

It is more difficult to glean information regarding *suinae* compared to the two other species. The teeth frequently fall from the jaw bones, for example (in contrast, sheep mandibles are usually complete). Consequently, usable data tends to be scarce and this quantitative handicap prevents us from developing detailed analyses.

Nevertheless, a general pattern emerges in which the majority of pigs were slaughtered at a young age: fourteen out of nineteen identified animals were less than 2 years old at the time of slaughter. Only a minority of animals exceeded 2.5 years of age (3 individuals identified on the basis of teeth).

Two slaughtering peaks are observed, one before the age of 4 months and the other between 17 and 23 months (fig. 27). It is common to find elements of the appendicular skeleton of new-born or very young *suinae*, which confirms the observations made on the basis of teeth. Since the natural juvenile mortality rate for domestic pigs can reach almost 20%, it is possible that some of the remains are those of animals that died naturally.

On the sites in the north-east of France, 40 to 50% of *suinae* were killed before the age of 13 months and animals kept over the age of 3 years are very rare. The slaughtering of pigs at Cuiry-lès-Chaudardes, therefore, follows the same rules as that governing the exploitation of this species in the LBK, with an overriding emphasis on meat production. However, a slight difference is evident (provided that the numbers are representative): we see a greater number of animals exceeding the age of 1 year on this site.

Data regarding the sex of the adult animals, based on bone measurements, reveals that neither sex was selected over the other since the proportions of males and female are identical.

Certain similarities can be observed in the processing of cattle and caprines: intensive slaughter

We are therefore dealing with a large body of sub-adult animals, with the number of adult females exceeding that of males.

Sheep were also slaughtered at an early age: 54 % before the age of two years, with a significant slaughter peak between three and six months. If we add individuals killed before the age of twenty-four months, without any other specification regarding age group, this figure rises to 65%. A second peak in slaughter occurred at three to four years and only 17% of animals exceeded this age. Several animals were maintained in the herd until they were killed at six to eight years of age (fig. 26).

If we compare these patterns with LBK sites in Alsace, we see that this region has two types of slaughter curve for caprines, more than a third of which were killed before the age of 18 months: the first involves animals that were slaughtered between the ages of 3 and 6 months, the second is made up of animals aged between 12 and 18 months. In the two cases we note a low incidence

was carried out at two points in time, one when the animals were very young (before 6 months for lambs and between 1 and 1.5 years for calves and juvenile bovines), and another once the animals had reached maturity at 2 to 3 years of age. Pigs also seem to have been killed at two points in their lifecycle, at 4 months and between 17 and 24 months. The variations in the processing between the species are evident from the proportions of the different slaughter age groups described above. However, it is worth synthesising these findings so as to be able to compare them to other sites with smaller samples. If the animals are classified according to the three main developmental stages identified in the archaeological samples (Habermehl, 1961; Silver, 1969; Payne, 1973; Jones and Sadler, 2012) *i.e.* very immature (less than 1 year), sub-adult (from 1 to 3 years, or 4 years for cattle) and adult (over 3 or 4 years) – the following slaughter pattern emerges for the site of Cuiry-lès-Chaudardes:

- for caprines, the three age categories are more or less equally represented, which, when compared to the situation for cattle, highlights higher proportions of lambs and adult animals;
- for cattle, the sub-adult category largely dominates over the very young and adult categories and the proportions of the latter two are similar;
- for *suinae*, the young and sub-adult categories predominate over the adult category.

For the three main domestic species, therefore, animal husbandry was geared towards the production of meat, probably with some exploitation of secondary products in the case of caprines.

The spectrum of wild fauna is very varied with numerous species being recorded. This fauna can be divided into two groups: a principal group consisting of large game and another composed of small game.

Four species of large game occur regularly: aurochs, wild boar, red deer and roe deer. The contribution of game to the meat diet was greater than might be thought on the basis of numbers of remains. At Cuiry-lès-Chaudardes, for example, cattle are followed in importance in terms of weight of meat by red deer, wild boar and pig (Hachem, 1995, fig. 59, vol. 2).

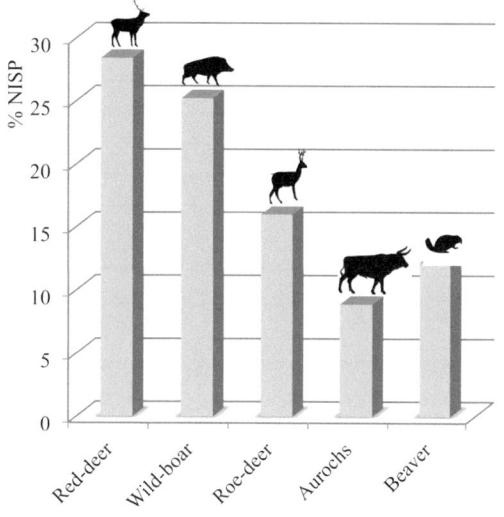

Figure 28
Proportions of the main wild species in the LBK of the Aisne valley (NR = 1134).

Among the large wild animals, wild boar and red deer occur in equal quantities (c. 27%) and between them they represent more than half of the remains (fig. 28). Roe deer comes in third position with 16%, while aurochs comes last with only 8% of the number of remains.

Other large wild animals, such as wolf, bear and horse, make up a very small percentage of the remains and are, therefore, categorised as rare animals. However, while this small number of bone remains indicates that the animals were probably not consumed, the bones nonetheless appeared to have a special status and it is possible that they were considered as "amulets" (see chapter 4).

There is considerable variety in the small game, which is principally composed of fur-bearing animals such as beaver (14 % of game remains (fig. 28) and badger: however, other species such as hare, fox, marten (or weasel), wild cat and even hedgehog are also present.

Birds and fish are also present. Frogs are frequently encountered, and while some of these may be recent intrusions, it is probable that many are Neolithic in date. In fact, they are associated with the other refuse,, display the same colouration as other bones and are frequently recorded on the LBK sites under consideration. They may have been consumed or may have been used for other purposes such as medicinal practices.

The breakdown of the age classes for the wild species is as follows:

- for red deer, the minimum numbers of young animals (27.9%) and sub-adult animals (6.9%) are not negligible, but adult animals remain predominant (65.1%).
- a significant proportion of roe deer also fall into the category of young animals (23.2%), but adult animals are much more common (76.6%) while sub-adult animals are absent from the sample.
- in contrast to the *cervidae*, young wild boars are less well represented (13.7%) than sub-adults (9.8%) and adults (76.4%).
- finally, in the case of aurochs, sub-adults (9.7%) and adults (87%) largely predominate.

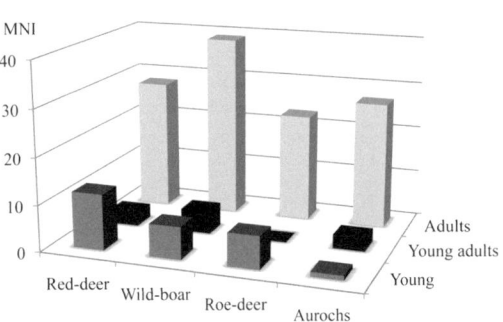

Figure 29: Mortality profile for game, Cuiry-lès-Chaudardes, LBK (after Hachem 2011a, fig. 69).

Slaughter ages for wild animals have been estimated on the basis of the stage of epiphyseal fusion in the bones of the appendicular skeleton because the number of mandibles is insufficient to provide accurate ages. The lack of tables listing the long bone epiphyseal fusion ages for wild animals, apart from tables for red deer and certain parts of the roe deer skeleton, prevents us from determining precise age groups. This is why individuals are divided into three broad categories, which give us an overall idea of the age structures of the hunted species (fig. 29). Since Cuiry-lès-Chaudardes yielded the greatest number of recorded remains, we have decided to present the data from this site but it should be noted that similar observations have been made for the other sites.

The minimum number of individuals obtained within each broad age category was calculated for each house. Differences become apparent between the age groups of the various species killed, but there is one point in common, namely the absence of very young animals. One possible explanation for this absence in the case of aurochs, roe deer and wild boar, could be potential confusion between the remains of young individuals of these species and those of juvenile cattle, sheep and pigs[9]. The three wild individuals identified here as being immature or sub-adult, were identified on the basis of fairly characteristic morphological and dimensional criteria. For roe deer, the longitudinal groove on the metapodials is a good indicator. For aurochs and wild boar, the total absence of epiphyseal fusion, or the presence of partial fusion, when the bone size of a young individual already exceeds that of an adult of the corresponding domestic species, appears to be a good indicator.

Furthermore, in the case of wild boar and aurochs, we observe a higher number of fully fused bones for female animals than for males. Hunting, therefore, appears to have targeted adult wild boar sows and adult female aurochs.

In conclusion, while we cannot determine the exact proportion of young wild animals killed, it appears, based on the *cervidae* age classifications (for which the bones of young animals are more easily identified than those of *bovinae* and *suinae)* and on comparison of the proportions of adult *suinae* and *bovinae*, that the hunting strategy employed at Cuiry-lès-Chaudardes targeted animals of reproductive age.

We have identified a minimum of 117 adult animals among the four large wild species, as against eleven sub-adults and twenty-seven young animals: 75% of hunting, therefore, targeted animals that had reached their mature weight.

Finally, we note two particularities: firstly, young *cervidae* (some of which were less than 8 months old) occur in greater numbers than young animals of the two other large game species; secondly, the majority of adult aurochs and wild boar were female (for the moment it is not possible to identify the sex of the *cervidae* on the basis of long bones).

All anatomical parts of the three domestic species are present in the lateral pits of every LBK settlement site, with an abundance of head fragments, ribs and limb bones, indicating that the animals were mainly used to produce meat. In the case of wild animals, however, the representation of skeletal parts varies according to species.

Deer hind limbs (tibia, femur) and antlers are abundant, indicating that this animal was exploited as a source of raw material for tools as well as for its meat. The aurochs remains, like those of wild boar, present particularities such as large numbers of metapodials and phalanges.

Beaver and badger are frequently represented by teeth, although other parts of the skeleton are also present. There is, however, a difference in treatment between the two species, as in general badger hind legs are absent. Traces of flint tools indicate skinning. These animals seem to have been used for fur, but they were certainly eaten as well. In fact, the same butchery practices were applied to the two species as has been described in the site of Cuiry-lès-Chaudardes (Hachem, 2011a, chap. II.1.2.2.b.).

This general description can be refined through the study of consumption of the seven principal species for each chronological phase of the RFBS (fig. 30).

In houses dating to the early phase of the RFBS, cattle were the primary source of meat in the diet and make up half of the sample of identified remains, caprines occupied second position and pigs came third with slightly fewer remains. The other species present are wild species, with red deer and wild boar representing equal proportions, followed by roe deer and then aurochs.

In houses belonging to the middle phase, cattle constitute a slightly larger proportion of the remains, while caprines and pigs were of equal importance. Red deer and wild boar also occur in equal proportions, followed in order of importance by roe deer and aurochs (fig. 31).

In the final phase, cattle comprise less than half of the remains and caprines occur in greater numbers than pigs. Red deer and wild boar occur in similar proportions, while roe deer is in third position and aurochs in fourth

Figure 32: Evolution of the main species proportions in the LBK of the Aisne valley (early, middle and late phases; data from figure 31). a- red deer, wild boar. b- beaver. c- aurochs and cattle.

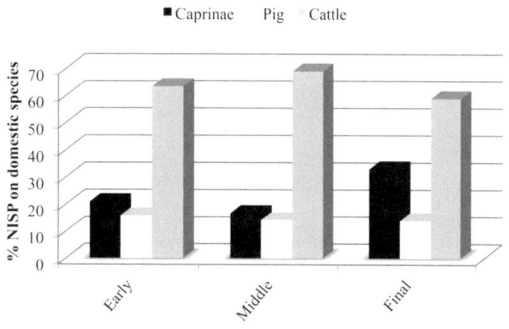

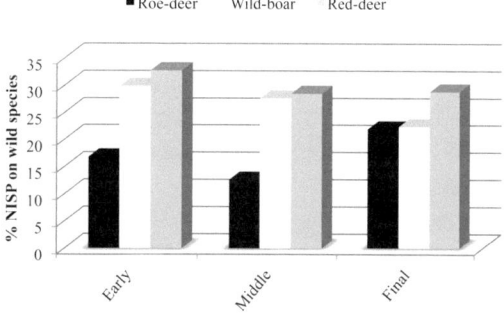

Figure 30: Evolution of the proportions of the main domestic species in the LBK of the Aisne valley (early, middle and late phases; data from figure 31). a- Domestic animals. b- Wild animals.

Figure 31: Number of remains of the ten main species by site and by chronological phase in the LBK of the Aisne valley. BCP: Berry-au-Bac "le Chemin de la Pêcherie", BVT: Berry-au-Bac "le Vieux Tordoir", CCF: Cuiry-lès-Chaudardes "les Fontinettes", BLF: Bucy-le-Long "la Fosselle", MDV: Menneville "Derrière le Village", MAC: Missy-sur-Aisne "le Culot".

Sites	Phase	Total remains	Total houses	NISP	Domestic	Wild	Cattle	Caprinae	Pig	Red-deer	Wild-boar	Roe-deer	Aurochs	Beaver
BCP	Early	5068	3	1868	1802	66	1176	371	255	30	6	13	11	0
CCF	Early	9774	6	3154	2290	748	1414	471	402	235	245	130	30	78
BVT	Early	662	3	324	264	51	169	68	27	18	7	2	22	0
Total	Early	15504	12	5346	4356	865	2759	910	684	283	258	145	63	78
MDV	Middle	2400	3	916	737	164	507	114	109	41	26	31	7	45
BLF	Middle	5029	6	1712	1551	149	1076	310	158	54	34	22	21	0
CCF	Middle	17349	10	5901	4757	1084	3275	749	726	303	326	123	113	134
Total	Middle	24778	19	8529	7045	1397	4858	1173	993	398	386	176	141	179
MAC	Final	3428	6	978	853	116	438	341	74	26	11	7	18	46
BLF	Final	4236	2	1427	1363	58	936	300	127	17	6	10	12	3
BVT	Final	1937	2	656	492	160	218	231	42	75	32	38	7	0
MDV	Final	5037	3	949	904	37	608	230	65	16	4	5	0	0
CCF	Final	21597	9	6512	5505	939	2877	1741	883	198	204	192	78	129
Total	Final	28571	22	10522	9117	1310	5077	2843	1191	332	257	252	115	178

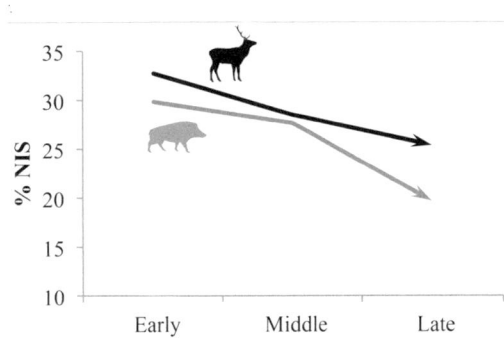

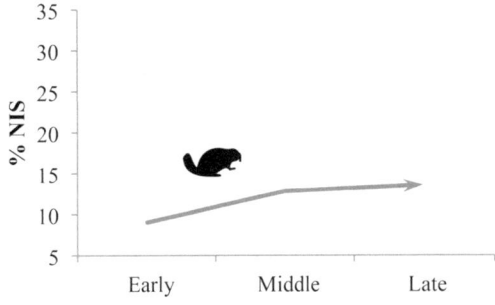

Figure 32: Evolution of the proportions of red deer, wild boar, beaver, aurochs and cattle in the LBK of the Aisne valley (early, middle and late phases; data from figure 31).

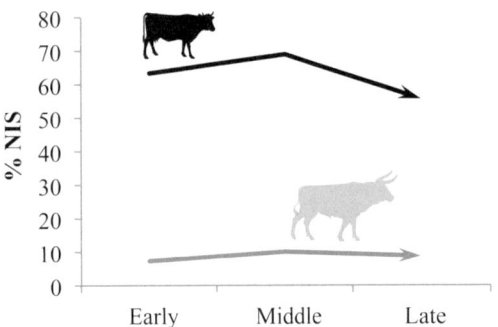

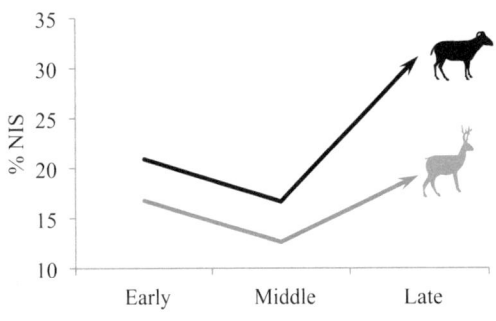

Figure 33: Evolution in the proportion of species between the beginning and end of the Aisne valley LBK sequence (data from figure 31): first the rise of caprines and roe deer, second the regularity of pig.

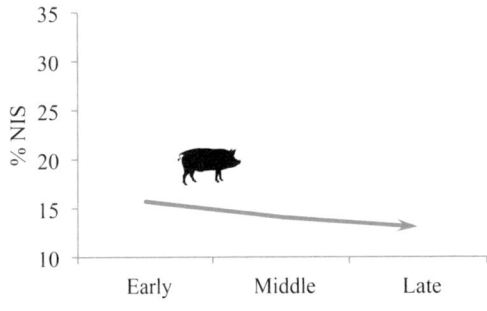

Taken as a whole, these observations indicate that significant changes occurred in the proportions of animals over the course of the chronological sequence.

If we consider the evidence on a village scale it emerges that the contribution from hunting declined over time. This is confirmed on the site of Cuiry-lès-Chaudardes, which has the advantage of spanning all three chronological phases, the sites of Menneville and Bucy-le-Long which have yielded evidence for only two phases. The proportion of wild animals drops from 16.4% in the early phase to 11.8% in the final phase (see appendices).

Over the space of the 100 years or so that a long-term settlement would have been occupied, numbers of the two wild species that yield the most substantial quantity of meat, namely red deer and wild boar, dwindle simultaneously indicating that the Neolithic population hunted more at the beginning of the occupation than they did at the end (fig. 32). There are numerous possible reasons for this decline. For example, hunting grounds may have shifted further and further away from the village as animals became wary of the human presence. A similar hypothesis has been forwarded for the low incidence of red deer hunting in central Germany and Poland where intensive grazing by herds of cattle may have degraded the natural environment of wild herbivores (Uerpmann and Uerpmann, 1997). Another possible explanation is a change in dietary choices: game may have become less and less essential in the diet as the domesticated herd grew in size.

However, it is interesting to note that roe deer does not follow quite the same pattern (fig. 33). Unlike red deer and wild boar, roe deer were often hunted more intensively in the final phase of the chronological sequence. This may have been due to the gradual disappearance of the other two species: the cultivated fields may have attracted roe deer that would have then taken the place of the other game animals.

Aurochs, the fourth species belonging to the large game category, follows quite a different pattern to the three other species as its relative frequency remains constant throughout the sequence (fig. 32). In fact a striking contrast between aurochs and the other species has been revealed at Cuiry-lès-Chaudardes and is confirmed on other LBK sites in the Paris Basin: our studies have revealed that auroch bones tend to be found in the assemblages with higher proportions of domestic animals, in particular cattle.

While the contribution of large game declines in importance over time, the incidence of small game remains constant throughout the sequence. This

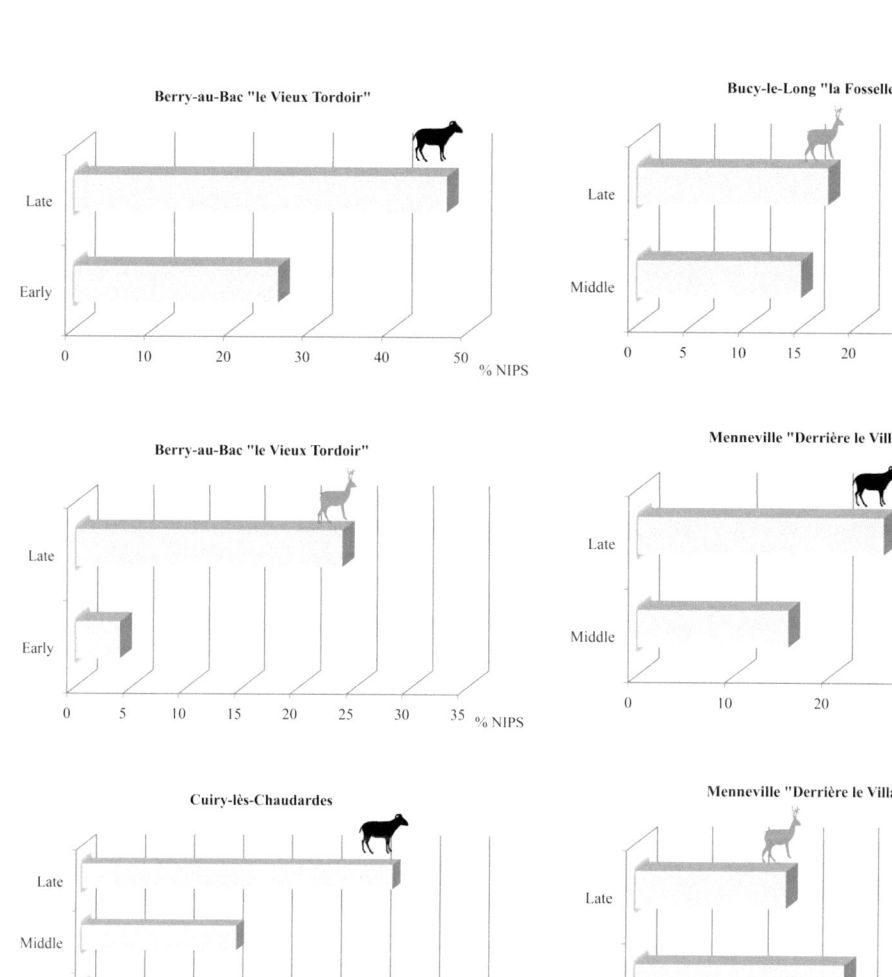
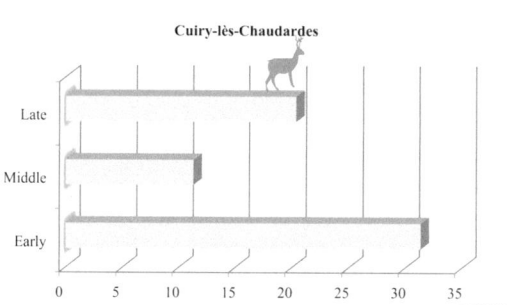
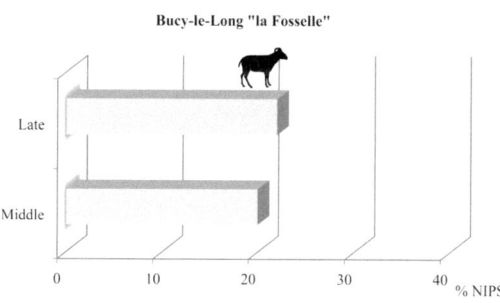
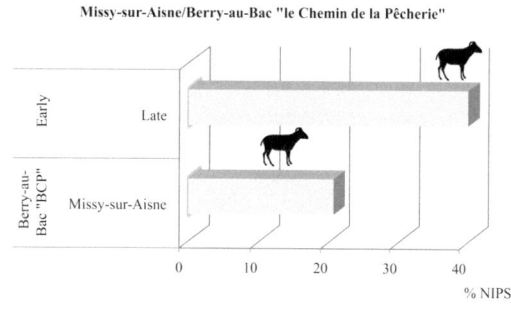
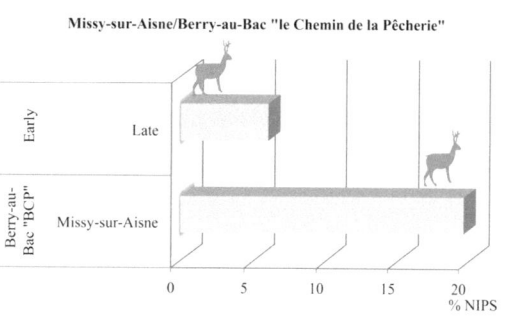

Figure 34: Proportions of caprines and roe deer between the beginning and end of the chronological sequence in six LBK sites in the Aisne Valley (data from figure 31).

tends to suggest that furs, which were probably used in exchanges between households, were an important resource over the entire lifespan of the village (fig. 32).

The decline in hunting was necessarily accompanied by an increase in livestock rearing. This growth is revealed by a marked increase in the numbers of caprines over cattle on certain sites (fig. 31), (fig. 33), (fig. 34).

The proportion of pigs remains stable over the entire sequence. Caprines and roe deer share a similar trajectory and the two animals may have been linked as we have demonstrated for the site of Cuiry-lès-Chaudardes (Hachem, 2011a).

It is worth noting that all types of finds point to a significant break towards the end of the RFSB. It is clear that this terminal part of the Aisne sequence is quite distinct from the others. Whether we consider lithic finds, faunal remains, architecture or ornament, clear changes are evident in acquisition territories, production and herd management. We can, therefore, discern two stages within the LBK of the Aisne, *i.e.* a first stage that sees the LBK establish itself and then develop *in situ* while retaining strong original characteristics, and a second, at the end of the sequence, which marks a clear break with the established tradition. We observe that this final step is characterized by an increase in the number of sites, and in the number and size of houses, within the valley, a phenomenon which is perceptible throughout the Paris basin and which prefigures a growth in the population.

The spectrum of fauna discovered to date on Blicquy/Villeneuve-Saint-Germain (BVSG) sites in Champagne is relatively limited: cattle remains predominate, followed by pig. In contrast, caprines are practically absent, which is indicative of a major dietary shift compared to the LBK. Wild boar, red deer, aurochs and roe deer are still present. Small game is generally absent, but this may be an artificial bias due to the small size of the assemblages (see appendices).

Of the forty-seven sites that have yielded faunal remains in the Aisne and Vesle Valleys, in the lower Marne valley, at the Seine /Yonne confluence, and in the Oise valley, twenty have been the subject of detailed zooarchaeological

analysis. Together they represent a very large corpus: about 120,000 remains were analysed (Bedault, 2012) during the preparation of a large-scale regional synthesis which the reader should consult for details regarding the faunal composition per house, per village and per region. In what follows we provide a summary of these results.

The chronological breakdown of the sites is as follows [10] (Bedault, 2012, 443):

- seven sites for the early phase (total number of remains = 53 345),
- twelve sites for the middle phase (total number of remains = 82 381),
- six sites for the late phase (total number of remains = 32 474),
- one site, located in the lower Marne valley, for the final phase (total number of remains = 99).

On all of the BSVG sites, the bulk of the meat in the diet was provided by domestic animals, *i.e.* between 70 and 90 % of remains. The species present are cattle, pig, sheep and goat. Cattle account for the highest proportion of remains (52%), followed by pigs (30%) and caprines (18%). Dog represents only a tiny proportion of the corpus.

Domestic cattle are, therefore, predominant. While on average they account for about half of remains, there is a significant degree of variation between sites: (between 21 and 86%). Cattle are almost invariably in the majority and in the rare instances where this is not the case, they hold an equal position with one of the two other main domesticated species.

Domestic cattle are relatively easily distinguished from aurochs. Their height at the withers remains the same as for the LBK, although metric data indicates that certain individuals were smaller and stockier in build than in the preceding period (Bedault, 2012). Sexual dimorphism is marked in the domesticated and wild species. Just as we have seen for the LBK in the Paris basin and on the sites in Alsace and Champagne, it seems unlikely that there was local domestication of aurochs or interbreeding between the two species. BVSG farmers appear to have maintained the practice of keeping the domesticated and wild domains separate, not only when they continued to occupy the same sites as their predecessors but also when they expanded westwards to occupy new territories.

For all of the BVSG sites, pigs are a little more plentiful than caprines but, depending on the settlements, one or other of the species may occupy second position after domestic cattle. While the average for all of the samples is about 26%, the frequencies can vary by a factor of one or two.

The available faunal corpora provide a particularly abundant frame of reference for distinguishing between wild and domestic forms of the *sus* genus and differentiation between the species is quite distinct. Their height at the withers remains the same as it was in the RFBS although, as was the case with *bovinae*, we observe a wider range of variation for the domesticated species and the presence of a few smaller individuals whose size is below that normally attributed to sows.

The proportion of caprines (on average 22%) can vary quite considerably.

A detailed study of the differentiation criteria for sheep and goats has revealed that sheep predominate. In contrast to cattle and pigs, there are no small-sized individuals and the stature is of the same order as that identified for the RFBS.

In some house assemblages the three domestic species occur in equal proportions, although this is never the case at site level.

Even though wild animal remains represent only a small proportion of the faunal corpus, they are, nonetheless, present on almost all of the sites. Their proportions vary between 3 and 33% with an average of about 9%. Hunting was principally focused on large game, with red deer in first position, followed in order of importance by wild boar, aurochs and roe deer. In general, red deer was the principal game animal hunted but on certain sites aurochs and wild boar surpass the *cervidae*. Roe deer is less well represented and is often absent from corpora. Examination of sexual dimorphism in the wild boars reveals a deficit of male animals while females occur on all of the sites: for the other wild species we note that females also tend to be more numerous.

Bear, horse and wolf are present but only in very small numbers.

Small game, which is present in small numbers on all sites, is principally made up of fur-bearing animals such as fox, badger, marten, weasel, and beaver.

Fish remains are only found on a few sites, but this may be due to a lack of sieving. The site of Trosly-Breuil has produced the largest corpus (7,255 remains). Quite a variety of species were fished, with a predominance of bream (*Abramis brama*) and pike (*Esox lucius*).

Bird species identified in the corpora are native to open forest and rivers, biotopes that probably existed in close proximity to the villages. Several varieties have been identified including blackbird (*Turdus sp.),* passerines

(*Turdidae*), duck (*Anas platyrhynchos*) and magpie (*Pica pica*). Their presence is never very pronounced within sites.

Analysis of the evolution of the proportions of domestic animals reveals a number of changes over time (Bedault, 2009, 2012).

Firstly, cattle decline in terms of the overall proportions but, nonetheless, remain predominant throughout the BVSG.

The high proportion of caprines observed at the end of the RFBS continues into the early phase of the BVSG and the beginning of the middle phase but the proportions drop somewhat over time. During the middle phase of the BVSG they decline markedly and are replaced by pig in the order of importance. This trend continues into the Late BVSG and can be observed on most sites, but there are a few exceptions which suggest that the chronological factor is not the only element to be considered when looking at species variations.

Cooking and eating within the home

The second factor determining the proportions of species is house size, which is linked to the type of house that the occupants inhabited.

A study of the house plans was conducted in order to identify clear distinctions between houses and to look for relationships between the faunal remains and the types of house; the aim in doing so was to throw light on the Neolithic social structure. In our study we refer to the description of house plans proposed by A. Coudart (Coudart, 1998) and more specifically to our formal classification (Hachem, 2000a, 2011a), which defined categories of house size depending on the number of bays (units) in the rear part of the house:

- Small houses with one rear unit (house length: 9.5 m to 15 m)
- large houses with two rear units (house length: 15 to 21 m)
- large houses with three rear units (house length: 21 to 39 m)

In fact, the houses respond to specific technical, social and identity rules (Coudart, 1998, 2015, 2009; Gomart *et al.*, 2015). Ethnography teaches us that

Figure 35:
Classification
of houses and
species at Cuiry-lès-
Chaudardes, based
on agglomerative
hierarchical clus-
tering (AHC) of
deviations from
the site average;
W: wild (hunting); M:
mixed; D: domestic
(stockbreeding).

the front part of the house (which may include the space outside, immediately in front of the house) is the transition between the interior and exterior domains, and signals the household's function, status, and identity. Generally separated from the exterior world by a "corridor", consisting of two closely spaced rows of posts, the central part of the Bandkeramik house seems to have been the place where daily activities took place and where visitors were received. The rear part is situated at the far end of the building and is separated from the central part of the house by a second "corridor". This was probably the most private part of the house. Unlike the central part, the number of units in the rear part is directly related to its length. This observation has led to the hypothesis that the size of the rear part is related to the number of people in the household (Dubouloz, 2008; Hamon and Allard, 2010; Allard et al., 2013; Coudart, 2015).

A recurrent pattern can be observed at Cuiry-lès-Chaudardes and several other LBK sites in the Aisne valley: each group of contemporary houses making up a settlement phase is composed of several small houses associated with one (or two) long houses. The rear parts of these large houses are characterized by a foundation trench. However, this feature disappears in the last settlement phase – apparently a change in building practice. At Cuiry-lès-Chaudardes the first chronological stage, which reflects initial occupation by migrants, includes just a few houses, while the last stage is characterized by a larger number of houses.

The search for relationships between fauna and house type was initially undertaken at Cuiry-lès-Chaudardes (Hachem, 1997, 2011a). This settlement can be taken as an example as it is representative of the other Paris Basin sites; it represents a coherent sample since there are numerous houses, preservation conditions are identical and the various houses belong to different occupation phases, which means that we have a good understanding of the chronology of the waste disposal.

Cuiry-lès-Chaudardes	House	Wild boar	Pig	Red deer	Roe deer	Caprines	Aurochs	Cattle	% Big game	% Domest.	Wild/Dom/Mixed
Group 1 of the AHC Pig/Wild boar	M425	++	++	+	-	--	---	+	+++	-	W
	M420	++	++	=	-	--	---	+	+++	-	W
	M635	+++	+++	--	-	--	+	=	++	-	W
	M126	+	++	+	=	+	---	-	+++	-	W
	M89	++	+	--	+	+	+	-	+	=	M
	M112	+	+	-	=	-	--	=	++	-	W
Group 2 of the AHC Red deer/Roe deer and Cattle or Pig	M640	--	+	+	+++	--	--	=	+	=	M
	M530	--	+	+	+++	--	--	-	-	=	M
	M330	--	+	++	++	-	--	=	=	=	M
	M580	=	+	++	--	-	---	=	+	=	M
	M45	--	+	++	-	-	--	=	+	=	M
	M390	-	+	+	--	-	++	=	=	=	M
Sub group of Gr.2 Caprines/Aurochs	M570	-	+	+	++	=	--	-	-	=	M
	M320	---	+	+++	+	++	-	-	++	-	W
	M360	-	--	-	-	+	+++	+	=	=	M
	M90	+	--	-	-	=	+	+	=	=	W
Group 4 of the AHC Caprines/Roe deer/Aurochs	M280	--	--	-	+++	++	+++	-	--	+	D
	M225	--	+	---	+++	+++	+++	--	--	+	D
	M245	-	--	+	=	+++	=	-	--	+	D
Sub group of Gr.3 Cattle/Wild boar	M400	++	--	-	--	--	+	++	+	=	M
	M440	++	=	-	--	--	=	+	+	=	M
Group 3 of the AHC Cattles/Aurochs and Red Deer	M500	---	+	+	--	-	+	=	--	+	D
	M11	---	-	+	+	-	++	+	--	+	D
	M450	---	--	+	-	-	+++	+	--	+	D
	M85	--	--	+	-	--	+++	++	--	+	D
	M380	-	--	+	--	--	+++	+	-	=	M
	M410	---	+	+	-	--	+++	+	-	=	M
	M520	---	++	+	-	-	+++	+	--	+	D
	M80	---	++	+	+	--	+++	+	--	+	D
Average of the site by species	%	36,1	15,9	34,7	18,5	23,41	10,8	60,7	14,6	85,4	

+++/--- Deviation ≥75% of the average of the site. ++/-- Deviation ≥25% of the average. +/- Deviation ≥ 5% of the average

According to the species, percentages and differences are calculated on the total of the domestic or on the total of the big game;
In bold the significant positive differences, because based on sufficient numbers to obtain reliable percentages

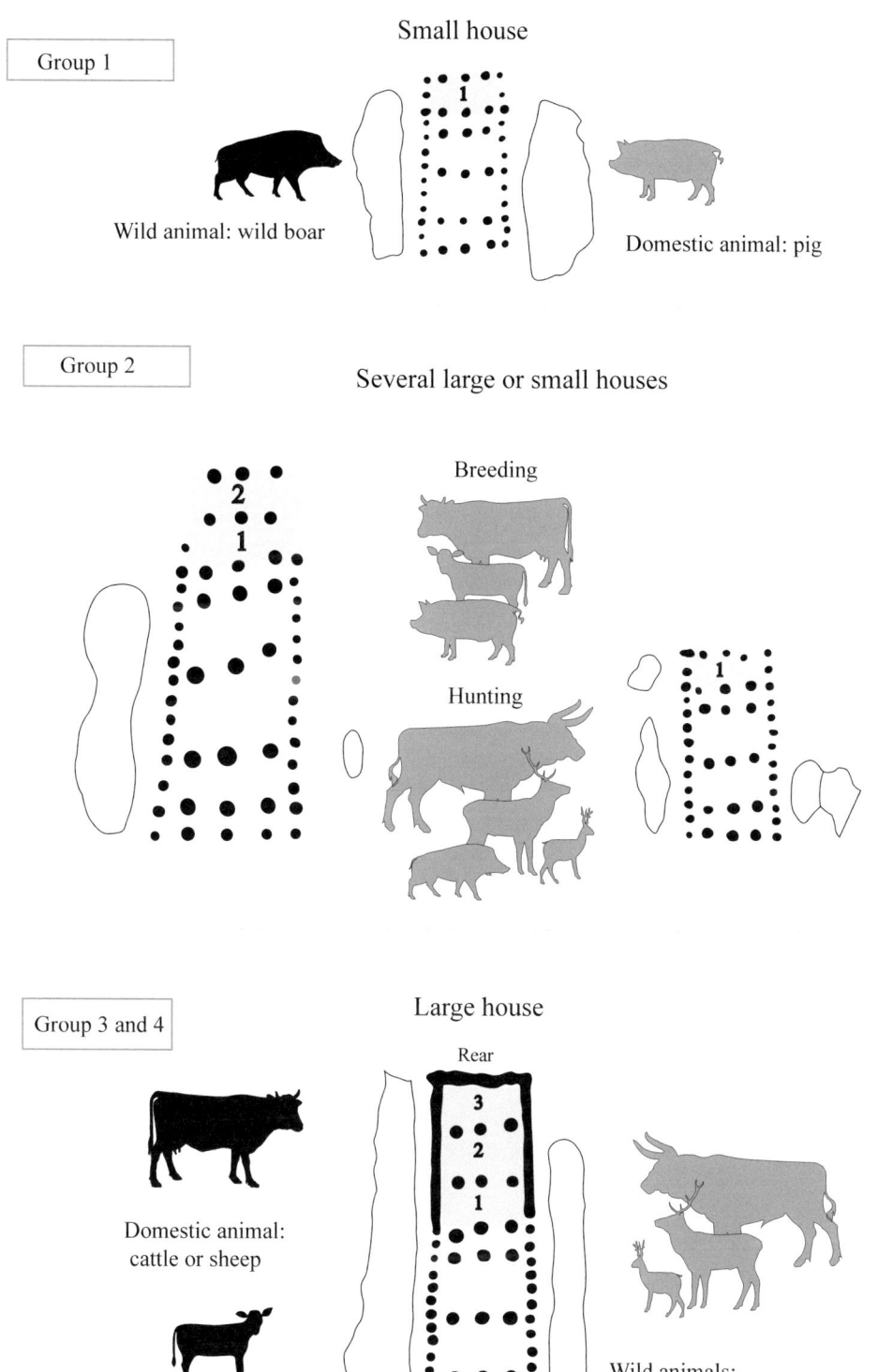

Group 1

Small house

Wild animal: wild boar

Domestic animal: pig

Group 2

Several large or small houses

Breeding

Hunting

Group 3 and 4

Large house

Rear

Front

Domestic animal:
cattle or sheep

Wild animals:
aurochs, red deer, roe deer

Figure 36:
Composition of a hamlet at Cuiry-lès-Chaudardes in the early and middle phases of the local LBK sequence: a large house with higher proportions of cattle or caprines, a small house with higher proportions of large and small game, especially wild boar, and several small and medium sized houses with no particular trend of this kind.

There is significant variation in the distribution of species per house relative to the site average.

Correspondence analysis data has revealed three clusters of houses grouped around dominant species. Three species offer particularly rich distributional variability for patterning: wild boar, domestic cattle and caprines. Two other species, pigs and to a lesser extent red deer, contribute marginally to the overall pattern. Pigs are particularly linked to hunted animals, especially wild boar. Red deer are always strongly linked to roe deer and show an overall opposition to cattle. We also note that red deer and wild boar are quite well opposed, as the two species are never abundant within the same house. It seems that houses characterised by a low rate of hunting have more red deer than boar, while, conversely, houses with a high rate of hunting have more wild boar than red deer. The aurochs is associated with livestock, particularly domestic cattle.

A coherent pattern emerges in which three categories of houses are found in each settlement phase (fig. 35). The first distinctive factor related to these groups of house units is the ratio of stock rearing to hunting identified in each house. Three different categories of household assemblages have thus been identified:

- The "herding" category, where stock rearing is more significant than the mean (a proportion of domestic animals between 91 % and 96 % of the total number of remains);
- the "hunting" category, where hunting is more significant than the mean (a proportion of wild animals between 23 % and 42 %);
- the "mixed" category, which groups together assemblages without particular proportions of animal remains (stock rearing between 60 % and 76 %).

Furthermore, the hunting rate closely depends on the house type of house (fig. 36).

The high hunting rate (> 23%) defines a class of small houses (with one rear unit), whereas a very high husbandry rate (> 90%) defines a class of large houses (with 2 or 3 rear units).

The large houses where the animal husbandry rate is very high, can be divided into two groups:

those with higher proportions of cattle, and those with higher proportions of caprines.

The game present in the pits of the small houses with a high hunting rate is characterised by higher incidences of wild boar than other hunted species.

The distribution of red deer by house type shows much less pronounced differences. It has been observed, however, that nine times out of eleven, red deer surpasses wild boar in large houses. Although there is no direct correlation between large houses and high rates for deer, it is nevertheless clear that deer was a particularly important large game animal for the bigger houses.

With the exception of beaver, the frequency of small game is not correlated to the frequency of large game.

In order to try and gain a better understanding of the link established between certain categories of house and certain consumed species, a further study was undertaken, this time on a larger corpus from several settlements (Hachem and Hamon, 2014). A combined analysis of animal remains and macrolithic tools, reflecting meat and plant food consumption, was conducted on six multi-phase LBK settlements in the Aisne valley, in order to further the exploration of variation between house units. In the end, twenty-six selected houses presented the most reliable contexts (complete plans, good preservation of the lateral pits and significant assemblages), and in many cases, precise chronological attributions. They were chosen in order to cross-compare house layouts with the various indicators of the subsistence economy.

Among macrolithic artefacts, the three main categories of tools were taken into account as evidence of the activities of grinding (querns and grinders), percussion (anvils and hammerstones) and abrading (polishers, abraders and so on) (Hamon, 2006, 2008). Grinding tools reveal the existence of food preparation, especially cereal processing. Abrading activities are more directly related to recurrent craft activities, such as the shaping of ornaments and bone tools (pointed tools, arrow shafts). Finally, percussion activities were linked to flint debitage and the maintenance of querns.

When all the results are combined, a coherent pattern emerges and a typology can be proposed (Hachem and Hamon, 2014, figs. 6-7). Three groups can be defined on the basis of the size of the houses, the activities represented by the macrolithic tools and the importance of the domestic fauna within the assemblage:

- in houses with one rear unit, evidence for hunting is abundant (with wild boar often predominant) as are abrading tools;
- in houses with three rear units, animal husbandry is massively predominant (90% sheep or cattle) and grinding tools are over-represented;
- in houses with one or two rear units, where neither of these trends (hunting or husbandry) is apparent, animal husbandry accounts for around 80% of the bone assemblage and macrolithic tools fall equally into three main categories, *i.e.* grinding, hammering and abrading tools.

Following on from this comparative analysis, which opened significant new possibilities for the interpretation of data, a third step was envisaged which would allow the construction of a new model for interpreting the organisation of the LBK house unit.

This model is based on a comparative analysis of three types of data of particular significance for dietary behaviour and production networks (fig. 37): faunal assemblages; macrolithic tools (Hamon 2006) and pottery production (Gomart, 2014; Gomart and Ilett, 2017). A novel approach in LBK research, this study has roots in the French school of social technology (Leroi-Gourhan, 1964; Latour and Lemonnier, 1994).

The three studies, each of which was subject to detailed descriptive and multivariate functional statistical analyses, made it possible to associate each house unit at Cuiry-lès-Chaudardes with a type of faunal assemblage, with a type of activity related to macrolithic tools and with one or more pottery making traditions. This approach also allowed us to characterise the waste assemblage from house unit in terms of subsistence and technical know-how. All of the data from the house units was cross-compared and sorted in order to identify the major trends.

It suggests a division of the house units into two groups (designated A and B), each with two sub-groups. In summary (fig. 38), Group A house units are characterised by homogenous pottery techniques, stock-rearing and cereal grinding, whereas Group B house units are defined by heterogeneous pottery techniques, a more substantial ratio of hunted animals and greater numbers of macrolithic tools associated with craft activities (Gomart *et al.*, 2015).

In chapter 5.1 we will look at how these results can be interpreted.

In the context of the LBK, each housing unit consists of the house itself as well as associated lateral pits: the latter were probably originally created for the extraction of clay required in the building of the house. The pits were subsequently

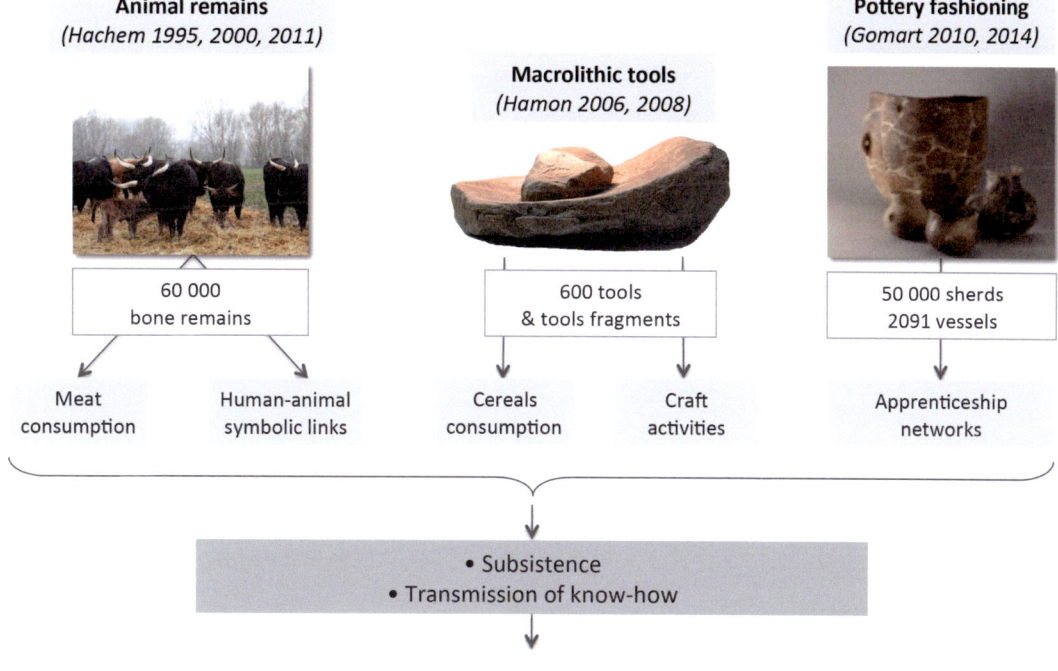

Figure 37: The three categories of data used in the analysis, with an indication of the number of faunal remains, macrolithic tools and vessels studied, and the range of evidence derived from these data (after Gomart *et al.* 2015, fig. 4, modified).

used for dumping waste material; this waste material, at least in part, reflects the activities that took place within the house and in its immediate surroundings (Soudský, 1969; Lanchon *et al.*, 1997; Coudart, 1998; Last, 1998; Allard *et al.*, 2013). Another view is that the pit-fills were affected by further depositional processes, particularly after abandonment of the house (Stäuble, 1997).

Several models have been proposed for the organisation of LBK domestic space and for over 40 years have provided the basis for reflection and discussion regarding LBK settlement, both in terms of domestic architecture and activity areas (Soudský, 1962; Modderman, 1970; Soudský and Pavlù, 1972; Coudart, 1998) and the organisation of space within the villages (Stehli, 1989; Claßen, 2005; Lüning, 2005, 1982; Květina, 2010; Pavlù, 2010, 2000, 2016; Květina and Končelová, 2013; Rück, 2013).

For the sites in the Aisne valley (and Paris Basin), we argue that the finds from the lateral pits are virtually unaffected by post-abandonment dumping or by intense post-depositional disturbance. First, and in contrast to most central European LBK sites, the density of houses is quite low, with no overlapping house plans and no really close-set houses. This reduces the likelihood that waste from "functioning" houses was discarded in the vicinity of abandoned houses. Second, stratigraphic sections through pit-fills show no evidence for recutting of pits after primary filling. Furthermore, as a rule the contents of the pits within a given house unit are homogeneous in terms of pottery decoration, suggesting a low degree of disturbance of pit-fill by intrusive finds. Pottery re-fitting also clearly shows that fragments from single vessels were discarded in the pits on either side of a given house (Allard *et al.*, 2013).

As we saw earlier, an analysis was conducted to assess variation in the relative frequency of species between house units.

The overall patterning of the remains was examined, firstly by assessing the duration of dumping in the pits, to see if it was possible to compare houses with one another (Hachem, 2011a; Allard *et al.*, 2013). First of all, we observed that the quantity of finds is not related to the volume of the pits, or to the size of the house unit and by extension probably not even to the number of inhabitants. While some variation can be seen from one pit or one house unit to another, there are no

GROUP A	GROUP B
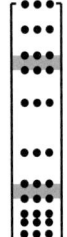	
* Long houses	* Short houses
* Dominance of herding	* Importance of hunting
* Dominance of cereal grinding	* Importance of craft activities (abrasion, hammering)
* Homogeneity and "conservatism" of pottery manufacturing practices	* Diversity of pottery manufacturing practices and appearance of new forming methods
Continuity and strong autonomy of production (with possible surplus)	Possible process of integration and economic maturation

Figure 38: Summary of the characteristics of group A and B houses with an indication of the main socio-economic hypotheses proposed (after Gomart *et al.* 2015, fig. 9).

major differences in the general composition of finds in lateral pits. There are hardly any cases of pits containing exclusively food refuse or exclusively waste arising from technical activities. In fact, most house units share the same kind of background noise as far as dumping is concerned.

Two kinds of data can be extracted from analysis of the faunal remains in order to address the question of the duration of the filling process. These are, firstly, seasonal indicators and, secondly, the quantity of meat consumed, based on an estimation of the minimum number of individuals (MNI). By applying these criteria to the faunal remains in the lateral pits adjacent to the houses, some marked trends can be observed.

Seasonal indicators (antlers, slaughter age) suggest that these pits were used for refuse disposal for at least one whole year, the most frequent indicators reflecting use from spring time through to autumn. Calculation of the minimum number of individuals provides additional information. The MNI is always the same for three quarters of the house units: the average numbers per house are three head of cattle, three caprines and two pigs, and less than four red deer, four roe deer and six wild boars. These

results do suggest, however, that the duration of pit use was identical for each house and that it was relatively short, these numbers being too low to represent a cyclical pattern of slaughter over a period of many years. Other evidence (pottery and macrolithic artefacts) suggests that the finds from these pits correspond to just a few years in the lifespan of the household – most probably the first years of occupation if one accepts that the pits were dug to provide raw material for building the walls of the adjacent house. A similar conclusion has been reached in other regions of LBK settlement (for example Domboróczki, 2009). Although it is difficult to estimate, the time that these pits remained in use must certainly be far less than the duration of the occupation of the average house: between 3 and 5 years maximum (Allard *et al.*, 2013).

Figure 39: Distinctions between houses at Cuiry-lès-Chaudardes based on the quantity of bone remains contained in lateral pits to the north (in grey) and to the south (in black) of the houses (after Hachem 2011a, fig. 18).

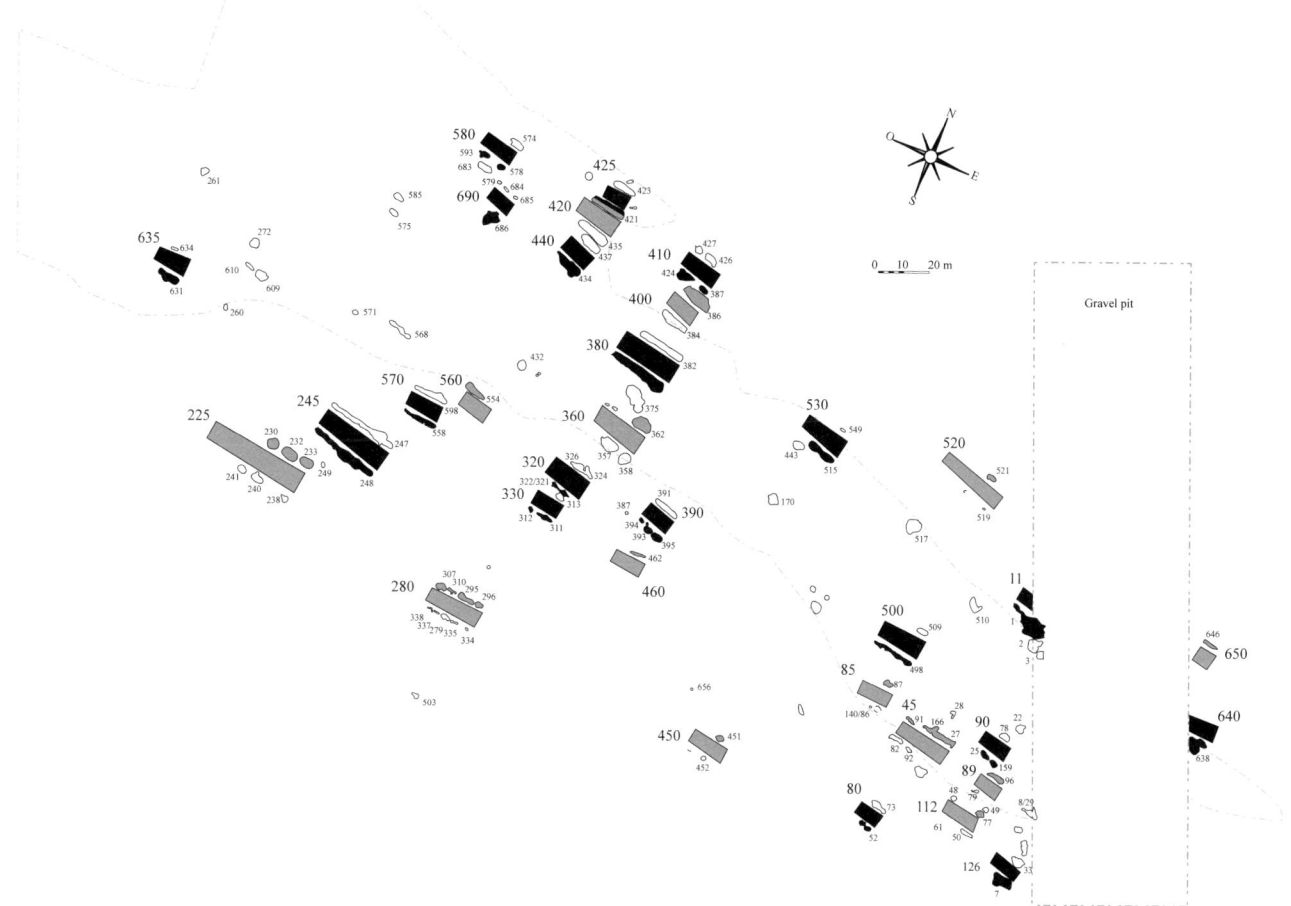

There is no standard amount of discarded animal bone per house; in fact, bone quantity varies widely from one house to another. Furthermore, quantity of animal bones is not directly correlated to house size, although it has been noted that the largest sets of faunal remains are associated with large houses.

The minimum for a normally preserved house is about 1000 fragments or 10 kg of bone (numbers below this threshold indicate major erosion of features) and the maximum is just under 7000 fragments or 65.5 kg of bone. This heterogeneity is not the result of taphonomy, and is simply a reflection of domestic activities. All of the houses display a preference for dumping to one particular side (either north or south (fig. 39). This pattern is not just restricted to animal bones,

but is also observed in the case of lithics and ceramics. In addition, there is a tendency for certain houses displaying southward dumping to produce more bone refuse than other houses.

In LBK contexts, the house plan and associated lateral pits form the house unit. Most probably the various pits associated within a unit were initially dug to extract subsoil materials needed for house construction and they subsequently served as refuse pits.

Isolated pits also occur within a 10 to 20 m radius of each house and these contain less waste material than the pits immediately adjacent to the buildings. However, they are similar in nature and can be associated with the houses (fig. 40).

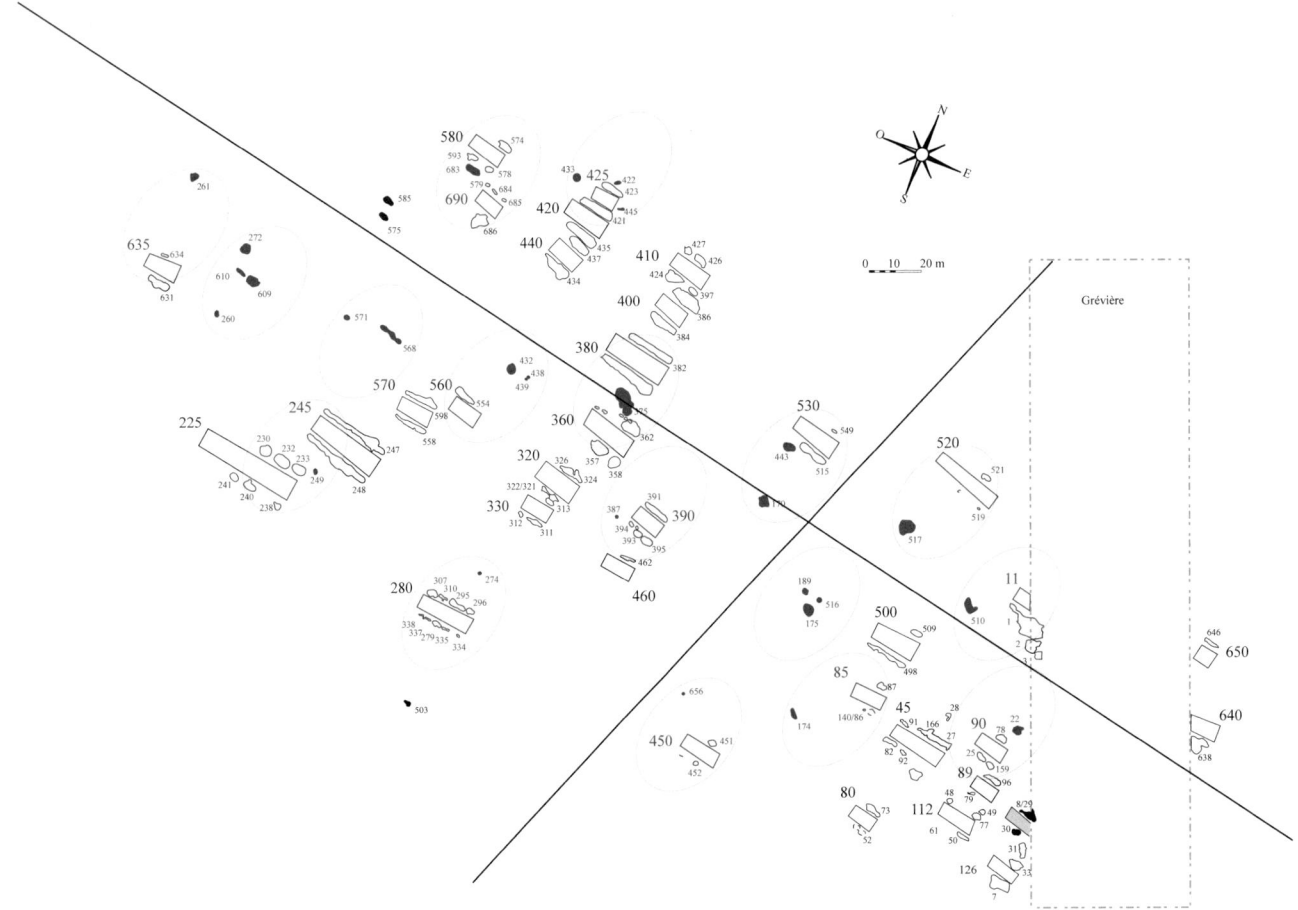

For the Blicquy/Villeneuve-Saint-Germain, the nature of the evidence differs from the LBK due to changes in the layouts of houses and settlements.

In addition, our knowledge of the internal organisation of BVSG settlements is incomplete as very few sites have been fully excavated: as a result we can only make suppositions based on fragmentary data.

The houses within the villages differ somewhat from those found on LBK sites. While lateral pits are still present, structural posts are more superficially set in the ground (Bostyn and Lanchon, 2003; Lanchon, 2006). The house size is more uniform and there is a general tendency towards longer buildings. In addition, small houses disappear from the record and the internal layout of houses changes radically with a reduction in the size of the rear sections of houses (fig. 41).

Village layout also changes, the houses being organised in long rectilinear rows and no longer dispersed, as can be seen at the sites of Poses (fig. 42) and Vignely, for example. We also observe the presence of "paired houses", as recorded on the sites of Jablines (Lanchon, 2008; Lanchon *et al.*, 2008), Echilleuses, Poses (Bostyn and Lanchon, 2003) and Balloy (A. Samzun unpublished).

The distribution of the number of remains per species reveals that pig and caprines vie for second position after cattle, and in terms of MNI, it is not rare for these latter two species to surpass cattle by a few individuals.

However, we no longer encounter houses with a high game rate; this shift is probably linked to the disappearance of smaller houses. Nevertheless, we do sometimes come across isolated pits containing a preponderance of game, as for example on the site of Buchères where, in an assemblage of 100 bone remains, only wild boar and red deer were present.

Forests, rivers and pastures: the natural resources

Another factor that is likely to influence the proportions of species within settlements is the local potential of agro-pastoral and forest resources in close proximity to the sites, as well as the relationship between the environment (geographical and topographical) and the exploitation of the domesticated herd.

While the sites are always situated on the valley floor, relatively close to river banks, there is nonetheless a degree of variability in terms of the siting of these LBK settlements (fig. 43).

An approach based on the theory of site catchment analysis was applied in the Aisne valley

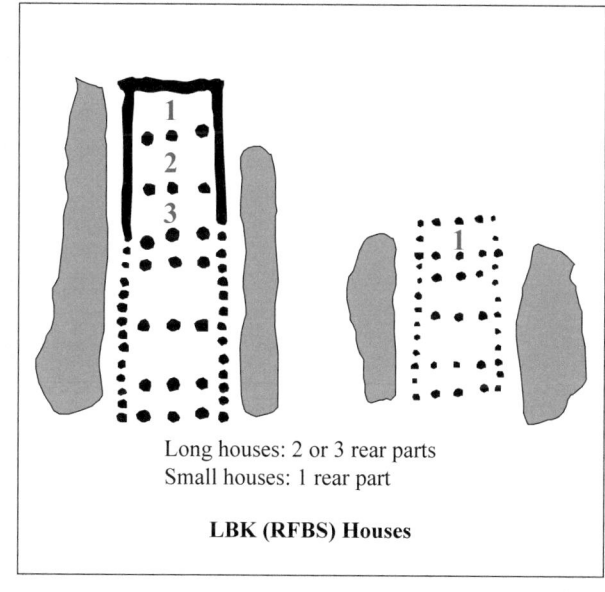

Long houses: 2 or 3 rear parts
Small houses: 1 rear part

LBK (RFBS) Houses

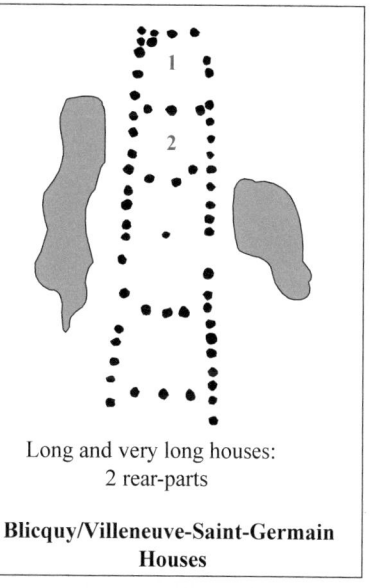

Long and very long houses:
2 rear-parts

Blicquy/Villeneuve-Saint-Germain Houses

Figure 41: Changes in the dimensions and internal layout of houses between the LBK and BVSG: lengthening of the buildings, disappearance of small houses and internal corridors. Left Cuiry-lès-Chaudardes "les Fontinettes" House 380 and house 400. Right Vignely "la Porte aux Bergers" (after Bedault and Hachem 2009, fig. 10, modified).

Figure 42: Poses "Sur la Mare" (Eure), plan of the BVSG site (after Bostyn 2003): the houses are laid out in parallel rows.

Poses sur la Mare (Eure)

using a classic analytical grid: *i.e.* in which territories were defined in terms of ten minute and one hour walking distances (Higgs and Vita-Finzi, 1972), but also including a calculation of the amount of agricultural land directly accessible within a 1 km (ten minute) radius from the settlement (Dubouloz, 2012a, fig. 4). The results were instructive as a fairly clear distinction emerged between long duration sites and short duration sites in terms of their access to the various environment categories (fig. 44). Long-term occupation sites are generally located in milieus characterized by a high proportion of land conducive to agriculture (>60%), while, in contrast, very short-term settlements occur in environments made up of more than 50% wetland. The availability of large, easily accessible expanses of land favourable for agriculture and livestock rearing was, therefore, an important factor governing the establishment of settlements; it ensured the stability of certain settlements over a long period and allowed the realization of long-term projects.

While LBK settlements were only established on valley floors, BVSG settlements were also located on plateaus. We might ask ourselves if these different choices were associated with different forms of exploitation of animal resources; questions also arise regarding the nature of exchanges between neighbouring villages and the sharing of territories. This question can be legitimately asked in the case of two contemporary Late BVSG sites located about 1.5 km apart at Bucy-le-Long in the Aisne valley

and displaying complementary characteristics. On the first site, "Le Fond du Petit Marais", there is a deficit of very young caprines, while on the second, "La Fosse Tounise", there is a deficit of adult and sub-adult animals (Bedault, 2012).

In the Lower Marne valley, BVSG sites also display differing durations of occupation, ranging from a single chronological phase to several phases. Characterisation of the topography within radii of 1 km and 2 km (*i.e.* within 15 minutes and 1 hours walk, respectively) of these settlements, indicates that long duration settlements, such as Vignely and probably Jablines (Lanchon, 2012, 2008), possessed more open land on the plain than shorter duration sites such as Changis (Lanchon *et al.*, 2008) or Rungis (Bostyn, 2002). Such configurations favoured the keeping of larger livestock herds and thus contributed to the permanence of the villages.

While the topography of the landscape had a significant influence on the duration of occupation of a site, it did not have a direct impact on the diet of Neolithic communities. In fact, slaughter profiles for domestic animals are not dependent on the choice of location of BVSG settlements, whether

Figure 43: Aerial view of Cuiry-lès-Chaudardes in the Aisne valley, 1994 (photo: F. Poitevin).

they are located low in the valley, on second or third terraces or on loamy plateaus.

However, in the case of wild species, certain specificities emerge (Bedault, 2012, fig. 270). Hunting of aurochs predominates at Fresnes-sur-Marne, red deer at Jablines and Vignely, and roe deer at Mareuil-lès-Meaux. The author of these studies suggested that such variations might reflect a degree of competition for access to wild fauna, or perhaps they simply reflect the composition of the local fauna. A more exhaustive study of finds (lithics, ceramics, etc.) from all of the sites in the Marne valley might throw further light on this subject (Bostyn and Lanchon, forthcoming).

On the basis of all of the above elements, a study was undertaken of the numbers of livestock reared within the LBK and BVSG settlements. Up to now the notion of "herd" has been little developed and only in theoretical and general terms. However, in order to adequately examine the issue of herd size in the Early Neolithic, it is our belief that we have to base our study on small-scale, concrete data. For this reason, we calculated the village MNI from the faunal remains associated with contemporary houses in a settlement.

In order to be able to make comparisons between houses, a study was initially carried out on the taphonomy and the accumulation times for bone waste in the lateral pits in the Aisne valley (Hachem, 2011a, 1995) and then in the Marne valley (Lanchon *et al.*, 1997; Hachem and Bedault, 2008). A similar analysis, based on an estimation of the numbers of ceramic vessels, grinding implements and other artefacts consumed and discarded per year, was carried out at Cuiry-lès-Chaudardes (Allard *et al.*, 2013) and at Jablines "la Pente de Croupetons": at the latter site, an occupation level

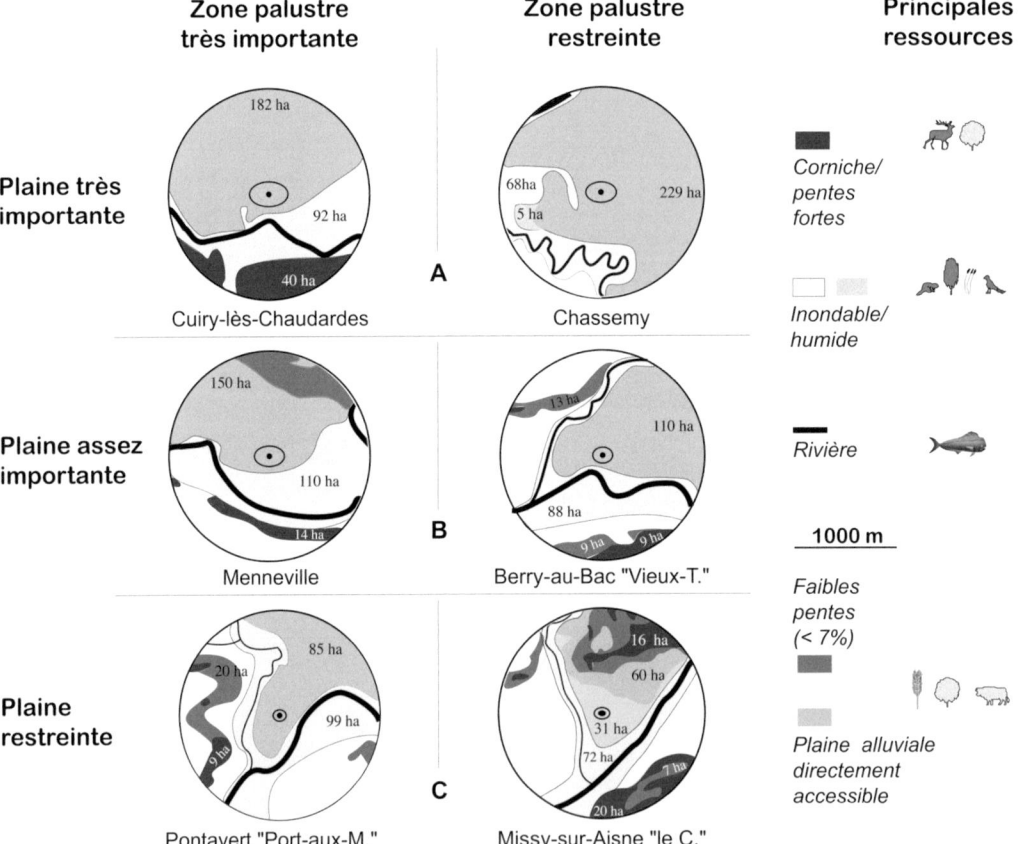

Figure 44: Analysis of the distribution of various LBK sites in the Aisne valley, in relation to wetland, arable zones and main ressources. A- Very important plain. B- Quite important plain. C- Restricted plain. Legend from top to bottom: ridges/steep slopes; floodable/wet; river; gentle slopes (<7%); directly accessible alluvial plain (after Dubouloz 2012, fig. 4).

synchronous with the lateral pits was excavated (Lanchon *et al.*, 1997).

At Jablines, the comparison of finds from the occupation level and from lateral pits revealed that, although the level contained 70% of all the settlement refuse, the finds from the pits did provide a representative picture of household activities.

These results, therefore, allow us to make comparisons between houses and to produce quantitative estimates of herd sizes, although it must be borne in mind that they do not reflect an absolute reality.

If we attempt an initial synthesis, we note that in the LBK, and indeed in the BVSG also, for most houses there is a standard maximum amount

Site	House	Cattle	Caprines	Pig	Wild boar	Red deer	Roe deer	Aurochs
CCF	11	3	2	3	0	2	1	1
CCF	45	4	2	4	1	2	2	1
CCF	80	3	1	2	1	0	1	1
CCF	85	3	1	1	1	1	0	1
CCF	89	4	4	3	2	1	2	1
CCF	90	8	5	3	2	3	3	1
CCF	112	5	4	7	3	2	2	1
CCF	126	4	6	5	5	4	3	1
CCF	225	10	19	12	2	1	3	2
CCF	245	6	11	3	2	1	1	1
CCF	280	7	7	3	2	1	2	1
CCF	320	2	2	2	1	1	1	1
CCF	330	3	2	2	1	1	2	1
CCF	360	7	8	6	6	2	3	3
CCF	380	16	9	5	4	4	3	3
CCF	390	3	2	2	1	2	1	1
CCF	400	6	1	2	3	2	2	1
CCF	410	3	1	2	0	1	0	1
CCF	420 (st 435)	4	2	3	3	3	2	
CCF	420/425 (st 421)	6	3	3	5	3	3	2
CCF	425 (st 423)	3	1	2	2	2	1	
CCF	440	7	4	5	3	2	1	1
CCF	450	4	3	1	0	1	0	1
CCF	460	2	0	2	1	1	0	
CCF	500	5	2	2	0	1	1	1
CCF	520	1	1	1	0	1	0	1
CCF	530	6	4	2	4	2	2	2
CCF	560	1	1	2	0	1	0	
CCF	570	5	6	6	2	2	2	1
CCF	580	4	2	2	2	2	1	1
CCF	635	1	1	4	3	1	2	1
CCF	640	4	3	3	2	2	2	1
CCF	650	1	1	0	0	0	0	
CCF	690	4	5	3	2	1	2	1
BLF	10	2	3	3	1	1	1	1
BLF	20	9	8	7	1	1	1	2
BLF	30	6	5	3	1	2	1	1
BLF	35	1	1	1	1	1		
BLF	40	2	2	2	1	1	1	
BLF	45	2	2	2	2	1	2	1
BLF	50	2	2	2	1			
BLF	90	4	1	1	1	1	1	1
MAC	75	1	1	1	1	1		
MAC	60	5	12	3	2	2	1	2
MAC	40	1	1	1		1		1
MAC	80	3	5	3		1		
BCP	195	16	13	8	3	1	1	1
BCP	200							
BCP	300	4	4	3	1	1	1	1
MDV	10	4-3	2	1	2	1	1	1
MDV	35	4-2	2	1	1	1	1	1
MDV	130	1- 2	1	2	1	1	1	
MDV	140	5- 4	2	3	1	2	1	1
MDV	185	4-4	2	3	1	1	2	1
MDV	200	9-3	3	2	1	2	1	
BVT	370	1	1	1	1	1		
BVT	585	3	3	1		1	1	1
BVT	590	3	3	1	2	2	2	1
BVT	620	2	1	1	1	1	0	2
BVT	630	2	2	1	3	2	1	1
BCM	125	14	12	6	1	1	1	5
BCM	165							

Figure 45: Minimum number of individuals for seven main species per house. The BCP and BCM combined cells correspond to the studies in which the houses are not separated. BCP: Berry-au-Bac "le Chemin de la Pêcherie", BVT: Berry-au-Bac "le Vieux Tordoir", BCM: Berry-au-Bac "la Croix Maigret", CCF: Cuiry-lès-Chaudardes "les Fontinettes", BLF: Bucy-le-Long "la Fosselle", MDV: Menneville "Derrière le Village", MAC: Missy-sur-Aisne "le Culot".

of refuse: it equates to five or less animals for a given domestic species and between four and six individuals for wild game (fig. 45). On the evidence from refuse discarded in lateral pits and other features, it appears from the Paris basin data that on average four to six bovines, four to six pigs and four caprines were consumed per house. It must be remembered that this is an average and that certain large houses have produced the remains of larger numbers of slaughtered animals.

Based on estimations made for sites whose full extent is known, LBK villages in the Paris basinwere made up of around five houses. In the case of the BSVG, we are not yet in a position to estimate minimum village size but it is very likely to have been similar.

This leads us to conclude that relatively high numbers of animals were slaughtered: between twenty and thirty individuals for each category of domestic animals (cattle, pigs and caprines) which means a herd size of 60 to 80 animals per village at a given time (fig. 46). As the live herd must have been larger than the number of slaughtered animals (taking into account sickness, predation and reproduction), it is certain that herd management and animal

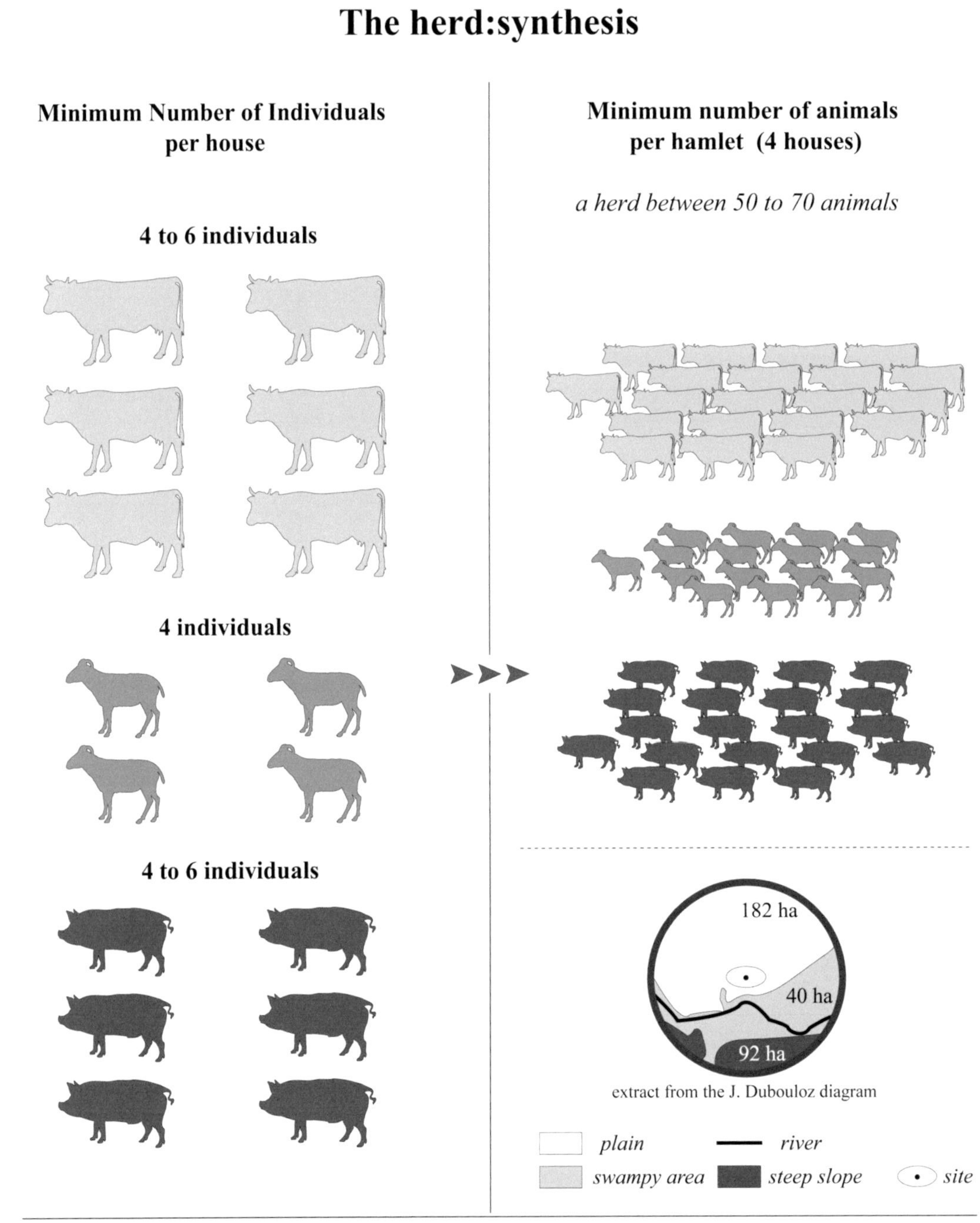

The herd:synthesis

Minimum Number of Individuals per house

4 to 6 individuals

4 individuals

4 to 6 individuals

Minimum number of animals per hamlet (4 houses)

a herd between 50 to 70 animals

182 ha

40 ha

92 ha

extract from the J. Dubouloz diagram

plain   river

swampy area   steep slope   site

Figure 46: Estimation of domestic herd size, according to the minimum number of individuals (MNI), per hamlet (5 houses) in the Paris Basin (after Bedault and Hachem 2008, fig. 16, modified).

rearing techniques were sufficiently developed to cope with such large numbers of animals.

This result gives rise to a series of questions regarding the potential need for winter fodder, the movement of livestock as a function of available land and the collective or individual management of herds. These issues have already been dealt with in a general way in a number of publications (*e.g.* Ebersbach and Schade, 2004; Thiébault, 2005) and a simulation has been carried out of the exploitation of a territory surrounding LBK sites in the region of Hesse (Wettereau), with a view to determining the minimum area required to feed a population who cultivated cereals and exploited available grass lands for cattle (Ebersbach and Schade, 2004).

A similar simulation has also been carried out based on MNI estimates from archaeozoological data for cattle herds on BVSG sites (Bedault, 2012).

We note that even in an upper limit hypothesis of territorial exploitation for pasture and agricultural use, *i.e.* 10 ha per head of cattle per year, an area within a 1 to 2 km radius around a site is largely sufficient to allow the herd to live normally. The low density of sites in the valleys suggests that there was no overlapping of the territories

associated with each village. However, leaves from trees were possibly used as winter fodder for livestock, which would imply access to extensive areas of forest.

A forthcoming study integrates palynological and carpological data in order to estimate the potential of natural resources favourable for the establishment of sites (Bostyn and Lanchon, forthcoming). We will thus be able to analyse the impact of environmental interactions on the exploitation of fauna- whether it be farmed, hunted or fished – and to examine the way in which the local potential of agro-pastoral resources (grazing zones, leaf fodder, etc.) can be exploited so as to maintain the herd and how this is co-ordinated between neighbouring villages.

The siting of the house within the village

A fourth factor influencing the proportions of species is the location of houses within the village (fig. 47).

The village of Cuiry-lès-Chaudardes serves as a good case study as it has been excavated in its entirety and has produced the most complete occupation sequence with thirty-three individual

Figure 47:
Reconstruction
of a BVSG house
at the Haute-Île
Archaeology Park,
Neuilly-sur-Marne
(photo: C. Hamon).

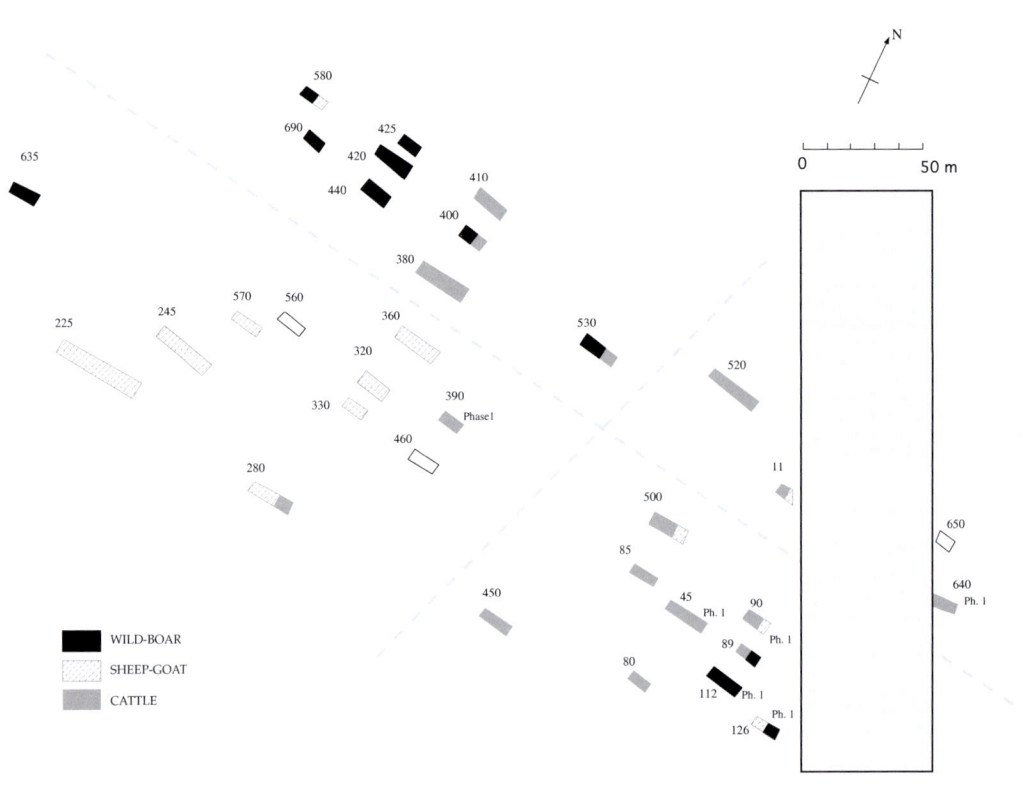

Cuiry-lès-Chaudardes « les Fontinettes »

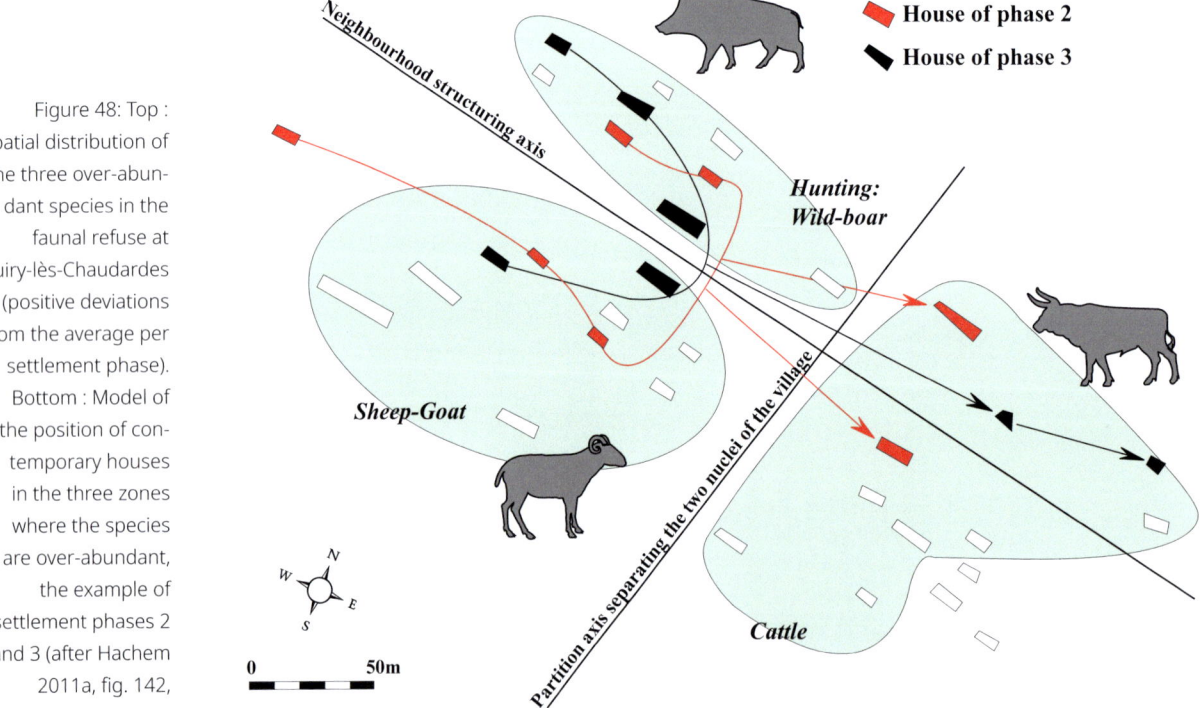

Figure 48: Top : Spatial distribution of the three over-abundant species in the faunal refuse at Cuiry-lès-Chaudardes (positive deviations from the average per settlement phase). Bottom : Model of the position of contemporary houses in the three zones where the species are over-abundant, the example of settlement phases 2 and 3 (after Hachem 2011a, fig. 142, modified).

Menneville "Derrière le Village",
plan of LBK structures

destroyed

anomaly, wind fall

archaeological structure

neolithic grave

surveyed area

archaeological structure
photographed
on aerial prospecting
geophysics

Departmental road n° 925

Figure 49: Menneville
"Derrière le Village"
(Aisne). The unexca-
vated parts of the
LBK enclosure ditch
were identified by
aerial photography
and geophysical
survey (after Coudart
and Demoule 1982,
Farruggia *et al.* 1996,
Thevenet 2014,
fig. 7.) (CAD:
C. Monchablon,
C. Thevenet).

Figure 50: Overlay of schematic plans of Cuiry-lès-Chaudardes (in light grey) and Menneville "Derrière le Village". The first LBK houses were established in the east of the settlements and then spread westwards.

houses investigated spanning several identifiable phases (Ilett and Hachem, 2001). The chronological sequence identified on the basis of ceramic decoration enables most of the houses to be placed within a chronological sequence (Ilett, 2012). The development of the village can be summarised as follows: the initial settlement involved the construction of a core cluster of houses in the eastern part of the site with a single house to the west; in the second phase, the pattern changes with the development of a larger concentration of houses at the west while the eastern cluster continues to exist. Over the course of the century or so that the site was occupied, the houses spread over an area of 6 hectares and the two settlement concentrations persisted. Archaeozoological analysis has shown that throughout the occupation of the village, contemporaneous houses were divided into three zones (Hachem, 1997, 2011a): one in

the east where there was an emphasis on the consumption of cattle, another in the south-west where more sheep were consumed and, finally, a zone in the north-west where more game, particularly wild boar, was consumed (fig. 48).

We have also analysed the plans of several LBK settlements in the Aisne valley, projecting them at the same scale and rotating them to the same orientation as Cuiry-lès-Chaudardes, in order to identify potential similarities in the initial layout of the LBK villages. The sites analysed were: Bucy-le-Long "la Fosselle", "le Vieux Tordoir" and "le Chemin de la Pêcherie" at Berry-au-Bac, Missy-sur-Aisne "le Culot", Menneville "Derrière le Village" and Bucy-le-Long "la Héronnière/la Fosse Tounise".

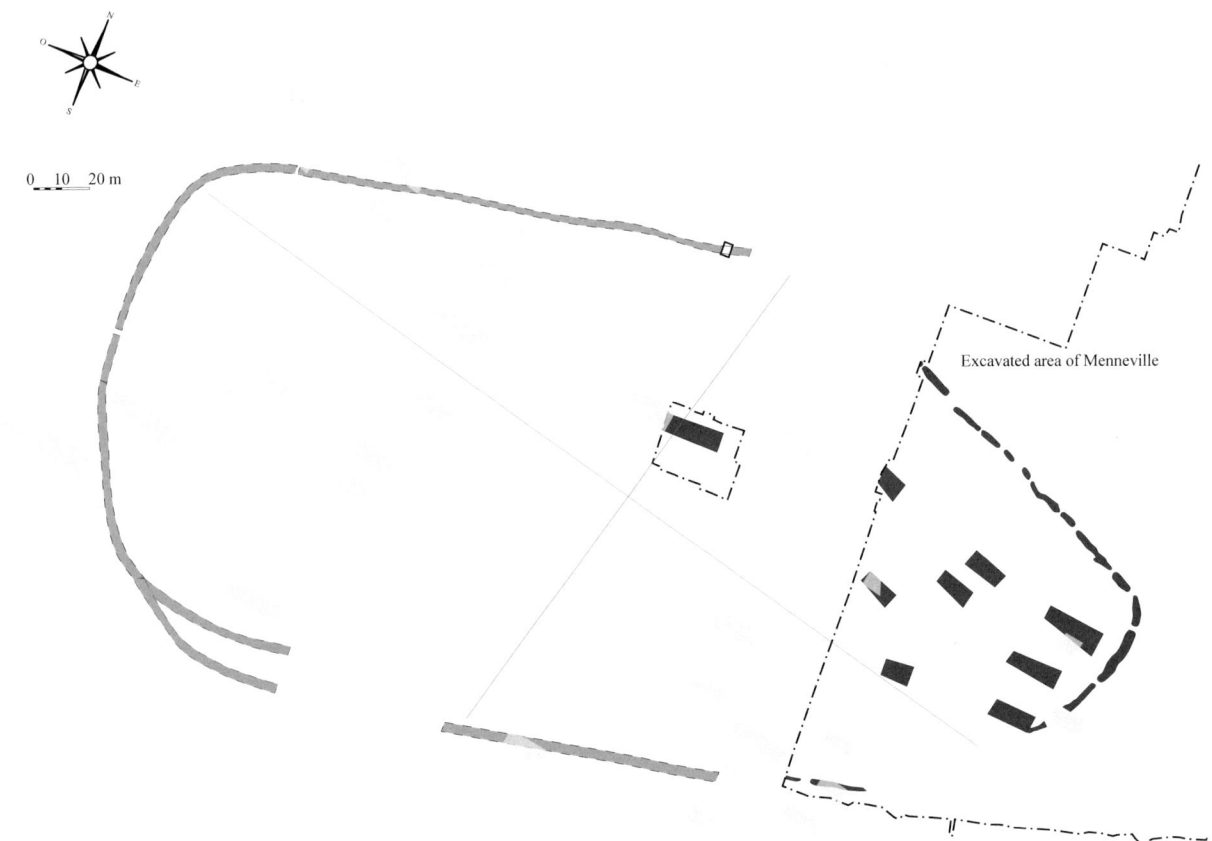

0 10 20 m

Excavated area of Menneville

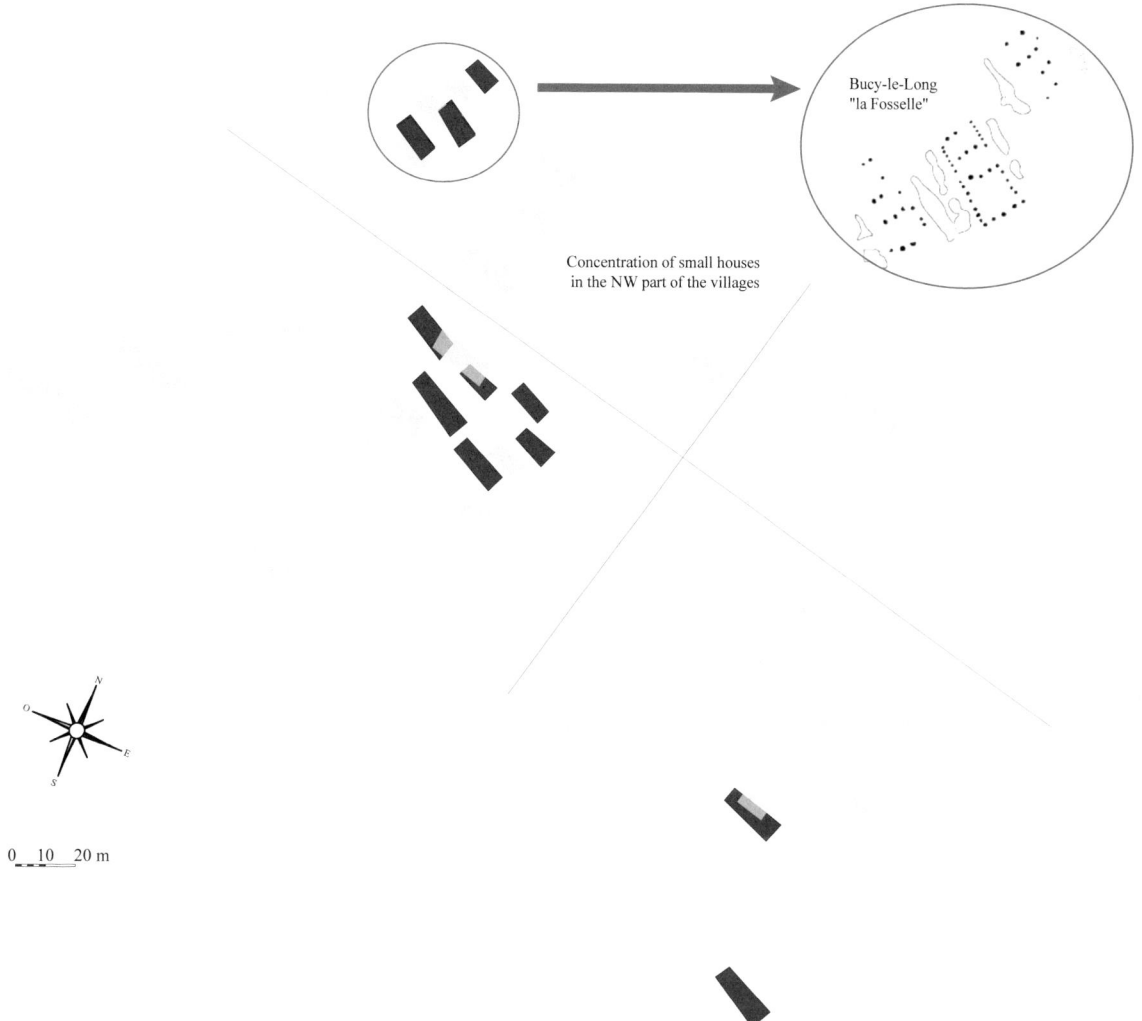

Concentration of small houses
in the NW part of the villages

Bucy-le-Long
"la Fosselle"

0 10 20 m

The results are encouraging as illustrated by two sites whose full extents are known and which are classified as large sites (6 ha).

The first, Menneville, had a long duration of occupation, and while it has not been excavated in its entirety, its full extent is known (fig. 49). By considering the plan of the enclosure (the western part of which has been identified through geophysical prospection) and the excavated areas in combination with detailed dating for the houses, a number of interesting observations can be made (fig. 50):

- the occupied area of the village did not exceed 6 hectares;
- the village seems to be expanding westward;
- finally, although it is difficult to demonstrate this because the entire surface of the village has not been excavated, it seems to us, according to the chronology of the houses uncovered, that they could be organized in clusters.

The second site, Bucy-le-Long "La Fosse Tounise, La Héronnière" (Ilett *et al.*, 1995), has yielded similar information (fig. 51):

- the occupied area of the village did not exceed 6 hectares;
- the village extended westwards;
- there is a concentration of small houses in a similar location to that at Cuiry-lès-Chaudardes.

According to our model, game should be found in small houses. So we tried in Bucy-le-Long to see if the small houses had game, but unfortunately the fauna is not conserved in this part of the village. In the future, we will make sure that we dig in the northwestern sector of Menneville to test the hypothesis of the presence of small houses to see if our model is validated.

Figure 51: Overlay of schematic plans of Cuiry-lès-Chaudardes (in light grey) and Bucy-le-Long "la Fosse Tounise/la Héronnière" (in dark grey). The pattern of houses at Bucy-le-Long appears to be similar to Cuiry-lès-Chaudardes and the houses in the north-western part of the site are also small (after Hachem 2017, fig.10).

HERDING

Rubané Moyen Champenois
(3 houses and several pits)

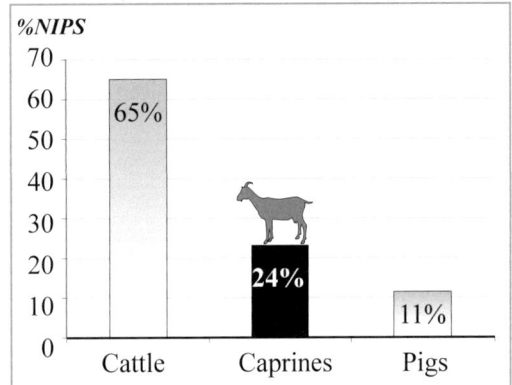

RFBS (LBK) early and middle phases
(38 houses)

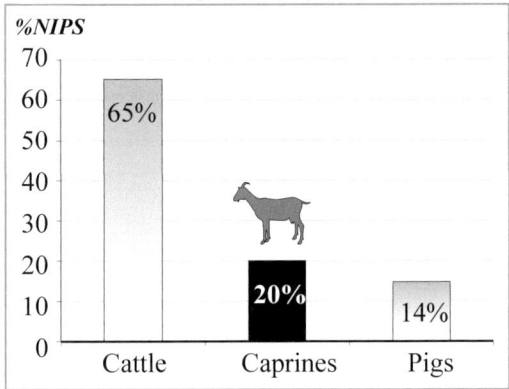

RFBS (LBK) late and final phases
(21 houses)

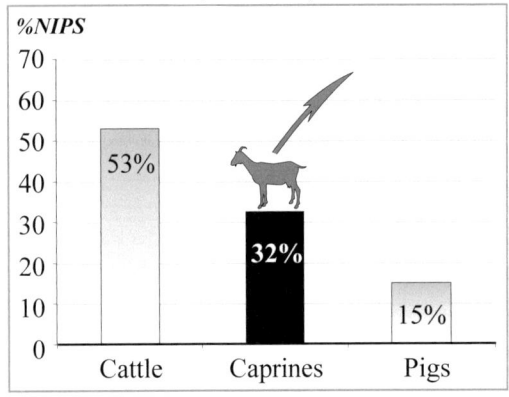

BVSG early phase
(7 houses)

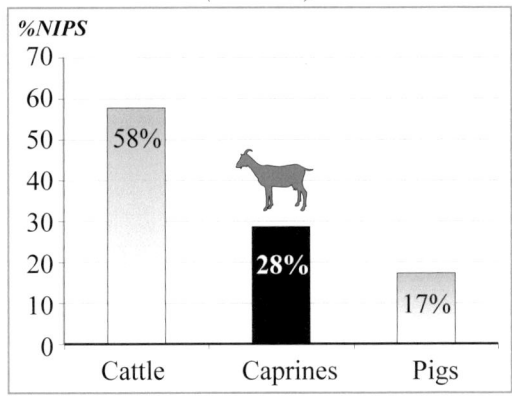

BVSG middle phase
(14 houses)

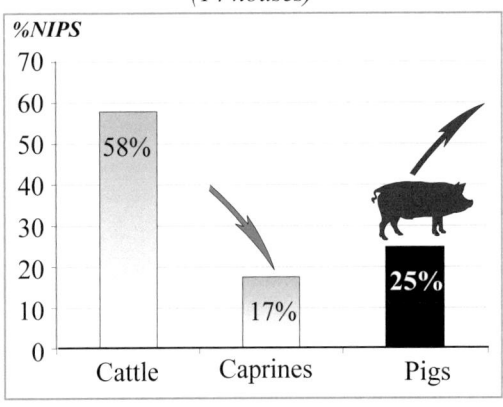

BVSG final phase
(3 houses)

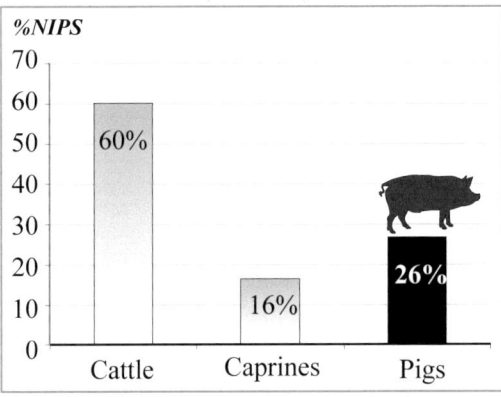

Figure 52: Animal husbandry: evolution of dietary patterns according to chronological phase in the Early Neolithic of the Paris Basin (LBK and BVSG) (after Bedault and Hachem 2008, fig. 8, modified).

An overview of the variation in species during the RFBS and BVSG

Analysis of the evolution of the proportions of the principal species reveals several developments. During both the LBK and BVSG, animal husbandry was central to cultural identity and cattle dominate the faunal spectrum; in the Paris Basin, cattle were reared specifically for butchering. Slaughter curves do not indicate that older animals were maintained within the herd nor do they indicate a mixed strategy including the production of milk; they thus support the findings of other studies on the exploitation of livestock in north-eastern France and western Germany. Nevertheless, the discovery of dairy fats on a number of ceramic sherds dating to the end of the 6th millennium in Poland, and the slaughter curves produced for certain LBK sites in Central Europe, tend to suggest that milk was in fact exploited (Gillis *et al.*, 2017), although probably not intensively. The milk may have been transformed into cheese which would have been more digestible for Neolithic populations who were lactose intolerant (Roffet-Salque *et al.*, 2017).

Work is currently under way to determine if milk traces survive on ceramics from Cuiry-lès-Chaudardes (Casanova *et al.*, 2020) and ongoing analysis is being carried out by R. Evershed and M. Roffet-Salque, in Bristol. Even if the presence of milk is proven, interpretation of its uses will, nonetheless, remain limited because, at the moment, we are unable to distinguish between cow's milk and ewe's milk. Moreover, it should be remembered that milk has many possible uses, such as the production of casein-based paint, for example.

In the Paris Basin, while cattle still predominate, their proportion tends to diminish from the latest phase of the RFBS, while that of sheep and goats tends to increase. More precisely, the proportion of caprines increases very significantly at the end of the LBK sequence. This high proportion of caprines persists through the early phase of the BVSG and into the beginning of the Middle BVSG, albeit to a lesser extent than for the Late RFBS (fig. 52). Their proportion begins to decline in the second half of the Middle BVSG in the valleys of the Aisne, Marne, Seine and Yonne. However, this was not the case in Champagne where caprines would long remain the second most important source of meat.

A change occurs in the choice of domesticated species exploited during the Middle BVSG when we see a clear increase in the rearing of pigs; this trend becomes more pronounced in the Late BVSG.

Compared with the Alsatian sites located further east in France, the second most predominant species is either pig or caprines (Arbogast, 1994). Later, in the Late LBK, the trend indicates a steady rise in the importance of pig.

How can this marked increase in sheep and goats during the final phase of the RFBS, and its persistence into the early BVSG, be explained? Mortality curves for caprines in the RFBS indicate that these animals were slaughtered between the ages of 3 and 6 months and between 3 and 4 years. Certain animals were kept within the herd until the age of 6 to 8 years (see chapter 2). While it appears that there was an emphasis on the rearing of animals for butchery, with the killing of lambs and young animals, we also see diversification in the use of caprines since a non-negligible proportion exceed the age of maturity. This could possibly attest to a desire to keep adult animals not only for breeding purposes but also for their secondary products, namely milk and wool. The significant increase in caprine numbers in the Final RFBS period (Aisne 3) could reflect either of these hypotheses, particularly since the oldest caprines at Cuiry-lès-Chaudardes were recorded in certain longhouses at the end of the sequence (Hachem, 2011a, chap. II.2).

In the case of milk production, difficulties in distinguishing between male and female remains limit our ability to draw definitive conclusions but it nonetheless remains a strong possibility.

Concerning the hypothesis of the exploitation of sheep wool and goat hair, archaeological evidence for the use of such materials to make clothing does not appear before the Bronze Age, although the use of fleeces before this period cannot be ruled out. In fact, due to its perishability, wool is never found in archaeological contexts, even in Scandinavia where conditions are favourable for the preservation of Neolithic textiles. However, indirect evidence, such as changes in the forms of certain tools, particularly spindle whorls and loom weights, around 3000 BC during the Corded Ware period in Switzerland has

HUNTING

Rubané Moyen Champenois
(3 houses and several pits)

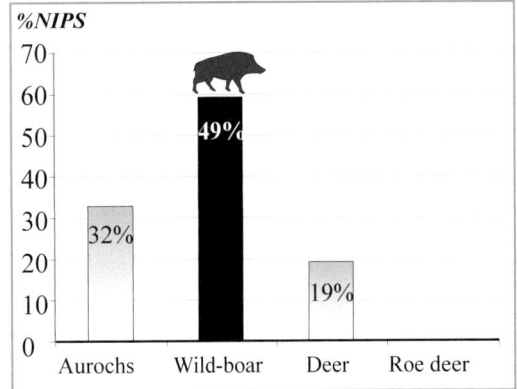

RFBS (LBK) early and middle phases
(38 houses)

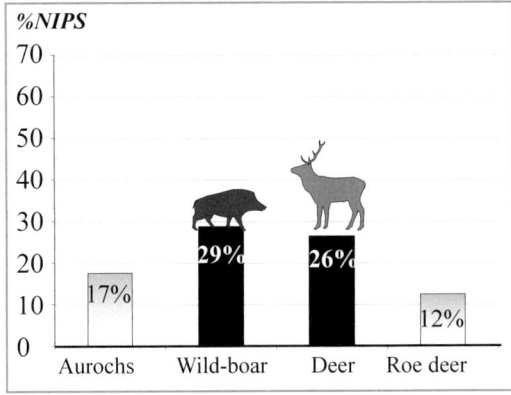

RFBS (LBK) late and final phases
(21 houses)

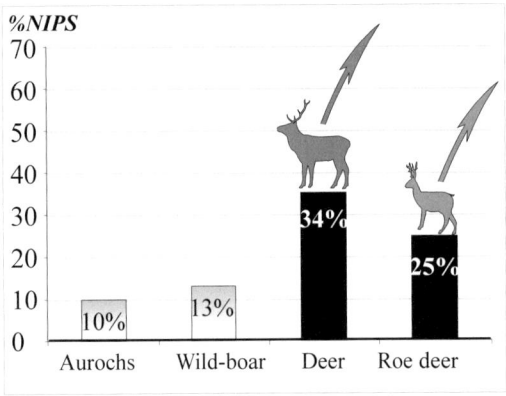

BVSG early phase
(7 houses)

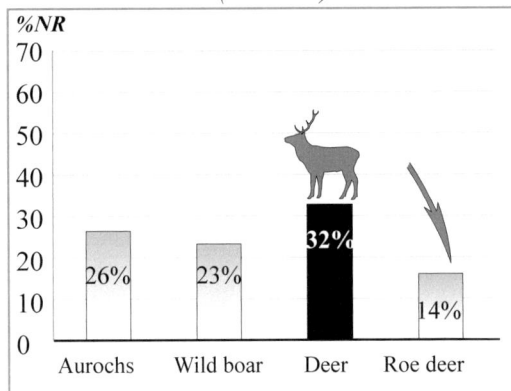

Small game = 9%

BVSG middle phase
(14 houses)

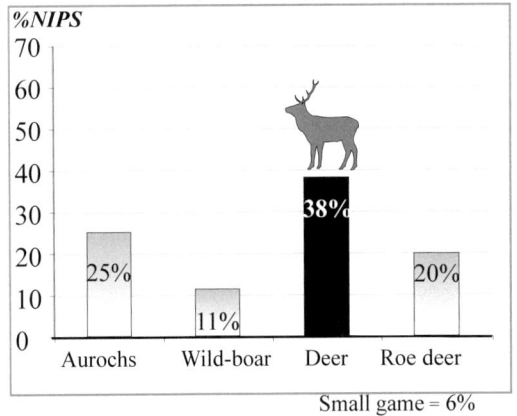

Small game = 6%

BVSG final phase
(3 houses)

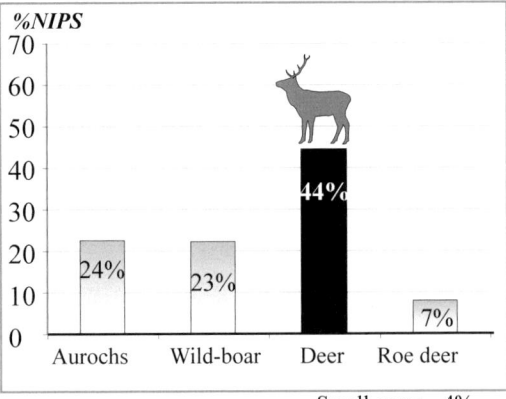

Small game = 4%

Figure 53: Hunting: evolution of dietary patterns according to chronological phase in the Early Neolithic of the Paris Basin (LBK and BVSG) (after Bedault and Hachem 2008, fig. 9, modified).

been interpreted as reflecting the development of woollen textiles (Rast-Eicher, 2014). Goat hair may also have been exploited in the Neolithic as attested to by evidence for the making of string, rope and braided mats in Turkey between 3000 and 2750 BC (Frangipane *et al.*, 2009). Primitive sheep breeds moulted in the spring; their outer coat was made up of stiff fibres that covered an undercoat of very fine wool. This was easy to pull off by hand once a year. The fact that this raw material could be harvested without the need for specialised shearing tools means that we could envisage the use of sheep fleece for purposes other than the manufacture of clothing. It is possible, for example, that wool was used on a small scale for stuffing, insulation or for making felt.

How can the shift from the intensive rearing of caprines to that of pigs in the Middle BVSG be explained? Two possibilities deserve further exploration. The first hypothesis is that this shift was a response to the dietary needs of a growing population. This is a legitimate possibility since the number of sites increases during the BVSG and this growth has already been demonstrated at the scale of the LBK (Lüning, 1998; Bocquet-Appel *et al.*, 2015). The observed increase in house sizes also supports this argument. The

population may have turned towards pigs as a way of significantly increasing the proportion of meat in the diet as they gain weight quickly, reproduce rapidly, and produce larger numbers of young more frequently than either cattle or sheep. However, we could also envisage a second hypothesis involving climatic deterioration and an adaptation by livestock farmers to the saturation or exhaustion of locally available pasture lands.

Hunting was practised throughout the early Neolithic. It developed to the same extent as livestock rearing, although not necessarily at the same pace.

Thus we see clear patterns in which the proportions of game can vary as a function of the following parameters:

- chronological phase: for example, hunting was more important at the beginning of the LBK sequence than at the end and this pattern is in line with observations made on other LBK sites in Europe (Uerpmann, 2001).
- the internal chronological sequence of a specific geographical entity (*e.g.* a valley): hunting was more important in villages dating to the first phase of the RFSB;

Figure 54: Hunting of roe deer increases at the end of the LBK sequence in the Paris Basin (photo: Lubos Houska, Pixabay).

- the internal chronology of a site: as shown in the Aisne valley, there is a phenomenon associated with the establishment of pioneer villages. Hunting is more important at the beginning of the occupation of a site than at the end; this is probably due to the fact that in the initial phase of occupation the domestic herd was not yet fully developed and the environment was largely uncleared which would have favoured hunting.

Four large game species are recurrent: aurochs, wild boar, red deer and roe deer. Despite the impression given by the actual numbers of remains, the nutritional input from game was far from negligible.

The proportion of game animals varies over time. When we compare LBK sites throughout Europe, both aurochs and red deer become more important in the Late LBK (Hachem, 1999, 331).

In the Paris Basin, during the RMC, there is an increase in the occurrence of wild boar; however, there is some doubt regarding representativity as the numbers of remains and archaeological sites are limited.

Subsequently, in the RFBS, we see a chronological trend in the proportions of species: in the early and middle phases of the RFBS, wild boar and red deer occur in equal proportions, but in the final phase of the LBK, red deer becomes the principal game species and the proportion of wild boar declines (fig. 53). This situation prevails throughout the BVSG.

We also note an increase in roe deer in the final phase of the RFBS, which occurs in parallel with a significant increase in the exploitation of sheep and goats during the same period (fig. 54). This change in the composition of game is evident

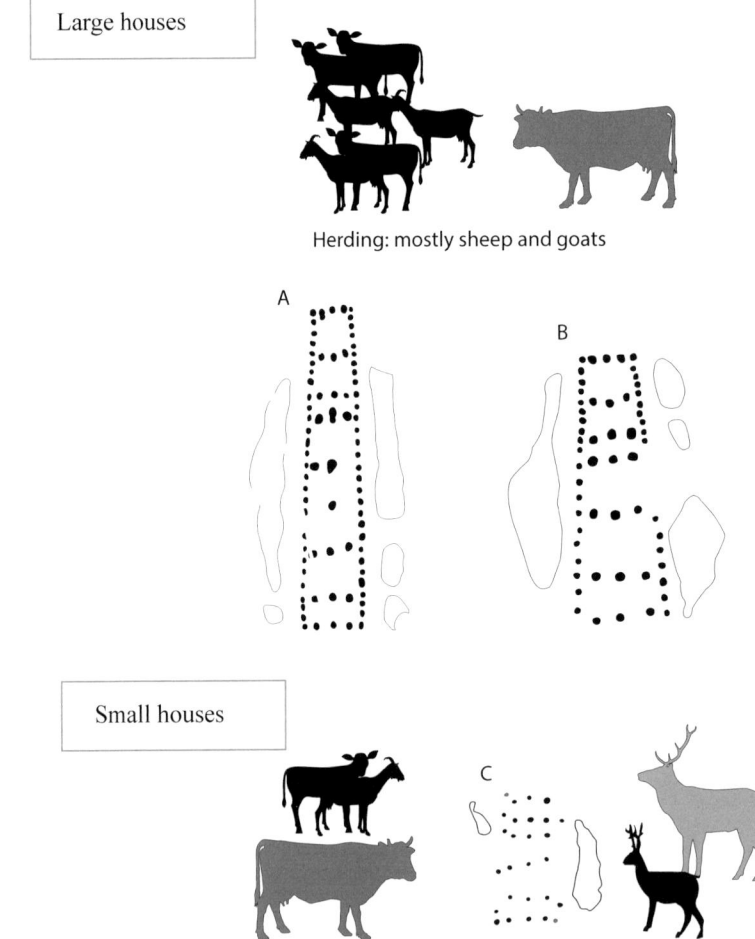

Figure 55: Composition of a typical Final LBK village in the Paris Basin: large houses with significant consumption of caprines, rarer small houses with hunting primarily focused on red deer and roe deer. A- House 630 at Berry-au-Bac "le Vieux Tordoir". B- House 200 at Menneville "Derrière le Village" C- House 89 at Cuiry-lès-Chaudardes.

before the change noted in the composition of livestock as it is only from the Middle BVSG onwards that we see an increase in the proportion of pig remains.

In the BVSG, the predominant wild species is almost invariably red deer; aurochs and wild boar occur in equal proportions. The proportion of roe deer decreases over time. Over the course of the BVSG, we also observe that the presence of wild boar goes hand-in-hand with high proportions of domestic pig, without chronology being a factor. This may be related to the groups of associated species identified on sites in the Aisne valley during the LBK.

Certain age and sex criteria governed the hunting of wild animals: Neolithic hunters targeted adult female animals. While numbers of very young wild boars and aurochs are probably somewhat under estimated, this result indicates that the selection exercised in the case of wild animal is the inverse of that adopted for domestic animals which were generally slaughtered at a young age. This pattern continues throughout the chronological sequence.

Another factor influencing the proportions of species consumed in an LBK household is house size. This variation is perceptible in the case of hamlets. In a settlement of five contemporaneous houses, for example, one house will tend to be larger with at least three rear units. In this case, consumption will be strongly orientated towards cattle or sheep. The wild animals present in the refuse will tend to be mostly aurochs, red deer and roe deer. In contrast, a second house will tend to be small, with a single rear unit and in this case there will be greater emphasis on hunting both large and small game, with a preference for wild boar. The dominant domestic species will be pig which is favoured for consumption. The other houses, which are generally small in size, will exhibit a profile without extremes, with consumption principally turned towards domestic animals, but not excessively so, and with the usual wild species also present.

A change occurs in the final phase of the RFBS: houses become longer, the number of inhabitants rises and sheep rearing increases in importance. Small houses become rare but certain examples still represent a contrast to the large houses, with a more pronounced presence of wild species, particularly red deer and roe deer, while wild boar becomes rare (fig. 55).

The siting of a house within a long duration LBK village also influences the composition of the species present. Depending on whether it is located to the east, south or north of the settlement, hunting is more or less developed and three species

are particularly representative of these variations: cattle, sheep (and goats) and wild boar.

At present, it is difficult to characterise the layout of BVSG villages because no site has been excavated in its entirety. However, we have gleaned a certain amount of information such as the fact that small houses disappear and that the houses are now laid out in rows, which probably reflects a change in social structure. Moreover, a degree of differentiation still exists with certain houses yielding a larger proportion of sheep or pig remains. Change is also to be seen in hunting, with a shift in focus towards red deer and the emergence of villages that favour one species over the others, which might indicate a degree of complementarity.

Lastly, the natural resources around Neolithic sites also play a role in shaping the makeup of the herd: open plain landscapes provide favourable conditions for the establishment of long-term villages and the development of herds, while more challenging environments tend to give rise to shorter-term occupation.

Comparisons with LBK sites elsewhere in Europe

Apart from the present overview, several others have been undertaken on fauna in the Linear Pottery culture (Müller, 1964; Bökönyi, 1974; Bogucki, 1988; Döhle, 1993; Arbogast, 1994; Arbogast et al., 2001; Tresset, 1996). In attempts to formulate a general model of animal exploitation, the authors have tended to combine the faunal data from individual sites at regional level. But the regional averages obtained in this way are problematic: they are often based on a geographical area that is artificially defined and are strongly determined by various contingencies (methodological, financial, and administrative) that have nothing to do with prehistoric reality. Furthermore, the smoothing of variabilities on the basis of "region" necessarily masks the particularities of individual sites. However, the characteristics of the individual settlement are essential to our understanding and must be examined in detail if we are to consider a site's significance at a larger scale.

The LBK sites grouped together by "region" in general present the average proportions of domesticated and wild animals. The main differences observed between regions tend to be interpreted in terms of cultural traditions. However, if we consider the extrinsic characteristics that shape the data, *i.e.* the site, it is apparent that this is a deformed vision of reality. A full review of published data was therefore undertaken in order to estimate the representativity of the series.

Because we are dealing with older excavations, of limited extent, it is tempting to think that the data presents too many weaknesses to be integrated, without hesitation, into a general interpretation.

In eastern Germany, for example, we might question the representativity of several faunal assemblages (Müller, 1964) which often yield very low percentages of game animals. Apart from the site of Eilsleben (Döhle, 1994), these corpora were retrieved from 15 sites that were excavated in the 1930s and 1950s. In general, the areas excavated were very small (one or two pits, on average) and even when this was not the case, as at Rosdorf (a village of fifty-two houses spread over 2.3 ha), for example, the faunal samples taken were very limited: at Rosdorf samples were taken from only 3 of the 8 pits investigated (Reichstein, 1977). Such limited samples cannot be taken as being representative of the entirety of a site

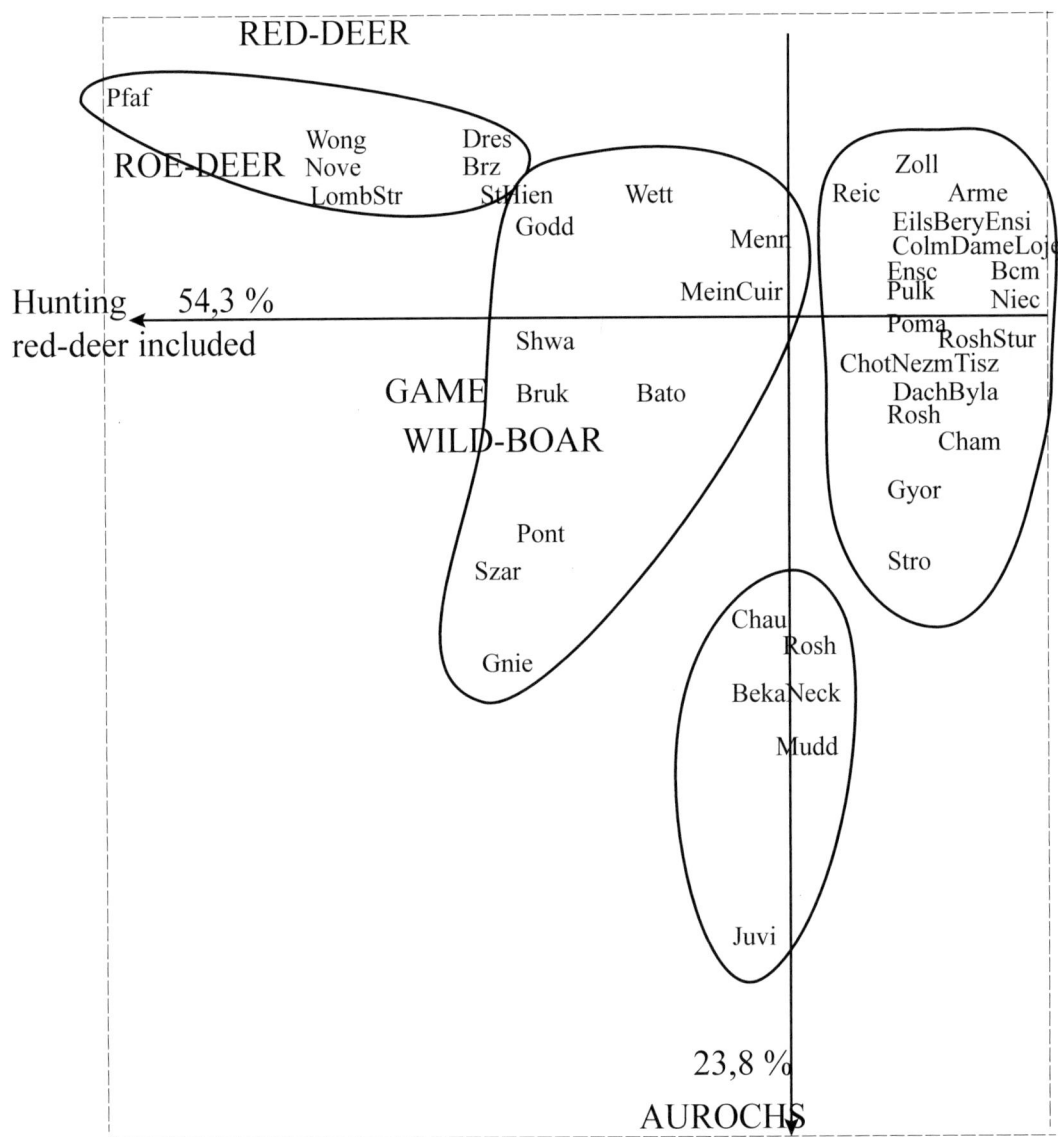

Figure 56: Correspondence analysis (carried out by J. Dubouloz) of the proportions of large game (red deer, roe deer, wild boar and aurochs) on 102 LBK sites. Abbreviations are composed of the first three letters of each site (see Hachem 1999, table 1 and appendix)

Number of overlapping dots : 21
Lamb(Nové) Stra(Brz) Brz(Arme) Eils(Bery) Grab(Bery) Hohl(Bery)
Lagi(Bery) Miku(Bery) Wett(Bery) Ens(Dame) Hatt(Colm) Trob(Colm)
Wett(Dame) Stur(Bcm) Larz(Bcm) DET(Pulk) Zale(Stuc) Larz(Chot)
Pili(Nezm) Dach(Byla) Rouf(Rosh)

and, therefore, the 7% average calculated for game remains in the region must be treated with caution.

The example of eastern Germany is repeated in Poland (Kujawie), where the average percentage for game animals is 5.5%. In this case there are additional issues which, when taken together with those mentioned above, lead to the conclusion that this approach gives a false impression. In fact, here we are dealing with a group of sites that are lumped together without taking the chronological sequence of the LBK into account. In reality, the earliest phase of the LBK in Europe is characterised by a high incidence of hunting, as has been demonstrated by M. Uerpmann (Uerpmann, 2001). The early LBK site of Gniechowice, for example, reveals a very high proportion of game (47%) (Sobocinski, 1978) which is masked by the 5.5% average.

For the Paris Basin, the smoothing of data by averaging is also an issue. The proportions of game species is 17%, which appears to be quite high compared to regions in central Europe. However, if we look at the faunal data at site level, we notice that the percentage of game at certain sites is 3%. The problem here is not the extent of the area excavated but rather the failure to consider the dates of the features or the size and location of the house within the village context. It should be remembered that at Cuiry-lès-Chaudardes, depending on whether we take a house belonging to the early or later phase, the proportions of wild fauna can vary between 41% and 4%, and variations are just as marked if we consider house type.

It is clear from these examples that the interpretation of the results of faunal studies is intimately linked to the archaeological context from which the samples are taken. Differences between regions can, of course, arise from particular cultural traditions, or from a particularly contrasted landscape (humid v chalky soil, for example). But it is essential to link these elements to the primary cultural context, *i.e.* the structure (house, pit), and its role or position in the settlement. If this is disregarded, then the fauna only provide a truncated view of the sample and cause us to miss what is essential, namely a true understanding of the household system.

It is therefore important to adopt an alternative approach to that of regional averages.

We have attempted to identify the elements that shape the variability in hunting and animal husbandry observed between LBK households. An alternative is not to simply divide sites into regional entities, but rather to seek the links between species composition and the context in which the remains were sampled, using different scales of analysis.

The sample provided in the literature is extremely heterogeneous and for the purposes of this analysis we have chosen to limit ourselves to settlements that have produced more than 100 pieces of bone; below this limit, the quantitative distribution of species is unreliable.

In total, we have examined data for large game (red deer, aurochs, wild boar and roe deer) from 102 published or accessible sites (fig. 56). An initial study was carried out on 95 published sites in Germany, Poland, Hungary, the Czech Republic, Slovakia, Austria, Belgium and France (Hachem, 1999)1999. This was subsequently augmented by the addition of a number of sites that had been excavated and published in the intervening period. Because the bibliographical references for all of these sites are too numerous to list in this work, we refer the reader to the following compilations: (Arbogast *et al.*, 2001; Schmitzberger, 2010; Bogaard *et al.*, 2017). For sites in the Paris Basin we refer the reader to the appendices of this volume and the archaeozoological reports we have written in the context of the Inrap excavations.

The percentage of identified remains is low and masks considerable disparities. The best assemblages are located in the Paris basin and in Alsace, while the poorest are located in Belgium and in the Dutch province of Limburg.

Despite these difficulties, analysis of the variations in the levels of hunting, taking all periods and regions together, reveals striking differences. Three categories of sites, which were already identified in a synthesis of LBK settlements (Döhle, 1994), can thus be highlighted:

- a majority of settlements where livestock rearing is overwhelmingly dominant (greater than 90% and often reaching 95%);
- a smaller number of sites where the level of hunting is quite significant (between 10 and 20%);

- a handful of sites where the incidence of wild animals is very high (greater than 20%).

We have seen how these variations in the proportion of hunting between houses can be the result of various factors. Let us now examine the first of these, chronology.

The earliest LBK ("älteste Linienbandkeramik") sites in Bavaria and Hesse have yielded faunal assemblages that are characterised by high proportions of game (Uerpmann and Uerpmann, 1997). According to the authors, this significant level of hunting may be due, in part, to pioneer-type behaviour. This is an attractive hypothesis and might well apply to the earliest LBK in other regions as well. Therefore, despite the relatively poor documentation (published sites also need to be well dated), it emerges that half of the early sites present a percentage of hunting greater or equal to 10%, regardless of the region in which they are located.

What is the situation in the later LBK? Settlements are more numerous and among the securely dated examples that have produced faunal assemblages, some three quarters reveal low levels of hunting, less than 10%. At the scale of a micro-regional sequence, as we have seen in the Aisne valley, for example, the same pattern, involving a drop in hunting, can be observed.

This decline in hunting between the beginning and end of the LBK sequence is accompanied by a shift in the species being hunted:

- in the early period ("älteste Linearbandkeramik"), faunal assemblages often display a preferential association of red deer with roe deer ;
- in the later period ("jüngere-Linienbandkeramik") we find much more aurochs and wild boar remains, particularly on sites where game represents more than 10% of the fauna.

However, this evolution is part of a general framework where certain characteristics, highlighted at Cuiry-lès-Chaudardes, appear to belong to the LBK cultural background as a whole.

Thus, all of the faunal assemblages show that red deer is the primary game animal, which is well represented in both early and late chronological phases (present on three quarters of sites). Like red deer, aurochs constantly occurs in the faunal records and particularly so in the later phases (present in almost ¾ of sites). Furthermore, these two animals are most frequently at the top of the list of wild animal remains found on sites. The presence of two of these animals is almost systematic on all LBK settlements. However, aurochs differs from red deer in that it is weakly correlated with high rates of hunting: it thus contrasts with other wild animals. Its particular status, which is certainly symbolic, has already been demonstrated by the types of bones found on sites in the Paris Basin: lower leg bones and skulls predominate which indicates a type of consumption and disposal that differs from other species.

However, wild boar only really appears to come to the fore on sites where hunting exceeds 10%, and it always occurs in equal proportions with one of the three other hunted species; it is poorly represented on other sites (*i.e.* the sites of Pomàz-Zdravlyàk, Pulkau, and Cuiry-lès-Chaudardes). We have, in fact, noticed that the presence of wild boar goes hand-in-hand with a high rate of hunting in certain small houses in the Paris basin and thus is representative of more intensive hunting.

Roe deer occurs in two configurations. At times, this animal, like red deer and aurochs, comes to the fore on sites where stock-rearing is significant (*e.g.* the sites of Dammendorf, Kothen-Geuz, and Tröbsdof). However, in cases where the rate of hunting is high, we see the same pattern as that observed for wild boar: the proportion of roe deer rises in conjunction with that of another animal, either red deer or aurochs (*e.g.* the sites of Wettolsheim, Gniechowice, and Pfaffingen). The "neutral" character of roe deer has been revealed at Cuiry-lès-Chaudardes: it more or less follows the same distribution as red deer in the houses and occurs in high proportions both in houses where hunting is pronounced and in those where stock-rearing is prominent. In cases where there is a marked focus on a single wild animal species – a phenomenon that is particularly evident on sites where hunting accounts for more than 30% of the remains – that species tends to be either aurochs (Neckenmarkt, Gniechowiche, Juvigny) or red deer (Hienheim, Straubing-Lerchenhaid). Apart from a single site, Pont-Sainte-Maxence, wild boar never follows this configuration.

While change in the exploitation of wild fauna over time appears to account for some of the variations that we have observed, it does not explain them all. In fact, some of the pioneering sites appear to ignore hunting while, in contrast, some of the later sites tend to favour it. In order to resolve this problem, we can introduce another of the aforementioned variables influencing the rate of hunting, namely house type.

Unfortunately, it is difficult to verify that wild faunal remains are preferentially found in small houses as the literature rarely provides a house-by-house breakdown of bone data. In order to partially overcome this dilemma we can limit the resolution of the field of investigation and simply match the overall rates of hunting with the absence or presence of small houses on each site for which the duration of occupation is known.

Of the thirteen sites that can be evaluated under these terms, and for which the hunting rate is high (over 15%), nine feature small houses. This high proportion would appear to be a promising indication of the consistency, at the scale of the entire LBK socio-economic system, of the model established at a micro-regional level: *i.e.* the association between house type and the fauna consumed.

Certain sites do not fit into the explanatory schemes presented here. However, most of the incoherencies identified can be explained by the quantitative or qualitative shortcomings of the samples or by the absence of accurate dating which hinders interpretation.

The Middle Neolithic

The various types of settlements

Over recent years, archaeological research has produced a wealth of new data on the Middle Neolithic, particularly in the North of France, and it has now been established that enclosures were a key element in settlement and territorial systems occupation over a large part of Europe at the time (Andersen, 1997; Dubouloz, 2018; Gronenborn, 2003).

The role of enclosures in the societies of the 5th millennium BC has been modelled by examining the interaction of numerous parameters. In particular we draw attention to a project that set out to explore this issue: a Franco-German project undertaken by the UMR "Trajectoires" and by the Römisch-Germanische Kommission-D.A.I. with the aim of producing data on a Europe-wide scale (Demoule and Lüth, 2010). A PhD thesis (Liétar, 2017), undertaken at the University of Paris I, which concerns the Paris basin was written on this

Figure 57: Pont-sur-Seine "le Haut de Launoy" (Aube), a circular building dating to the Middle Neolithic 1 (Cerny) (photo M. Ilett, University of Paris 1, Panthéon-Sorbonne).

occasion, and several other articles *e.g.* (Dubouloz *et al.*, forthcoming; Hachem and Maigrot, 2019; Höltkemeier, 2010; Manolakakis and Giligny, 2011).

Approximately the fauna of thirty occupation sites have been analysed, spanning the period from the Cerny Culture to the Balloy Group, *i.e.* from 4200 BC to 3800 BC. They are located in the Departments of Aisne, Oise, Marne, Seine-et-Marne, Nord, Somme, Pas-de-Calais, and, further west, in the Orne (fig. 9).

In the 1990s, a number of works were published on the food economy of the Middle Neolithic in northern France (Arbogast, 1989; Hachem, 1989; Tresset, 1996). Then, in the 2000s, analyses were undertaken based specifically on the archaeological contexts (Guthmann, 2010; Hachem, 2011b; Höltkemeier, 2013). More recently again, a number of important sites, excavated as part of rescue operations, have further expanded the initial corpus and have added to our knowledge concerning the fauna of northern France. For the initial period of the Middle Neolithic (designated the Middle Neolithic I), these sites include the settlement in Conty (Bostyn *et al.*, 2016) and the Fleury-sur-Orne necropolis (Ghesquière *et al.*, 2019a); the following period (Middle Neolithic II) is represented by enclosures found in Passel (Cayol, forthcoming), Carvin (Monchablon *et al.*, 2011), Escalles (Praud, 2015; Praud and Panloups, 2015) and Villers-Carbonnel (Bostyn, 2014).

We will look at these sites in greater detail in the following sections; the actual data is provided in the appendices or is available as lists of fauna per site (Arbogast, 1989; Hachem, 1989; Tresset, 1996).

For the Middle Neolithic II, Chasséen occupation layers, such as those at Louviers (Giligny *et al.*, 2005) and Bercy (Lanchon and Marquis, 2000), have been interpreted as accumulations of waste originating from nearby settlements and thus the bone refuse they contain can be considered as having resulted from everyday consumption.

The Middle Neolithic I

For the Middle Neolithic I period (Cerny and Late Rössen), only a small number of sites have yielded faunal remains and these are very heterogeneous in terms of their archaeological contexts: ditched enclosures, domestic pits, ovens, specific pits, occupation layers, graves and funerary monuments (fig. 57). This means that each site should be regarded as singular. Nonetheless, here we will attempt to characterise the fauna associated with the sites presumed to refuse arising from everyday consumption; animal bones from funerary contexts and communal consumption contexts within enclosures will be dealt with in later chapters.

One settlement of the Cerny Culture with buildings has been discovered at Conty "ZAC Dunant" in Somme (Bostyn *et al.*, 2016). As it stands, Conty is currently the only documented settlement site featuring buildings. Its corpus is also much larger than that of other sites

Figure 58: Cuverville "le Clos du Houx" (Calvados), an oven dating to the Middle Neolithic 1 (after Fromont 2016, fig. 15).

(NISP=1185), except the Balloy enclosure. The proportion of domestic animals is significant (90.4%). Cattle still prevail, but in a relatively smaller proportion (46.2%), and the place of pig increases (36%). Caprines remain under-represented (4.9%). The limited proportion of red deer is matched by that of wild boar (2.5% each).

An oven (fig. 58) was found at Cuverville "le Clos du Houx", Calvados (Hachem, 2016). Bone remains (NISP=44), which had accumulated in the backfill of the oven and its access pit, indicate that domestic animals were numerically dominant: they include the remains of eighteen cattle, one pig and one caprine. We also note the presence of red deer in the form of bone remains and antler fragments.

Sites attributed to the Late and Final Rössen consist of enclosures, which will be dealt with in the chapter on communal consumption (chapter 3).

The Middle Neolithic II

For the Middle Neolithic II, sites are more numerous than for the preceding period. We observe the spread of groups belonging to a shared cultural background, with distinctive regional variants including sites belonging to the Michelsberg, the northern Chasséen, the Noyen Group and the Balloy Group.

The "open" Michelsberg settlement at Cuiry-lès-Chaudardes, in the Aisne, is composed of 80 pits spread out over an area of 6 ha, surrounded by a palisade; twenty of these pits have yielded faunal

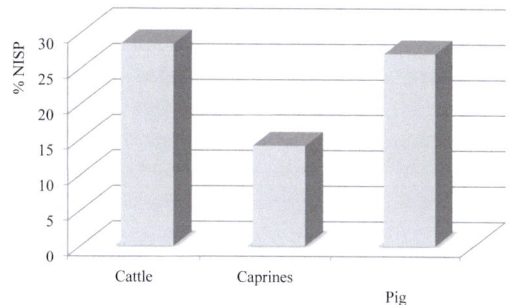

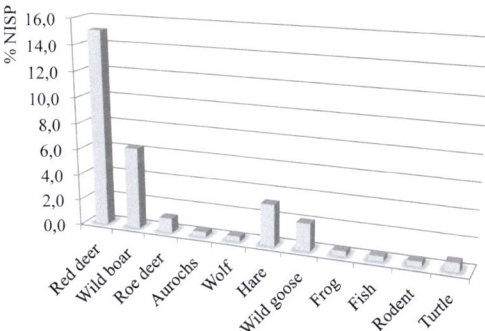

Figure 59: Proportion of species on the Middle Neolithic 2 (Michelsberg) settlement at Cuiry-lès-Chaudardes "les Fontinettes" (Aisne) ; an example of an "open" or unenclosed site composed mainly of pits. Up: domestic species. Down: wild animals.

Figure 60: Louviers "la Villette" (Eure), an example of an occupation layer dating to the Northern Chasséen: concentration of wood, including worked pieces, in a palaeochannel (photo F. Giligny, University of Paris 1, Panthéon-Sorbonne).

remains, corresponding to an NISP of 295 remains (fig. 59, Hachem, 2011b).

The proportion of game is high (30 % of NISP), and a wide variety of species are present. We note the importance of red deer and wild boar, but also of small game including hare. Inversely, the contribution from livestock rearing, principally represented by cattle and pigs (28% respectively), is relatively low.

By comparison, the situation at the site of Mairy in the Ardennes is very different even though this site also features numerous pits (along with evidence for buildings and an enclosure). In fact, dozens of pit-silos were filled with very particular faunal remains, with no game at all; a single red deer bone was recorded among the thousands of bones present (Arbogast, 1989). Cattle (78.9 % of NISP) and caprines (16 %) make up the bulk of the fauna, while pig only occurs occasionally (4.4 %). In this case we are probably looking at a site used for cult activities.

Very few remains associated with the Noyen Group have been recorded.

An occupation layer excavated at the site of Noyen (assemblage 5) yielded 105 identified remains (Tresset, 1996). Game is present in high proportions (17.1 % of NISP). Among the domestic animals, cattle make up 61.9% of the remains, pigs 20 % and caprines only 1%. Game is principally represented by red deer, with wild boar in second position.

A site featuring numerous pits has also been discovered at Châtenay "LP" (St 12) and it has yielded 115 identified remains (Tresset, 1996). Domestic animals predominate (92.1 %), but cattle are relatively poorly represented (45.2 %), while pigs are well represented (36.5 %) and sheep are relatively rare (10.4 %). Red deer is the main game animal present.

Northern Chasséen sites are found in the regions of Ile-de-France, Hauts-de-France, and the department of the Eure.

A number of levels, interpreted as occupation layers, have been discovered at Paris Bercy ("Channel" and "Layers 12, 13, 15") (Tresset, 1996) and at Louviers "la Villette" in the Eure (Tresset, 2005). Consumption remains are very numerous (total number of remains: Bercy = 11377; Louviers = 3676), but have sometimes suffered disturbance, in particular at Bercy. At the latter site, domestic animals are strongly represented (90.9 % of NISP), and we observe a slight difference in the proportion of domestic animals between Bercy "Channel" where cattle occur in high proportions (73.9%), followed by pigs (10.5 %) and caprines (4.8%) and Bercy "Layers 12, 13, 15" where cattle are less numerous (56.3%), followed by pigs (18.8%) and caprines (15%) in almost equal proportions. The game animals are largely dominated by red deer (7.1 % and 6.9 % respectively); the other wild species present are wild boar, aurochs and roe deer.

At Louviers 86 % of the determined remains are from domestic livestock. Cattle largely predominate (74.3 % of NRD), while pigs (7.7 %) and caprines (4.1 %) are much more poorly represented (fig. 60).

The presence of dog is quite important (1 %).

Wild animals are represented, in order of importance, by red deer (8.7 %), aurochs (2.4 %), wild boar (1.3 %) and roe deer (0.3 %). Rare animals such as wolf and bear are present, as are small game animals such as beaver and badger. The presence of a complete male aurochs skull on the site is also noteworthy.

The slaughter curves for cattle on the two sites show the presence of young animals that were slaughtered for their meat and of older or castrated animals which are interpreted as dairy animals and draft animals.

At Louviers, the hunting of deer and aurochs targeted adult animals but was indiscriminate in terms of sex since there are equal numbers of males and females.

Only a single site composed of pits is recorded, Limay in Yvelines (Gasnier *et al.*, 2014; Hachem, 2011c), but the sample here is too small to allow conclusions to be drawn (NISP: 16).

An occupation layer at the site of Châtenay ("LB") yielded a faunal assemblage of the Balloy Group (NISP = 102) (Tresset, 1996). The proportion of domestic animals present stands at 85.2 %, with a relatively low proportion of cattle (54.9 %), followed by 12.7 % pigs and 9.8 % caprines. Game is represented solely by red deer.

Sites belonging to the Spiere Group are still very rare and only the faunal assemblages from Carvin and Escalles, located in Pas-de-Calais, have been studied. We will deal with this material in the chapter on communal consumption (chapter 3).

Ten features making up a building (n° 6), and associated pits, were discovered on the site of Saint André-sur-Orne "la Delle du Poirier", Calvados (Ghesquière *et al.*, 2016).

The site yielded sixty bone remains, fifty of which could be determined, which is an excellent ratio compared to other samples from Neolithic settlement sites in Normandy which tend to be significantly eroded.

Meat consumption was largely dependent on domestic animals. Cattle represent almost half of the remains (48%); sheep occupy second position with 32 % and pigs come third with 20 %. Wild animal bones were absent and although red deer antler was made into tools, we cannot be certain that the animal was actually hunted.

The minimum number of cattle is a single individual (MNI), but it is possible that two were present because of the different diaphysis dimensions observed on the long bones. It was not possible to determine the slaughter age but these

were adult or sub-adult animals. Three caprines are represented: a very young individual, a juvenile and an almost adult individual. The minimum number of pigs is also three (MNI): a very young individual, a juvenile and an adult or sub-adult.

The Late Neolithic

An imbalance in favour of funerary contexts

Between 2001 and 2008, a joint research programme (PCR financed by the Minister for Culture) brought together researchers from the CNRS and the Inrap, as well as students from various universities, with a view to exploring the end of the Neolithic in the Centre-North of France (the regions of Pas-de-Calais, Picardy, Champagne-Ardenne, Ile-de-France, Centre and Bourgogne). The work of compiling an inventory of sites and gathering information was complemented by new excavations, new dating and the re-examination of finds. The work provided an up-dated overview of the material culture which

Figure 61: Beaver was greatly sought after, from the Early to the Late Neolithic (photo Skeeze, Pixabay).

in turn led researchers to propose at least three regional facies and three chronological phases for the Late Neolithic, *i.e.* the second half of the 4th millennium BC, in the Centre-North of France (Cottiaux *et al.*, 2014b).

The settlement sites whose faunal assemblages were studied are located in Seine-et-Marne, Hauts de France, Champagne and Nord-Pas-de-Calais. The archaeological contexts are diverse and include occupation levels, enclosure ditches, and pits with various functions. The particular status of deep pits, isolated tombs and monumental complexes, such as that at Pont-sur-Seine in the Aube will be considered in the chapters on collective consumption and funerary contexts (chapter 3 and chapter 5).

Different types of occupation
The sites mentioned above were the subject of an archaeozoological analysis which involved a relatively substantial sample[11]. A further site, a pit at Luzancy "le Pré-aux-Bateaux", Seine et Marne (Lanchon *et al.*, 2013), has been omitted because of the small size of the assemblage (NISP = 11).

The site of Mareuil-lès-Meaux "la Grange du Mont" revealed a large pit containing eighty-three remains, seventy of which could be determined. The species present are exclusively domestic. Pigs predominate (54 remains, MNI =2 individuals), followed by cattle (15 remains, MNI = 1); a shed red-deer antler was also found.

The excavation of the site of Mareuil-lès-Meaux "les Lignères" revealed an occupation layer dating to the Late Neolithic which had been miraculously preserved from erosion by a large later building, possibly dating to the Post-Bell Beaker period or, more probably, to the Early Bronze Age (fig. 10). This archaeological level yielded 100 pieces of animal bone, 63 of which were determined. Five species have been identified, all domestic apart from red deer. In terms of the numbers of remains, cattle occur in the same proportion as pigs (% NISP= 44.4 % and 41.3 %), but in terms of minimum number of individuals, the latter predominate: we observe two pigs (a young animal and an adult) and a single head of cattle (adult). Caprines are also present (NISP = 9.5 % MNI = 1), as is dog (NISP = 1).

At the site of Vignely "la Noue Fenard", a pit (feature 264) dating to the early phase of the Late Neolithic contained 115 bone remains, 36 % of which have been determined. The assemblage is made up of several layers of bone waste that include at least five wild boars, three aurochs, five beavers (fig. 61) and two red deer, which together represent a considerable volume of meat; to these we can add isolated bear, otter and fox bones which bear cut marks attesting to the removal of the animals' skins. The analysis shows that the assemblage is the result of a single event of relatively short duration; it reflects a number of technical activities indicated by the presence of lithic tools and worked bone and antler. At present, we have no parallels for this pit with its predominance of game in the data for the Neolithic of the North of France.

The site of Cuiry-lès-Chaudardes "les Fontinettes" featured a Late Neolithic pit containing 175 animal bones, 118 of which have been determined. The sample is predominantly made up of pig bones (NISP = 112; MNI = 5 individuals), the other species represented being cattle (NISP = 1) and caprines (NISP = 2). Red deer is very discretely represented by part of a long bone and two fragments of antler; fish vertebrae are mentioned in the site report but they were not located during our study. It appears that the contents of the pit represent two types of discarded bone: the first, corresponds to "normal" consumption, principally of pig (presence of diverse carcass parts, bones fractured to allow extraction of marrow, specific cut marks on the ribs), the second involves the disposal of four very young piglets (3 were perinatal) which were probably unsuitable for consumption.

Topsoil removal at Pont-sur-Seine "le Haut de Launoy; phase 3" uncovered several Late Neolithic features which yielded a total of 124 bone remains. They were exclusively cattle bones (at least three individuals), and no other domestic or wild species were identified.

The faunal remains recovered from a distinctive feature, which took the form of a rectangular stone-paved surface, are comprised of 60 fragments of bone, three quarters of which have been determined. The remains are virtually all from cattle (at least two individuals).

Finally, five bones were recovered from a ditch. These are the remains of a horse but it is not possible to tell whether the animal was domestic or wild. Two complete coxae, a right and a left, were from the pelvis of a single individual. An almost complete metatarsus and the core of a vertebra were also recorded. The intact nature of these bones leads us to suggest that this was a special disposal or perhaps even a deliberate deposit.

Taken together, this data indicates a different composition for the bone refuse depending on the type of feature that it was placed in and for the moment it is not possible to use it to build a clear image of the Late Neolithic diet. If we base our reasoning on the three sites that have produced some evidence for consumption (*i.e.* Mareuil-lès-Meaux "la Grange du Mont" and "les Lignères", and Cuiry-lès-Chaudardes "les Fontinettes"),

Early Neolithic (5200-4700 BC)

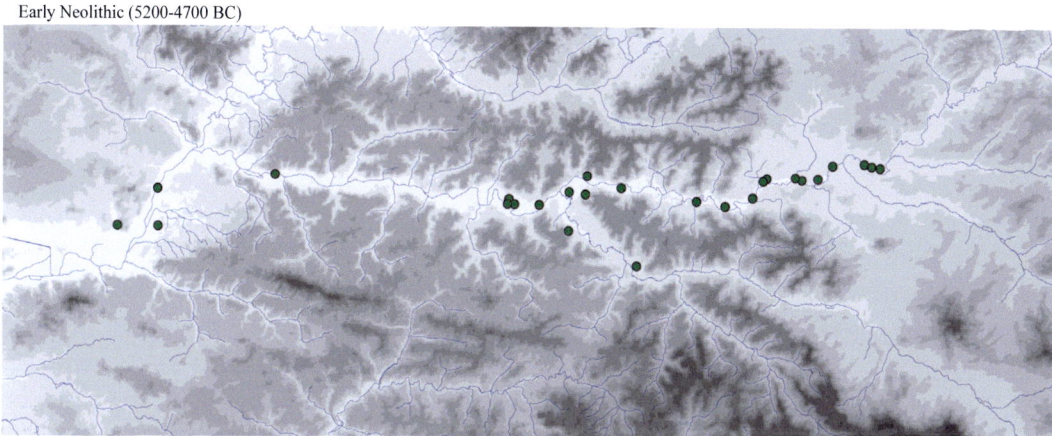

Middle Neolithic (4700-3950 BC)

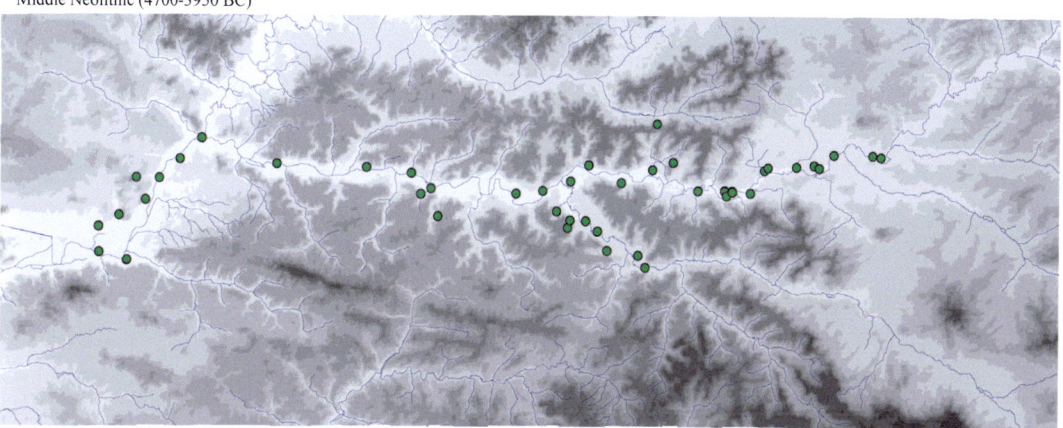

Late/Final Neolithic (3950-2900 BC)

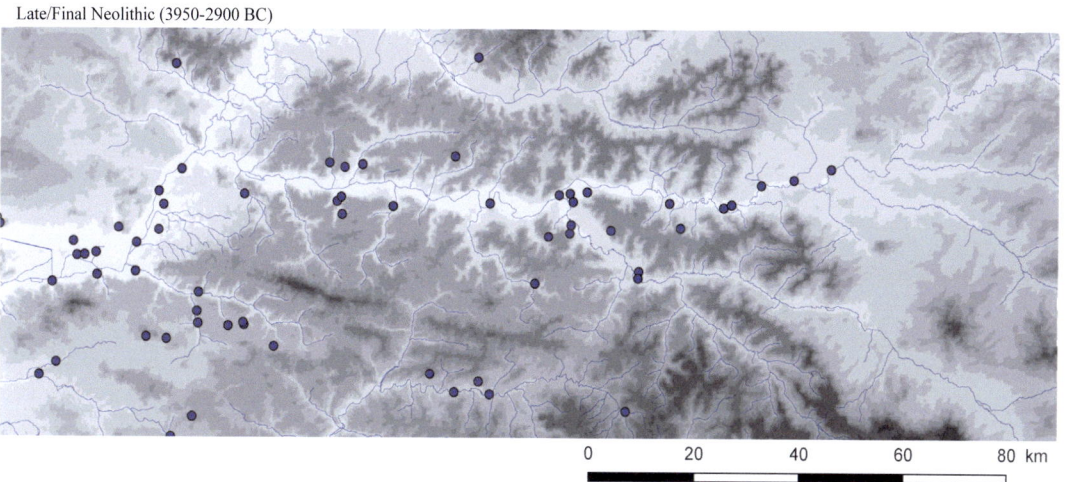

Figure 62: Sites with evidence for settlement in the Early (top), Middle (middle) and Late (bottom) Neolithic in the valleys of the Aisne and Oise (after Liétar and Giligny 2016, fig. 1)

0   20   40   60   80 km

then it would appear that pigs were the primary resource and cattle secondary. Caprines were marginal while red deer is present in each case. Red deer antler was an important raw material in the bone-working industry and numerous antler sleeves have been recorded (Cottiaux and Salanova, 2014).

The other assemblages are unique, due to the very different nature of the features in which they were found. We note the almost exclusively cattle remains from the features at Pont-sur-Seine while the isolated pit at Vignely only contained game species.

The Final Neolithic

Fresh settlement evidence

Recent discoveries made in the context of rescue archaeology have greatly added to our knowledge of the 3rd millennium BC, particularly in the departments of the Nord, Pas-de-Calais and the Somme. A striking example is the vast structure, featuring a palisade and numerous buildings, that was uncovered at Houplin-Ancoisne "le Marais de Santes" in the Nord, and which dates to the beginning of the 3rd Millennium (Praud, 2015).

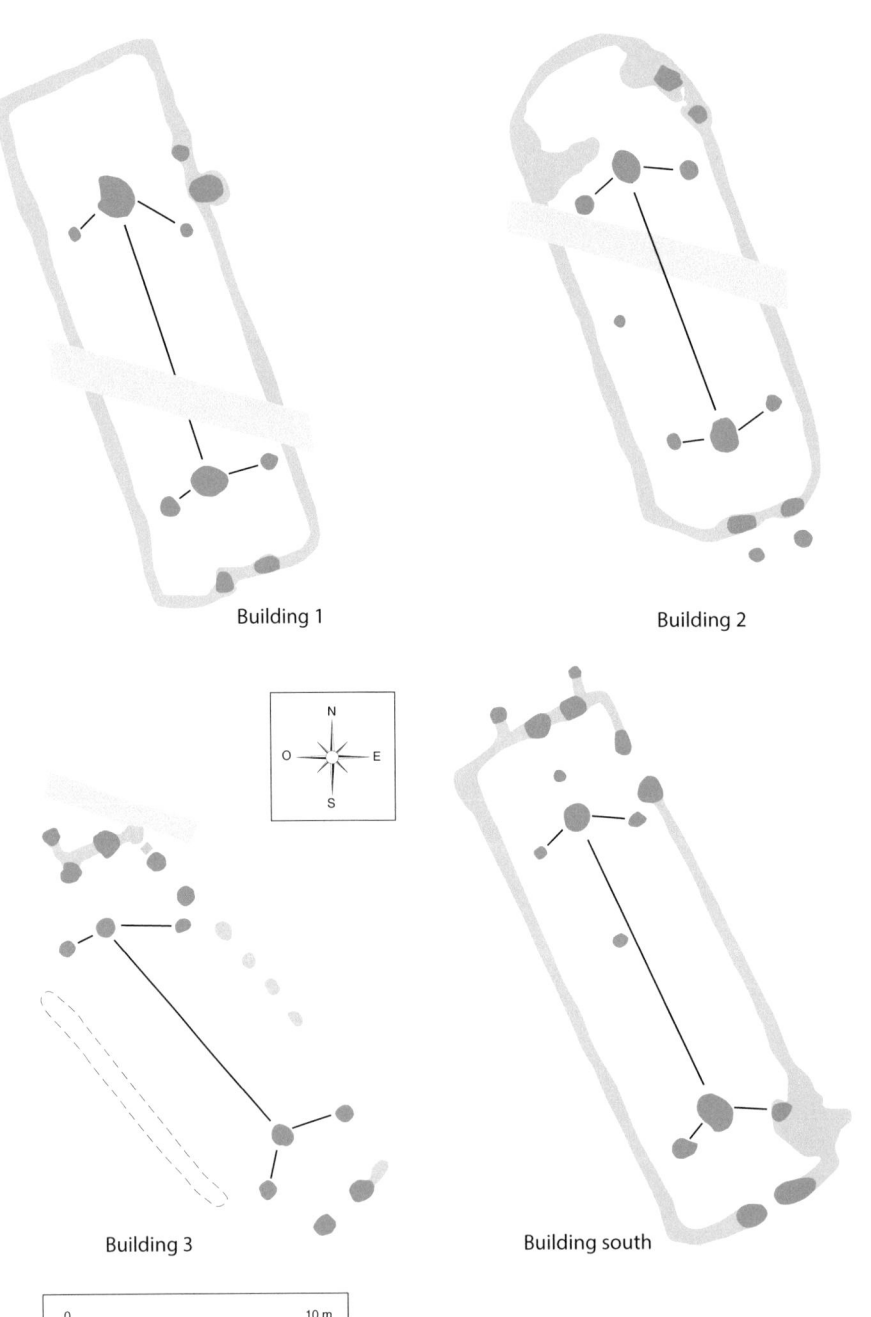

Figure 63: Glisy "Zac Jules Vernes" (Somme), buildings dating to the Final Neolithic (after Joseph 2008, fig. 2 modified).

A number of sites have been discovered in Picardy but three quarters of all known settlement sites in this region have been found in the Oise (fig. 62), where they are principally located on alluvial terraces (Dubouloz *et al.* 2005). These sites almost always take the form of a "trapped" archaeological level, preserved to a greater or lesser degree. The quantities of finds yielded by these sites are greater than those found on Late Neolithic sites and they show that certain sites extended over very large areas. Also in this region, clay spindle whorls make their first appearance and attest to textile production. Lithic finds show that a variety of raw materials were exploited, many of which were transported over long distances.

Data for the Bell Beaker period is much rarer than data for the preceding periods.

Faunal remains from settlement sites

Faunal remains dating to the Final Neolithic are quite rare on known sites; in fact, only a handful of archaeozoological analyses are available to allow us to make some preliminary observations regarding the beginning of the 3rd millennium, and there is a total lack of information for the end of the millennium. We are mainly dealing with remains retrieved from the postholes of buildings. A full list of the sites is provided in the footnotes[12].

Other faunal remains have been gathered during testing carried out by the Inrap (Hachem unpublished): the find contexts are remnants of occupation levels, or specific features as at Meaux (Seine-et-Marne), Gonesse (Val d'Oise) or Pont-sur-Seine (Aube), but in general the assemblages are too small and their dating too imprecise (Late/Final Neolithic) to allow them to be considered.

However, the excavation of a ditch at Houplin-Ancoisne "le Marais de Santes", (Nord-Pas-de-Calais), has yielded a more substantial sample which we will look at in the chapter on collective consumption; the site of Saint-André "la Delle du Poirier" (Calvados) will also be looked at in the same chapter (see chapter 3).

The faunal assemblage from the site of Glisy "Zac Jules Vernes" was retrieved from the post holes of three buildings and corresponds to consumption waste (fig. 63). The remains are poorly preserved with 157 bones remaining undetermined. The consumed fauna (NISP = 143) is made up of 41.3 % caprines, 30.8 % pigs and 21.7 % cattle. The wild fauna, which is very scarce (3 %), is made up of red deer (1.4 %), horse (0.7 %) and hedgehog.

Burnt bones were uncovered in the southern and central parts of a building which the plans suggest may have contained two hearths.

At Marquion, faunal remains were collected from the post holes of three buildings belonging to the Final Neolithic/Bell Beaker period. Preservation was very poor and only about twenty fragments were recovered, six of which could be determined. Nonetheless, the remains deserve mention because they come from well-dated structures. The identified remains consist of five fragments of caprine bone and one fragment of pig bone; there is a strong probability that the undetermined bones are also from caprines and/or pigs due to their size. The presence of burnt remains in the foundation trench of the building at Marquion supports the hypothesis that there was an internal hearth, as has also been suggested at Glisy.

The animal bone from Meaux "Route de Varreddes" comes from an occupation layer and is therefore poorly preserved. The assemblage is comprised of 1073 pieces of bone, almost half of which have been determined. The remains are dominated by domestic animals, namely cattle (56.9 % NISP), caprines (21.6 % NISP), pigs (15.7 % NISP) and dog (0.6 % NISP). As regards wild animals, we observe red deer (3.5 %), horse (0.6 %) and possibly a wild boar, along with two birds.

The management of the cattle herd, represented by ten individuals, shows a predominance of juveniles and young adults which indicates a focus on meat production. The same is true in the case of the pigs. The age distribution of the caprines, however, does not reveal a specific focus on the production of meat, milk or wool.

The bone remains from the site of Compiègne "le Gord" (NISP = 276) were recovered from an occupation layer associated with a floor surface; they are thus poorly preserved (160 undetermined bones which equates to 58 % of the total remains). While domestic fauna are predominant (63.5 % of NISP), wild fauna are significant and varied (36.5 % of NISP). Cattle are in the majority but the overall proportion is low (35 %); next come pigs (18 %) and caprines (11.5 %); we also note the presence of one piece of dog bone. Wild boar (14 %) and red deer (13 %) are well represented and occur in similar proportions. Roe deer (3%), aurochs (1 %) and horse

(1 %) are more discrete. Small game is represented by beaver (0.5 %), and birds and fish complete the list of wild species.

The domestic animals were slaughtered at a young age (the majority were killed before the age of 3 years), but older cattle (between the ages of 5 and 9 years) have been identified.

The Metal Ages

This overview examines the issue of livestock management in the Aisne valley, between Osly-Courtil and Menneville, and in the valley of its tributary, the Vesle, between Chassemy and Reims (Aisne et Marne). We also look at sites located on the Troyes Plain, in the middle Oise valley, between its confluence with the Aisne in the north-west and the Thève in the south-west (Oise), in the department of Seine-et-Marne, particularly in the Seine-Yonne interfluvial zone and in Bassée (the name given to the middle valley of the Seine between Montereau-Fault-Yonne and Nogent-sur-Seine, at the Seine/Yonne confluence). Consideration is also given to a number of sites located in the Marne valley, in the Crould Basin (spanning the eastern part of the department of Val d'Oise and western part of the department of Seine-Saint-Denis), in the Caen Plain (on the western fringe of the Paris Basin, Calvados), and on the plateaus located in the interfluvial zone between the marshy valleys of the Somme and the Avre to the south-east of Amiens. Significant corpora of faunal remains have been studied in all of these regions allowing us to adopt a diachronic approach to the subject.

The division of these faunal assemblages into broad chronological units highlights the small proportion of Bronze Age/Early Hallstatt assemblages compared to the Late La Tène. Looking at the situation in greater detail, the Bronze Age period is particularly well documented for the Late Bronze Age/Early Hallstatt (Hallstatt A2/B1/B2/B3 and Hallstatt C, *i.e.* between 1150 and 630 BC), although in most cases the sizes of the corpora are limited. Farm sites dating to the Middle Hallstatt (640-530 BC) have produced meagre assemblages while those from Late Hallstatt/Early La Tène sites are more substantial. While Middle La Tène sites (La Tène C1, 270-180 BC) are generally poor in terms of faunal remains, numbers of assemblages from sites dating to the Late La Tène increase considerably throughout the geographical areas under consideration.

Bone remains from the various study areas have been subject to a certain degree of post-depositional deterioration but, on the whole, the skeletal material tends to be well preserved. For this reason, it is possible to envisage comparisons between individual sites and regions.

The Bronze Age

Regardless of their location within northern France, most farmstead sites produce little meat consumption waste. One of the reasons for this is the often faint footprint left by human occupation on the landscape, but also the management of waste on the margins of the settlement areas, to which must be added the undoubtedly low proportion of meat consumption within the diet of Bronze Age people generally. In certain areas it is still not possible to propose "models" for animal husbandry and to evaluate the contribution of hunting activities to the diet of Bronze Age populations in the period between 1500 and 1000 BC. Despite these obvious difficulties in assembling a statistically viable corpus for the successive phases of the Late Bronze Age, and thanks to the intensity of excavations and the growing number of small assemblages, now provide enough evidence for us to make suggestions regarding social organisation on the basis of bone refuse (Auxiette *et al.*, 2015; Auxiette, 2017a; Auxiette *et al.*, 2020). The decision has been made to divide the descriptions of the various sites belonging to this period into two categories: farms of modest size, on the one hand, and large sites, on the other.

Production and consumption on farmstead sites

The most commonly identified farming settlements take the form or sites composed of a number of pits and, in the best cases, the remains of a dwelling. Evidence for material culture is generally scant: the finds are limited to no more than a few hundred pot sherds and bone fragments. Nevertheless, a small number of sites stand out from the rest in terms or their extent and the fact that they feature several buildings which, in some cases, cans be analysed from the perspective of their synchronicity or diachrony (e. g. Changis-sur-Marne; Lafage, forthcoming; Lafage *et al.* 2006, 2007).

In southern Picardy, which is one of the areas to have produced the best evidence, the farm sites principally occur on the alluvial plain of the Aisne valley (*i.e.* the sites of Bucy-le-Long "le Grand Marais", Limé "les Fussis", Berry-au-Bac "le Chemin de la Pêcherie", "Derrière le Village" and "la Bourguignotte" at Menneville) and on the valley slopes (*i.e.* Pasly "les Côteaux" and "Derrière Longpont") (Auxiette, 1997; Desenne, 2017). Analyses of the assemblages from these sites highlight the singularity of each farm in terms of the dominant domestic species and variations in the proportions of wild species, variations which may be linked to the size of the corpus, among other things. Among the best documented sites, pig predominates at Menneville "Derrière le Village" located in the east of the Aisne valley, while sheep and goats predominate at the neighbouring site of Menneville "la Bourguignotte". As regards wild fauna, the proportions can vary from very low, as at "Derrière le Village" (1%), to quite significant, as at "la Bourguignotte" (8.4%), and to quite high as on the two sites at Pasly (18.2 and 21% respectively). For sites that have yielded good evidence for slaughtering age, the sites of Menneville "Derrière le Village" and Pasly "les Côteaux" indicate that pigs were preferentially slaughtered between the ages of 1 and 2 years and caprines between 1 and 3 years. These animals were raised for consumption; the corpus lacks evidence for the slaughter of lambs and calves which might have indicated the production of milk. Furthermore, data is lacking for cattle.

At Compiègne "Fond Pernant" (Oise), pigs and caprines occur with equal frequency and there is a very limited presence of wild fauna, approximately 1% (Méniel, 1984).

Farms sites are well recorded in the Bassée area, situated at the confluence of the Seine and the Yonne in Seine-et-Marne. Such sites have generally produced faunal assemblages ranging from several dozen to several hundred specimens: assemblages of 1000 or more bones are rare (Hermetey, 1995). The size of the farming settlements also varies: two examples have a layout that indicates the presence of several domestic units which may be synchronous or, alternatively, which may reflect a household's move to a new house, instead of repairing the old. To date, seven sites have been recorded: "Motteux" and "le Grand Canton" in Marolles-sur-Seine, "les Méchantes Terres", "Ferme d'Isle", "les Champs Pineux" and "les Roqueux" in Grisy-sur-Seine, and Barbey "Chemin de Montereau". As before, the livestock composition of each farm is unique. As regards the domestic fauna, cattle predominate at Grisy-sur-Seine "les Champs Pineux" and at Marolles "Motteux", while sheep are predominant at Grisy-sur-Seine "les Roqueux" and at Barbey. The proportion of wild fauna varies between 5 and 10%. In the three farms that have produced the best evidence for herd management, we see a preference for keeping cattle into maturity (from 4 years old to more than 12 years old) for the production of secondary products, probably

including large quantities of milk. It is rarer to find animals that were slaughtered before the age of 3 years. In the case of caprines, the approach differs between the farm at Barbey, where the flock was reared for secondary products, and Grisy where the slaughter curve is spread out over time suggesting the slaughter of lambs and kids to stimulate milk production based on the studies carried out by D. Serjanston (Serjeantson, 2007). On the left bank of the Loing, to the north of Nemours, the large corpus (several thousand bones) from the site of Grez-sur-Loing "l'Epine" (Seine-et-Marne; Late Bronze Age IIb, faunal studies, Legoff in Valéro 2008) is notable for the very high proportion of pig present (79%) and the very low proportion of wild fauna (1.6%) (Valéro, 2008). Almost half of the pigs were slaughtered before the age of 2 years. In the case of cattle, adult animals and juveniles are equally represented. Caprines were preferentially slaughtered before they had reached the age of 2 years. An estimate of the Minimum Number of Individuals (MNI) indicates the presence of more or less forty pigs, four cattle, six caprines, numbers which place this site among the larger settlement sites. The high incidence of pig may be a reflection of collective meals. This settlement also provides clear evidence for the consumption of horses. The site is composed of a number of houses existing together in a hamlet type cluster; the ceramic evidence allows us to estimate that the site was occupied for one or two generations (Valéro, 2008).

In Champagne-Ardenne, in the Aube Department, 15 km north of Troye, the farm at Villemaur-sur-Vanne "les Gossements", which was probably made up of several domestic units, has yielded an assemblage of several thousand bone fragments, the largest proportion of which are pig bones (54%) with wild fauna bones not exceeding 2% (Hermetey, 1994). A study of the ages of the pigs (NISP=31) reveals a peak in the slaughter of animals between 1 and 1.5 years; 78% of individuals were slaughtered between the ages of 6 months and ± 2 years. In the case of cattle, juvenile (aged between 1 and 2 years), mature and culled animals are recorded in similar proportions. Caprines were principally slaughtered before the age of 2 with almost half of these being killed between 1 and 2 years. This indicates the selection of animals who had not yet attained their optimal weight but whose meat was of better quality. Milk production was not especially sought after.

Regardless of whether we are dealing with farms scattered along the Aisne valley or those situated at the confluence of the Seine and Yonne, the two best documented areas, the scarcity of data (ranging from ten to a few hundred bones) is an obstacle when it comes to developing models for herd management in the Late Bronze Age IIIb/Early Hallstatt (Auxiette, 1997, 2017a; Hermetey, 1995). The faunal spectra are very varied even though they are principally composed of domestic species that are commonly found on farm sites (cattle, caprines and pigs). Apart from rare examples (*e.g.* the site of Pasly), evidence for hunting activity is scarce for this period. There was a preference for hunting red deer over wild boar, and these are followed in importance by roe deer and occasionally aurochs, which disappears from the faunal record at the beginning of the Iron Age: to these main species we can add a number of small fur-bearing mammals, such as fox and hare, and certain birds.

Two sites in Ile-de-France stand out from the others: the settlement site at Grez-sur-Loing, which has produced evidence for pig consumption on a scale surpassing that of a simple farm, and Grisy-sur-Seine, where the management of the caprine herd reflects the production of milk and wool.

The best documented Late Bronze Age faunal assemblages recovered from the communes of Ifs and Mondeville (Basse-Normandie) located further west on the Caen Plain in Normandy (Auxiette, 2000b, 2011a), highlight the scale of caprines and beef consumption (caprines 54 %, Cattle 32 %) and the much lower proportion of pig (10%). Herd management, particularly of sheep and goats, shows clear evidence of a preference for quality meat production, which is reflected in the preferential slaughter of juvenile and young adult animals (aged 1 to 2 years); a portion of the herd was made up of older animals that were kept to produce wool and milk. In the case of cattle, we see evidence for the consumption of calves as well as of young animals. Pig was preferentially produced from animals approximately 18-20 months old. The three principal species were reared *in situ*. Evidence for hunting is virtually non-existent.

Predating the establishment of these farms, a Middle Bronze Age double enclosure at Mondeville "ZI Sud" (Chancerel A., Marcigny C., 2006) has yielded a corpus of about 1000 bones (31% of which can be identified to species) within which cattle bone predominates (65%); in this particular context, caprines and pigs each represent only a small proportion of the corpus (18% and 8.5% respectively) (Arbogast in Chancerel *et al.* 2006, 166-168). As in the case of the Late Bronze Age farm settlements, cattle are the most numerous species, but the predominance of cattle and horse remains singles this settlement out from the others (*cf. infra*).

These patterns in the animal husbandry practiced on farms on the Caen Plain partially agree with the findings for small farm sites in Aisne (Auxiette, 1997; Auxiette and Méniel, 2005b) and Ile de France (Hermetey, 1995) with a notable difference being that pigs were far less significant in the westernmost areas. The proportion of meat obtained through hunting activities is even smaller in these areas than elsewhere.

Production and consumption on large sites in upland areas and plains

Several sites fall outside the category of simple farmsteads. They are distinguishable by their layout and by the quantity and quality of the surviving remains. These settlements were established on hilltops or on the plains, close to water courses. Some of these sites will be dealt with again in the chapter on collective meals (see chapter 3).

For the upland sites in Ile-de-France, the partially excavated fortified hill settlement of Boulancourt "le Châtelet" (Seine-et-Marne), dating to the Hallstatt B2/B3 (Late Bronze Age IIIb), yielded 33,000 faunal remains, 8000 of which have been identified (24%, 37 taxa). The faunal remains come from a ditch. The faunal spectrum is overwhelmingly dominated by domesticated mammals (approximately 95%) (Bălăşescu *et al.*, 2008). Pigs represent 61.4 % of the corpus, cattle 19.1%, and caprines 12.5%. Red deer and wild boar are the principal wild animals present (3.5%) within a wide spectrum. Age analysis highlights the large proportion of young animals slaughtered before they had reached their optimal weight (see chapter 3).

In Picardy, fortified hill settlements are as rare as in Ile-de-France. Catenoy "Le Camp de César" in the Oise has produced a relatively large corpus of faunal remains although there is some doubt regarding the chronology of the assemblages, which renders the data somewhat unreliable. The study of ± 3500 remains (out of 7600) has highlighted the predominance of cattle (51%), half of which were slaughtered before the age of 4 years. A significant portion of the pig population (± 34%) was killed before the age of 9 months but those slaughtered around 1.5 years predominate. Caprines represent only a small proportion of the meat diet (±10%). The contribution from hunting is low (4%, 1.9% red deer out of ten taxa) (Méniel *et al.*, 1987). While it is clear that the production of meat was a central concern, the presence of a few older cattle and equids indicates that a portion of the livestock herd was maintained to provide other products and services.

Belonging to the category of large plain sites, the settlement of Choisy-au-Bac "le Confluent" in the Oise has yielded 7200 identified bones. The faunal sample from the Early Hallstatt level indicates a predominance of pigs (62.8%) with a slaughtering age centred on young individuals aged between 6 and 12 months. For the most part, caprines were slaughtered between 12 and 24 months. More than half of the cattle population was killed after the age of 48 months (Méniel, 1984, 43-45). Representing only 0.6% of the identified remains, wild fauna made up a very insignificant portion of the diet (Méniel, 1984, 41). The predominance of pig sets this area of the site apart and reflects a level of pig consumption that exceeds that normally encountered on simple farm settlements.

On the same site of Choisy-au-Bac, in an area known as "Canal Seine-Nord" which dates to Hallstatt B3/Hallstatt C, Late Bronze Age IIIb and Early Hallstatt (c.12,500 bones, 6000 of which have been identified), domesticated animals represent some 97% of the assemblage. The incidence of pigs (33.2%) is slightly higher than that of cattle (27.6%), whereas caprines make up only a small proportion of the total livestock (10.7%). While goats were certainly present, only a very small number of bones were determined with certainty, suggesting that they were a minor element of the livestock population alongside sheep. Horses and dogs barely register (0.6%) compared to the species already mentioned. Analysis of slaughter age for pigs reveals that there is a division between young animals (almost 56% were killed before the age of 24 months) and adults (44% were killed after the age of 24 months, of which 15% were killed after the age of 36 months). Cattle represent an important meat source: a little more than half of the herd was killed before reaching the age of 4 years, with a concentration of individuals aged between 24 and 30 months. Slaughtering appears to have occurred in the months of May to June and November to December. It is clear, therefore, that cattle rearing was primarily orientated towards the production of meat. Sheep and goats were killed, either in their first year with the selection of young animals aged between 5 and 10 months,

that is between August and January (18%), or much later after the age of 4 years (48-72 months, ±30 %). The flock was kept for mixed production of meat and secondary products: the killing of lambs may be linked to the stimulation of milk production. Culled animals could exceed 6 years (Auxiette and Bedault, 2015).

The site of Osly Courtil "la Terre-Saint-Mard" (Aisne; Early Hallstatt B2/B3, Late Bronze Age IIIb), located on the right bank of the Aisne and the only site in the corpus situated downstream from Soissons, was partially excavated and is characterized by an abundance of storage features, an absence of houses, and the presence of significant quantities of pottery (at least 480 vessels). The site was densely occupied compared to simple farming settlements. On the basis of the density and distribution of features, it seems likely that the site extended further westwards. The site, therefore appears to have occupied a special position within the settlement network as a centre for the storage and redistribution of excess cereal production (Le Guen *et al.*, 2005). The nature of the site leads us to include it within the category of large sites. Compared to the two sites discussed above, the corpus of faunal remains is very limited, comprising only 2000 remains (±1230 bones studied; Auxiette *in* Le Guen *et al.* 2005): pigs and caprines are present in similar proportions (32% and 33%, respectively). The proportion of wild fauna is amongst the highest in the study area (17 %). Meat consumption at the site was clearly focused on young sheep, principally slaughtered between the ages of 4 and 12 months, with killing taking place between the months of August and March.

The site of "les Pétreaux", located on a bend of the Marne at Changis-sur-Marne (Seine-et-Marne), was occupied on a perennial basis from the Early Neolithic to the Middle/Late La Tène and has produced detailed records spanning the entire Late Bronze Age sequence. The settlement is composed of clearly identifiable domestic units that provide an insight into the organisation and shifting of households (Lafage, forthcoming; Lafage *et al.*, 2007, 2006) over the Late Bronze Age IIIa (or Hallstatt B1) and the Early Hallstatt (Hallstatt C). A total of eleven domestic units developed over an area of five hectares and analysis has shown that, in certain cases, two synchronous

Late Bronze Age

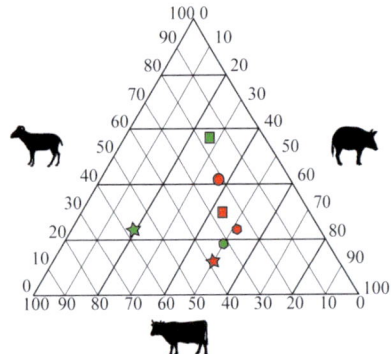

Aisne

- ● Pasly "les Côteaux de Pasly"
- ■ Menneville "la Bourguignotte"
- ★ Beaurieux "les Grèves"
- ● Osly-Courtil "la Terre-Saint-Mard"
- ● Cuiry-lès-Chaudardes "les Fontinettes"
- ■ Menneville "Derrière le Village"
- ★ Vasseny "le Dessus des Groins"

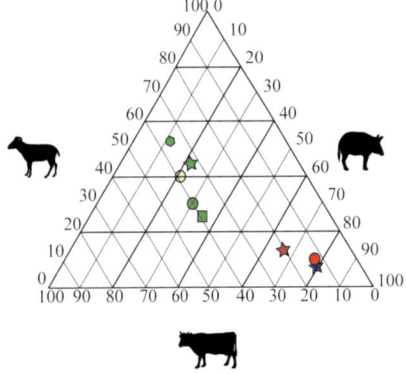

Seine-et-Marne

- ★ Boulancourt "le Châtelet"
- ● Villiers-sur-Seine "le Gros Buisson"
- ■ Grisy-sur-Seine "les Champs Pineux"
- ★ Grisy-sur-Seine "Enceinte des Roqueux"
- ● Barbey "Chemin de Montereau"
- ● Marolles-sur-Seine "Motteux"
- ★ Grez-sur-Loing "l'Epine"
- ○ Changis-sur-Marne "les Pétreaux"

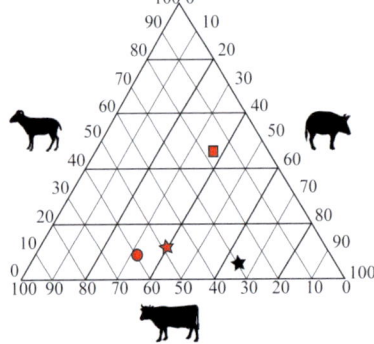

Figure 64: Late Bronze Age, frequency of the main domestic species in the selected regions.

Oise

- ★ Choisy-au-Bac "le Confluent"
- ★ Choisy-au-Bac "Canal Seine-Nord"
- ● Catenoy "le Camp de César"
- ■ Compiègne "le Fond Pernant"

units existed. Each unit included at least one pit and one or two silos, depending on the size of the household. Each of these farms housed a family unit for about a generation. The faunal corpus, composed of more than 7000 remains (41.7% determined), includes 19 taxa. The meat component of the diet of the inhabitants of Changis-sur-Marne was dominated by bovids (cattle 34%, sheep/goats 35%) with pigs making up only a very small portion. Domestic animals were reared on site and we see a broad range of slaughter ages for pigs and sheep and, to a lesser extent, for cattle where it is more difficult to discern. In the case of sheep, young animals (5-10 months of age) slaughtered in the autumn and winter perhaps to stimulate milk, occur along with both young adults (c.2 years) killed during the spring and mature animals aged between 3 and 5 years. Sub-adult and adult animals predominate. As was the case for sheep, approximately half of pigs were killed during their first few months of life and the other half were killed once they reached their mature weight. In the case of cattle, we have little evidence for the slaughter pattern, although the youngest animals were killed in the spring. Estimates of the minimum number of individuals indicate that sheep predominate (MNI=20) while cattle are in the minority (MNI=10); pigs (MNI=15) fall in the middle. So these MNI estimates provide a rather different image from the calculations based on numbers of remains, with a livestock herd largely composed of caprines and pigs.

However, cattle undoubtedly provided a greater volume of meat than all of the smaller mammals put together.

Hunting activities represent a variable portion of the corpus depending on the species but the overall contribution is marginal (6%) with red deer and wild boar being the principal species hunted. The presence of other wild species is minimal and may reflect hunting for the purpose of acquiring furs (hare, badger, beaver, cat, and fox). Each family had its herd of livestock, comprised principally of caprines (between two and five animals), a few pigs (one to three animals) and one or two head of cattle, all reared on the site and slaughtered either when they had attained their mature weight or when they were older and had supplied the community with services and products over a number of years. The estimation of the MNI for dog and horse, together with the evidence for the distribution of body parts, reveal only one horse and one dog for contemporaneous units. This suggests that these animals were probably held in common by the various households.

The exceptional site of Villers-sur-Seine "Le Gros Buisson" (c. 800 BC) (Seine-et-Marne) was established on the floor of the Seine valley. Extending over an area of 2 hectares, the site is delimited by four concentric ditches and a palisade to the west, the Seine to the south and a palaeochannel to the north. Faunal remains are particularly well-preserved: of 36,000 bone remains recovered, 24,000 have been analysed (69% determined, thirty two taxa, total weight, 252 kg). Wild faunal remains, which represent 15.5% of the corpus, are principally composed of red deer (11%) and wild boar (3.5%). Red deer occupies second place in the overall faunal spectrum after pig; cattle and caprines occupy a secondary position. There is a preference for juvenile animals consumption over adults. This indicates that meat quality was the primary focus and that the quantity of obtainable meat was secondary. In total more than 30,000 kg of meat were consumed. The structure of the site and the nature of the remains strongly suggest that the site was communal in nature (see chapter 3) (Auxiette *et al.*, 2020, 2015; Peake *et al.*, 2009).

The settlement site of Buchères/Saint-Léger-Près-Troyes/"Parc Logistique de l'Aube" is situated further east, in Champagne-Ardenne (Aube), more specifically on the Troyes Plain, which marks the transition between the chalky plain and the low hills of the Côte des Bars. Located on the banks of the Seine, the site has yielded a corpus of about 11,500 bones from dozens of pits, 5,600 of which were sampled (twelve pits, 36% determined) for the purposes of further study. Here, sheep and pigs were an important part of the meat diet (31.2% and 21.1% respectively). The proportion of wild animals in the sample is 5.4% and red deer is the predominant species (Auxiette, 2014a). In the case of pigs, there was a preference for killing immature animals (fifteen out of twenty three individuals were less than 16 months old). Therefore meat quality was probably primary concern. The management of caprines was clearly orientated towards

the production of secondary products with a large proportion of animals being killed after reaching the age of 3 or 4 years. The data is less complete for cattle but the killing of animals over 5 years is prevalent and is in line with the type of management identified for caprines. The total absence of individuals under the age of 4 years seems to support the hypothesis that cattle were acquired "on the hoof" from other communities.

Lastly, the livestock herds associated with farms in Ile-de-France and Basse-Normandie indicate a form of animal husbandry focused on the rearing of bovids (cattle and sheep), in contrast to Picardy where, apart from rare exceptions, pig tends to dominate the assemblages. The consumption of pig is particularly marked on sites with the largest assemblages, such as Villiers-sur-Seine, Grez-sur-Loing and Boulancourt in Seine-et-Marne, Choisy-au-Bac in the Oise, Buchères/Saint-Léger-Près-Troyes and Villemaur-sur-Vanne in the Aube. In the case of the Aisne, only the site of Osly-Courtil could be included in the same category, albeit with a lower proportion of pig (fig. 64, see appendices).

Pigs seem to have been particularly associated with practices of collective meals in which wild animals played a complementary role of variable importance. Farms with a predominance of pigs and caprines are only recorded in Picardy (the sites of Compiègne and Menneville "la Bouguignotte").

On the farm sites, age analysis highlights a form of herd management involving the slaughter of juvenile and adult animals in more or less equal proportions. The primary aim appears to have been the production of meat. Nonetheless, there is some evidence for the stimulation of milk production, reflected in the killing of very young caprines within the herds at Glisy and Changis.

The selection of very young animals that had not yet attained their mature weight may be explained in the context of particular supra-domestic practices which involved episodes of consumption that were exceptional and perhaps regulated. The most relevant evidence for this comes from the large sites located on the plains and in upland areas. Culled bovids are rare with the exception of farm sites in Bassée (Hermetey, 1995).

The presence of dogs and horses is minimal, with MNIs that do not exceed two individuals per farm, and it is very difficult to detect evidence for their consumption.

Among the wild animals most regularly encountered in the assemblages from all of the sites studied, red deer was without doubt the species most frequently hunted, before wild boar, roe deer and aurochs. Fox, beaver and weasel, which have been identified in the corpora from several farms, were probably hunted for their fur although no evidence has been found of specific cut marks that might support this hypothesis. There is a similar lack of clear evidence for their consumption. Bear, which occurs only very rarely in the record, is almost always represented by isolated bones or canines (*cf.* the sites of Buchères "Parc Logistique de l'Aube", Changis-sur-Marne "les Pétreaux" and Villiers "le Gros Buisson" in Seine-et-Marne).

The Iron Age

Different scales of interpretation are adopted in light of the correlations between the nature of the sites and the choice of husbandry within local areas, territories, and, more broadly, within cultural entities. When the data permits, our approach is based on the hierarchical ranking of farmsteads (Ranks 1, 2, 3, 4) as defined by François Malrain and Yves Menez (Malrain *et al.*, 2002; Menez, 2008). This classification of sites allows us to analyse inter-site relationships within a given locality in the geographical areas with the best records, such as the Aisne and Oise valleys, the Caen Plain, and the confluence of the Seine and the Yonne (Gransar *et al.*, 1999; Issenmann, 2009; Le Goff, 2009; Malrain and Pinard, 2006; Pion *et al.*, 1996, 1990). This in turn enables us to discuss relationships of reciprocity and interactions in the best documented areas and to put into perspective the variation observed for the various settlement categories (resemblances and dissemblances). Differences in herd composition between the east and west and between the north and south of the Paris Basin have already been identified (Méniel *et al.*, 2009; Zech-Matterne *et al.*, 2013). Significant faunal corpora have been recorded in all of the regions for the Late Hallstatt, Early La Tène and Late La Tène periods. The Caen Plain, located in Basse Normandie, has yielded larger assemblages than anywhere else for the Middle La Tène period.

Taking all of the sites together, the geographical distribution of the frequencies of the five main domestic species reveals a clear distinction between the north-eastern and the north-western quadrants of the Paris Basin. The rearing of small domesticated animals is favoured in the north-eastern quadrant whereas, inversely, cattle predominate in the north-western quadrant. At the same time, we see a clear evolution in the husbandry choices between the Late Hallstatt/ Early La Tène, on the one hand, and the Middle-/ Late La Tène on the other.

In the Late Hallstatt/Early La Tène, the distribution of the five main domestic species varies markedly from one region to another. In the Aisne, Oise and Seine-et-Marne, which have produced abundant records, we see a clear tendency towards the rearing of small livestock, with caprines and pigs being predominant. In the vicinity of Reims, livestock rearing was focused on bovids. However, analysis of an area further to the west has produce quite a different picture with the preferential rearing of cattle, followed by pigs and, to a lesser extent, caprines. Equids also seem to be slightly more common in this area than elsewhere (Auxiette *et al.*, 2005; Auxiette and Méniel, 2005a, 2005b).

During the Middle La Tène, the beginnings of a change can be seen in the north-eastern part of the Paris Basin, with the decline of caprines and an increase in sites with a high proportion of cattle and pig. In the Late La Tène, on the wealthiest enclosure sites of the Aisne, Oise and Seine-et-Marne, we see a predominance of pig and cattle over caprines even though the latter enjoyed a prominent position between the 7th and 5th centuries BC. A decline in the incidence of this species is already discernible in the Middle La Tène. The sites of the Caen Plain stand out from the sites just mentioned by virtue of the omnipresence of cattle in the herds, the increasingly intense rearing of which forms part of a continuum with Late Hallstatt practices. Dogs and horses have a more marked presence in the herds than was the case in the Bronze Age and the rearing of these species clearly intensifies between the Late Hallstatt/Early La Tène and the Late La Tène.

Consumption of animals on Late Hallstatt/Early La Tène farms (530-325 BCE)

During the Late Hallstatt/Early La Tène in the Aisne valley, farms are characterised by relatively large faunal assemblages (Auxiette, 1997; Auxiette *et al.*, 2003)[13]. Six of these sites fall within the category of highly structured sites, with many being surrounded by a palisade. We observe the importance of bovid rearing (66% of domestic species), particularly sheep rearing (38% average), on farms established on the alluvial plain of the river Aisne, between the communes of Menneville to the east and Villeneuve-Saint-Germain to the west.

As regards cattle, we can distinguish between animals slaughtered before the age of 4 years and those slaughtered after the age of 4: the frequency of these two broad classes is more or less equal. As regards sheep, which represent the majority of the caprines, 76% were consumed before the age of 3 years and a non-negligible proportion (40%) were killed before reaching their optimal weight. With the exception of "les Grèves" in Beaurieux (Auxiette, 2019a), none of the sites exhibits a predominance of pig. In the case of "les Grèves" we are dealing with a mass of bones, with little fragmentation and originating from the carcasses of two pigs. This assemblage has been interpreted as the remains of a collective meal. Evidence for the slaughtering of lambs in order to stimulate milk production has been identified at the farm sites of Villers-en-Prayères, Menneville, and Bucy-le-Long.

For the most part, the farms feature small herds composed, on average, of three head of cattle, five sheep, four pigs, a dog, and a horse. On sites with better records, we can interpret the assemblages at the scale of the domestic unit, in which case the average herd is composed of a single head of cattle, two pigs and three sheep. Goats are rare. Chicken (*Gallus gallus*) appear, but in relatively low proportions. Poultry also feature in funerary rites where they were deposited in the graves of certain individuals (see chapter 5). The site of Villeneuve-Saint-Germain "les Grèves" (Debord and Desenne, 2005), which is enclosed by a substantial ditch, has been only partially excavated but has yielded a large assemblage of faunal remains: the proportions of caprines (± 40 %) and pigs (± 40 %)

are greater than those recorded on other sites while cattle are poorly represented. These differences are interpreted as an indication of status and mark this settlement out from the other farmsteads. This evidence, corroborated by the quality of the ceramic assemblage (particularly fine, decorated vases and platters), allows us to identify this as a high status site and, once again, pig appears to be a marker of this status.

The place of horse and dog within the herd varies considerably from farm to farm (1.5% to 7.5% and 1% to 5.7%, respectively). It is likely that horses and dogs formed part of the diet but evidence for their consumption remains tenuous. The proportions of wild mammals vary between 0.4% and 4.2% and cervids, particularly red deer, are predominant.

In the Oise valley (Méniel, 2006), bovids are in the majority and, of these, caprines are predominant, but pigs occupy a slightly more prominent position than on farms in the Aisne valley (Auxiette and Méniel, 2005a). They predominate on the settlement sites of "le Bois d'Ageux" and "la Butte de Rhuis" in Longueil-Sainte-Marie, Verberie "les Moulins", and Compiègne "le Fond Pernant". The trends observed in animal husbandry are more contrasted than in the Aisne valley. In the case of cattle, we observe radically different approaches to herd management: on the farm at Houdancourt, a significant proportion of calves and young animals, aged less than 2 years, were slaughtered. To a completely different end, at Compiègne, the majority of cattle were killed after the age of 9 years (Méniel, 2006, 189). Most pigs were killed before reaching the end of their second year and at Compiègne in particular we see a significant proportion (60%) of animals aged less than 6 months (*op. cit*, fig. 143, 191). The slaughter age for caprines varies from site to site: at Houdancourt old animals occur in significant proportions, while at Compiègne the slaughter of young animal is favoured, probably as a measure to stimulate milk production (*op. cit*, fig. 144, 191). We have no information regarding the rearing of horses and dogs. Wild mammals never exceed 2% of the corpus.

Turning to the Marne, in Reims and its hinterland[14], the assemblages from farm sites generally indicate preferential rearing of bovids (cattle 39%

Late Hallstatt/ Early La Tène

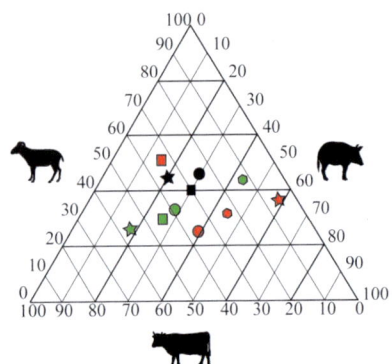

Aisne

⭐ Beaurieux "les Grèves"
★ Bazoches-sur-Vesle "les Chantraines"
⭐ Condé-sur-Suippe "le Déprofundis"
🟩 Limé "la Prairie"
🟥 Limé "la Fosse aux Chevaux"
⬛ Menneville "Derrière le Village"
🔴 Villers-en-Prayères "les Mauchamps"
🟢 Bucy-le-Long "le Grand Marais"
⚫ Bucy-le-Long "le Fond du Petit Marais"
🔴 Villeneuve-Saint-Germain "les Etomelles"
🟢 Villeneuve-Saint-Germain "les Grèves"

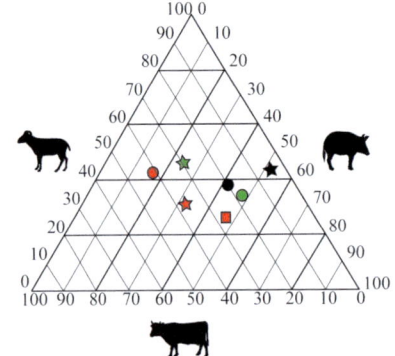

Oise

⭐ Houdancourt "les Esquillons"
⭐ La Croix-Saint-Ouen "les Longues Raies"
★ Longueil-Sainte-Marie "la Butte de Rhuis"
🔴 Ponpoint "les Prés Véry III"
🟢 Compiègne "le Fond Pernant"
⚫ Longueil-Sainte-Marie "le Bois d'Ageux"
🟥 Verberie "les Moulins"

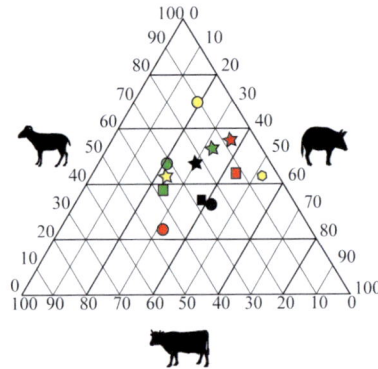

Seine-et-Marne

⭐ Ville-Saint-Jacques "le Bois d'Echalas 1"
⭐ Souppes-sur-Loing "le Poirier Métais"
★ Changis-sur-Marne "les Pétreaux"
⭐ Ecuelles "Malassis et Charmoy"
🔴 Egligny "le Bois de la Pêcherie"
⚫ Varennes-sur-Seine "Beauchamps"
🟢 Changis-sur-Marne "les Pétreaux"
🟥 Varennes-sur-Seine "Beauchamps"
⚪ Varennes-sur-Seine "Volstin"
⬛ Larchant "les Groues"
🟩 Ville-Saint-Jacques "le Bois d'Echalas 1"
🟡 Ville-Saint-Jacques "le Bois d'Echalas 2"

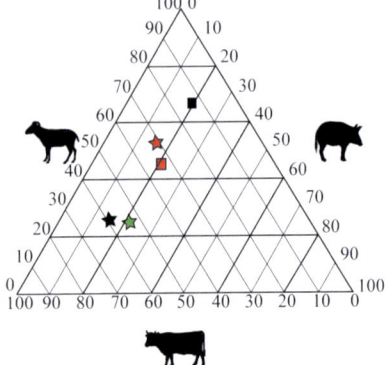

Marne

⭐ Reims "Croix Blandin"
★ Reims "les Hauts des Nervas"
⭐ Reims "Thillois"
🟥 Cernay-lès-Reims "la Borne-Saint-Laid"
⬛ Cernay-lès-Reims "le Puisard"

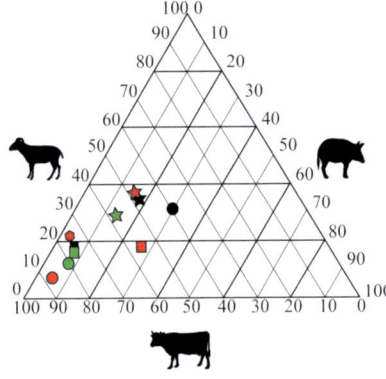

Calvados

⭐ Hérouvillette "les Pérelles"
★ Ifs "AR67"
⭐ Ifs "Object'Ifs sud"
🟥 Saint-Martin-de-Fontenay "Le Chemin de May"
⬛ Fleury-sur-Orne "les Mézerettes"
🟩 Eterville "les Prés du Vallon"
🔴 Bourguébus "la Main Delle"
⚫ Condé-sur-Ifs "la Bruyère du Hamel"

Figure 65: Late Hallstatt/Early La Tène,
frequency of the main domestic species.
Top: Aisne, Oise, Seine-et-Marne. Bottom:
Marne, Calvados.

and caprines 34% on average), with a lesser focus on pigs (18.5% on average). The ages of cattle are not well documented but it appears that slaughter age is divided equally between animals aged less than 4 years and animals aged over 4 years. Caprines were almost invariably slaughtered before the age of 3 years (90%, of which 45% were killed in their 1st year) and 87% of pigs were killed before the end of their 2nd year. The presence of lambs in the assemblages from

"Thillois" and "Haut des Nervas" in Reims and from Caurel "le Puisard", may correspond to slaughter for the stimulation of milk production. Taken as a whole, the evidence from the settlements indicates the presence of small herds of livestock comprised, on average, of a single cow, three sheep, two pigs, a dog and a horse.

The place of horse and dog within the herd varies greatly from farm to farm (1.5% to 5% and 0.3% to 3%, respectively). The proportions of wild mammals vary from 0.5% to 4.4%, with red deer and hare predominating.

In Seine-et-Marne, settlement sites are geographically dispersed[15]. In the three best-documented sites (with between 1000 and 2100 remains), namely the farms of Changis-sur-Marne, Ecuelles and Larchant, we once again observe the importance of bovids and, in particular, of caprines. Taking all of the farms as a whole, the proportions highlight an abundance of caprines (41%), followed by pigs (27%), with cattle occupying third position (22%). The data for cattle ages is poor, but there are equal proportions of animals under 4 years and over 4 years. In the case of caprines, estimates indicate a preference for slaughtering young animals (80% before the age of 3 years, 40 % of which were killed before the age of 1 year): the majority of pigs were slaughtered before the age of 2 years (76%, 38% of which were killed under the age of 1 year). The farm at Ecuelle is the only site to reveal evidence for controlled milk production.

Similar to the situation observed in the Aisne valley, the farms here also feature small herds which, on the basis of the minimum number of animals slaughtered, would have been made up, on average, of a single cow, three sheep, three pigs, a dog and a horse. The place of horses and dogs varies considerably from farm to farm (0.2% to 7.1% and 0.1% to 14%, respectively). In the case of dog, the greatest incidence is recorded on the Ecuelles farm site where there is clear evidence that dogs were eaten. The proportions of wild mammals vary between 1% and 5%, with cervids predominating.

On the farm sites in the Crould catchment area, we see a predominance of pig (35% to 56%; Gonesse "Zac des Tulipes II", Saint-Denis "Nozal Chaudron", Saint-Denis/Saint-Ouen "EDF" and, "le Château" and "le Dessus de la Rayonnette" in Roissy-en-France) (Auxiette and Jouanin, 2018). It is difficult to deal with the question of herd management as the available data is quite poor. Cattle were preferentially slaughtered before the age of 4 years. On the sites of Gonesse and Saint-Denis/Saint-Ouen, pigs were principally killed at the end of their 1st year and sheep were generally slaughtered between 1 and 2 years. None of the farms appear to have focused on specific products. The place of horse and dog within the livestock herds remains poorly documented.

The Caen Plain and surrounding areas (Calvados) constitute a veritable storehouse of data for understanding animal husbandry. Archaeozoological studies have focused on fifteen farm sites.

For the Late Hallstatt / Early La Tène[16], the faunal spectra on the farms are largely dominated by cattle (51% on average), which were the principal source of meat and secondary products. The assemblages are also characterised by a significant presence of caprines (28% on average) while pigs play a minor role in meat production (14% on average). The consumption of horse is clearly attested although it remains marginal (4.4% on average). Indicators for the consumption of dog are very limited (2.4% on average). We see some variations in the frequencies of the principal domestic species between the various phases of Early La Tène settlement: there is a marked increase in the proportions of sheep and pig relative to cattle at the end of this period. The slaughter of cattle centres on animals less than 3 years, sometimes less than 1 year, and 71% are less than 4 years. The slaughter of very old animals is an unusual feature at Soulangy.

We observe a spread in the ages of caprines, with a relatively significant proportion of slaughtered animals being less than 1 year old: the killing of such young lambs may have been aimed at diverting milk production for consumption by humans (at Ifs and Eterville, more than half are less than 1 year old, 20% on average). Juveniles and young animals (62.5%), and a number of mature caprines, were retained within the herd for the production of secondary products, particularly at Ifs (17.5% on average). Pigs were generally killed before reaching the end of their 2nd year (75%).

On average the livestock herds are composed of three head of cattle, eight sheep, three pig, a dog and a horse.

In the area around Amiens, Hallstatt period farms remain poorly documented; the farm at "les Terres de Ville" in Glisy (Méniel in Gaudefroy *et al.* 2000) is characterised by a herd largely composed of pigs (41%) and caprines (31.3%). Much further north, herds on the farms at "Marais de Dourges" in Dourges and at "Pièces à Liard" in Etaples-sur-Mer are principally made up of bovids, of which cattle predominate. There is a wide range in the proportions of horse and dog. Wild species never represent more than 1.5% of the assemblages.

The livestock herds associated with farms in Ile-de-France and Picardy indicate a form of animal husbandry focused on the rearing of sheep and pig in contrast to Basse-Normandie and the Marne where, apart from rare exceptions, cattle (and sheep) tend to dominate the assemblages (fig. 65).

On the farm sites, age analysis highlights a form of herd management involving the slaughter of juvenile and adult animals in more or less equal proportions. The primary aim appears to have been the production of meat. Nonetheless, there is some evidence for the stimulation of milk production in all regions, in the form of the killing of young animals within the herds at Villers, Menneville, Bucy, Compiègne, Reims, Caurel, Ecuelles, and Ifs. In the case of cattle in the Oise, we observe radically different approaches to herd management: a proportion of calves and young animals were slaughtered at Houdancourt and very old cattle at Compiègne.

Chicken appears at this time in faunal assemblages. Sometimes we have clear evidence that dogs were eaten. From the mid-5th century BC, farming practices focus to a much greater degree on cattle rearing, a marked shift that continues until the Late La Tène.

The fauna of Middle La Tène farms

Middle La Tène farms are unevenly documented in the area between the valleys of the Aisne and the Oise and the Seine-Yonne interfluvial zone and in Bassée. In contrast, farms are numerous and well documented in the Caen Plain.

Farms well-dated to the Middle La Tène are rare in southern Picardy: a single example is recorded in the Aisne (Chambry "Zac du Griffon", (Audebert, *et al.*, 2013) and just two in the Oise (Chevrières "la Plaine du Marais" and Verberie "le Buisson Campin", (Méniel, 2006). In certain cases, such as the farm at Longueil-Sainte-Marie "le Vivier des Grès", there is some doubt regarding attribution to the La Tène C1 and/or C2. The frequencies of the principal domestic species vary: cattle predominate in the Oise and caprines in the Aisne. Milk production is attested to on the farm at Chambry.

In Seine-et-Marne, four farms have yielded relatively modest assemblages (Jossigny "Pré aux Chênes", Changis-sur-Marne "les Pétreaux", Varennes-sur-Seine "la Justice" and "le Marais du Colombier)" (Auxiette, 2011b, 2013a; Lafage, forthcoming; Séguier *et al.*, 2008) which provide a diversified picture of herd composition. Caprines are in the majority at Changis and Jossigny (even

though at Jossigny the proportions of the three main domestic species are fairly equal), cattle at the settlement of Varennes-sur-Seine "la Justice", and pigs at "Marais du Colombier". The question of herd management cannot be examined in detail, due to the extremely fragmented nature of the cattle bones on the four sites. The caprines were most often killed between the ages of 1 and 3 years and pigs between the ages of 1 and 2 years. Livestock rearing was principally orientated towards meat production.

In the Crould Basin, the faunal assemblages from five Middle La Tène farm sites (which yielded between one hundred and several hundred bones), located at Villiers-le-Bel/Gonesse "Déviation RD 10-370", Le Mesnil-Aubry "le Bois Bouchard-Carrière REP-Véolia", Gonesse "Zac des Tulipes nord", Roissy-en-France "Zac de la Demi-Lune", and Fontenay-en-Parisis "la Lampe", reveal the predominance of cattle (between 40 and 60%) along with pig (between 18 and 36%): Gonesse and Roissy are the exception with a predominance of caprines (Auxiette and Jouanin, 2018). On the probable high-status farm at Fontenay-en-Parisis (Daveau and Yvinec, 2001), which was partially excavated, the slaughter of pigs focused on animals aged between 18 and 24 months. In the case of cattle, 60% were killed before the age of 4 years and we see preferential slaughter of caprines aged between 1 and 2 years (42%). It is clear that meat of very high quality was consumed on this site (Daveau and Yvinec, 2001, 97). These trends differ from those observed for other, smaller farm sites where bovids tend to be older (>4years).

In the area around Amiens, two farms at Glisy "Les Quatorze, ZAC de la Croix de Fer/Pôle Jules Verne, site C" and "les Terres de Ville" (Auxiette, 2015; Méniel, 2000a) yielded very different assemblages (in terms of quantity and quality). In the first, the faunal spectrum is dominated by caprines (32%) and cattle (26%); herd management is based on the mixed slaughter of juvenile and mature animals with no particular focus on either.

On the second site, which is very well documented, cattle and pigs share first position (33%). Most of the slaughtered pigs, which are very numerous (more than ninety), are juveniles (about fifty were between 0.5 and 1.5 years old). Out of slightly more than fifteen cattle, more than half were slaughtered between the ages of 1 and 2 years and it emerges that a significant proportion of the caprines were killed before the age of 1 year even though the ages are spread out. This settlement is believed to have been a high ranking farm and taken together the particularities of the assemblage undoubtedly indicate a high quality meat diet, probably associated with occasional feasting. In addition, we cannot rule out the possibility that some of the caprines were used for the production of milk (Méniel, 2000a).

Outside this area, the herd associated with the farm at Méaulte "Zac du Pays des Coquelicots" (Auxiette, 2014b), located c. 20 kms to the northeast of Amiens, is largely composed of bovids (33.8% cattle, 28.9% caprines); pigs are in the majority in terms of MNI and were almost invariably killed before reaching the age of 1 year. The caprines were preferentially killed between the ages of 1 and 3 years, with a significant number of individuals being only a few months old, which suggests that ewes were being used to produce milk for human consumption. These characteristics may also reflect particular propitiatory ritual practices involving the killing of piglets and lambs[17]. Cattle were slaughtered before reaching their 4th year. These are unusual trends that suggest that the residents enjoyed a certain degree of affluence.

At Ifs, located in the Caen Plain, cattle predominate (57%) and are followed in importance by caprines (23%) and pigs (11.7%) (Auxiette, 2000b). Estimations of the slaughter ages for cattle allow us to distinguish two broad groups: the first is made up of animals aged between 6 months and 1 year, and the second groups together animals aged between 1 and 1.5 years. These animals were slaughtered during autumn and winter. After an almost complete cessation of slaughter between 1.5 and 3 years, another portion of the cattle herd, aged between 3.5 and 4 years, was slaughtered in the autumn. These results reflect herd management for the production of meat from juvenile and sub-adult animals, a choice that suggests a certain level of affluence among the population who could forego the significant mass of meat that might be obtained if the animals were maintained in the herd until they had reached their mature weight. In the case

of caprines, we observe a broad spread in the ages at which animals were slaughtered with a significant proportion of lambs killed for consumption. Among this population, it can be seen that more than half were lambs aged 4-5 months, which were slaughtered in summer, *i.e.* during the months of July and August. This selection pattern corresponds to a particular taste for tender meat obtained from lambs, which were perhaps killed to stimulate milk production, and from juvenile and young adult animals, principally killed in spring and winter (12/24 month interval).

Pigs, which were raised solely for meat production (and exploitation of by-products), were principally killed between the ages of 1.5 and 2 years and particularly between 17 and 25 months: a few individuals were consumed before the age of 1 year, at about the age of 8 months (slaughter in January/February or August/September if we follow the hypothesis of two litters per year).

The faunal assemblages from several enclosed farmsteads situated at Mondeville "l'Etoile" support the findings from the Ifs farmsteads (Auxiette, 2009a). At Mondeville I (Phase 2), cattle are predominant (40%) and constitute the greatest proportion of meat consumed, substantially supplemented by sheep (30%). The proportions of pig do not exceed 14%. We observe a significant percentage of horse bone (8%) but proof that the animals were actually eaten is elusive.

In the main, cattle were killed around the age of 40 to 50 months, but the curve stretches between 5/6 months, equivalent to slaughter in October and November, and some animals were more than 5 years old when slaughtered. These animals had, for the most part, attained their optimum weight; they were preferentially killed in spring or autumn. Age estimates for the caprines indicate that slaughtering was spread out between 3 months and more than 30 months; half of the animals had not reached their first year and the other half had exceeded their 2nd year. When superimposed on a calendar, we observe that most of the animals were killed between the end of the winter and the end of the summer. This reflects a real desire to consume high quality meat from young sheep alongside a wish to maintain part of the herd for the purpose of producing secondary

Middle La Tène

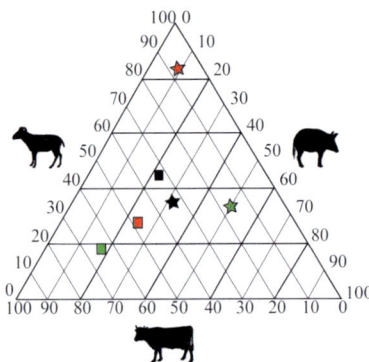

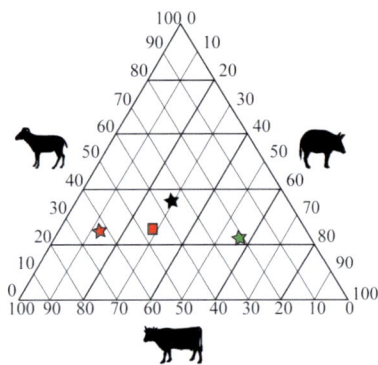

Figure 66: Middle La Tène, frequency of the main domestic species in the selected regions.

Seine-et-Marne
⭐ Changis-sur-Marne "les Pétreaux"
★ Jossigny "Pré aux Chênes"
⭐ Varennes-sur-Seine "le Marais du Colombier"
🟥 Varennes-sur-Seine "la Justice"
⬛ Chatenay-sur-Seine "les Sécherons"
🟩 Marolles-sur-Seine "le Grand Canton"

Calvados
⭐ Ifs "Object'Ifs Sud"
★ Mondeville "l'Etoile"
⭐ Mondeville "MIR"
🟥 Fleury-sur-Orne "ZL7/CD120

products. Finally, pigs, which were very much in the minority in these herds, were essentially killed around the age of 20 months and had almost always attained their optimal weight: a few juvenile animals, aged 6 to 8 months, were also identified.

The estimate of the minimum number of individuals (14 head of cattle, 24 sheep, 18 pigs, 4 horses and 3 dogs) for Site I, Phase 2, reveals a picture that differs somewhat from that obtained on the basis of the number of remains: sheep occupy first place with pigs not far behind; nevertheless, beef was undoubtedly the primary resource.

On the site of Mondeville III, the frequency of the three main domestic species is similar to that already observed for site I: 67% of the bones come from bovids (34 and 33%) and 26% are from pigs, a frequency that is twice that recorded on the Site I farm. The caprines were, for the most part, killed in spring and summer, cattle preferentially in winter and spring, and pigs of all ages probably throughout the year.

Middle La Tène farms are unevenly documented in the different areas. We observe heterogeneity in the make up of herds (fig. 66) with faunal assemblages providing a diversified picture of herd composition. The principal domestic species is either cattle or caprines on most sites, and only occasionally pig. The importance of pig consumption seems to have been linked to particular events or specific roles.

The farms of the Caen Plain are well documented: at Ifs, cattle predominate and are followed in importance by caprines. Estimations of the slaughter ages for cattle reflect herd management

for the production of meat from juvenile and sub-adult animals.

The fauna of Late La Tène farms and oppida (180-20 BCE)

For Late La Tène, our overview deals with the various tribal territories separately. We will begin with a brief introduction to the farmsteads before turning our attention to the herd management.

In the vast alluvial plains of the Aisne and Vesle valleys, between Reims and Soissons, the faunal corpora are rich and well documented. In the territories of the Remi and Suessiones tribes, several relatively large farmsteads and four oppida have yielded assemblages that vary greatly in size, from just a few bones to several tens of thousands of bones.

Four farmsteads have been recorded within the territory of the Remi[18]. The faunal corpora, which vary considerably in size, do not allow us to adopt multiple approaches to herd management analysis in all cases. Differences in data quality are linked to the nature of the site investigations, which can be partial, as in the case of Berry-au-Bac "le Chemin de la Pêcherie" and Damary "le Ruisseau de Fayau" (Haselgrove and Lowther, 1992). At Damary, the faunal remains come from a finds-rich occupation layer (>4000 remains, Auxiette unpublished). In most cases, the nature of the farm determines the quantities of finds. In the case of the oppida, which have been only partially excavated, the quantities of faunal remains produced vary greatly between Condé-sur-Suippe/Variscourt (>60 000 remains) (Auxiette, 1994; Méniel, 1984; Paris, 2016) and in Reims-*Durocortotum* "Rue d'Anjou" (c.1400 remains) (Auxiette, 2003). Also associated with the Remi tribe but situated 20 km north of the Suessiones capital on the border between the territories, a group of farms have been excavated near Laon. The sites are located at Barenton-Bugny (sites "L" and "M") and Chambry ("ZAC du Griffon") (Audebert, *et al.*, 2016; Audebert *et al.*, 2016; Auxiette, 2016, 2014c).

Within the territory of the Suessiones, farm sites are more numerous and better documented. Their diversity allows us to define their nature and status: to date, ten farmsteads and two oppida have been identified[19].

The numerical differences between the faunal corpora are again determined by the nature of the sites and the conditions of the excavations: partial excavation of the farmsteads at Mont-Notre-Dame and Villeneuve-Saint-Germain, for example. The nature of the subsoils, which are sometimes acidic and not conducive to the preservation of bone, as at Limé "les Sables-Sud", may also be responsible for the low counts and differential preservation.

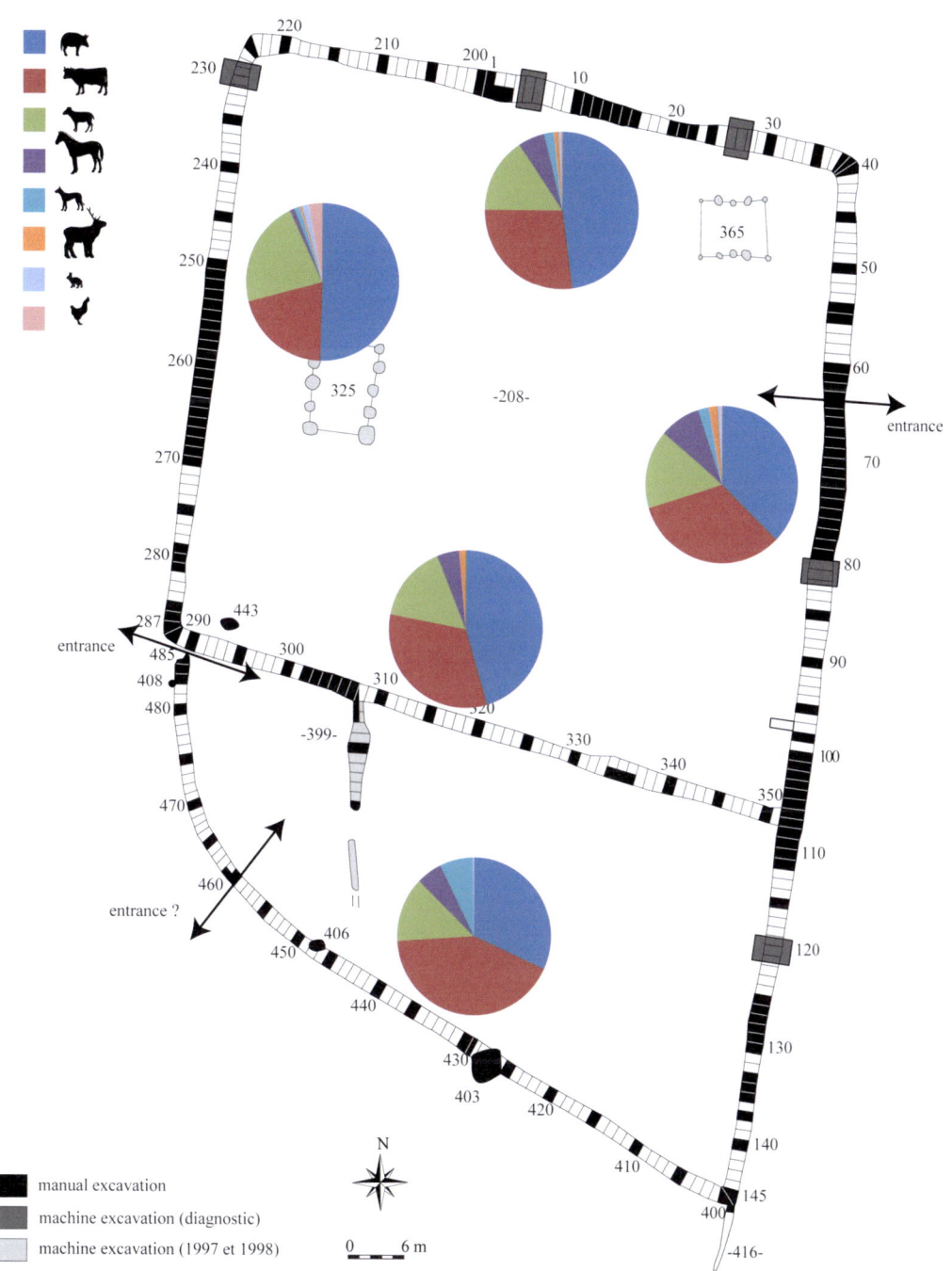

Figure 67: Braine "la Grange des Moines", frequency of the main domestic species.

However, small assemblages cannot always be attributed to erosion, preservation conditions or site areas excavated: for example, the site of Bazoches-sur-Vesle "la Foulerie" (Auxiette *et al.*, 1995), where there were no issues with preservation and which was excavated in its entirety, produced very few faunal remains. These "deficits" may be linked instead to prevailing dietary factors or to waste management practices that differ from those normally encountered, perhaps involving discard outside the settlement area. The farms at Bazoches-sur-Vesle "les Chantraines" (Gransar and Pommepuy, 2005), Braine "la Grange des Moines" (Auxiette and Desenne, 2017), and Villeneuve-Saint-Germain "les Etomelles" (Hénon *et al.*, 2012) have yielded the richest corpora (1950/>10 000 remains) and correspond to the category of high status farmsteads (Malrain, 2000; Menez, 2008). The oppidum of Villeneuve-Saint-Germain "les Grèves" (± 70 000 studied bones), which was established

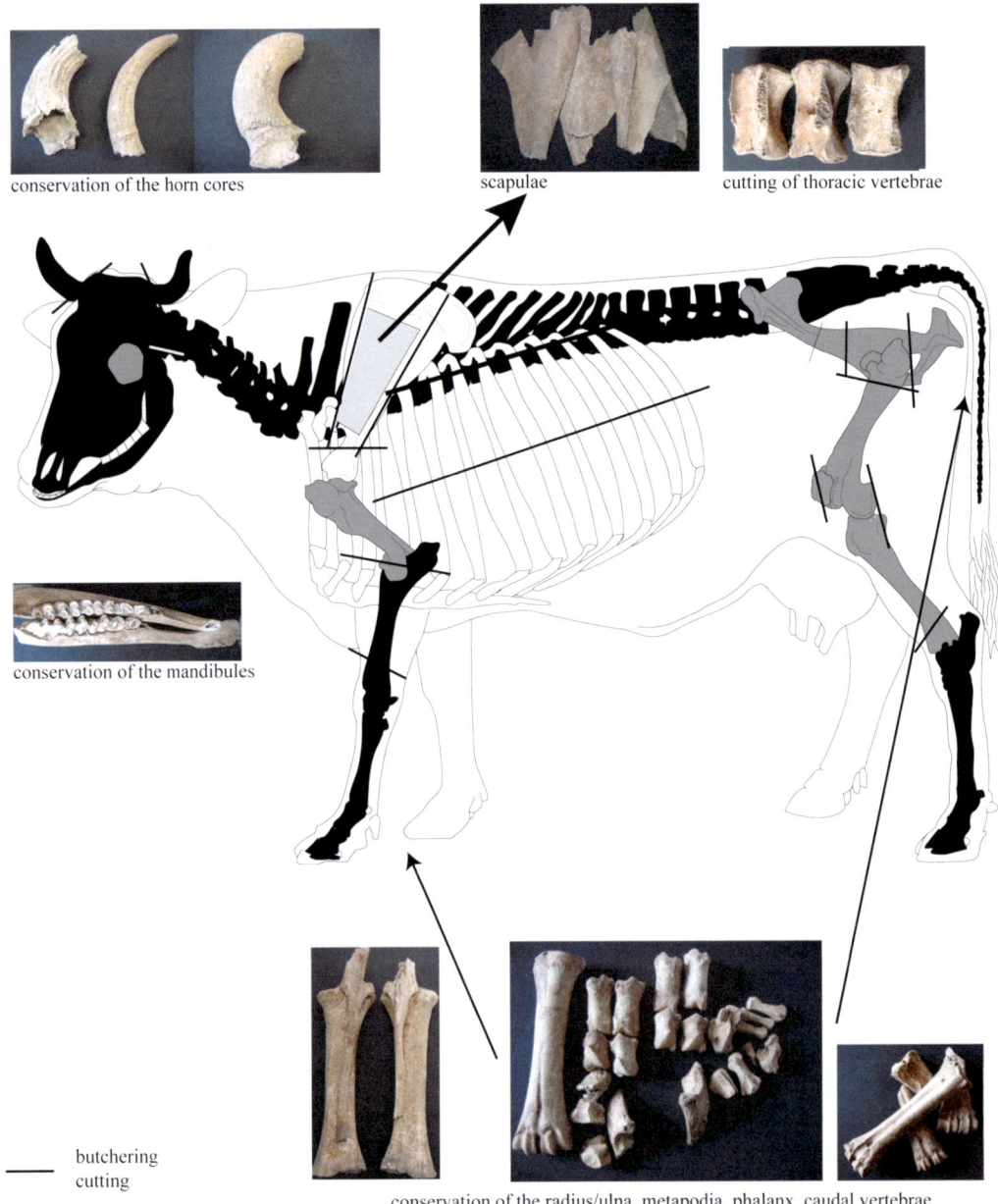

conservation of the horn cores

scapulae

cutting of thoracic vertebrae

conservation of the mandibules

conservation of the radius/ulna, metapodia, phalanx, caudal vertebrae

—— butchering cutting

■ parts removed at the time of carcass preparation

▦ crushed bones of the pieces of most important meat

Figure 68: Villeneuve-Saint-Germain "les Grandes Grèves" (Aisne), Late La Tène, oppidum, cattle butchery in the craft sector (photo: G. Auxiette).

during the second half of the 1st century BC after the demise of the three high ranking sites mentioned above, is divided into clearly identifiable zones (Auxiette, 1996; Constantin and Debord, 1982; Debord, 1995, 1993, 1990, 1982; Paris, 2016; Ruby and Auxiette, 2010). At present the oppidum of Pommiers"l'Assaut" remains too poorly documented to allow us to comment on its organisation and patterns of meat consumption (Auxiette, 1994).

The bone refuse on Late La Tène (LTD1/D2) farms is almost exclusively found in ditches that enclose relatively large spaces; the sites of Bétheny "les Equiernolles" (Auxiette, 2002), close to Reims in the east, and Villeneuve-Saint-Germain "les Etomelles" near Soissons to the west, are the exception. At this stage of our overview, it is worth turning our attention to the nature of the find contexts, which are principally ditches; this will entail (re)discussing the impact of the nature of features on the nature of the domestic waste. In the 1990s, debate regarding the "nature and differential preservation of bone" drew attention to the preferential selection of waste with the recording of a raised frequency of bones from large mammals (Méniel, 2000b, 1997) and its corollary, a higher frequency of identified bones, from ditch contexts. The width of the ditches and their relatively long durations of use, were deemed responsible for this; but this explanation fails to take account of the recurrent episodes of cleaning-out which would have regularly emptied the ditches of waste material. Furthermore, numerous studies have shown that spatial analysis greatly remedies this overall view of the omnipresence of large mammals by highlighting sectors with concentrations of bone from small animals, particularly in the vicinity of houses (Auxiette and Desenne, 2017; Auxiette and Jouanin, 2018; Méniel, 2006). Spatial

analyses carried out on the site of Braine "la Grange des Moines" provide a good example. The faunal remains retrieved from the fill of the western enclosing ditch immediately adjacent to the largest house appear to have been the result of feasting and are largely composed of pig, while the faunal remains from the southern segment of the ditch are dominated by cattle (fig. 67; Auxiette and Desenne, 2017, 175).

Whether we are dealing with the territory of the Remis or that of the Suessiones, the faunal assemblages provide a diverse picture of herd composition.

Within the Remi territory, caprines dominate the herd at Berry-au-Bac (MNI) while the herd at Damary is principally made up of pigs and cattle, but especially caprines and pigs on the basis MNI. On the Betheny site, pigs are in the majority (MNI) while the number of caprine remains is higher. At Barenton-Bugny and Chambry, caprines surpass all other species. Cattle predominate largely on the site of Reims "les Hauts des Nervas".

In the case of the Suessiones territory, the farms at Limé (with small faunal assemblages) and the two sites at Bazoches-sur-Vesle in the Vesle valley are characterised by a high frequency of cattle, while at Mont-Notre-Dame and Braine the herds are dominated by pigs. The frequency of horse is not insignificant (> 7 % at Limé, Bazoches-sur-Vesle "les Chantraines", and Braine; 11 % at Bazoches-sur-Vesle). At the same time, the frequency of horse on the farm at Mont-Notre-Dame is twice as low (3.2%). The faunal spectra of the farms located closest to Soissons tend to be dominated by caprines (in MNI at Sermoise and Ciry-Salsogne), while at Villeneuve-Saint-Germain pigs are in the majority.

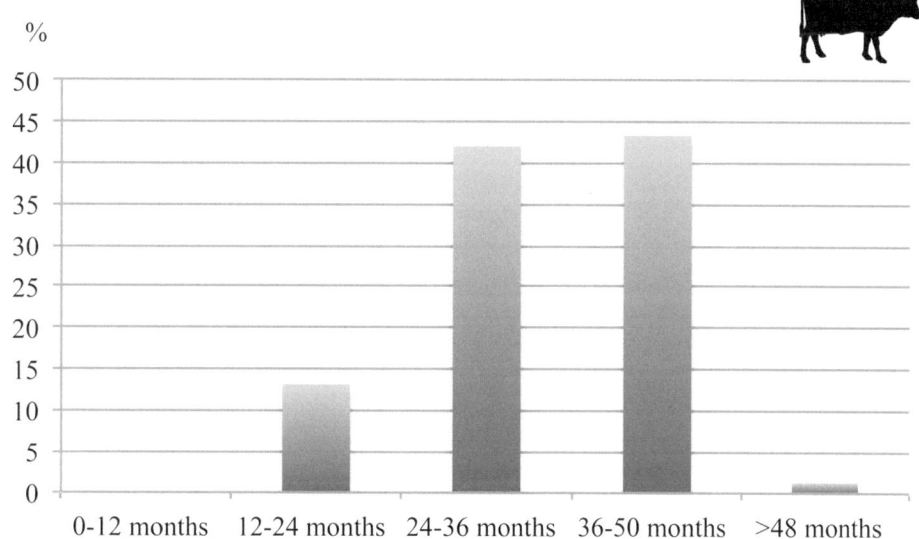

Figure 69: Villeneuve-Saint-Germain "les Grandes Grèves" (Aisne), Late La Tène, oppidum, cattle slaugthering (mandibules).

Small-sized farms, with herds composed of only a few individuals per species, are part of a typical bovid rearing tradition that was already in evidence in the Late Hallstatt/Early La Tène period (Auxiette, 1997; Hachem and Auxiette, 2006). Herd composition is fairly uniform, with a predominance of small livestock (caprines and pigs), supplemented by a few head of cattle and a horse or two. The three high-status farms (Bazoches-sur-Vesle "les Chantraines"; Braine "la Grange des Moines"; Villeneuve-Saint-Germain "les Etomelles") clearly stand apart in terms of the size of their herds. The first two were excavated in their entirety[20]; in addition to the fauna, they are characterised by large quantities of diverse finds (vases, amphorae, metal objects and personal ornaments, evidence of metal working, etc.), imposing buildings, and rare objects. The sites were therefore undoubtedly ostentatious in character. The third high-status site, which is apparently not enclosed by a ditch (at least, no ditch is apparent today), was only partially excavated. It nonetheless produced remains that are clearly indicative of its high status: specific elements of its layout, remains that suggest communal feasting, conspicuous finds, all of which are comparable to elements found on the two sites described above. Bazoches and Braine have yielded faunal spectra that are similar to each other: on both sites cattle and pigs are the principal

species consumed. At Villeneuve-Saint-Germain, pigs dominate the assemblage, which makes it comparable to the faunal spectrum for the more recent oppidum of the same name.

Archaeozoological studies (±70 000 bones) carried out on the Villeneuve-Saint-Germain oppidum have revealed the co-existence of different patterns of meat consumption within the same site. In the area of the site associated with craft working, the fauna attests to standardised butchery activity for the production of large quantities of cattle with systematic processing of carcasses (fig. 68). This was probably obtained in part from herds reared on farms within the territory controlled by the oppidum (Auxiette and Paris, 2017); the peak in slaughtering between the ages of 36 and 48 months is the main indicator of this (fig. 69). These studies have also revealed the slaughter of very gracile animals (Auxiette, 1996, 76-77; Hachem and Auxiette, 2006; Paris, 2016, 33-39).

This area of the site also produced evidence for the processing of pelts, particularly those of dogs which varied greatly in size with some individuals being very small (Auxiette, 1996; Paris, 2016; Yvinec, 1987). In the residential area, the importance of pig rearing (between 49 and 63% depending on the domestic unit), which were principally slaughtered between the ages of 18 and 22 months, as well as caprines (14 to 24%) is undoubtedly a consequence of the necessity to feed the inhabitants of the oppidum who were probably more particular in their choice of meat than the inhabitants of the farms (fig. 70) and (fig. 71).

Thus, the contribution of pig to the domestic diet eclipsed that previously provided by caprines (Auxiette, 1997). The slaughter age for sheep shows a wide distribution over time with a peak

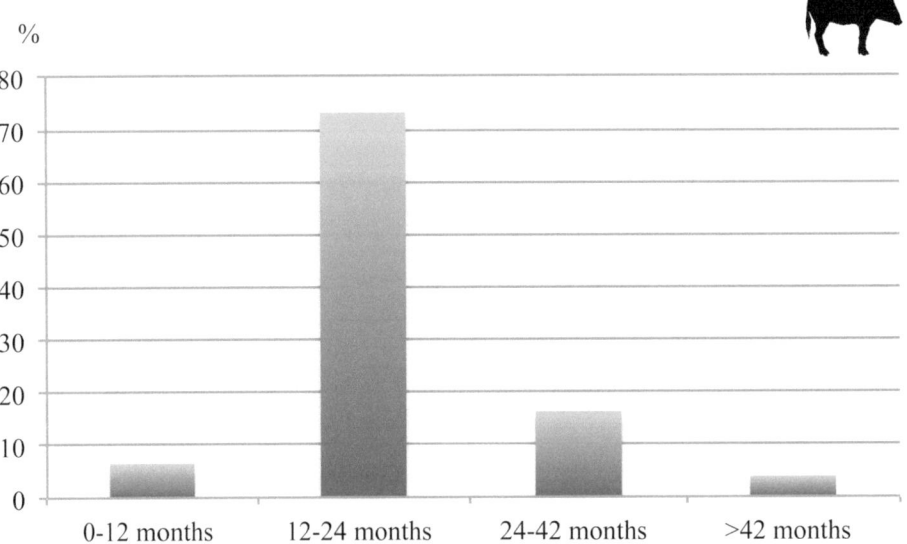

Figure 70: Villeneuve-Saint-Germain "les Grandes Grèves" (Aisne), Late La Tène, oppidum, pig slaugthering (mandibules).

around 2 years. The joints of beef consumed in the residential area of the site were shared amongst the domestic units, a pattern that contrasts with the meat from smaller animals, which was not shared between households. To summarise the evidence for the residential area, each family owned a herd principally composed of pigs and caprines. Cattle, on the other hand, were common property with their secondary products and, ultimately, their meat being shared out among the inhabitants. At the same time, large numbers of cattle were killed in an area of the site dedicated to slaughtering and butchering in order to feed a constantly increasing population.

Taking all of the studied areas of this oppidum together, estimations of the minimum number of individuals and of the corresponding weight in meat are staggering: 280 head of cattle representing an estimated 74,200 kg of beef, 380 pigs representing c.28,500 kg of pig, 212 sheep corresponding to 3,800 kg of sheep, 24 horses corresponding to 8,400 kg of meat and, lastly, 101 dogs that would have provided 3,800 kg of meat.

We can thus identify the emergence of large scale meat production, where cattle played an increasingly important role and which gave rise to a form of production management akin to that of a town, with highly organised butchery and market-like concentrations of butchers' stalls within a delineated area (markets) (Auxiette, 1996; Auxiette and Paris, 2017; Paris, 2016).

In the vicinity of Reims, the study of the fauna from the farm at Bétheny reveals a herd containing a high proportion of pigs, indicating that the site ranked somewhere between a simple farmstead and a high-status site. On the farm site of Reims "les Hauts des Nervas" (situated outside the oppidum

of Reims-*Durocortorum*), there appears to have been an emphasis on the rearing of cattle and horses. However, this assemblage has been heavily impacted by the acidity of the subsoil, as attested by the large proportion of isolated teeth recovered and by the preferential preservation of larger, more solid bones (Auxiette, 2004). As a result, species representativity is undoubtedly biased and the proportion of smaller livestock is minimised. Finally, the quantities of pig recorded at "Rue d'Anjou" inside the oppidum prefigure the increase in this species within the oppidum (Auxiette, 2010).

The study of a significant corpus of bone (± 70 000 studied bones) from several clearly planned areas of the oppidum of Condé-sur-Suippe (Auxiette, 1994; Méniel, 1984; Paris, 2016; Pion *et al.*, 1997) has revealed a predominance of pig (between 52.5 and 70.7%), with bovids accounting for about 20% of the assemblage. Several zones have been identified, including areas dedicated to craft working and butchery, indicating large scale economic development from the La Tène D1a onwards. We can distinguish areas of carcass preparation and processing from others clearly characterised by domestic consumption. There are also areas with evidence suggesting the use of bone in metallurgical activities.

Slaughter age analysis reveals that a significant proportion of cattle were killed between the ages

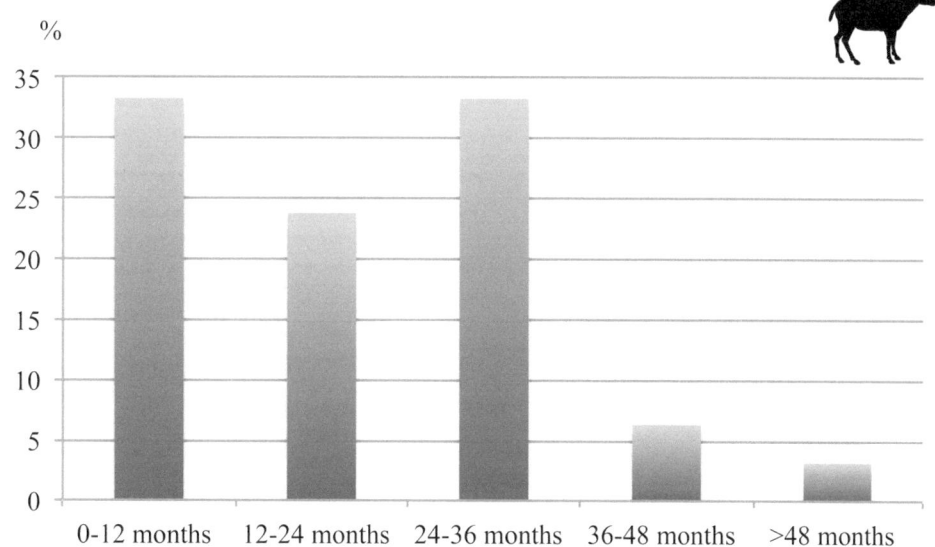

Figure 71: Villeneuve-Saint-Germain "les Grandes Grèves" (Aisne), Late La Tène, oppidum, sheep/goat slaugthering (mandibules).

of 15 and 36 months: this includes animals that had not yet attained their optimal weight along with mature or old animals. As for pigs, there is a peak in slaughtering at around 18 to 21 months and the killing of a non-negligible proportion of animals around the age of 12 months. Finally, the slaughter of caprines is less focused but we nonetheless note a peak around the age of 2 years. The estimated MNI and associated meat weights are considerable: 68 head of cattle with an estimated meat weight of 17,200 kg, 238 pigs with a meat weight of about 17,500 kg, 139 sheep representing 2,400 kg of meat, 31 horses for 10,850 kg of meat, and finally 35 dogs corresponding to 700 kg of meat.

On farms dated to the Middle La Tène/ early Late La Tène located in the vicinity of Laon, on the edge of the Remi territory, the majority of slaughtered sheep were juveniles and at Barenton-Bugny

Site L they were killed at about 10 months. Thus they were killed during the winter, between December and March (lambs are born in March/ April). In order to identify the period of the year when slaughtering took place, the ages of the different individuals are superimposed on a birth calendar; for example, animals aged 5/6 months (born in March/April) were killed in August and September, while those aged 15/16 months were killed during the months of June and July. The fact that sheep only lamb once a year allows us to make a simple projection of the slaughter ages of the animals during their first year. By preference, pigs were killed between the ages of 1 and 2 years. These results are in line with those obtained for the Chambry "Zac du Griffon" farm site, where lambs and young adult sheep dominate the animals consumed. The clustering of ages in two groups, the

Site	Field	Date	MNI					Ages cattle		Ages sheep/goat			Ages pig		
			Cattle	Pig	Sheep/ Goat	Horse	Dog	Cattle less than 4 years	Cattle more than 4 years	Sheep/ Goat less than 1 year	Sheep/ Goat between 1 and 3 years	Sheep/ Goat more than 3 years	Pigt less than 1 year	Pig between 1 and 3 years	Pig more than 3 years
Villeneuve-Saint-Germain	les Etomelles	LTD1b	9	24	15	2	4	2	4	2	0	11	1	18	1
Ciry-Salsogne	le Bruy	LTC/D	3	7	2	1	1	2	1	4	3		1	1	
Vasseny	Dessus des Groins	LTD													
Limé	les Sables sud	LTD1	4	3	3	5	3	3	1	1	1	1	2	1	
Bazoches-sur-Vesle	la Foulerie	LTD1/D2	2	1	1	2	1	1		2			1	1	
Bazoches-sur-Vesle	les Chantraines	LTD1a	14	18	32	8	2	7	7	11	11	10	1	17	
Braine	la Grange des Moines	LTD1b	53	106	41	18	5	13	26	8	6	20	10	73	19
Braine	la Grange des Moines	LTD2a	9	4	5	2	1	4	2	1	2	3		2	2
Berry-au-Bac	la Pêcherie	LTD1a	2	3	5	1	1	2			3	2			
Damary	le Ruisseau de Fayau	LTD1	12	14	13	6	3	3		3	4	1	4	4	1
Mont-Notre-Dame	Vaudigny	LTD1b	2	9	9	inc	inc	2			8	1		5	4
Reims	les Hauts des Nervas	LTD1	4	4	3	2	0		4	1	1	1			
Betheny	les Equiernolles	LTD1	5	15	9	3		1	1	2	1	6		11	1
Barenton	Site L	LTC2/D1	3	2	8	2	3			8			1	1	
Barenton	Site M	LTC2/D1	2	2	3	1	1								

first around 5-6 months and the second around 12 months, undoubtedly reflects the fact that slaughtering took place seasonally within two distinct episodes: the first in spring (animals aged 12 months) and the second in autumn (animals aged 5-6 months). This pattern in no way corresponds to the opportunistic slaughter inherent in typical domestic herd management, but rather reflects selection within a framework of very specific events. At the very end of the Late La Tène, there is a radical change in the direction of caprine rearing with sheep and goats being slaughtered between the ages of 4 and 6 years (Audebert *et al.*, 2016). It was rare for animals aged less than 2 years to be selected for consumption. The rearing of pigs was practically non-existent while cattle rearing, though not negligible, was limited. The predominance of late slaughter of caprines, which suggests that husbandry was orientated towards milk and wool production, might be indirect evidence of the woollen textile production for which the Remis were renowned (cf. Strabo who refers to the export of *saga*- short cloaks- during the reign of Augustus (Gonzalez-Villaescus, 2010): this trend is absent in other assemblages known for the period. The absence of evidence for weaving suggests that bales of wool were exported to workshops located away from the production zones.

The breakdown of slaughter ages within broad groups shows how preferential choices made in terms of species are related to site categories (fig. 72).

On simple farmsteads, cattle tended to be killed around the age of 4 years or older and, with the exception of farms located on the periphery of Laon, sheep were generally slaughtered over the age of 2 years. On high ranking farms, these trends continue to prevail. When examined in more detail it is clear that some of the pig herd was preferentially slaughtered between the ages of 17 and 23 months at the site of Villeneuve-Saint-Germain "les Etomelles" (twenty mandibles) and between the ages of 18 and 24 months at the sites of Braine "la Grange des Moines" (50%, 106 mandibles) and Bazoches-sur-Vesle "les Chantraines" (eighteen mandibles) (Pommepuy and Gransar, 1998). As regards caprines, the preferred slaughter age was between 24 and 48 months at

Villeneuve-Saint-Germain (fifteen mandibles), between 24 and 36 months at Braine (thirty four mandibles), and in equal proportions according to the three broad classes at Bazoches-sur-Vesle (*i.e.* less than 1 year; between 1 and 3 years; over 3 years; thirty two mandibles). In the case of cattle, slaughter ages vary between 30 and 48 months and between 12 and 15 years at Braine (forty), between 30 and 50 months at Villeneuve-Saint-Germain (five mandibles), and in equal proportions according to the two broad classes at Bazoches-sur-Vesle (*i.e.* less than 4 years; greater than 4 years; fourteen mandibles).

What conclusions can be drawn from this evidence? It appears that each farm had its own pattern of herd management. This is expressed through the various choices made regarding the livestock: we observe both opportunistic slaughter, which would have satisfied changing needs, and episodic slaughter, which was probably determined by seasonal demands. In reality the two patterns may be intertwined and difficult to dissociate. The mandibles of juvenile or young adult animals are particularly useful for examining the issue of seasonal slaughter since estimations of age are more easily situated within a calendar of births. Thus, the assemblages from Barenton-Bugny Site L and from Chambry "Zac du Griffon" have enabled us to identify, with certainty, two periods for the slaughter of sheep: one in spring and the other in autumn. At Braine, we can identify a single episode in spring. At Bazoches sheep tended to be killed at the end of spring and during the summer, while at Betheny killing took place at the end of the winter and during the spring. Pigs were preferentially slaughtered between the autumn and the start of spring at Villeneuve-Saint-Germain "les Etomelles" and at Braine. While the slaughtering of calves and lambs may have formed part of a deliberate strategy for milk production, certain slaughter episodes could also be related to collective feasting associated with specific festivals in the farming calendar. We can also envisage collective meals on the occasions of weddings, births, funerals rites of passage or the inauguration/decommissioning of houses... there are, in fact, a multitude of reasons for which people might come together to feast (Auxiette and Desenne, 2017; Hénon *et al.*, 2012).

Among the Late La Tène sites in the middle Oise valley, within the territory of the Bellovaci, we can distinguish between simple farms – "la Plaine d'Herneuse I" (LTC2/D1) and "la Plaine de Saint-Germain" (LTCD) in Verberie, Lacroix-Saint-Ouen "le Pré des Iles" (LTD) (Méniel, 2006) and Jaux "le Camp du Roi" (LTC2D1) (Malrain *et al.*, 1996) – and high status farms, namely Longueil-Sainte-Marie "le Vivier des Grès" (LTC1C2) and Verberie "la Plaine d'Herneuse II" (LTC2D1 et D2) (Méniel, 2006; Méniel *et al.*, 2009). The faunal assemblages vary in size from a few tens of bones to several thousand. The proportions of the species relative to each other also vary: the incidence of pig varies between 22 and 50%, cattle between 28 and 45 % and caprines between 28 and 45%. Pig predominates at Jaux (although the actual numbers are low) and at Verberie "la Plaine d'Herneuse II". Cattle predominate at Lacroix-Saint-Ouen, Longueil-Sainte-Marie and on the site of Verberie "la Plaine de Saint-Germain". Caprines rarely predominate on these farms but it is worth noting that they take precedence over pigs at Verberie "la Plaine de Saint-Germain".

The high ranking farm at Verberie "la Plaine d'Herneuse II" has yielded the richest faunal assemblage (about 10,000 bones). Spatial analysis of the bone refuse shows variation between the various sectors of the site and enables us to interpret more finely the relative frequencies of species. Domestic mammals were preferentially slaughtered at intervals that reveal the consumption of immature animals which had not yet reached their optimum weight, thus indicating that good quality, tender meat was being sought: cattle were mainly killed between 0 and 2 years, pigs between 0.5 and 1 year, and caprines between 0 and 2 years with half under the age of 1 year (Méniel, 2006, 142-144). At least 30 (thirty) head of cattle, about 140 pigs and 80 (eighty) sheep/goats were consumed on site. The fauna from the farm site of Longueil-Sainte-Marie "le Vivier des Grès" (c. 4800 bones) stands out in that one of the pits yielded very large quantities of cattle bone: in this case the remains of butchered cattle carcasses are suggestive of collective rather than domestic consumption. In another area of the site, pigs predominate. The slaughtering of young cattle and pigs attests to the consumption of high quality meat.

For the most part, cattle slaughtered on the site of Verberie "la Plaine de Saint-Germain" were rather old, or indeed very old, and the management of the caprine herd seems to have followed the same pattern (Méniel, 2006, 142). Evidence for slaughtering from the other farms indicates a greater spread of slaughter ages without any distinctions.

Within the territory of the Senones, at the confluence of the Seine and Yonne (Seine-et-Marne), many enclosed farmsteads dating from the end of Late La Tène C1 to the La Tène D2, some of which have only been partially excavated, have produced faunal assemblages that are in some cases quite considerable in size (Auxiette, 2013a; Gouge and Séguier, 1994; Horard-Herbin *et al.*, 2000; Séguier and Auxiette, 2006). These include the sites of Cannes-Ecluses "le Petit Noyer" (partially excavated, La Tène D1 and La Tène D2), Marolles-sur-Seine "le Grand Canton" (La Tène C2/La Tène D1), Balloy "les Défriches" (La Tène D2), Bazoches-lès-Bray "le Tureau aux Chèvres" (La Tène C2/D1 transition), Bazoches-lès-Bray "la Voie Neuve" (LTD1), Balloy "la Fosse aux Veaux" (LTD), Villiers-sur-Seine "le Gros Buisson" (LTC2) (Auxiette, forthcoming)(Auxiette forthcoming), Varennes-sur-Seine "la Justice" (LTD1/D2) (Auxiette, 2013b) and the settlement known as "le Marais du Pont" (LTD2) (Séguier, 1996) also in Varennes-sur-Seine. In the Loing valley, the upland site of Souppes-sur-Loing "à l'Est de Beaumoulin" (Séguier and Auxiette, 2006) completes the inventory of Late La Tène sites. The areas enclosed within these sites vary as do the sizes of the faunal corpora: the assemblages are made up of several hundred to several tens of thousands of bones.

In the two farm sites in Bazoches-lès-Bray, cattle clearly dominate the faunal spectra, in Cannes-Ecluse pig predominate, and in Marolles-sur-Seine cattle are in the majority. Analysis carried out on the two sites in Varennes-sur-Seine reveals a high proportion of pig (45%). The site of Varennes-sur-Seine "la Justice" stands out from other sites by virtue of the quantities of meat consumed, the seasonal slaughter practices, the probable preservation of meat, the occasional but large scale ritual/cult deposits and the consumption of wine. On the basis of these characteristics, the site can be interpreted as a "central place" with considerable economic and social power and probably closely associated with the lowland clustered settlement located at "le Marais du Pont" (Séguier, 1996).

On the site of Varennes-sur-Seine "la Justice", cattle (forty one) were preferentially slaughtered between 4 and 6 years (82.5 %). This tendency to slaughter animals over the general age for beef cattle indicates that herd management was orientated towards milk production. There is no evidence for the consumption of veal or for natural mortality among calves. A few young adults are recorded and these were killed during the autumn and winter (between 15 and 18 months, then between 30 and 33 months).

The majority of pigs (n = 107) were killed between 1.5 and 2 years (58%) and older (23.3%). Juvenile animals are rare with just a few instances of pigs being killed at 7 to 8 months. The principal objective of the pig rearing on this site was the production of the greatest possible quantities of meat – hence the almost exclusive slaughter of animals that had reached their optimum weight-rather than the production of tastier, tenderer meat. The virtual absence of evidence for natural mortality raises the issue of *in situ* rearing.

Study of sheep slaughter (n=102) produces results in line with those obtained for cattle: there is an emphasis on the exploitation of secondary products with animals being preferentially slaugh-tered between the ages of 3 and 6 years (60.8%). A few were killed at around 2 years (20.5%) and, in rare cases, much younger. There is no evidence for the killing or consumption of lambs.

In conclusion, there was no consumption of calves, piglets or lambs. There is very little evidence for on-site rearing of animals. Consumption by the inhabitants and/or visitors depended in most cases on the cooking of adult animals: bovids were preferentially consumed after a life of service to the community (Auxiette, 2013b).

The incidence of horse apparently varies from region to region, with the highest values occurring to the east of the Seine-Yonne confluence (> 8%) (Horard-Herbin *et al.*, 2000).

The amount of hunting, as shown by cuts of venison and wild boar, seems to be an indicator of farm status (Auxiette, 2013a; Horard-Herbin *et al.*, 2000, 197-198) and particularly so when

combined with the presence of certain other categories of finds such as weaponry, amphorae, large quantities of ceramics, metal objects or imported wares. In the case of lower status sites, such as Cannes-Ecluse, Grisy-sur-Seine and the two sites at Bazoches-lès-Bray, when wild mammal remains (especially wild boar) occur, the proportions never exceed 1%. The high ranking farm at Varennes-sur-Seine "la Justice" has produced accurate data on the consumption of red deer (2.2% of large game). In fact, the joints of meat identified come from specific parts of the animal; they present all the characteristics of having been prepared from pieces of meat from 10 animals. The cuts consumed principally come from the shoulder (humerus/radius) and from the haunch (fig. 73). All require preparation, most often marination, prior to cooking. The absence of cuts of meat removed in the initial stages of carcass preparation, immediately after the killing of the animal (axial skeletons -including antlers- and lower legs are completely missing), suggests that it is most likely that the cuts intended for consumption were brought onto the site.

The enclosed settlement of Varennes-sur-Seine "le Marais du Pont", the full extent of which remains unknown, was excavated over an area of c. 1.5 ha in the 1990s and produced over 40,000 animal bones (Horard-Herbin *et al.*, 2000). In terms of actual numbers, the faunal spectrum is dominated by pig (42.3%) but cattle predominate in terms of the amount of meat generated. The type of herd

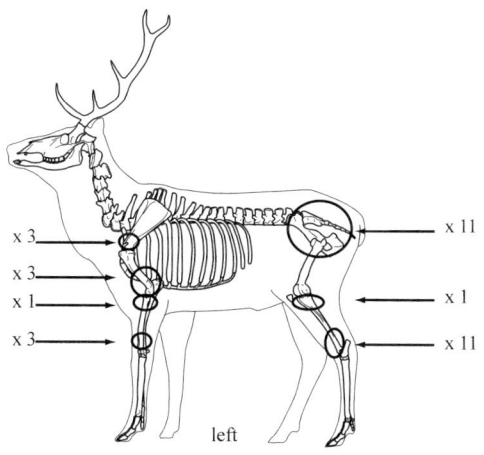

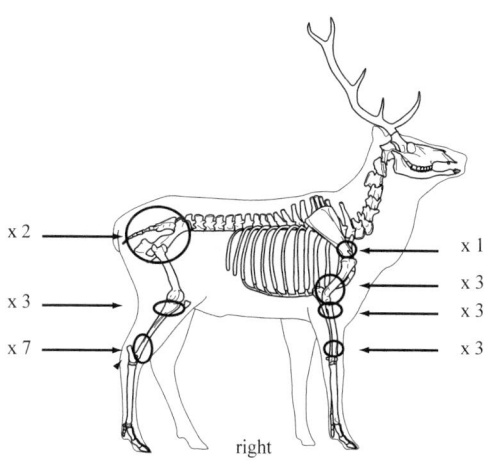

Figure 73: Varennes-sur-Seine "la Justice" (Seine-et-Marne), Late La Tène, red deer butchery.

management practiced compares well with that identified at the "la Justice" farm site, located less than 1 km away, with slaughtering centred between autumn and beginning of spring based on the hypothesis of two litters per year (Auxiette, 2013b). In fact the slaughter age curve for pigs indicates three peaks, the first at 8 to10 months, the second at 16 to 20 months, and the third at 22 to 24 months. When the ages are aligned on a birth calendar featuring two litters, it emerges that slaughtering was principally centred on the winter months. In the case of a single litter per year (which is an unlikely scenario (Auxiette *et al.*, 2020), slaughtering spans the entire year with the exception of spring time.

This curve attests to strict herd management and probably reflects seasonal killing. The two sites differ in terms of the quality of meat produced with the consumption of juvenile pigs being characteristic of "le Marais du Pont". Most of the cattle were killed before the age of 4 years for meat production. In both cases, the slaughter curves highlight the mixed exploitation of the cattle herd with a portion of the animals being reared for meat production (slaughtered between the ages of 3 and 4 years) and a portion (more than 4 years) being kept for secondary products. At "le Marais du Pont", the majority of the caprines were reared for meat production and slaughtered between the ages of 3 and 4 years, with an older population (28% killed over 4 years of age) indicating parallel production of milk and wool (Méniel and Horard-Herbin, 1996).

One of the most notable differences between the two sites involves the presence of new-born and fœtal pigs and of dogs and horses which attest to *in situ* rearing of domestic species on the "Marais du Pont" clustered settlement: these elements are absent in the assemblages from the "la Justice" farm site.

Spatial analysis of the faunal remains from segments of the enclosing ditch at Varennes-sur-Seine "la Justice" has revealed several zones of preferential discard, which are interpreted as domestic refuse deposits. Thus in the east, on either side of the eastern entrance, we find a predominance of pig; in the west, close to some buildings, cattle remains are slightly more numerous than those of pig (38 and 33%, respectively); lastly, in the south, on either side of the southern entrance, pig and cattle remains occur in roughly equal proportions (37 and 38%). The incidence of small mammals does not change greatly with the nature of the context.

Among the other farms in the Bassée, the site of Bazoches-les-Bray "la Voie Neuve" (dating to the second half of the 2nd century BC), which is located about 20 km to east of the sites at Varennes-sur-Seine, is noteworthy. In fact, the site shares several aspects in common with the farm at Varennes-sur-Seine "la Justice", including the occurrence of large quantities of bowls and amphorae, but it differs in terms of the relatively limited number of faunal remains (±1700 bones). Study of the fauna has revealed three prime areas of discard, each with its own particularities, including a notable concentration of refuse near the entrance. As regards management of domesticated animals, there was a preference for slaughtering pigs either at the age of 12 months or, more commonly, at 24 months or older. More than half of the cattle were killed before the age of 4 years (almost 40% between 0 and 2 years and approximately 20% between 2 and 4 years). The majority of caprines were slaughtered between 2 and 4 years, but never before 1.5 years. As at the site of Varennes-sur-Seine "la Justice", there is no evidence for the actual rearing of animals on site.

The site of Villiers-sur-Seine "le Défendable" combines a number of characteristics that set it apart from lower status farms. The faunal corpus is largely dominated by cattle (40%) and pigs (34%). Taken together, age analyses of the principal domestic species allow us to identify favoured slaughtering periods for cattle and pigs. These episodes took place either at the end of spring or over the winter. Age matching between the three principal domestic species is not "perfect" but the closeness of the temporal ranges points to a certain degree of correlation. We estimate the MNI as follows: sixty cattle, hundred pigs, almost fifty caprines, fifteen horses and ten dogs (see chapter 3).

The site of Souppes-sur-Loing "A l'Est de Beaumoulin" is by far the most exceptional site of all. The faunal assemblage consists of a high percentage of identified bone (>80%) composed of large mammals (cattle and horses) and small mammals, which also played a large role in meat production; several wild species, including red deer, complete the corpus. The entrance area of the site yielded high concentrations of faunal remains, particularly skulls and various portions of skeleton, both articulated and disarticulated, which appear to be deposits. Estimations of the slaughter ages for the various species reveal that seasonal collective feasts took place here on a very large scale (see chapter 3).

In the vicinity of the city of Amiens-*Samarobriva* (Somme), several Late La Tène (La Tène C2/La Tène D2) farms reveal diverse animal husbandry practices that nonetheless focus either on cattle rearing (Amiens-Renancourt "Zac de Renancourt" site 5, and Pont-de-Metz "l'Hôpital" for La Tène D1; Amiens-Renancourt "Zac de Renancourt" site 2 for La Tène D2) or pig rearing on the other (Méaulte "Zac du Coquelicot" and Glisy "les Champs

Tortus" for La Tène C2; Amiens-Renancourt "Zac de Renancourt" site 5, for La Tène D2) (Auxiette, 2017b, 2014b, 2011c). The incidences of the principal domestic species vary considerably: cattle or pigs always predominate, never sheep. The evidence thus differs considerably from the two Middle La Tène sites where pigs made up only a small part of the total livestock. From La Tène C2 onwards, pigs constituted a significant portion of the meat consumed, surpassing the contribution of sheep. Animals were principally reared for meat with the exception of caprines on the farm at Glisy "les Champs Tortus": at this site the wide distribution of slaughter ages suggests that only some portion of the sheep were raised for their secondary products. This farm falls into the category of high status farms and was the site of collective feasting with a particular emphasis on the consumption of pig (75% of pigs were killed between 1 and 2 years). The main slaughtering periods occurred during the winter and spring. The composition of the livestock herd has been calculated as follows: twenty cattle, forty sheeps/goats, ninety pigs, fifteen dogs and a few horses. Farms "2" and "5" at Amiens-Renancourt "Zac de Renancourt" are characterised by high levels of meat consumption, with a particular emphasis on cattle and sheep on the former site and on pigs and cattle on the latter: in both cases meat animals predominate. The number of individual animals reaches 100 on Site 2 and exceeds this number on Site 5. It is clear that the status of the

site governs the level of importance of pigs within the herd.

In the Crould Basin, faunal assemblages from five Late La Tène farms located at Villiers-le-Bel/Gonesse "Déviation RD 10-370" (Jouanin and Robin, 2010), Le Mesnil-Aubry "le Bois Bouchard-Carrière REP-Véolia" (Jouanin and Touquet Laporte-Cassagne, 2013), Gonesse "Zac des Tulipes Nord" (Auxiette, 2012), Roissy-en-France "la Rayonnette" and at Pierrefitte-sur-Seine "les Tartres" (Frère, 2012), reveal variability in the dominant species. Pigs and caprines predominate at Mesnil-Aubry and Roissy-en-France (LTC2), cattle and pigs at Villiers-le-Bel (LTC2 and LTD1), cattle at Gonesse and Pierrefitte (LTD), thus relegating caprines, with rare exceptions (e.g. a silo deposit at Mesnil-Aubry), to third position in the order of importance. The high status farm site at Pierrefitte, which features a very large herd of livestock (65 head of cattle, 132 caprines and 91 pigs), provides valuable information regarding herd management and is particularly unusual in terms of the slaughter pattern for cattle, with rare evidence for the killing of calves along with a high proportion of animals aged between 2 and 4 years. The exploitation of caprines is divided between the production of quality meat (animals killed between 0.5 and 2 years) and secondary products. Pigs were generally killed before the age of 2 years, and the proportion of piglets is not negligible (15%). In addition, this farm has also yielded a significant proportion of

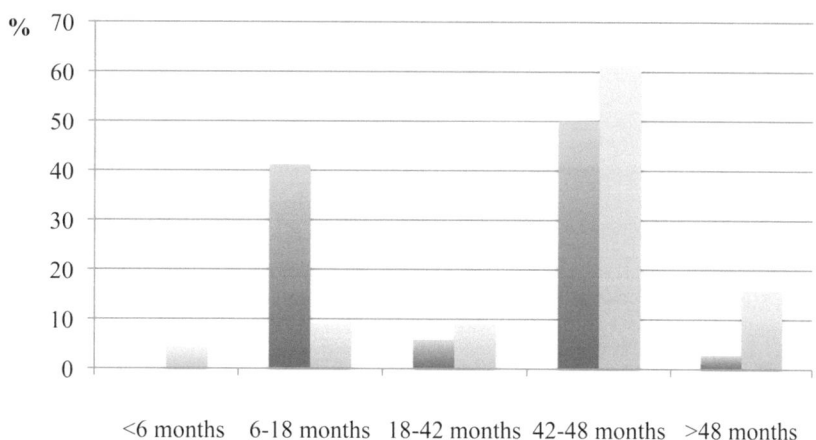

Figure 74: Ifs "Object'Ifs-Sud" (Cavados), Middle La Tène/Late La Tène, cattle slaugthering.

older animals. At Mesnil-Aubry, pigs were mainly slaughtered between 0.5 and 2 years (>60%), while the killing of caprines focused on two age groups: sheep were killed either between the ages of 0.5 and 2 years (a similar pattern to that observed at Pierrefitte) or between 4 and 6 years. Both sites have evidence for the slaughter of calves and lambs which would suggest deliberate stimulation of milk production.

At Ifs "Object'Ifs-Sud", in the Caen Plain, the consumption of young cattle identified in the Middle La Tène (thirty four mandibles) becomes marginal in Late La Tène (forty four mandibles, fig. 74) most of them were slaughtered between the ages of 2.5 and 4 years at Late La Tène (Auxiette, 2000b).

Sheep rearing was focused on meat production with most animals being killed between the ages of 2 and 2.5 years. As in the Middle La Tène, we still have evidence for the consumption of lambs but the proportions are drastically reduced. Sheep were rarely killed after the age of 4 years. Little difference can be seen in the consumption of pig between the Middle La Tène and Late La Tène: there is still evidence for the slaughter of juveniles but the majority of animals were killed between 1.5 and 2.5 years.

The slaughter ages were aligned on the birth calendar for each species, i.e. March-April for caprines, May-June for cattle, and March-April and/or August-September for pigs (depending on whether there were one or two litters per year). The following patterns emerge: during the La Tène D1/D2 two slaughtering periods are identified, in spring and in summer (25 to 28 months); in the La Tène C1, caprines were preferentially killed in two episodes, in the summer (juveniles aged 4-5 months and adults aged 25-28 months) and in the winter/early spring (animals aged 21-24 months).

The killing of cattle over the course of the autumn and winter is well documented for the entire chronological sequence (5 to 7 months, 17/18 months, 30 months and 40 months). Evidence for the killing of very young animals, which is recorded for the La Tène C, is absent in the La Tène D1/D2 assemblages. Finally, there was a preference for slaughtering pigs in spring time (March to May) or over the winter (at 1 month, 7/8 months and around 19/23 months).

On the farm at Mondeville "l'Etoile" site III, cattle bones dominate the assemblage (50%), while sheep occupy a substantial second place (22%) and pigs constitute a minor element (12%). Horse represents 15% of the remains, which is the largest proportion recorded for all of the Mondeville farm sites. Cattle slaughter ages range from 5/6 months to

over 50 months. Juvenile and young adult animals account for most of the slaughter carried out in September/October (16/17 and 30/31 months, 40 months) and in the springtime (around 24 months). The corpus did not contain any culled animals: most of the evidence points to the consumption of young animals, which suggests that the inhabitants of the site enjoyed a relatively high standard of living. Sheep were generally killed around the age of 2 years, during spring/summer (25 to 28 months), which does not preclude the slaughter of young animals during the winter (21/24 months) and the killing of adult animals (>2.5 years). Above all else, this reflects a dual aspect to the sheep rearing with a desire to produce not only meat but also secondary products such as milk and wool. The killing of young pigs attests to a taste for tender meat (aged between 4 and 8 months, killed in July-November and/or December-April) which was substantially supplemented by animals aged between 1.5 and 2 years (October-February and/or March-July, with a preference for animals aged 19-21 months killed between October and December or between March and May). Although there is indisputable evidence for the consumption of horses, the evidence from the dog bones is insufficient to prove that dog meat was eaten as well.

The two Late La Tène sites at Fleury-sur-Orne "les Mézerettes" and "ZL7/CD120" reveal a clear preference for rearing bovids, and particularly cattle (53%) (Baudry, 2018). Cattle are a little less frequent on the farm site of Saint-Martin-de-Fontenay "la Grande Barberie" (47%). Data on herd management suggests variable practices with some similarity between the two Fleury-sur-Orne sites where over 50% of cattle were slaughtered before the age of 4 years, and more specifically between the ages of 2 and 4 years at "ZL7/CD120". The proportion of culled cows differs on the two Fleury-sur-Orne sites, and there is a clear absence of young individuals aged less than 1 year which is in contrast to the data from Saint-Martin-de-Fontenay. The caprine age patterns clearly indicate that there were two approaches to the management of the flocks: at Saint-Martin-de-Fontenay, significant proportions of animals were slaughtered between the ages of 2 and 12 months (the majority between 6 and 12 months) and between 24 and 36 months; on the Fleury-sur-Orne

Late La Tène

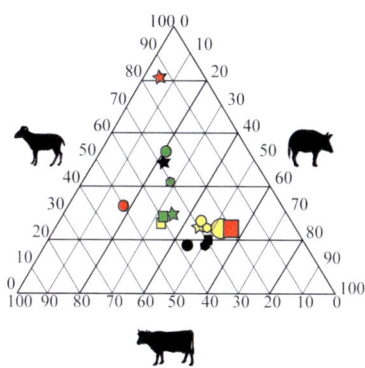

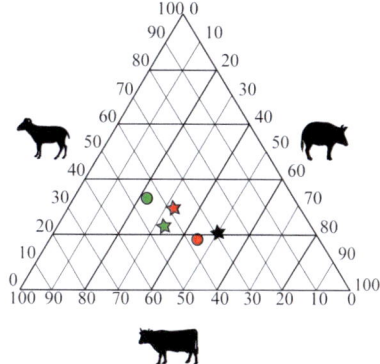

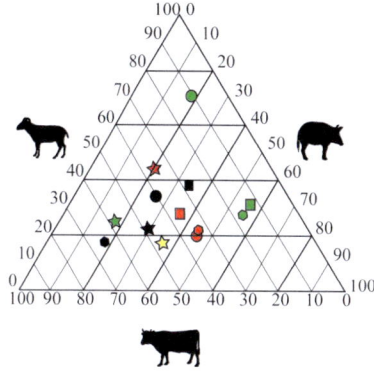

Aisne

🟥 Condé-sur-Suippe/Variscourt - oppidum
🔴 Ciry-Salsogne "le Bruy"
⭐ Barenton "Zac du Griffon-site L"
⭐ Juvincourt et Damary "le Ruisseau de Fayau"
🟩 Bazoches-sur-Vesle "les Chantraines"
🟢 Berry-au-Bac "le Chemin de la Pêcherie"
🟢 Bucy-le-Long "le Fond du Petit Marais"
⚫ Braine "la Grange des Moines"
⬛ Mont-Notre-Dame "Vaudigny"
★ Barenton "Zac du Griffon-site G"
🔴 Villeneuve-Saint-Germain "les Etomelles"
⚪ Bucy-le-Long "le Fond du Petit Marais"
⚪ Barenton "Zac du Griffon-site M"
☆ Sermoise "les Prés du Bout de la Ville"
🟡 Braine "la Grange des Moines"
🟡 Villeneuve-Saint-Germain "les Grèves" - oppidum

in red La Tène C2D1, in green La Tène D1a,
in black La Tène D1b, in yellow La Tène D2

Oise

⭐ Lacroix-Saint-Ouen "le Pré des Iles"
⭐ Longueil-Sainte-Marie "le Vivier des Grès"
★ Verberie "la Plaine d'Herneuse II"
🟢 Verberie "la Plaine de Saint-Gemain"
🔴 Jaux "le Camps du Roi"

Seine-et-Marne

☆ Villiers-sur-Seine "le Gros Buisson"
🟢 Marolles-sur-Seine "le Grand Canton"
🟢 Grisy-sur-Seine "les Méchantes Terres"
⭐ Bazoches-les-Bray "Près le Tureau aux Chèvres"
🟩 Bazoches-les-Bray "la Voie Neuve"
⚫ Marolles-sur-Seine "le Grand Canton Sud"
⬛ Balloy "les Défriches"
★ Egligny "le Bois de la Pêcherie"
⚫ Souppes-sur-Loing "A l'Est de Beaumoulin"
🔴 Varennes-sur-Seine "la Justice"
🔴 Varennes-sur-Seine "le Marais du Pont"
⭐ Changis-sur-Marne "les Pétreaux"
🟥 Poincy "Près le Pont de Trilport"

Marne - Late La Tène

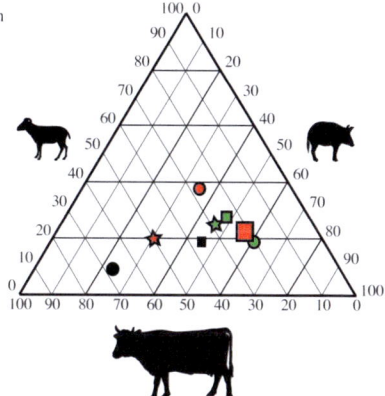

🟥 Condé-sur-Suippe/Variscourt - oppidum
🔴 Betheny "les Equernolles"
⭐ Reims "les Hauts des Nervas"
☆ Reims "Rue d'Anjou"
🟩 Reims "Villa des Capucins"
🟢 Reims "Rue Chanzy"
⚫ Reims "Rue Carnot"
⬛ Reims "Rue Rockfeller"

In red La Tène D1, in green La Tène D2, in black La Tène D2/augustéen

Calvados

🟢 If "Object'Ifs sud"
🟢 Mondeville "l'Etoile"
⭐ Mondeville "l'Etoile"
🟩 Saint-Martin-de-Fontenay "la Grande Barbarie"
🔵 If "Object'Ifs sud"
⚫ If "Object'Ifs sud"
⬛ If "AR67"
★ If "Object'Ifs sud"
⚫ Bretteville-l'Orgueilleuse "le Bas des Prés"
🔴 If "Object'Ifs sud"
🔴 If "Object'Ifs sud"
⭐ Fleury-sur-Orne "les Mézerettes"
🟥 Fleury-sur-Orne "ZL7/CD120"
☆ Saint-Martin-de-Fontenay "la Grande Barbarie"

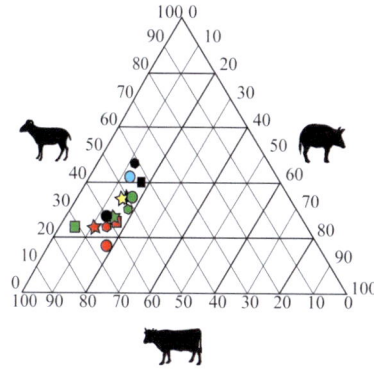

Figure 75: Late La Tène, frequency of the main domestic species: Aisne, Oise, Seine-et-Marne, Marne, Calvados.

sites, the animals were generally aged between 12 and 48 months at the time of slaughter. The slaughter profiles for pigs can vary (95% were killed before the age of 2 years) but they were never killed before they were 6 months old.

For Late La Tène, it is now clear that assemblages linked to consumption vary significantly between farms, even when they are located relatively close together. The most obvious differences are geographical : significant proportions of pigs and sheep on the sites of the Aisne, very few caprines on sites around Reims (Marne), a lot of cattle on sites in the Caen Plain (Calvados), more disparate herd compositions in the Oise and Seine-et-Marne (fig. 75).

The sites that exhibit the most similarities, with pigs playing a predominant role, are well documented in the regions under consideration: for the La Tène C1, "les Terres de Ville" at Glisy in the Somme and Fontenay-en-Parisis in the Val-d'Oise; for the La Tène C2, "les Champs Tortus" at Glisy in the Somme, Villiers-sur-Seine (Seine-et-Marne) and Pierrefitte (Val d'Oise); and for the Late La Tène, the sites of Souppes-sur-Loing, Varennes-sur-Seine and Poincy (Seine-et-Marne), Pierrefitte (Val d'Oise), "les Etomelles" and "les Grèves" at Villeneuve-Saint-Germain, Braine (Aisne), Verberie, Longueil-Sainte-Marie and Beauvais (Oise).

Oscillations are particularly clear between cattle and pig (fig. 75); caprines are most often relegated to third position apart from some rare exceptions (Barenton-site L in the Aisne and Grisy-sur-Seine in Seine-et-Marne). While the place of horse varies significantly from settlement to settlement, the greatest proportions are found on high-ranking sites. We should remember that while there is evidence that horse was eaten, it is probable that not all individuals were consumed. Dog was also eaten and used for its skin (*e.g.* at the oppidum of Villeneuve-Saint-Germain), and becomes more and more common over time.

In our opinion, the distinctions between simple farms and high-ranking farms are expressed by several criteria: herd size, the ages of animals selected for slaughter (even though this can be opportunist with animals being killed throughout the year), the quantities of meat consumed (which can attain several thousands of kilograms) and the hunting of wild game, even though the products of such hunting only represent a small proportion of the overall weight of meat consumed. Recent studies have highlighted the fact that the cutting up of carcasses into large pieces was also a feature of sites where shared consumption took place on various scales. The consumption of very young animals, however, does not appear to be a discriminating criterion. In almost every case, except in

the territory of the Bellovaci, the consumption of meat is accompanied by the consumption of wine, attested to by the presence of *amphorae* (except at Beauvais "les Aulnes du Canada"). The size range of the animals appears to be linked to site status (hierarchical rank). In fact, from the La Tène D1b, larger than average cattle and horses, and in variable proportions depending on the geographical area, are more characteristic of high-ranking farms, clustered settlements and *opidda*. The coexistence of several morphotypes is recorded from the La Tène D1 onwards. A recent study has shown that changes in the morphology of cattle are not linked to the Roman conquest and that greater variation is observed on urban sites. This variability might be the result of local production of several cattle morphotypes (Nuviala, 2016, 2015). Heterogeneity in the sizes of horses is a characteristic of equine populations of the Late Iron Age (Arbogast *et al.*, 2002a, 45; Méniel, 1984); as is the case for cattle, this tendency becomes more marked from the La Tène D1 onwards. Pauline Nuviala (Nuviala, 2015) postulates that specific animal husbandry practices were developed to improve the stature of the animals and to respond to evolving needs. This new hypothesis does not exclude the possibility that horses were imported from elsewhere.

Endnotes

1   Orconte "les Noues" (Arbogast, 1994); Norrois "la Raie des Lignes" (Poulain,1986); Larzicourt "Champ Buchotte" (Poulain unpublished); Larzicourt "Ribeaupré" (Poulain unpublished).

2   Pont-sur-Seine "Marnay/La Gravière" (Fournand et al., 2010; Fournand, 2012); two houses and pits; Lesmont "Les Graveries", two houses and pits; Juvigny (Arbogast, 1991; Meunier, 2013), several pits; Bréviandes "Zac St Martin" (Laurelut, 2010, 2017); NISP= 16,192 remains from sevral houses and pits ; Buchères "les Bordes" (D 11, D12 et D32), one house and several pits; Buchères "Le Clos II", one pit; Buchères "Parc Logistique de l'Aube" ("PLA" D 39), one house; Buchères "Parc Logistique de l'Aube" ("PLA" D 43-44), one house. About 100 remains were retrieved from both "Le Clos II" and "PLA" in Buchères, almost 600 from Buchères "les Bordes" and Lesmont, and almost 1,300 remains from Pont-sur-Seine "Marnay/La Gravière".

3   Buchères "les Bordes" (D 11, D12), Buchères "Parc Logistique de l'Aube" ("PLA" D 43-44).

4   Bucy-le-Long "la Fosselle" (Hachem et al., 1997, 1998a); Missy-sur-Aisne "le Culot" (Charier, 1986; Farruggia and Constantin, 1984) (Farruggia and Constantin, 1984; Charier, 1986); Berry-au-Bac "le Vieux Tordoir" (Allard et al., 1997); Berry-au-Bac "le Chemin de la Pêcherie" (Ilett and Plateaux, 1995); Menneville "Derrière-le-village" (Farruggia et al., 1996; Hachem et al., 1998b); Cuiry-lès-Chaudardes "Les Fontinettes" (Ilett and Hachem, 2001; Hachem, 2011a).

5   Bucy-le-Long "la Fosse Tounise" (Aisne) (Ilett et al., 1995); Bucy-le-Long "le Fond du Petit Marais/le Grand Marais" (Aisne) (Constantin et al., 1995); Tinqueux "la Haubette" (Marne) (Hachem, 2003; Hachem et al., 2007); Trosly-Breuil "les Obeaux" (Aisne) (Bréart, 1991). It should be noted that the site name Bucy-le-Long "la Héronière/la Fosse Tounise" is applied to both an LBK settlement and a subsequent BVSG settlement on the same site.

6   Changis-sur-Marne "les Pétreaux" (Lanchon et al., 2008); Vignely "la Porte aux Bergers" (Bostyn et al., 2018b); Jablines "la Pente des Croupetons" (Bostyn et al., 1991); Mareuil-les-Meaux "les Vignoles" (Cottiaux et al. unpublished); Fresne-sur-Marne "les Sablons" (Brunet et al. 1992).

7   The fact that it is impossible to determine whether the tali belong to cattle or young aurochs is due to the nature of this bone, which has only one centre of growth. Therefore, if the aurochs are young and have not finished growing, it is impossible to distinguish these bones from those of domestic cattle on the basis of measurements.

8   For the early phase of the RFBS (Aisne 1), 15,649 remains from twelve houses, i.e.: three houses at Berry-au-Bac "le Vieux Tordoir"; three at Berry-au-Bac "le Chemin de la Pêcherie"; and six at Cuiry-lès-Chaudardes "les Fontinettes". For the middle phase of the RFBS (Aisne 2), 24,778 remains from nineteen houses, i.e.: three houses at Menneville "Derrière le Village"; six at Bucy-le-Long "la Fosselle"; and ten at Cuiry-lès-Chaudardes. For the final phase of the RFBS (Aisne 3), 35,129 remains from twenty houses, i.e.: two houses at Berry-au-Bac "le Vieux Tordoir"; three at Menneville "Derrière le Village"; two at Bucy-le-Long "la Fosselle"; four at Missy-sur-Aisne "le Culot"; and nine at Cuiry-lès-Chaudardes "les Fontinettes".

9   In order to compensate in part for the consequences of a differential determination between domestic and wild young suinae and bovinae, which would artificially accentuate the deficit in the latter category, we have examined the suinae and bovinae metric data derived from bones where the epyphisis has fused. The conclusions show that the number of adult wild boar bones is much higher than those of domestic pig, regardless of the part of the skeleton examined and that the number of aurochs bones with fused epyphises is higher than that of domestic cattle. The hunting strategy, therefore, seems to target adult wild boar and aurochs while livestock rearing focused on the slaughter of young pigs and cattle.

10 The Early BVSG comprises: Bucy-le-Long "la Fosselle", house n°20 ; Tinqueux "la Haubette"; Oise: Pointpoint "le Fond du Rambourg"; Changis-sur-Marne "les Pétreaux"; Vignely "la Porte

aux Bergers" ; Villeneuve-la-Guyard "les falaises de Prépoux"; Saint-Pierre d'Autils "Carrière GSM". The Middle BVSG sites are: Trosly-Breuil "les Obeaux"; Tinqueux "la Haubette"; Longue-il-Sainte-Marie "la Butte de Rhuis" II and III; Vignely "la Porte aux Bergers"; Mareuil-les-Meaux "les Vignoles"; Jablines "la Pente de Croupetons"; Fresne-sur-Marne "les Sablons"; Marolles-sur-Seine "le Chemin de Sens"; Marolles-sur-Seine "les Prés Hauts"; Poses "Sur la Mare »; and Aubevoye "la Chartreuse". The Late BVSG sites are: Bucy-le-Long "la Fosse Tounise"; Bucy-le-Long "le Fond du Petit Marais/le Grand Marais"; Maurecourt "la Croix de Choisy"; Vignely "la Porte aux Bergers"; Marolles-sur-Seine "les Prés Hauts"; Passy "la Sablonnière". The Final BVSG site is: Vignely "la Porte aux Bergers".

11 Mareuil-lès-Maux "la Grange du Mont" (Cottiaux *et al.*, 2014a); Mareuil-lès-Meaux "les Lignères" (Brunet *et al.*, 2014; Brunet and Irribarria, 2018); Vignely "Noue Fenard" (Lanchon *et al.*, 2006); Cuiry-lès-Chaudardes "les Fontinettes" (Claude Constantin *et al.*, 2014); Pont-sur-Seine "Les Hauts de Launoy" (Hachem, 2015a; Peltier and Fournand, 2015a).

12 Glisy "Zac Jules Vernes" (Hachem, forthcoming; Joseph, 2008); Marquion "Sauchy-Lestrées" (Hachem unpublished); Meaux "la Route de Varreddes" *et al.*, 2004); Gord (Méniel, 1985).

13 Beaurieux "les Grèves", Barenton "Site N", "les Chantraines" and "la Foulerie" in Bazoches-es-sur-Vesle, Braine "la Grange des Moines", Berry-au-Bac "le Chemin de la Pêcherie", Bu-cy-le-Long "le Grand Marais", Ciry-Salsogne "la Bouche à Vesle", "la Prairie", "la Fosse aux Chevaux" and "le Gros Buisson" in Limé, Menneville "Derrière le Village", Sermoise "les Prés du Bout de la Ville", "les Etomelles" and "les Grèves" in Villeneuve-Saint-Germain

14 Romain "Cense Sauvage", "la Borne Saint Laid", "les Champs Virés", and "le Puisard" in Cernay-lès-Reims, Caurel "le Puisard", "les Hauts des Nervas", "Thillois" and "Croix Blandin" and "Zac Dauphinot" in Reims, Gueux "les Batailles".

15 Changis-sur-Marne "les Pétreaux", Ecuelles "Malassis et Charmoy", "Jardins de la Méridi-enne", "les Perpignans", "Zac de la Pyramide" and "lot 934" in Lieusaint, Ville-Saint-Jacques "le Bois d'Echalat", Egligny "le Chemin de la Pêcherie", "Beauchamps" and "la Justice" in Varennes-sur-Seine, Poincy "Près le Pont de Trilport", Saint-Mard "Zac de la Fontaine aux Bergers" and Larchant "les Groues".

16 "Object'Ifs-sud" and "AR67" in Ifs, Brette-ville-l'Orgueilleuse "le Bas des Prés et Résidence du Parc", Hérouvillette "les Pérelles" (studies by G. Auxiette), Saint-Martin-de-Fontenay "le Chemin de May", Fleury-les-Ornes "les Mézer-ettes", Eterville "les Prés du Vallon", Bourguébus "la Main Delle" and Condé-sur-Ifs "la Bruyère du Hamel" (after Baudry 2018).

17 *cf.* Bernard Sergent who refers to the sacrifice of piglets in Greece during Thesmophoria (an important autumn festival dedicated to Demeter Thesmophoros, goddess of cereals). Thrown into underground pits, they were abandoned and allowed to rot; three months later the bones were gathered and mixed with cereal seed to ensure a good harvest (Sergent, 1999, 25).

18 Berry-au-Bac "Le Chemin de la Pêcherie", Damary "le Ruisseau de Fayau", Betheny "les Equiernolles" and Reims "les Hauts des Nervas" and two oppida, Condé-sur-Suippe/Variscourt and Reims-*Durocortotum* "Rue d'Anjou".

19 To the east, Bazoches-sur-Vesle "les Chantraines " and "la Foulerie", Limé "les Sables-sud", Mont-No-tre-Dame "Vaudigny", and Braine "la Grange des Moines". To the west of Braine, many settlements were excavated at Ciry-Salsogne "le Bruy", Missy-sur-Aisne "le Culot" (no bone), Sermoise "les Prés du Bout de la Ville", and Villeneuve-Saint-Germain "les Etomelles". Oppida unequally explored are situated at Villeneuve-Saint-Germain "les Grandes Grèves" and Pommiers "l'Assaut" (*see* bibliography in Auxiette and Desenne dir. 2017) (Auxiette and Desenne, 2017)

20 The issue of the time span over which the animal remains accumulated is important: if we follow the hypothesis that the ditches were cleaned out on a regular basis over several decades (recutting the stratigraphy), emptying them of accumulated refuse, then we can reasonably estimate that the waste present at the time when the features are archaeological-ly excavated represents a relatively short, or indeed very short, time span.

3. COLLECTIVE MEALS

"Feasting is recognised as critical in reaffirming notions of power and identity, and it has a central role in sustaining social systems and inter-community relationships (Dietler and Hayden, 2001). Anthropological research has demonstrated common features in ceremonial food sharing, with prescribed behaviours relating to consumption, movement or action (Twiss, 2008). Many systematised feasting practices cannot be reconstructed archaeologically and, consequently, the identification of feasting is often based on criteria such as large quantities of remains, low-level of processing (such as no marrow extraction) and a dominance of meat-bearing elements." (Madgwick and Mulville, 2015, 629).

The Neolithic

Self-sufficiency and sharing between households in the Early Neolithic

Analysis of the various types of finds on LBK settlements in the Aisne Valley suggests that households could have been relatively self-sufficient in terms of food supply (cereals and meat) and pottery making (Hachem, 1995, 2011a; Hamon, 2006; Gomart, 2014).

The refuse assemblages from the lateral pits of the houses invariably contain animal bones mostly from domestic species, macrolithic equipment for cereal processing and a broadly equivalent proportion of coarse and fine ware pottery. The subsistence economy of each house is thus largely based on the exploitation of products originating from agriculture and stock-keeping (Hachem and Hamon, 2014). In our view, this apparent self-sufficiency does not rule out exchange between houses. In fact, some variability can be seen in the composition of the refuse assemblages: (i) as regards faunal remains, some house units have above-average values for wild animals and for certain domestic species; (ii) as regards macrolithic tools, the number of querns varies from one house to another and some house units have more evidence for craft activities involving abrasion and percussion; (iii) as regards pottery, differences can be observed between houses in the techniques and methods of vessel manufacture.

Certain households stand apart. Thus, for example, at the site of Cuiry-lès-Chaudardes, the lateral pits of two houses from the middle and late phases contained the remains of numerous animals less than 1 year of age. House 380 (where cattle remains are predominant) revealed a minimum of seven calves out of a total of sixteen bovines, three lambs out of ten caprines and two piglets out of five *suinae*; and House 225 (where caprine remains are predominant) has yielded a minimum of six lambs out of a total of twenty caprines, three calves out of ten bovines and eight piglets out of twelve *suinae* (Hachem, 2011a). This particularity might be explained by the very high number of remains (see appendices) but this explanation is insufficient because in other houses with a high number of remains, or where the proportion of livestock exceeds 90 %, on average only one or two young animals are recorded per domesticated species (on the basis of MNI). On the evidence of other categories of finds, such as ceramics, flints and grinding tools, these two houses have been interpreted as communal meeting houses because of the specificities of their assemblages (Gomart *et al.*, 2015). In such a context, the very high number of slaughtered young animals might suggest that the building was the venue for communal meals for the village, a hypothesis that does not exclude the possibility that the building was permanently occupied by one or more families.

Furthermore, these two long houses (and a third, House 360), yielded the largest numbers of aurochs remains (respectively 24 and 63) and the only examples of aurochs long bones (humerus, radius, tibia, femur) in the entire settlement.

In all the other houses, aurochs bones are present in small quantities (four bones on average) and, in addition, they appear to be selected and are comprised mainly of foot bones (metapodials and phalanges). The quantitative and spatial distribution of the aurochs remains follows a logic that differs from that of red deer, roe deer and wild boar: the aurochs bones are systematically present, as indeed are red deer bones, but in small proportions. The aurochs remains are concentrated in certain buildings, as are those of wild boar, but without any particular spatial patterning at site level, which is not the case with wild boar. The probability that a whole aurochs was consumed in each of the houses is low because of the small number of remains and the fact that the bones most often recorded do not yield much meat. We can therefore hypothesise that the three long houses that have yielded the most aurochs remains (House 225 even yielded an aurochs bucranium) were meeting places where the members of the village community came together and feasted on the meat of an aurochs after the hunt. Certain parts of the carcass, such as the feet (metapodials and phalanges) were then carried back to each house as a physical reminder of what had been shared together.

In this scenario the aurochs plays both a nutritional and a symbolic role (fig. 76).

Middle Neolithic enclosures as places of assembly

We consider the enclosures as collective places because archaeologists have long emphasized the monumentality of the features, which refers to the joint efforts of one or more communities to build these monumental works. Meals were eaten within the enclosures, as evidenced by the faunal remains. It is not so much the quantity of remains that makes us think that they are collectively shared meals, as the nature of the refuse. As regards faunal remains, two types of deposition are recognized in the enclosure ditches. Both types of deposition relate to very little diversified species, mainly domestic, slaughtered at specific ages. The first takes the form of consumption waste and is made up of fragmented bones often occurring in concentrations near the ditch interruptions that are considered to be the passage-ways in and out of the enclosure. The second type takes the form of deposits of articulated or semi-articulated bone, bucrania and antlers.

Figure 76: The aurochs played both a symbolic and nutritional role in the Neolithic. Tentative reconstruction of this extinct species, site Nature Arlaines, SARL Desmarest, Pontarcher, Vic-sur-Aisne (photo: B. Robert).

The Middle Neolithic I

For the Late to Final Rössen period, only two sites can be considered: the enclosures at Osly-Courtil "la Terre Saint-Mard" (Dubouloz, 2003b; Hachem, 2011b) and Berry-au-Bac "la Croix-Maigret" (Méniel, 1984; Dubouloz *et al.*, 1991) (Dubouloz *et al.*, 1991; Méniel, 1984) both located in the Aisne valley.

Animal husbandry predominates at the two settlements, but cattle consumption is comparatively low (41.8% in Berry-au-Bac and 56% in Osly-Courtil). The proportion of pigs is significant (34.3% and 29.5%, respectively), contrary to that of caprines (4.6% and 9.8%, respectively). However, game is much more common in Berry-au-Bac (18%) than in Osly-Courtil (2%), with a predominance of red deer followed by aurochs.

Six enclosures have been recorded for the Cerny period:

- Balloy and Châtenay in Seine-et-Marne (Tresset, 1997) ; Barbuise-Courtavant (Tresset, 1997) and La Saulsotte in Aube (Hachem, 2015b); Maisons-Alfort in Val-de-Marne (Hachem, 2000b); Gurgy in Burgundy (Bedault, 2007; Meunier *et al.*, 2012) (see appendices).

These Paris basin enclosures reveal a very high proportion of domestic animal remains (more than 90 % of total remains), with a predominance of cattle (over 70%). Pigs appear as the second source of meat (between 15 and 19%), while the proportion of caprines is almost insignificant (between 2 and 6%). The proportion of game is low, red deer being the principal species hunted. Nonetheless, the proportion and composition of the game may vary, as in the case of the enclosure in La Saulsotte "le Vieux Bouchy" where a significant proportion of wild animals was found.

Apart from the enclosures, there are singular pits in which remains of connected animal skeletons indicate unusual consumption.

A number of isolated, deep pits dating to the Cerny and containing particular deposits, have been recorded on several sites : Escalles "le Mont d'Hubert" in Pas-de-Calais (Hachem and Chombart, 2014), Vitry-sur-Seine "Rue du Génie " in Val-de-Marne (Hachem, 2015b). Their dating is dependent on 14C, and there is no way of verifying these dates because the features are devoid of any other associated finds.

These isolated pits contain peculiar faunal assemblages in the form of deposits of articulated animal remains with certain parts missing.

Two very deep pits (st 546 and st 495) were discovered at the site of Escalles, which is located on a headland overlooking the sea. The pits were found within an enclosure dating to the Middle Neolithic II. The first contained two distinct deposits: at the base of the feature was the skeleton of a young male roe deer, aged five months, which was missing its hind quarters; the second deposit, contained in the upper fill of the pit, consisted of

Figure 77: Escalles "le Mont d'Hubert" (Pas-de-Calais), pit 495 dated to the Middle Neolithic 1 (Cerny); deposition of the fully articulated remains of a young calf (after Praud *et al.* 2014, fig. 117; photo: Inrap-CG62).

a two large red deer antler palms (fig. 77). At the bottom of the second pit were the disarticulated but almost complete remains (a few bones were missing) of a young calf, probably male and about five months old. The young deer and the calf had probably been slaughtered in the autumn.

The pit at Vitry-sur-Seine, discovered during archaeological testing, contained selected bones. Ten red deer ribs, seven of which were complete, were placed at the bottom of the pit accompanied by the upper part of a red deer antler, three caprine bones and two worked bones, a point and a scraper made from a *suinae* tooth (fig. 78).

The Middle Neolithic II

The western border of the Michelsberg Culture lies in the Oise where sites correspond to northern Chasséen settlements. The four Michelsberg enclosures discussed here are:

- Bazoches-sur-Vesle "le Bois de Muisemont ", Aisne (Hachem, 1987, 2011b; Dubouloz *et al.*, 1997),
- Crécy-sur-Serre "la Croix Saint-Jacques", Aisne (Naze, 2014; Hachem, 2015c),
- Maizy "les-Grands Aisements", Aisne (Le Bolloch *et al.*, 1986; Hachem, 1989),
- Vignely "la Noue Fenard", Val-de-Marne (Lanchon *et al.*, 2001; Hachem, 2011b).

The Bazoches-sur-Vesles enclosure was excavated in its entirety (fig. 8) as well as the Crécy-sur-Serre enclosure except the inner area. While the first occupies a larger area, its ditches yielded less bone waste than the second partly due to the more calcareous sediment that has had a significant impact on the faunal remains.

Figure 78: Vitry-sur-Seine "108 Rue du Génie" (Ile-de-France), pit dated to the Middle Neolithic 1 (Cerny); deposit of red deer antler in the upper part of the feature (after Durand 2017, fig. 17).

The composition profile of domestic fauna is rather similar on all four sites (fig. 79), with a low proportion of cattle accounting for less than half of the remains found (on average 43% of the NISP). The proportion of pigs is high (23% on average), while caprines are not well represented (10% on average). The proportion of game is variable with major discrepancies between sites. The frequency of game is lowest at Bazoches-sur-Vesle (4%) and highest at Crécy-sur-Serre (20.8%). The values obtained in Maizy and Vignely (15% and 12.4%, respectively) lie between these two extremes.

Comparing the various carcass parts uncovered in Crécy-sur-Serre, for example, we observe that, in general, these are varied for the principal meat producing animals, *i.e.* the three domestic species. We find all elements of the skeleton, from the skull to the feet, which means that these animals were probably slaughtered close to the site and that all of the carcass was used. The presence of perinatal individuals, which indicates that livestock were reared in the immediate vicinity, supports this theory.

Most of the enclosures of the "Noyen Group" are found in the Bassée part of the Seine valley, and the Pont-sur-Seine enclosure is located in the same valley, about 25 km upstream of the Bassée.

The faunal remains from the four enclosures in the Bassée-Gravon, Châtenay, Grisy and Noyen "Fd" and "F" (Mordant, 1992)- are quite similar to the remains just described from the Michelsberg enclosures, with the usual pairing of cattle and pigs (Tresset, 1996). Wild animals are not numerous (10 %).

The enclosure at Pont-sur-Seine "Ferme de l'Ile", in the Aube (Dugois and Loiseau, 2019) has yielded a large quantity of animal bones; in total 1693 remains have been studied, 47% of which have been determined (Auxiette and Hachem, 2019). The proportion of game is very high in comparison to the enclosures of the Michelsberg, Chasséen and Spiere Groups; in fact, the numbers are the highest of all the sites studied (30.9 % of NISP). Even though they are the primary source of meat, the proportion of cattle is low (37 %). The second most important source of meat among the domestic animals is pig (16.2 %), with caprines coming a distant third (4.9%).

Large game is well represented by red deer (21 %), followed in equal measure by wild boar and aurochs (6 % and 6.9 %). Roe deer (1 %) and horse (0.5 %) complete the wild species. Small game such as beaver and badger is present.

Most of the northern Chasséen sites discussed here are located in the Oise and the Somme departments. Broadly speaking, the proportion of cattle is about 66% of the NISP, while pig is about 15% and caprines around 11%.

Post-Rössen enclosures

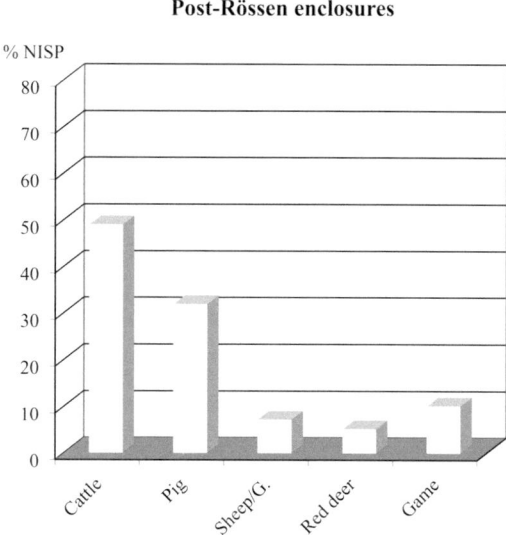

Cerny enclosures

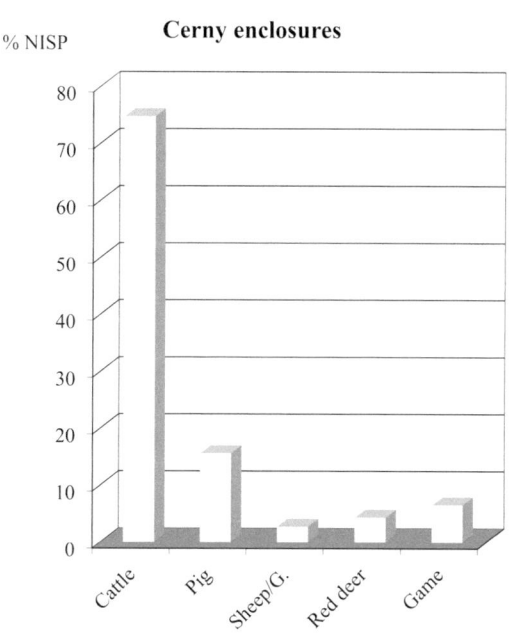

Michelsberg enclosures

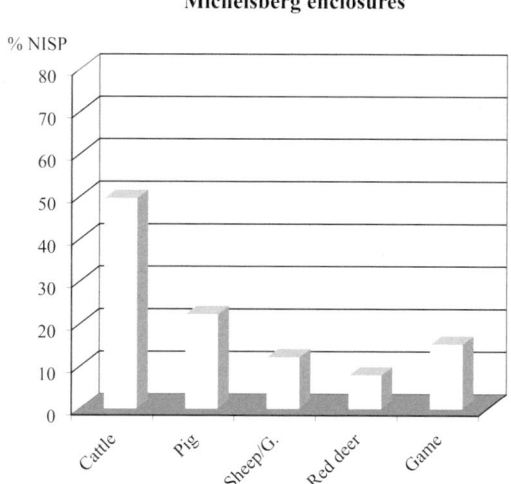

Northern Chasséen enclosures
(without Passel and Villers-Carbonnel)

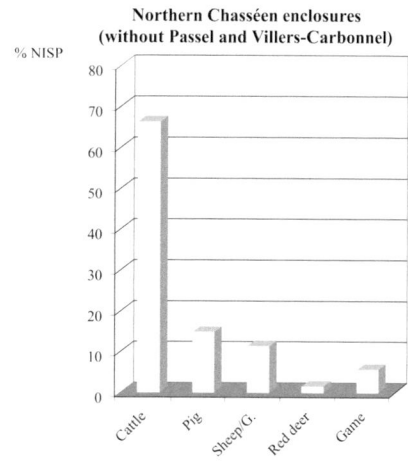

Noyen Group enclosures

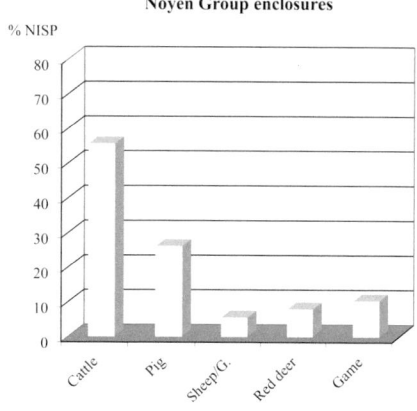

Balloy Group enclosures

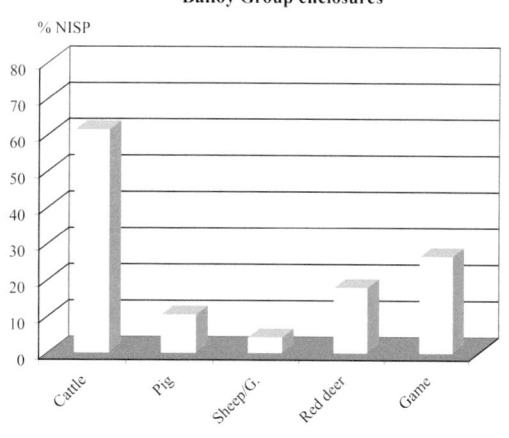

Pont-sur-Seine enclosure (late Middle Neolithic)

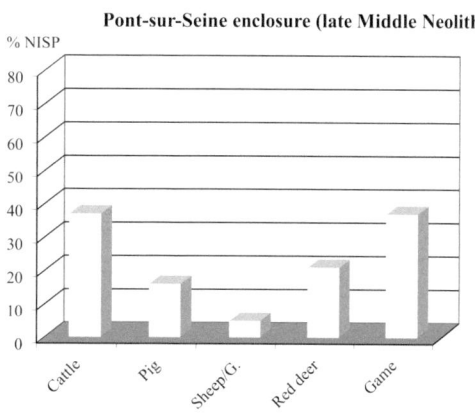

Figure 79: Proportions of the principal domestic and wild species in the Middle Neolithic 1 and 2 of the Paris Basin. (NISP = 36046).

We will look at the case of the Boury-en-Vexin deposit a little later, in the chapter devoted to caprine deposits (chapter 4).

Compared with the Michelsberg and Noyen Group sites, the Chasséen sites, as at Boury-en-Vexin (dump) (Méniel, 1984) for example, are characterized by higher proportions of cattle and caprines. In all cases pigs outnumber caprines. Game, however, is found in comparatively low proportions (around 9%), except at the Maisons-Alfort enclosure site where it reaches 16% (Cottiaux *et al.*, 2008; Hachem, 2011b).

Two large-scale sites have significantly contributed to our knowledge of the northern Chasséen Culture, namely the enclosures at Passel (Oise) and Villers-Carbonnel (Somme) (Hachem *et al.*, 2016). These sites complement the available archaeozoological record for the Middle Neolithic II in northern France; until the discovery of these two sites, the corpus numbered 49,142 identified faunal remainsrecovered from 25 sites (see appendices).

Large quantities of faunal remains were found on these two sites: 8571 identified remains at Passel (Hachem *et al.*, 2017) and 2747 at Villers-Carbonnel (Hachem and Bedault, 2014; Hachem *et al.*, 2016). They both display quite similar faunal spectra with a proportion of domestic animals in excess of 95% (fig. 80). If we only consider the domesticated trio, we find that cattle are better represented at Passel (71.5%) than at Villers-Carbonnel (63%), while the opposite is true for caprines (2.7% against 14.8%).

However, the proportions of pigs are close for both sites (22% and 26 %, respectively).

The significance of pig as the second source of meat is confirmed, but in proportions that are higher than those previously recorded for the northern Chasséen. Indeed, these values are closer to those encountered in the Michelsberg and Noyen Group enclosures.

As regards wild species, their proportion is very low at both sites, which appears to be a characteristic of Chasséen enclosures, whereas game proportions are slightly higher for other site types of the same period. Nevertheless, the sites always feature the four species usually consumed, *i.e.* red deer, wild boar, aurochs and roe deer. In Passel, wild boar is the primary game resource (48% of wild animal remains found). The individuals are rather large and a significant proportion are males. On the other sites, red deer usually comes first, as in the case in Villers-Carbonnel (38.9% of wild animal remains) where wild boar comes second (29.6%). Aurochs is well represented with a higher proportion in Passel than in Villers-Carbonnel (30.9% vs. 9.3%, respectively). Smaller game species (wild cats and birds) only account for a very small number of remains, as is usual on such sites.

At Villers-Carbonnel, as at Passel, the mortality profiles of the three domestic species reveal the preferential slaughtering of juvenile/sub-adult animals and young adults (fig. 81). This trend is especially visible for cattle, slaughtered between the age of 2 and 3 years, and had previously been observed at other Chasséen sites such as Catenoy, Boury-en-Vexin (dump area), Louviers and Bercy (Méniel, 1984; Tresset, 2005). This selection of young animals is particularly evident in the case of cattle at Passel. A second slaughtering peak identified for older animals at Villers-Carbonnel has also been identified at Catenoy (Méniel, 1984). In addition, an ongoing osteometric analysis is hinting at the presence of castrated individuals at Passel and

Figure 80: Proportions of domestic species in two major Northern Chasséen sites. a- Passel "le Vivier" (Oise). b-Villers-Carbonnel "la Sole d'Happlincourt" (Somme).

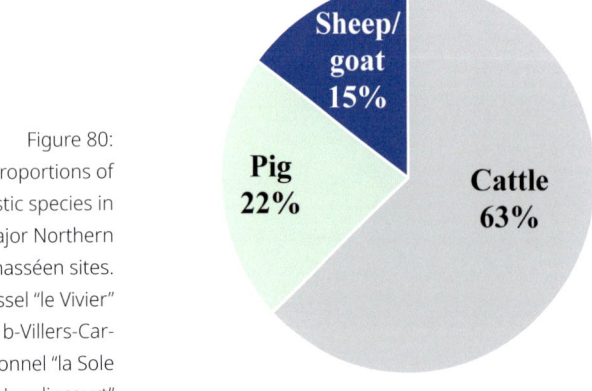

Villers-Carbonnel

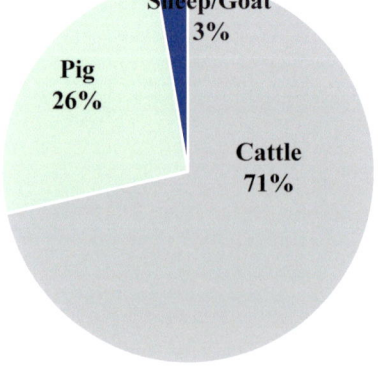

Passel

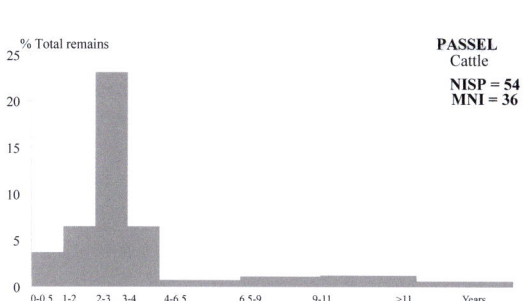

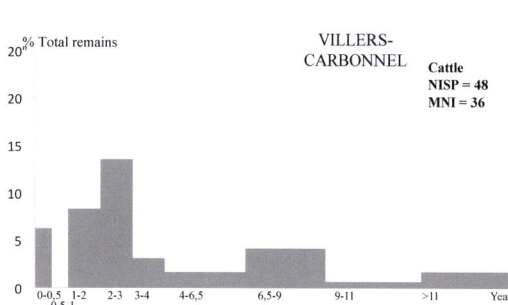

Figure 81: Slaughter pattern for cattle, Northern Chasséen enclosures, Middle Neolithic 2 (after Hachem *et al.* 2016, fig. 8, modified). a- Passel "le Vivier" (Oise). b-Villers-Carbonnel "la Sole d'Happlincourt" (Somme).

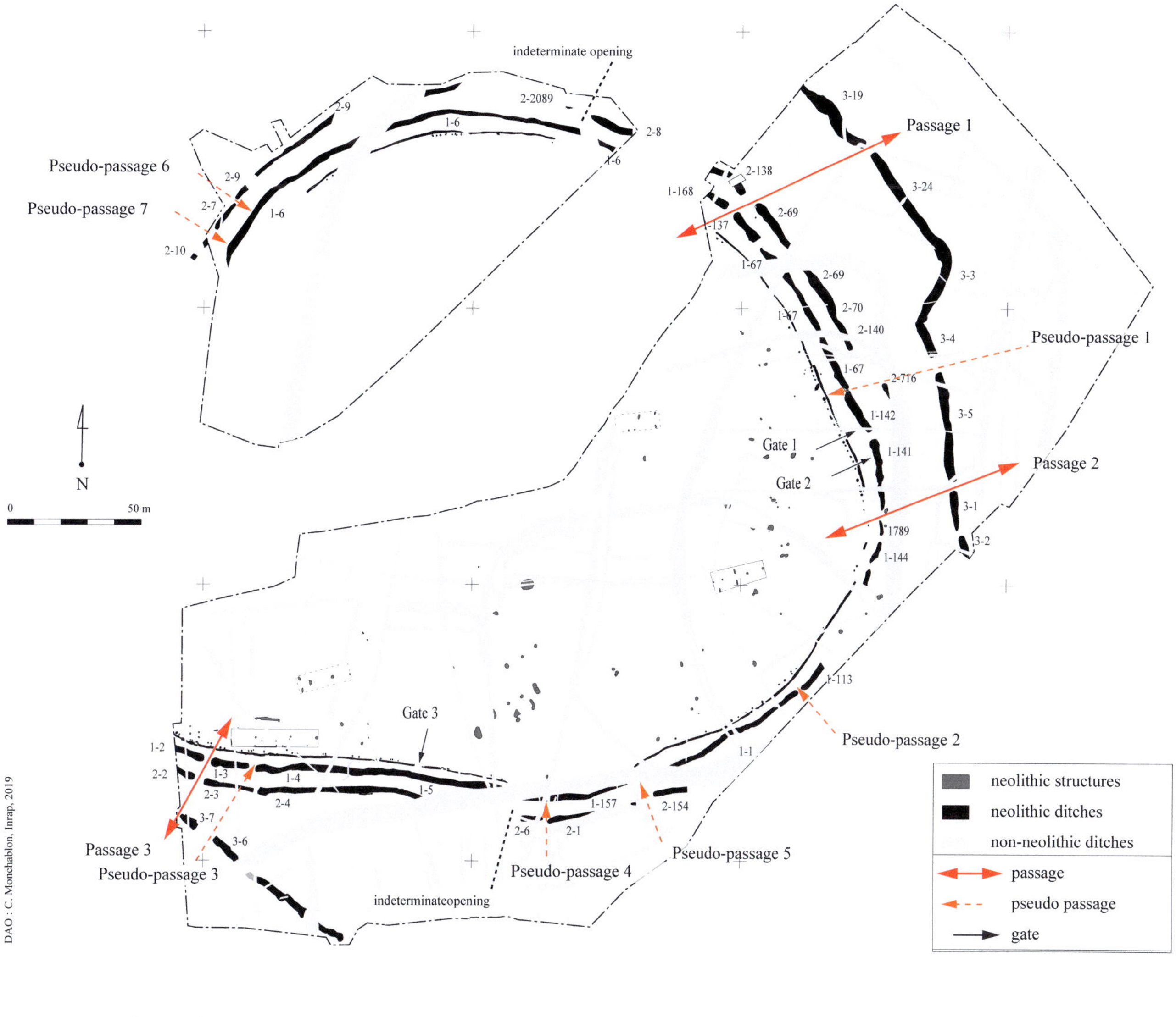

Figure 82: Carvin "la Gare d'Eau" (Pas-de-Calais), plan of the Spiere Group enclosure, Middle Neolithic 2 (after Monchablon *et al.* 2011, fig. 3 modified and Monchablon *et al.* 2014; CAD: C. Monchablon).

Villers-Carbonnel, which suggests that cattle were being used as draught animals. This data clearly shows that livestock may have been put to a variety of uses during the Chasséen period.

At both Passel and Villers-Carbonnel, most of the pigs were slaughtered at a young age (between 6 months and 2 years), even though some individuals were allowed to live longer. A similar trend has been observed for the Chasséen enclosures at Catenoy and Boury-en-Vexin. In the case of caprines, two slaughtering peaks are recorded: a first peak corresponding to an age of between 6 months and 2 years, and a second at 4 years and over. However, due to the partial nature of the available data, no conclusion can be drawn for the cultural entity as a whole.

Overall, this data very clearly emphasizes that breeding strategies were aimed at producing meat, without excluding the possibility that animals were also being used as draught animals or to provide milk or fleece. We will see in chapter 6 the interpretations we can draw from these results.

Domestic animals represent between 70 and 75% of faunal remains at the two Balloy Group enclosures (designated as "LM" F and "LM" FA) in Châtenay (Tresset, 1996). The amounts of cattle (60%) and pig (between 8 % and 12 %) remains are higher than caprine (2% to 6%).The proportion of game animals is high (27%), and red deer is the main animal hunted.

Recorded sites belonging to the Spiere Group are rare and to date only the faunal remains from the sites of Carvin and Escalles have been studied. These sites are located in the department of Pas-de-Calais.

At the first site, the enclosure at Carvin "la Gare d'Eau" (fig. 82), only small amounts of faunal remains have been found (Monchablon *et al.*, 2011; Monchablon, 2014) even though this is an imposing enclosure and all segments have been excavated (NISP = 298). Poor bone preservation is certainly a significant factor, but it is probably not the only explanation; indeed, a similar dearth of bone refuse has already been documented for the Bazoches enclosure. Domestic species predominate (94 % of NISP as is the case in other Middle Neolithic II enclosures, but, contrary to what has been observed in all other instances, the proportion of caprines (20.1 %) exceeds that of the pig (12.4%) (Hachem, 2014). However, caprine bones are more fragile and therefore should not be so well represented if poor preservation was the only factor behind the rarity of faunal remains

Wild fauna is particularly rare, with red deer being the most frequent game animal. For the first time, to our knowledge, there is a clear difference between the treatment applied to cattle discarded bone on the one hand, and to caprine and pig bone waste on the other; most of the bones from the latter two species have been burnt. Small heaps of charred bones are found throughout the first two ditches of the enclosure. The third ditch contains two types of faunal deposit that have also been observed in Michelsberg enclosures (see chapter 4); they feature four different species – dog, caprines and, to a lesser extent, cattle and aurochs.

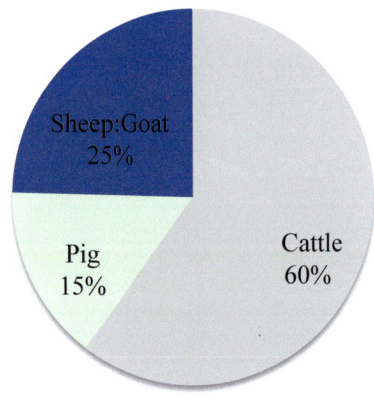

Carvin

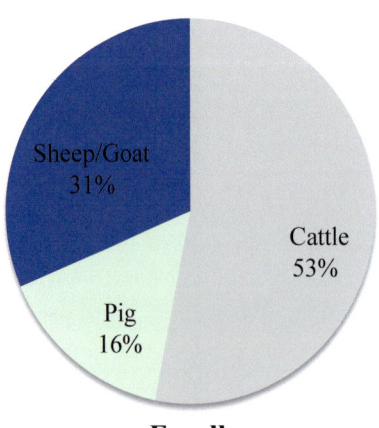

Figure 83: Proportions of domestic species at two Spiere Group enclosures. a- Carvin "la Gare d'Eau" (Pas-de-Calais). b- Escalles "Mont d'Hubert" (Pas-de-Calais), Middle Neolithic 2.

Escalles

The second site, Escalles "le Mont d'Hubert", Pas-de-Calais (Praud, 2015; Praud and Panloups, 2015), currently stands a little apart in cultural terms since the pottery exhibits both Spiere Group and Chasséen characteristics (Colas *et al.*, 2016). This is an unusual site because of numerous human bones, often bearing cut marks, that are found mixed with the animal remains in the enclosure ditch.

Contrary to Carvin, the number of bones found is very significant despite the fact that only four segments of the ditch have been excavated (NISP = 9472). The faunal composition of the Escalles enclosure is similar to that of Carvin (fig. 83), with a comparatively low proportion of cattle (NISP=52.5 %), a high proportion of caprines (29 %) and small amounts of pig (15.4 %) (Hachem and Chombart, 2014). Also, caprine remains are treated here in a way that sets them apart from other species, as they are laid out in ritualized, partial articulation. Dog, which is strongly associated with the symbolic world of the Middle Neolithic, is remarkably well represented at both sites. However, what really separates Escalles from all of the other sites is the near-exclusivity of domestic fauna in the faunal remains. The proportion of game is extremely low (0.7 %). Only two other sites are totally lacking evidence for game and they are deemed exceptional on account of their faunal composition: Boury-en-Vexin, deposit, (Méniel, 1984) for the Chasséen and Mairy refuse pits (Arbogast, 1989) for the Michelsberg. It is highly likely that Escalles belongs to this category of extraordinary sites, all the more so because of the presence of human bone with cut marks (Praud *et al.*, 2015a).

The slaughter patterns for all three domestic species show an identical trend and reveal that the primary intent was to obtain tender meat. Specific age groups were selected: between 2 and 12 months for caprines, between 5 and 14 months for pigs and between 5 and 12 months for cattle.

The surprisingly large number of individuals suggests that they came from one large herd, or several of them. The very narrow age brackets possibly indicate that mass slaughtering took place over a short period of time and during a particular season.

The hypothesis of cattle husbandry for the purpose of milk production – as is presumed to have been the case during the Late Neolithic in northern France (Tresset, 1996) may be envisaged but cannot be clearly established. Milk production implies specific livestock management practices such as the maintenance of lactating cows within the herd and the mass slaughtering of 5-9 month old calves (post-weaning) (Balasse and Tresset, 2002). This slaughtering peak is observed in Escalles, but it cannot be demonstrated in Carvin on account of the scarcity of available data.

The maintenance of older animals within the herd could also be an indication of the exploitation of livestock for draught purposes. The presence of castrated animals corroborates this hypothesis. There are hints of animal castration in Escalles that need to be confirmed by more thorough investigations, but measurements clearly show that there is a group of individuals that are distinct from the males and the females. The practice of castration is attested at the Michelsberg site of Mairy and the contemporary site of Hetzemberg in Germany (Stephan, 2008). There may be several purposes behind this castration: to increase the animal's mass and thereby to obtain more meat, or to use the animal for draught purposes, for ploughing for instance.

The Late Neolithic

At the site of Pont-sur-Seine "le Haut de Launoy" (fig. 84), excavation has revealed a Late Neolithic settlement made up of two enclosures, one inside the other, two monumental structures and rectangular buildings (Desbrosse and Peltier, 2010).

A first enclosure takes the form of a narrow palisade ditch which defines an area of 1.4 ha and which yielded an accumulation of faunal remains. A second surrounds two buildings that are striking in terms of their size and trapezoidal ground plans. Both have an entrance at the east end, flanked by large antenna-like trenches that form a funnel-like feature leading to the entrance. Subsoil features within the largest structure yielded 19 tonnes of stone. Two deposits containing lithics, animal bone and ceramics were excavated at the extremity of the southern antenna of the larger of the two buildings.

The faunal remains from the site have yet to be studied in detail but preliminary observations indicate the presence of cattle, caprines and pigs, with occasional wild boar and red deer remains in some features.

The Final Neolithic

Two sites provide evidence for communal consumption. These are:

- Houplin-Ancoisne "le Marais de Santes", Nord-Pas-de-Calais (Martial and Praud, 2007; Praud *et al.*, 2015b),
- Saint-André-sur Orne "la Delle du Poirier", Calvados (Hachem, 2017a; Ghesquière *et al.*, 2019b).

The site of Houplin-Ancoisne "le Marais de Santes" is one of a group of sites dotted along the right bank of the Deûle. It is situated at the foot of a slope on the edge of a marshy area and has been dated to the first half of the 3rd millennium BC, *i.e.* the Final Neolithic. The site includes a curvilinear palisade that encloses a monumental post-built building measuring 43.5 m in length and 12.8 m in width (fig. 11). The excavations also revealed several other timber structures and a refuse discard zone preserved in a paleo-channel of the Deûle which filled up over the course of the Holocene. The excavated area is 1 ha in extent but

Figure 84: Pont-sur-Seine "le Haut de Launoy" (Aube), aerial view of the Late Neolithic site: conjoined enclosures, monumental structures and rectangular buildings (Desbrosse and Peltier 2010, fig. 3; photo: f.canon@ vertical-photo.com).

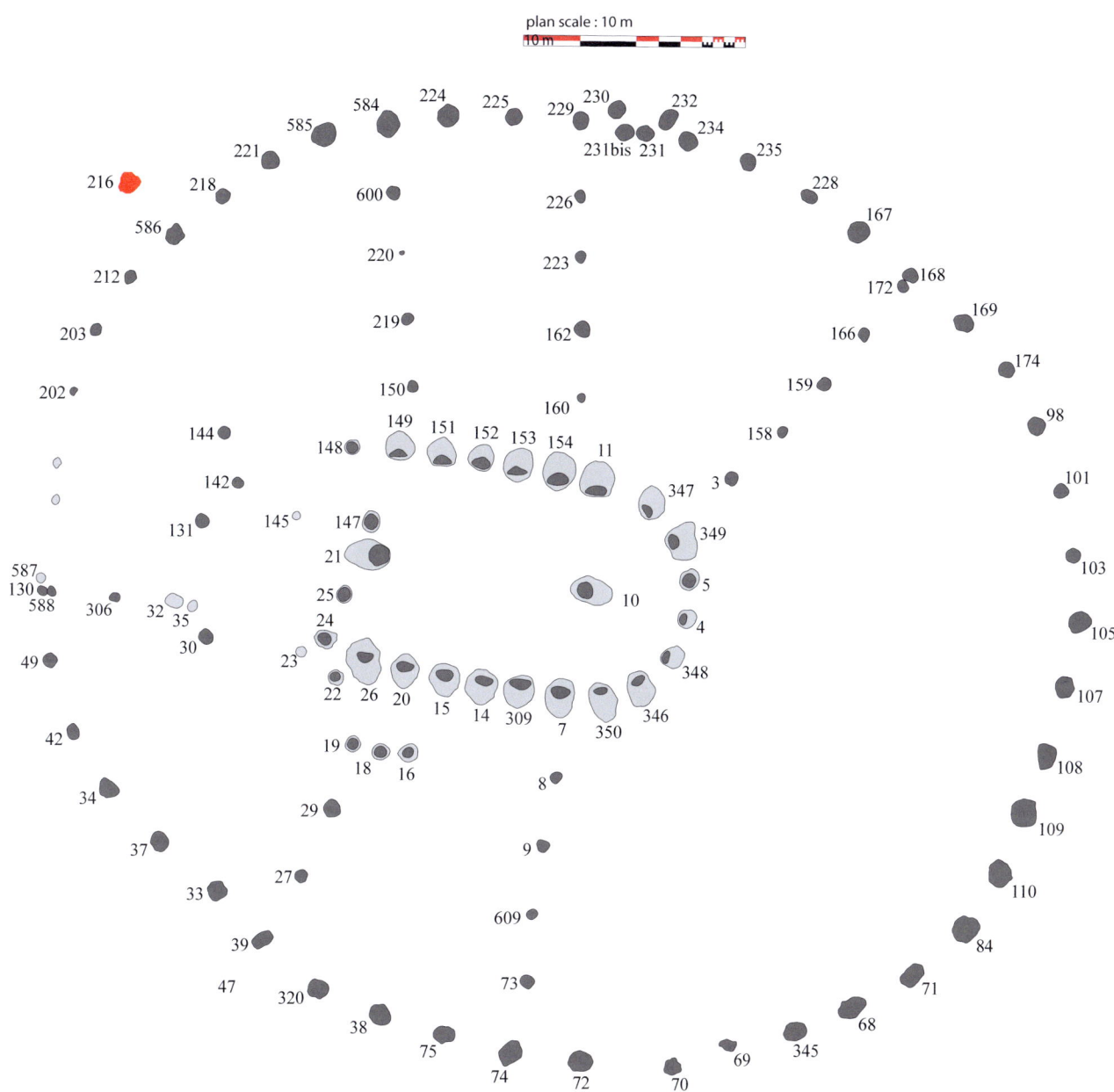

the total area of the enclosure interior is believed to be about 2.5 ha. The good preservation conditions on site meant that a substantial amount of evidence was recovered on the environment, economy and material culture of the Deûle-Escaut group.

The faunal material was retrieved from several locations, including a meander that yielded 2,952 determined and well-preserved remains (Braguier in Praud *et al.*, 2015b).

Domestic animals predominate (98 %) with cattle and pig making up the bulk of the assemblage (45.3 % and 36 % of NISP, respectively), followed by caprines (16 %). Some of the cattle bones, such as metapodials and phalanges, display traces of stress which suggests that some animals were used for draft purposes.

Dog is well represented with 27 remains (two individuals). The wild fauna consists of wild boar, which is the most numerous species, red deer, aurochs and roe deer. Other recorded species include wolf (3 individuals), fox, horse, badger and a few birds.

Pig has a predominant role in the livestock, especially if we consider the MNI which reveals the presence of a large number of individuals. This confirms a study carried out at Houplin-Ancoisne "la Rue Marx Dormoy", a contemporary site located about 700m from "Marais de Santes" (Martial and Praud, 2007).

The importance of pig finds a parallel in contemporary sites, attributed to the Artenac Culture, in the

Figure 85: Saint André-sur-Orne "la Delle du Poirier" (Calvados), Final Neolithic: plan of Building 1 and its associated enclosure (Ghesquière *et al.* 2019, fig. 1 modifed; CAD: E. Ghesquière).

Centre-West of France, where pigs also predominate over cattle and sheep (Braguier, 2000). The same pattern is observed in sites such as Chalain 3 in the Jura (Pétrequin, 1997). However, on the latter site, hunting (principally of red deer) rivals livestock as a source of meat.

It was not possible to define the status of each building at Houplin-Ancoisne "Marais de Santes" but the spatial distribution of activities reveals three broad zones (Praud *et al.*, 2015b): the large building is characterised by evidence for the transformation of plants and the production of sandstone tools; the area between the building and the palisade shows little evidence for activities; on the exterior, three pits contain tools associated with grinding and the preparation of hide. In addition, certain activities associated with textile production are evidenced in various sectors of the site.

Houplin-Ancoisne represents the northernmost extent of the distribution of these Final Neolithic monumental buildings and is also one of the earliest manifestations of the phenomenon.

The domestic nature of the site appears to be confirmed by the range of activities that took place there. In this respect, it is comparable to other sites with monumental architecture dating to the 3rd millennium BC. Certain researchers interpret these long buildings, which were constructed by the various Final Neolithic groups, as communal houses because of the significant social effort and resources invested by the community in the building of these gigantic structures, in terms of the quantity of wood consumed and the labour force required. At "Marais de Santes", for example, the building required substantial weight-bearing timber elements such as a central post that alone is estimated to have weighed 8 tonnes.

The faunal remains from the site of Saint-André-sur-Orne were retrieved from the post holes of a building (n°1), from the enclosing palisade and from rows of postholes that radiate from the building to the palisade (fig. 85).

The determined remains number (87); the indeterminate remains (197), mainly consist of burnt fragments; indeed, more than half of the bones recovered bear traces of fire. The non-burnt bone takes the form of small fragments, to the extent that no complete epiphyses survive, which is generally rare for faunal remains found on settlement sites.

Figure 86: Saint André-sur-Orne "la Delle du Poirier" (Calvados), Final Neolithic, photos and drawings of both faces of the sandstone stele found in Posthole 147. The associated *bovinae* scapula is represented at the same scale (after Ghesquière *et al.* 2019, fig. 8; photos and CAD: S. Giazzon, E. Ghesquière, Inrap).

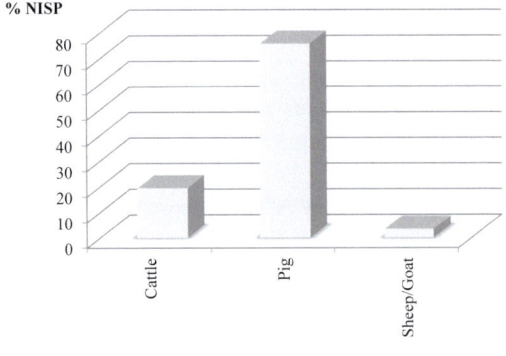

Figure 87: Saint André-sur-Orne "la Delle du Poirier" (Calvados), Final Neolithic, Building 1, proportion of domestic species.

Only a single complete bone is recorded and it comes from a posthole (st. n° 147) at the entrance to the building; it is a cow scapula that was placed upright against a trapezoidal sandstone stele. This arrangement suggests that the bone had a certain symbolic significance (fig. 86).

The bones result exclusively from the consumption of domestic animals (fig. 87). Pig is by far the predominant species (76 % of NISP), followed by cattle (19.5 %); caprines are rare (3.4 %). No wild animals have been recorded in the bone assemblage although both worked and unworked red deer antlers were discovered.

The Bronze Age and the Iron Age

Exceptional consumption: sheep bones in silos

Archaeozoological studies improve our understanding of how consumption evolved over millennia but also reveal how it diversified as societies became more complex.

Analysis of a large number of faunal corpora, of various sizes, has led us to reconsider certain assemblages as more than simple domestic refuse. In most cases bones are retrieved from excavated features that acted as the ultimate destination for the mixed and very fragmentary waste arising from domestic activities, including the preparation and consumption of meals. The presence of bone in these assemblages is, therefore, rarely the result of a deliberate act. However, certain assemblages retrieved from very identifiable features, namely grain storage pits, appear to combine a number of unusual distinctive characteristics.

The specificities of these assemblages include unusual quantities of bone combined with a high degree of preservation and the selection of a particular species, in these case sheep. Taken together, these characteristics clearly distinguish such assemblages from the disorganised and composite refuse found in most other contexts (pits and ditches).

The function of the features, principally silos, which produce this bone material therefore appears to determine the nature of the refuse contained within them. Usually pits and ditches yield relatively small amounts of bone remains compared to the available volume, which when taken together with the *ad hoc* species associations and the very fragmentary nature of the carcass parts, clearly indicates the casual nature of the disposal. However, in contrast, the composition of the assemblages from the silos appears to reflect deliberate acts.

Moreover, these silos, which are located within the perimeters of the farms on which they depend, tend not to be located in the immediate vicinity of houses, but rather in areas that are almost always devoid of domestic features, a fact that renders their role as receptacles all the more significant.

It required several years of study, covering a wide chronological span, to draw our attention to the assemblages from these features, which are so dispersed both in time and space. Ultimately it was the recurrence of certain characteristics that allowed assemblages from these contexts to be isolated from more "classic" faunal assemblages.

This study focusses on 40 or so sites, distributed over a wide geographical area encompassing the departments of Calvados, Aisne, Marne, Seine-et-Marne, Val-d'Oise, Pas-de-Calais and Somme. The chronological span covers the first millennium BC, from the Late Bronze Age IIIb (950/800 BC) to La Tène D (1st century BC). While the practice begins at the start of the 1st millennium BC, we observe an increasing number of these assemblages between the 6th and 3rd centuries BC (Auxiette, 2017c).

The seven principal characteristics of the assemblages that caught our attention are the following: the state of preservation, the degree of fragmentation, the number of remains, the carcass parts represented, the nature of the butchery marks, the ages of the animals, and lastly, the concentrations of bone.

The bone surfaces are smooth, lack rodent/carnivore teeth marks and are very pale beige in colour (fig. 88). The state of preservation of the bones is very similar to that of remains found in funerary contexts (faunal deposits in La Tène graves (see chapter 5). These preservation characteristics are thus evidence that the remains were rapidly buried and sealed.

Compared to the more random weights of bones in "classic" assemblages, which are often affected by various taphonomic processes (alteration and dissolving of bone surfaces, chewing by animals, etc.), the average weight of the "silo" bones is another criterion for distinguishing the differential preservation of these assemblages. The average weight of a sheep bone from a "classic" domestic context is 4 g but is 7 g in the case of a "silo" context (Auxiette, 2017c; Viand *et al.*, 2008). The assemblages range in size from a few tens to several hundreds of bones and are invariably from sheep (99.9%). These assemblages are characterised by a high incidence of ribcage remains (vertebrae and ribs), scapular and pelvic girdles, and long bones. In most cases these bones are preserved in their entirety but the carcasses are never complete: it is difficult to interpret the presence or absence of certain anatomical parts. In most cases we are confident that almost all of the bone material within these features ends up in the laboratory thanks to careful manual excavation. The cuts of meat are very varied: head, shoulder, saddle, haunch, rack, loin chops, and filet. Thus all kinds of cut of meat were apparently chosen. Butchery marks are recurrent on the skull, scapula, humerus, ulna, coxae and also on the vertebrae and ribs (fig. 89). The bone may be sectioned (*e.g.* vertebrae), abraded (*e.g.* on the ribs), or defleshed (defleshing of long bones and girdles). These marks are very difficult to spot and their recording requires particular vigilance. The numerous marks indicate that the animals slaughtered on these occasions were dismembered and cut up. The cuts of meat may take the form of entire joints or half-joints (for example, a humerus can be whole or cut in two).

Figure 88: Milly-la-Forêt "le Bois Rond" (Essonne), sheep bones, pit 2132 (photo: G. Auxiette).

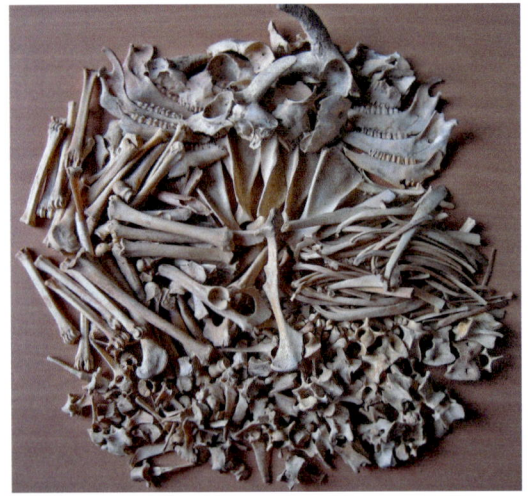

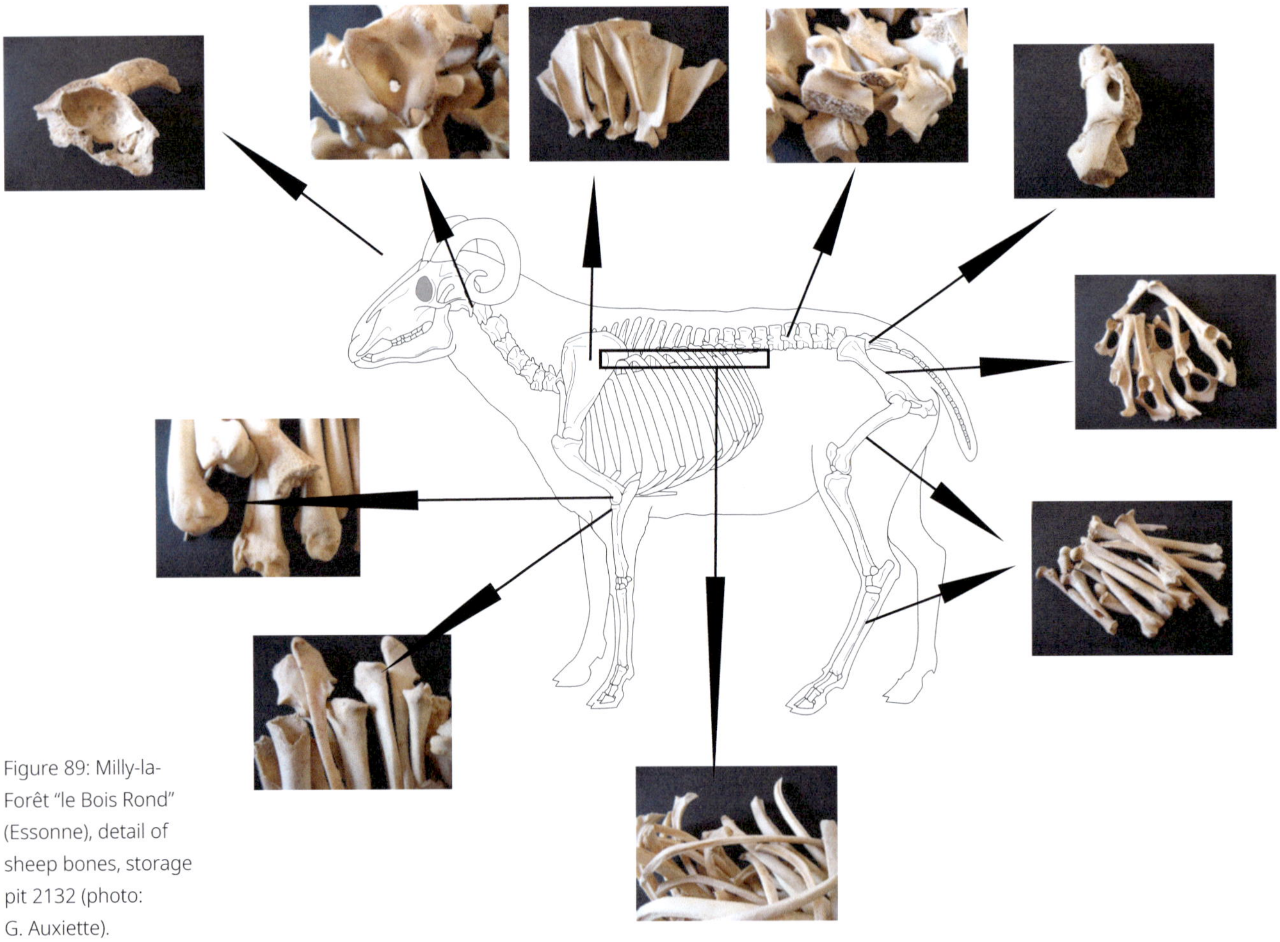

Figure 89: Milly-la-Forêt "le Bois Rond" (Essonne), detail of sheep bones, storage pit 2132 (photo: G. Auxiette).

Furthermore, the bones display little of the anthropic fragmentation usually observed in composite assemblages from more "classic" domestic contexts: in the case of household refuse, mandibles are broken open and the shafts of long bones and foot bones are fractured using percussion in order to prepare certain dishes or to remove marrow.

On certain sites, the quality of the corpus allows us to identify the slaughter ages through careful examination of tooth surface wear: observation of the degree of epiphyseal fusion in limb bones does not provide the same degree of aging accuracy. This analysis reveals the clear dominance of immature and young adult animals (less than 12 months old) among the individuals selected for consumption. Older sheep do occur and represent a non-negligible proportion of the assemblages. In order to identify the period of the year when slaughtering took place, the ages of the different individuals are superimposed on a birth calendar. In the best documented cases, it has been possible to identify two age blocks, one around 5/6 months and the other around 12 months, for example at the site of Chambry "Zac des Griffons". This projection undoubtedly reflects two distinct slaughtering episodes, a first at the end of summer or start of autumn (animals aged 5/6 months) and a second in the spring (animals aged 12 months, killed in March/April). This pattern does not indicate opportunistic slaughter governed by on-going demand, but rather reflects the sacrifice of immature and very young adult sheep in response to the organisation of specific events.

All of these assemblages correspond to identical consumption practices. The axial skeleton is well represented within these deposits, along with a very significant incidence of ribs, while heads, girdles and long bones occur in variable quantities. All occur in well defined fill layers inside the features. In general, the composition of the assemblages attests to the consumption of several sheep, including both mature and immature animals, of various ages. Comparison of the ages at slaughter tends to indicate that, in most cases, deposition of the bones occurred as two distinct episodes within a single pit. The episodes in question were at the beginning of spring and at the end of summer/beginning of autumn. We have been able to infer that specific choices and gestures were involved. Abundant and recurrent cut marks indicate that the animals were eaten. The bones were covered over with soil immediately after their deposition in the silo.

Analysis of the assemblages reveals that the bone refuse originated from meals consisting of pieces of meat that varied in size but that were generally large. The choice of the feature, in most cases a silo that had been emptied of grain and reused as a receptacle for bones and/or pieces of meat, cannot be arbitrary.

The animals selected, the quantities of meat involved, and the quality of the cuts of meat are clear evidence of a deliberate desire to consume more than normal on a given occasion. This implies removal from the herd of several animals, including some very young individuals: this would have deprived the community of a potentially large quantity of meat had the animals been allowed to reach their mature weight. Possibly these sheep were in fact selected from several flocks within the wider community.

An important question remains: why were sheep selected for this exceptional consumption, particularly when pigs were omnipresent within the livestock herds of these communities? It is reasonable to question the status of this species within Iron Age agricultural societies, where different rules governing consumption might have existed alongside those strictly related to domestic consumption. In fact, in all of these cases, the remains of sheep consumption follow a different pattern from that of normal meat consumption waste and indicate particular practices associated with particular consumption. This is the expression of perennial practices established on the farms. Over a period of almost a thousand years, these practices were widely adopted by communities throughout the regions studied here. The quantities of meat involved are quite significant in most cases and represent the slaughter/sacrifice of several animals on a scale that exceeds normal household requirements.

Finally, this practice of collective feasting leads us to reflect on the social dimensions of eating habits.

In his publication *Cooking, cuisine and Class*, Jack Goody (1984, 30) provides an overview of various anthropological approaches to this issue and cites, among others, Robertson Smith who wrote that according to an ancient belief, those who drink and eat together are thereby bound to each other through friendship and mutual obligation; the action of drinking and eating together is the solemn and immutable expression of the fact that all those who share the meal are brothers and that all obligations of friendship and brotherhood are implicitly acknowledged by this common act.

Questions arise regarding the implementation of the rules of consumption that prevailed in these Bronze and Iron Age societies; did they apply to all cuts of meat from a given species, or to all species? The observations made for sheep suggest that there were codes specifically governing their consumption, rules that we have not yet identified, or been able to identify, for other species. Waste from cuts of pig for instance does not follow the same pattern as for caprines. The degrees of preservation are similar to those found in funerary contexts and the studied assemblages are incomplete from an anatomical point of view. We therefore have to envisage the consumption of the missing carcass parts. Perhaps the missing parts were shared out among the families after the collective meals, and consumed differently, in other circumstances and in other locations. Sharing is an attested practice in the context of funerals in these societies from as early as the 6th century BC (Desenne *et al.*, 2009b). Indeed, in many inhumation and cremation burials, various quantities of meat from animals sacrificed in the course of the ceremony were deposited next to the deceased, probably as sustenance in the afterlife. We might ask what happened to the non-deposited pieces of meat. The most likely hypothesis is that they were shared out among guests on the occasion of a communal meal, but the archaeological evidence for this is elusive. However, it is evident that these modes of consumption were perpetuated during the first millennium BC. It is reasonable to propose that soil fertility rites formed part of farming feasts, which reinforced social cohesion and which were tied in to the rhythm of the agricultural calendar. Why sheep were chosen for these feasts remains a mystery.

Collective meals from the Bronze Age (Hallstatt A and C) to the Late La Tène (La Tène C2/D2)

Certain sites among the hundreds studied stand out on the basis of the total numbers of remains produced (in some cases several thousands of bones), the minimum number of individuals (several tens of individuals per species), the slaughter ages, the preparation of the meat in the form of large cuts, and by extension, the quantities of meat consumed, which runs into thousands of kilos. These criteria were defined thanks to hundreds of analyses carried out on various categories of sites, thereby allowing the singular nature of certain sites to be highlighted. The practices take different forms over time from Late Bronze Age to the end of Late La Tène.

The environmental contexts, which are often similar (consisting of alluvial plains in most cases), enable us to compare the corpora in terms of preservation. The origin of the food refuse, which is principally recovered from ditches (with the exception of Late Bronze Age assemblage from Villiers-sur-Seine), means that we can make comparisons regarding the products and modes of consumption.

For the majority of farm sites, the period of occupation does not exceed 50 years. Therefore, when projected for a period of a decade, the quantities of meat involved remain substantially above the normal meat consumption expected for an extended family, particularly when we consider that meat was probably not eaten every day.

Fauna from high status Bronze Age enclosure sites

Faunal studies for eight Early/Middle/Recent Bronze Age sub-circular enclosures located at Abbeville "Mont à Cailloux Sud", Etaples-sur-Mer "le Chemin des Près" and Guînes "Jardins du Couvent 2" (Somme and Pas-de-Calais) (Auxiette, 2017d, 2013c, 2006), Blainville-sur-Orne "Terres d'Avenir-Site 1A" (Auxiette, 2019b), and Mondeville "MIR", "l'Etoile", "Z.I. sud" and "Rue Nicocéphore Nièpce " located in Calvados (Chancerel A., Marcigny C., 2006) (Auxiette, G, 2019d) have allowed us to develop hypotheses regarding meat consumption that is strikingly different from what is observed in domestic contexts. Despite the small quantities of faunal remains recorded when compared to Iron Age farm sites, the corpora can be characterised on the basis of preservation, species selection and the cuts of meat included. The specific characteristics of these assemblages include an overwhelming proportion of cattle (more than 65 %) relative to small livestock and the diversity and quality of the joints of meat represented, which are generally large pieces that have not been subdivided.

At Abbeville (Early Bronze Age/Middle Bronze Age, half of the seventeen cattle were slaughtered before the age of 30 months (the youngest was 7/8 months old); the other half of the cattle had reached weight maturity, *i.e.* 3.5 to 4 years. At Etaples-sur-Mer (Early Bronze Age/Middle Bronze Age), 7 cattle were slaughtered before the age of 48 months (between 24 and 48) and 2 were less than 1 year old.

At Blainville-sur-Orne "Terres d'Avenir-Site 1A" (Early Bronze Age), cattle were slaughtered at different ages: under 30 months, or over 10 years. At Mondeville "l'Etoile" (Late Middle Bronze Age), most of the cattle are juveniles and at Mondeville "Rue Nicocéphore Nièpce " (Late Bronze Age) (Auxiette in Besnard 2019) cattle were slaughtered at different ages.

Furthermore, the sites of Etaples and Mondeville revealed human bones, which must be significant.

The distinguishing characteristics shared by these sites include the number of cattle involved, the weight of meat produced and the preparation of large cuts of meat for cooking.

On the basis of this evidence, it seems likely that the sites acted as temporary community gathering places. These gatherings produced waste from selected cuts of meat (fig. 90), which attest to collective meals: the meat for these meals would probably have been sourced from animals from several herds, which were butchered specifically for the occasion. Furthermore, wild animals do not feature in these "special" meals.

These principal characteristics allow us to distinguish these five assemblages from those encountered on simple domestic sites: for example, the site at Rœux "Château d'Eau" (Pas-de-Calais), revealed frequencies for the main domestic species that are very different; c.40 % cattle, 22 % pigs, and 36 % caprines (Desfossés *et al.*, 1992a).

Therefore, the evidence from these Bronze Age ditched enclosures in the North and West of France (*i.e.* the overwhelming proportion of cattle relative to small livestock, associated with the diversity and quality of the selected cuts of meat) suggests that

these sites were dedicated to collective feasting practices that were very different to those identified through faunal studies of simple farm sites. The assemblages correspond to a desire to acquire tender, good quality meat for a form of consumption that surpasses the ordinary domestic framework. We can suggest that these sites were the venues for periodic gatherings and that the pieces of meat consumed were probably selected from several herds and brought on site for these occasions.

The fully excavated, emblematic Late Bronze Age site of Villiers-sur-Seine "le Gros Buisson", located within the Paris Basin, illustrates these feasting practices through a faunal corpus that is rarely surpassed for this period. A second, partially excavated site located at Boulancourt and situated on a spur in a different sedimentary context, also produced a significant corpus. However, in this case the assemblage has undergone significant alteration due to tapho-nomic processes caused by the acidity of the soil.

At Villiers-sur-Seine "le Gros Buisson" (fig. 91 and fig. 92), the domestic species are dominated by pigs (64%) and game represents a staggering 15.5

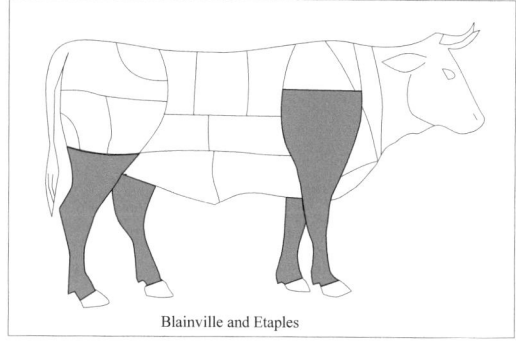

Blainville and Etaples

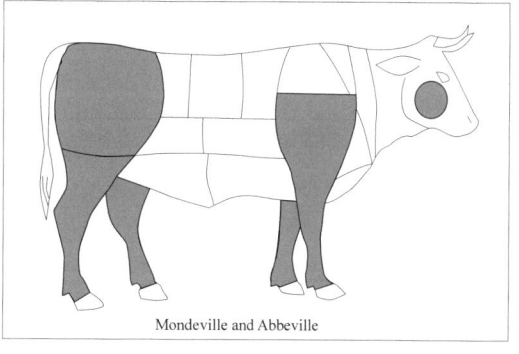

Mondeville and Abbeville

Figure 90: Bronze Age enclosures, main pieces of cattle meat.

% of the assemblage (see chapter 3). Red deer is the second most frequent species in the assemblage after pig (Auxiette *et al.*, 2020).

The analysis of 251 mandibles was used to determine the slaughter ages of pigs, whereby 18 % were slaughtered between the ages of 0 and 6 months, 50 % between the ages of 6 and 12 months, 15 % between 12 and 18 months, 10 % between the ages of 18 and 24 months and less than 6.5 % over the age of 24 months (fig. 93).

The slaughter of animals before they reached optimal size or weight underlines the wealth and ostentation associated with meat consumption. This practice would be completely unsustainable with just a single herd of pigs. The cuts of meat are also quite standardised with a marked preference for shoulder and hams, with other anatomical parts being partially represented. The butchering of red deer carcasses is also stand-ardised with favoured cuts being shoulder, hock,

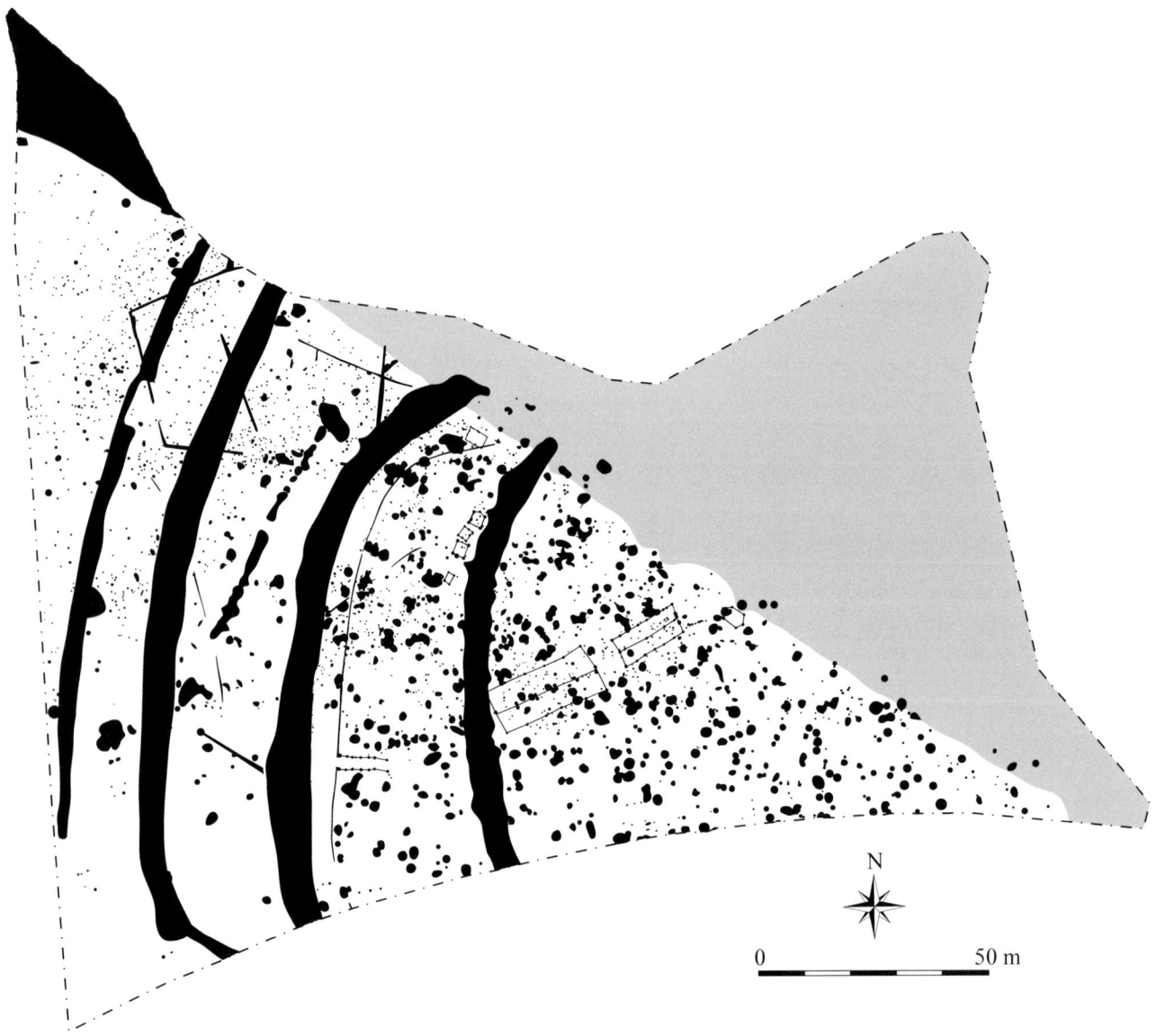

0   50 m

leg, ribs and neck, in very similar sized portions to those of the pig cuts. Therefore, we seem to be witnessing a very structured way of consuming meat, with standard portions for everyone. The study has revealed a concordance of age groups within the same contexts. In many cases the analysis of several mandibles has revealed the deposition of large amounts of refuse at specific moments in the year.

If a pit contains mandibles from both 6 month- and 18 month old pigs, correlating this information with when the pigs were born allows us to confirm that they were slaughtered at the same time: this presupposes that pigs have two litters per year (fig. 94).

The repetition of this phenomenon in several pits can be interpreted as collective feasting, mainly taking place during the winter and spring months (based on the age of the animals when slaughtered). Large quantities of meat were consumed, a conservative estimate being 30,000 kg of meat in total (fig. 95).

These results, highlighting the exceptional nature of meat consumption, are supported by studies of other types of find. The site is indeed remarkable for the large quantity of plant remains, found, the unusually large percentages of specific plant species such as broomcorn millet, lentil and

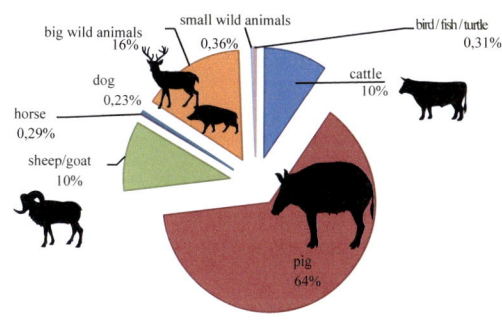

Figure 92: Villiers-sur-Seine "le Gros Buisson" (Seine-et-Marne), Late Bronze Age, frequency of species.

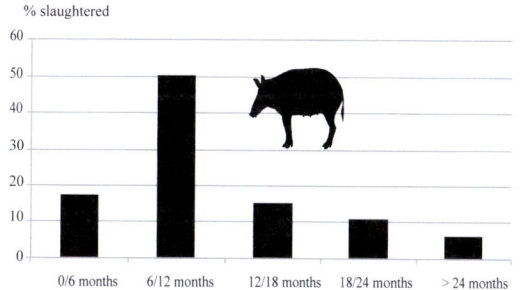

Figure 93: Villiers-sur-Seine "le Gros Buisson" (Seine-et-Marne), Late Bronze Age, age at death for pig.

		Winter			Spring			Summer			Autumn		
Year 1		J	F	M	A	May	June	J	A	S	O	N	D
Litter 1	Theorical age for pigs				1	2	3	4	5	6	7	8	9
	Number per month				9	4		14	20	17	34	62	28
Litter 2	Theorical age for pigs									1	2	3	4
	Number per month									9	4		14
Year 2		J	F	M	A	May	June	J	A	S	O	N	D
Litter 1	Theorical age for pigs	10	11	12	13	14	15	16	17	18	19	20	21
	Number per month	20	18	25		12	4	14	15		11		17
Litter 2	Theorical age for pigs	5	6	7	8	9	10	11	12	13	14	15	16
	Number per month	20	17	34	62	28	20	18	25		12	4	14
Year 3		J	F	M	A	May	June	J	A	S	O	N	D
Litter 1	Theorical age for pigs	22	23	24	25	26	27	28	29	30	31	32	33
	Number per month		16		7		4		2				
Litter 2	Theorical age for pigs	17	18	19	20	21	22	23	24	25	26	27	28
	Number per month	15	11	17	16	7	4	2					

Figure 94: Villiers-sur-Seine "le Gros Buisson" (Seine-et-Marne), number of occurrences per month in a theoretical schedule of births to one or two annual litters, spring and/or autumn ; after Auxiette *et al*. 2020. In certain cases (outlined) the age range covers two months (theoretical ages in months; yellow – litter 1 ; green – litter 2).

poppy and, finally, the quantity and the diversity of transformed foodstuffs.

Plants consumed on the settlement of Villiers-sur-Seine include the full range of plants cultivated in this area during Late Bronze Age, apart from foxtail millet, new glume wheat, flax and gold-of-pleasure (Auxiette *et al.*, 2020, 2015; Toulemonde, 2013, 2010). Moreover, the large quantities of fine pottery used and then immediately thrown into refuse pits, as well as the many little-used grindstones discarded on the site, could also be products of these collective meals.

At Boulancourt "le Châtelet" the domestic species are dominated by pigs (61.4 %). Red deer and wild boar are the principal wild animals present (Bălăşescu *et al.*, 2008). Age analysis highlights a large proportion of young animals slaughtered before they had reached their optimal weight. The slaughter age of pigs varies between 2 and +60 months (3 groups): between 6 and 12 months, (45% of the assemblage, 28% of which were aged between 8 and 10 months), between 12 and 18 months (30% of the remains), and between 20 and +60 months (20% of the assemblage). More than half of the cattle were killed between the ages of 6 months and 2 years (61%) and the majority were between 1 and 2 years old. The young cattle were not slaughtered as part of strategy to stimulate milk production since they had already passed weaning age (Balasse and Tresset, 2002). Caprines were preferentially slaughtered between the ages of 6 and 12 months.

The meat was provided by the settlement's domestic herds or brought in from other settlements as was the case for pig. Hunting also constituted a significant source of meat, but the choice of large game, namely red deer and wild boar, indicates that hunting was practised more for prestige than out of necessity. The consumption of large quantities of meat on the site followed precise rules as can be seen from the manner in which the carcasses were butchered, the association of different species (pig and red deer) and the

standardisation of the animal bone assemblages. Also, the systematic slaughtering of young animals which provide tenderer meat is not sustainable by one herd and it is evident that the animals or the meat were imported into the settlement from outside. Analysis of the age groups of slaughtered animals indicates mass consumption of meat mainly in winter and spring which could be interpreted as a collective event during which feasting took place.

Fauna from high status Iron Age sites

On the high ranking Iron Age farms, there are several lines of evidence for large-scale meat consumption. First of all, the assemblages include large numbers of animals from particular species, representing large quantities of meat. Furthermore, the cuts of meat are of high quality and, lastly, there is a clear selection by age and sex for the animals killed.

The proportion of large mammals (cattle and horse) is particularly high in these corpora and the bones are generally preserved in their entirety indicating that the cutting up of the meat into pieces rarely involved cutting the bone. Small mammals were also significantly involved in these practices and pig played an essential role. The spectrum of species is completed by wild species, particularly red deer, which seems to have had a special role in these collective meals.

On certain sites, the age estimations for the various species indicate seasonal slaughter, suggesting a periodicity for the occupation of the site. The large numbers of animals involved (in the hundreds) suggest consumption patterns that were part of a calendar of special seasonal events. Together with the food remains we also find evidence for the deposition of meat close to the monumental site entrances and for the display of skulls, often in large numbers, as if they functioned as markers of wealth or status.

This communal feasting would have played a central role in maintaining social cohesion and would have reaffirmed the power of elites within a complex economic network of exchange and circulation of goods, in accordance with well-established territorial rules.

They find parallels in certain practices involving the consumption of liquids described by anthropologists in other areas (Dietler, 2001, 1990). These ethnographic studies reveal that the sharing of drinks plays a fundamental and essential role in the forging of social relations expressed within the framework of hospitality. This close association with hospitality, and its ritual and symbolic resonance, is indicative of the powerful social value of the act of sharing drinks.

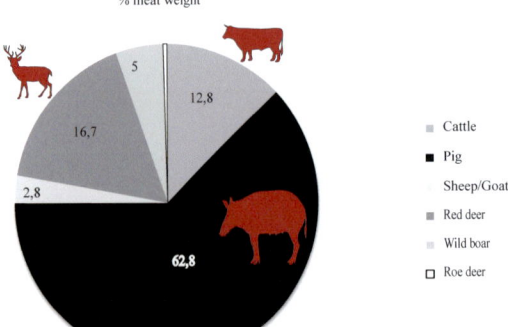

Figure 95: Villiers-sur-Seine "le Gros Buisson" (Seine-et-Marne), Late Bronze Age, % meat weight.

In order to best characterize the assemblages, we must first consider the various sites defined as "banqueting sites" (Poux, 2004): we will identify their shared features that constitute the expression of fundamental factors common to a number of Gaulish tribes. Among these groups, consumption and deposition rituals took various forms depending on the location and they differ from the cult practices observed in sanctuaries (Arcelin and Brunaux, 2003; Brunaux and Meniel, 1983; Méniel, 1999, 1991).

The capacity of utensils, whether they are of metal or ceramic, is another criterion because this capacity often surpasses that of vessels from domestic contexts; the use of the cauldron is particularly evocative of such communal meals. The same is true of amphorae in areas that maintained commercial links with the Mediterranean. As regards table wares, vessels that are found in their hundreds undoubtedly correspond to serving vessels for food and/or drink. Among other utensils that may be connected to feasting are fire brackets, flesh hooks, grills, ladles and grinding stones if they are present in large numbers. Next come concentrations of food consumption refuse and in the sites examined here this is primarily made up of faunal remains, since other organic materials are rarely preserved. These feasting practices sometimes went hand-in-hand with other more ritualised practices which are revealed by the deposition of certain categories of objects, such as the broken necks of amphorae to cite an obvious example (Poux, 2000, 219-220). The display of animal heads, sometimes in large numbers, also emphasises the ostentatious nature of the feasting sites.

We will now turn our attention to a number of case studies relating to the Iron Age. The La Tène C2 farm (200-130 BC) at Villiers-sur-Seine "le Défendable", located in the upper Seine valley (Seine-et-Marne), is made up of two systems of ditched enclosures, one of which is stirrup-shaped in plan and encompasses an area of 4100m^2 (Auxiette, forthcoming). Study of the faunal corpus, which is composed of 8000 bones (67% identified) with a total weight of 146 kg, reveals preferential consumption of cattle (40.2%) and pigs (33.8%). Caprines are present in much smaller proportions (14%), while horses (7.7%) and dogs (2.5%) make up the remainder of the faunal spectrum. The frequencies of the latter two species are relatively high compared to those observed on lower status farms. Wild species are almost absent (1%). A fair number of cattle were slaughtered around the age of 2 or 2.5 years and others were killed between 40 and 50 months (MNI = 30). Within the same slaughtering episode cows aged between 2 and 2.5 years can be slaughtered with cows between 4 and 4.5 years. These estimations suggest that slaughtering took place in two distinct episodes, namely at the end of spring (May to June) and/or at the beginning of winter (November to December). This pattern distinguishes the farm at Villiers-sur-Seine "le Défendable" from other farms where we observe more opportunistic livestock management with killing taking place when the need arose. The slaughter (sacrifice?) of young adult animals suggests that there existed strong social motivations behind the selection of such young animals from the herds. Half of the pigs were killed at 1.5 to 2 years (MNI = 100). This selection follows the same logic as that revealed for cattle. In fact, pigs of different specific ages were killed simultaneously. The ages of these animals are mutually compatible: within the same slaughtering episode, 1 year old pigs can be slaughtered with pigs aged 1.5 years and also animals aged 2 years, based on two litters per year: in March to April and in September to October. Thus, two slaughtering peaks can be observed: one at the onset of winter, around the month of November, and another at the end of this period, at the end of February or beginning of March. The majority of the animals were killed around the age of 1.5 years in the case of the first episode, and around 2 years for the second to which are also added pigs aged about 1 year. The animals killed between the months of May and July cannot be included in this pattern of seasonal slaughter linked to special events.

Analysis of about 40 caprine mandibles has mostly revealed the selection of young animals around the age of 2 years, supplemented by about 10 older animals aged around 3 years and, more rarely, young individuals less than 1 year old. According to these results, we suggest that there were two slaughtering periods: one in the spring for animals aged 1, 2 and 3 years and a second in winter for animals aged 0,6 years (lambs are born during spring). It follows that animals aged between 12 and 21 months may have been slaughtered during either of these two periods and it is impossible to determine in which. The same applies to all animals over 2 years where the one-year age step prevents their slaughter being attributed to an exact point in the calendar. It is only in the case of animals killed around the ages of 10 to 12 months and between the ages of 21 to 24 months that we can say with certainty that they were killed during the spring. This episode of springtime slaughter of sheep coincides with that already identified for cattle, while the winter episode coincides with episodes identified for cattle and pigs.

The available data enables us to estimate the minimum numbers of individuals as follows: 60 head of cattle, 100 pigs, a little less than 50 sheep, 15 horses and 20 dogs. The bone refuse points to large scale consumption on certain occasions during the year, namely in spring and at the start of winter. The selection of certain categories of animals, probably from several herds (the evidence suggests that it would not have been possible to assemble these animals unless they were selected from several herds belonging to several communities), reflects modes of consumption that were out of the ordinary. If we base our interpretation on the hypothesis that all of the animals were consumed on site, then we can extrapolate that approximately 10,000 kg of meat were eaten (fig. 96).

We now turn our attention to the site at Braine "La Grange des Moines" (Aisne) (Auxiette and Desenne, 2017). This La Tène D1b site, situated on the valley floor on the right bank of the Vesle,

is centred on a vast, sub-divided, quadrangular enclosure covering about 6000m². The site is defined by a ditch, with a monumental entrance at the east. The faunal corpus consists of ± 10,000 bones (54.8% of which are identified) with a total weight of 506 kg. The bones come from the ditches. The faunal spectrum is dominated by pig (fig. 97; 43.7% of the remains).

The meat diet also included a large proportion of cattle (30.2%). The incidence of caprines (18.1%) is very low compared to cattle and pig, while that of horse is relatively high (7.9%): there is incontestable evidence for the consumption of horses while the consumption of dogs was very marginal (2.6%). Wild fauna are not well represented in the assemblage (1%). Age analysis of the cattle (fifty in total, the majority of which were cows) reveals that a little less than a third of the animals were under three years when slaughtered; none were under the age of 1 year. Most of the individuals were over 4 years and had thus reached their mature weight. The absence of very young animals and calves along with the preference for slaughtering adult cattle supports the theory that the aim was to produce sufficient meat to feed a large number of people. When matched with a birth calendar (calves born in May/June) the killing of cattle less than 4 years appears to be concentrated in spring and winter. The pig slaughter age estimates were based on a population of over hundred animals (n=106). The consumption of pig focused on the selection of young adults aged approximately 1.5/2 years (72%), the rest consisting of very young animals or piglets and a handful of very mature animals.

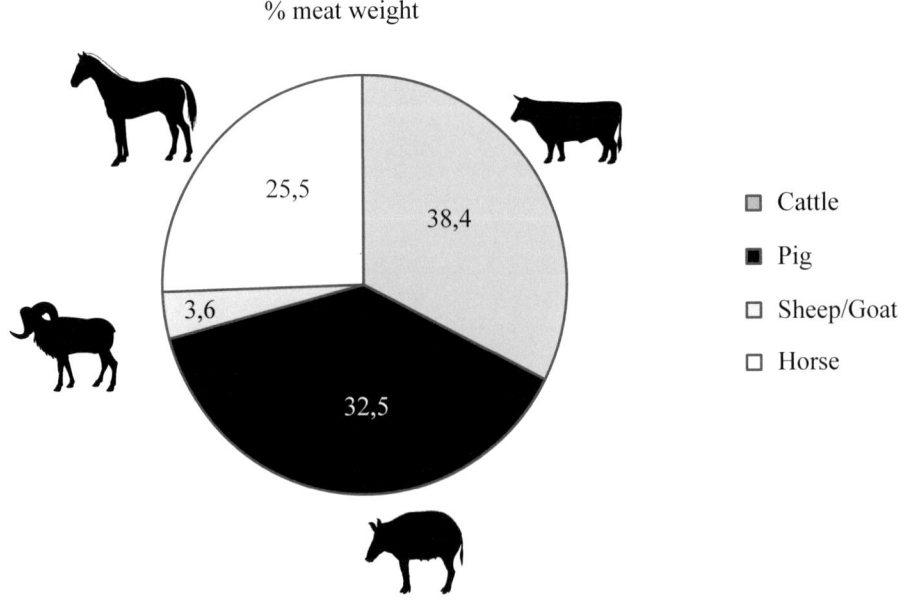

% meat weight

25,5

38,4

3,6

32,5

■ Cattle
■ Pig
□ Sheep/Goat
□ Horse

Figure 96: Villiers-sur-Seine "le Défendable" (Seine-et-Marne), Late La Tène, % meat weight.

If we assume that pigs have 2 litters per year, one in February/March and the other in August/September, then the number of pigs killed between October and March appears to be high; in fact, we find animals of different ages associated with the same episodes of slaughter and consumption in the case of fresh meat. However, pigs were being killed throughout the year. Out of thirty four caprine mandibles, three quarters are from animals aged 2 years and over. There is no evidence for the preferential consumption of lambs in this context. The projection of their ages on a theoretical calendar suggests that there was a preference for slaughtering during the spring.

The estimation of the slaughtering age for the three principal domestic species therefore

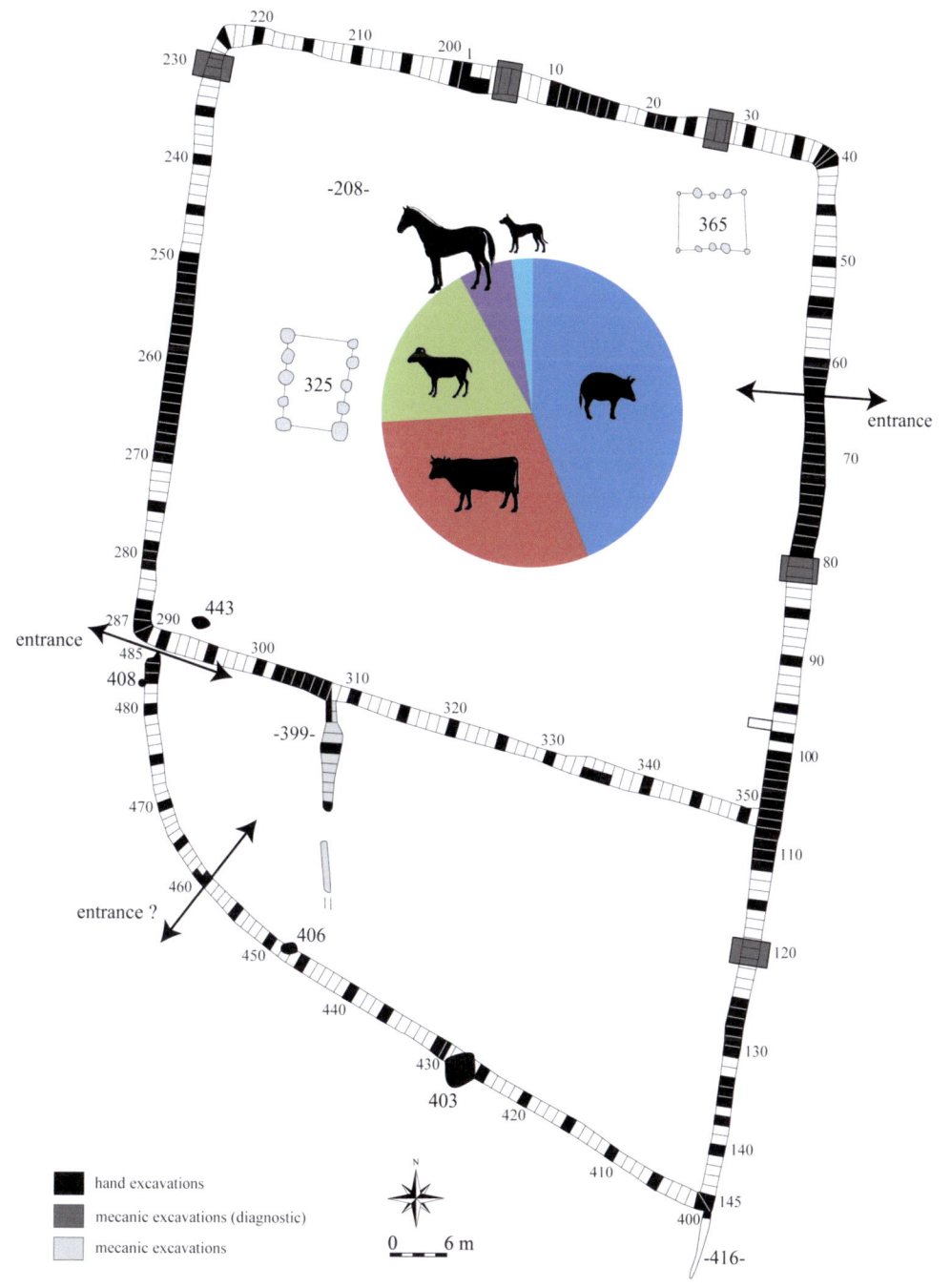

Figure 97: Braine "la Grange des Moines" (Aisne), Late La Tène, frequency of main species.

reveals the consumption of meat from animals that had, for the most part, reached their mature weight. The theoretical projection of ages over the calendar year indicates preferential consumption in spring and winter for cattle, in spring for caprines, and throughout the year for pigs (based on the hypothesis of two litters per year). Taken together, the seasonality, the number of animals involved, and the quantity of meat generated suggest communal eating habits that far exceed normal consumption in a domestic context. This pattern of consumption probably forms part of a calendar of special events associated with agricultural practices (harvest, seasons). From the minimum number of individuals calculated, we can estimate that the total mass of meat consumed exceeded 12,000kg divided between cattle, pigs, caprines and horse (fig. 98 and fig. 99): the respective contributions of cattle and pigs, in terms of the weight of meat generated, were more or less equal. Recorded food waste was particularly concentrated in front of the largest buildings and around the entrance of the enclosure.

To the west of the Middle Oise valley, the site of Beauvais "les Aulnes du Canada" (Oise) which was partially excavated and interpreted as a "banqueting site" with ritual elements (human skulls and metal objects), is in fact the only site with a faunal assemblage similar to that from Enclosure B at Braine: we observe similar ratios

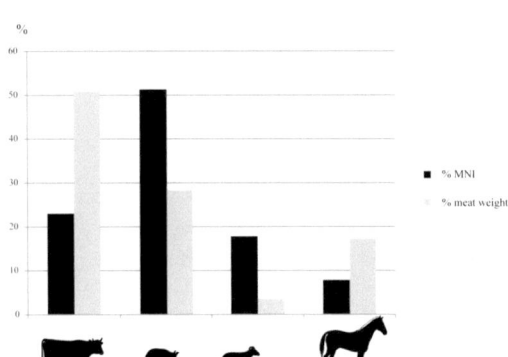

Figure 98: Braine "la Grange des Moines" (Aisne), Late La Tène, % meat weight.

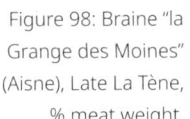

Figure 99: Braine "la Grange des Moines" (Aisne), Late La Tène, comparison between MNI and meat weight.

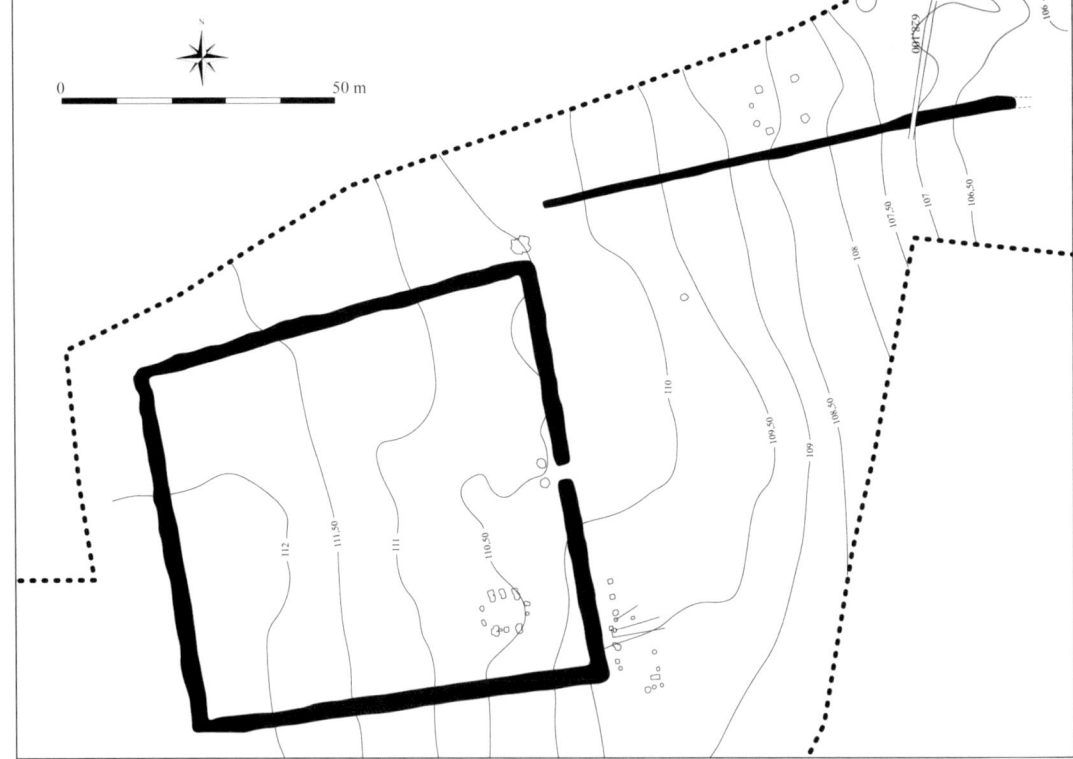

Figure 100: Souppes-sur-Loing "À l'Est de Beaumoulin" (Seine-et-Marne), Late La Tène, enclosed settlement (after Séguier, Auxiette 2006).

between the various species with pig predominant and sheep occupying a minor position after cattle and pigs; there is evidence for the eating of horses on both sites; the slaughter pattern is identical for cattle and pigs; the MNIs are high and are, to some extent, comparable between the two sites (Méniel, 1990; Woimant, 1990).

Another exceptional site is that of Souppes-sur-Loing "A l'Est de Beaumoulin" (Seine-et-Marne, La Tène D2b) (Séguier and Auxiette, 2006). The main feature of the site is a trapezoidal ditched enclosure (ca. 5000 m²) with a monumental covered entrance at the east; the enclosure itself is part of a wider system of ditches (fig. 100 and fig. 101).

The site is located on a limestone plateau overlooking the valley of the river Loing. The bones come from the ditch. The faunal corpus is made up of ± 10,000 bones (84% identified) with a total weight of 364 kg. Domestic species form the vast bulk of the assemblage (94%). The faunal spectrum is largely dominated by cattle (54.3%) with a non-negligible proportion of pigs (15.8%), while caprines make up only a small proportion (5%, mainly sheep). There is good evidence for hippophagy (9.3%) while cynophagy (1.9%) is attested but remains marginal relative to the overall consumption. The contribution of meat from wild species is not insignificant; red deer and roe deer are the principal sources. Of the 105 head of cattle represented, more than half were slaughtered around the age of 3.5/4 years, a significant number around the age of 5 years (optimum weight) and a small number of culled animals were over 10 years old. There is little evidence for the consumption of very young cattle and none for the consumption of calves. There is no doubt that we are looking at the selection of meat animals that were killed and eaten to fulfil specific needs. On the basis of the mandibles found, we estimate that about seventy sheeps were consumed. The ages obtained from teeth show that slaughtering was rather spread out with a substantial number of very young animals aged less than 12 months (25%) on the one hand and a preponderance (60%) of animals aged between 2 and 3 years on the other. As for pigs, ninety-one mandibles indicate that animals were slaughtered at various ages. However, a substantial proportion of the remains are of piglets or very young animals (48%) and these can be divided into two groups: the first is made up of animals aged between 0 and 6 months

Figure 101: Souppes-sur-Loing "À l'Est de Beaumoulin" (Seine-et-Marne), Late La Tène, concentration of animal bones near the entrance porch (photo: C. Valero).

(26.3%) and the second of animals aged between 6 and 12 months (22%). Almost 90% of the remains come from animals aged less than 2 years. It is clear that cattle and horse provided most of the meat in the diet.

The shoulder heights of the cattle range between 1 m and 1.24 m (n = 79), with a main cluster between 1.10 m and 1.16 m.

The analysis carried out involved crossing the Smalest part of the Diaphysis and Greatest Lenght measurements (codified measurements Van den Driesch and Forest & Rodet-Belarbi) (Driesch von den, 1978; Forest and Rodet-Belarbi, 2002, 294) and values for the gracility index (Gr-Iet, following Chaix and Méniel) (Chaix and Méniel, 2001, 80). The results highlight the dispersion of values that obscure the distinction between the sexes on the one hand (cows/bulls) and castrated animals on the other.

Above all else, these observations highlight the heterogeneity of the cattle populations and tend to support the hypothesis that the animals came from a number of different herds and were transferred to the site for collective feasting. The mixing of cattle populations has been previously proposed for the cattle found at "le Marais du Pont" in Varennes-sur-Seine, located about 40 km to the north-east of the Seine/Yonne confluence (Méniel and Horard-Herbin, 1996).

Analysis of data concerning the metapodes reveals the absence of large cattle (> 1.20 m), like those identified on the sites of "le Marais du Pont" et "la Justice" in Varennes-sur-Seine (Auxiette, 2013b; Méniel and Horard-Herbin, 1996). Recent studies have revealed the presence of large animals on most of the Senones sites, while, in contrast, very gracile animals are characteristic of the Suessiones (Auxiette, 1996; Duval *et al.*, 2012, 89). The absence of large animals at Souppes-sur-Loing could thus be interpreted as the selection of animals that responded to certain criteria and specific needs, in this case collective meals, as has been proposed by other authors (Nuviala, 2016, 604).

The faunal remains from Souppes-sur-Loing attest to large scale production of meat for consumption by a large number of people. Regarding small mammals, there is ample evidence for the consumption of lambs and piglets. This reflects a certain level of comfort in the procurement of

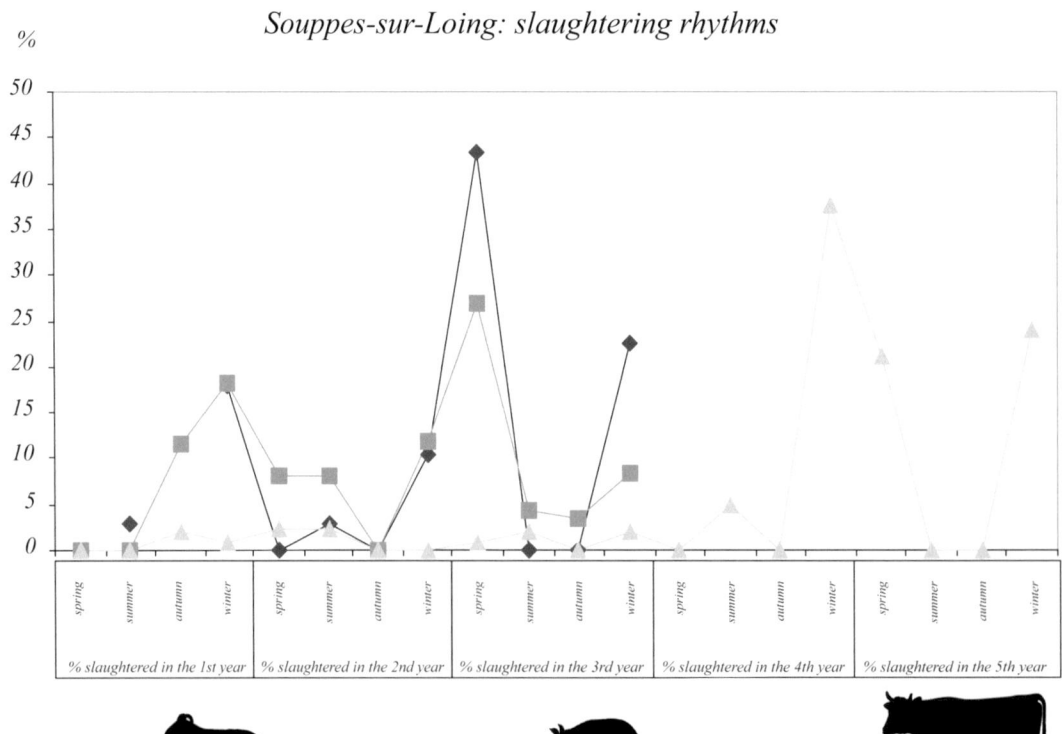

Souppes-sur-Loing: slaughtering rhythms

Figure 102: Souppes-sur-Loing "À l'Est de Beaumoulin" (Seine-et-Marne), Late La Tène, slaughtering seasons.

food. The evidence from the slaughter patterns suggests that these large-scale selections were made from numerous herds, and that they were intended to satisfy specific demands arising from the organisation of large-scale feasting. It is unlikely that all of these animals were reared on site as the slaughtering curves do not reflect sustainable herd management. In order to reveal the seasonal slaughtering patterns, the ages of the cattle, pigs and sheep were projected onto birth calendars for the three species. In fact, while the slaughtering period for cattle aged between 36 and 50 months (or over) is difficult to position in time due to the absence of smaller age-intervals, that of caprines and pigs reveals some degree of biannual periodicity. Hence, the majority of animals were killed in spring and during the winter, *i.e.* around April-May and December-January (fig. 102).

By extrapolation, we suggest that the less accurate slaughter curve for cattle (mainly bulls) can be made to line up with the same periods of the year. This analysis indicates that there was massive consumption of animals during specific, possibly recurrent, seasonal events, which brought together large numbers of people. On the basis of the calculated minimum number of individuals we can estimate that over 21,500 kg of meat were consumed (fig. 103).

We have to envisage these seasonal events within a continuum of more normal meat consumption since the settlement was undoubtedly occupied and maintained outside of the seasonal festivals, hence the "background noise" discernible in the spread of slaughter ages between the two seasons of large-scale killing. However, it should be stressed that this reasoning is only valid in the case of the consumption of fresh meat (immediately following slaughter), which is generally the case in temperate regions where fresh meat does not keep more than a few days; but we cannot exclude the consumption of salted or smoked meats. We can consider that Souppes-sur-Loing is a "banqueting site".

In conclusion about communal consumption during Metal Ages, it is clear that the sites examined share a number of striking similarities: the predominance of one species over the others (sometimes less significant once the MNI is calculated); seasonal slaughter patterns; the enormous quantities of meat involved in these communal meals; the preparation of cuts of meat that are markedly different from those cooked in a domestic context.

Consumption was essentially focused on the main domestic species identified on all sites in the north of France, *i.e.* cattle (*Bos taurus*), pig (*Sus scrofa domesticus*), caprines (*Ovis aries/Capra hircus*), horse (*Equus caballus*), and dog (*Canis familiaris*). Wild mammals represent a negligible proportion of

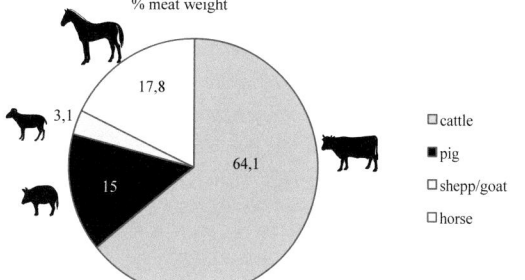

Figure 103: Souppes-sur-Loing "À l'Est de Beaumoulin" (Seine-et-Marne), Late La Tène, % meat weight.

the faunal corpora, with the exception of the Late Bronze Age site of Villiers-sur-Seine "le Gros Buisson", which is one of the earliest sites examined. Red deer (*Cervus elaphus*) is the most frequently occurring wild species, followed by wild boar (*Sus scrofa*), and roe deer (*Capreolus capreolus*); small game occur in tiny proportions (*Lepus europaeus, Vulpes vulpes, Castor fiber etc.*). In the Iron Age, backyard livestock are principally comprised of chickens (generic term, *Gallus gallus*) and geese (*Anser anser*).

Most of these farm sites are characterised by very large faunal assemblages made up of thousands of bones. However, for the Bronze Age, five early sites (Mondeville x 4, Blainville, Abbeville, Guînes and Etaples) stand out from the others by virtue of their small assemblages which, nonetheless, tell us a great deal about the nature of consumption on these sites. These enclosed sites dating to the Early/Middle Bronze Age have revealed a preference for the consumption of cattle on the occasion of these communal meals. In contrast, pig and red deer were the preferred species on the Late Bronze Age site of Villiers-sur-Seine. In the Middle La Tène, cattle and pig were the preferred species for these "exceptional" consumptions; horse occurs in much smaller quantities but is nonetheless non-negligeable, while dog is fairly marginal. Calculation of the Minimum Number of Individuals enables us to estimate the numbers of animals involved: the large numbers obtained imply that live animals were selected from several herds. Obviously these estimations do not permit us to affirm that all of the meat generated was consumed during the communal meals: however, the MNI discrepancies between the various carcass parts are sometimes such that it seems probable that a portion of the meat was redistributed, or that cuts of meat were specially brought to the site by guests as a contribution to the festivities. In addition to the numbers of animals involved, and the quantity of meat generated, these assemblages are characterized by preferential slaughter at certain times of the year -sometimes masked by background noise – and by food preparation practices that differ from those observed in the assemblages from ordinary farms. Estimation of slaughtering ages reveals an uneven distribution between the various age groups, which also varies depending on the species and site.

In the case of cattle, juvenile animals are selected less frequently than young adults or mature adults. On the three earliest sites, juvenile and adult animals were selected. The 24 to 30 months age bracket is fairly well represented; these animals were therefore killed between June and December.

For pigs, the estimates highlight a particularity of the Late Bronze Age site of Villiers-sur-Seine where a high number of animals were killed between the ages of 2 and 12 months and a large proportion of these were between the ages of 6 and 12 months (42%). The estimated slaughter ages for pigs on Iron Age sites is marked by recurrent choices: the 12 to 24 month slaughtering interval, reduced to 18 to 24 months (Villiers-sur-Seine, Beauvais, Braine, Souppes-sur-Loing), is the best documented; in addition, we see a high proportion of piglets on the site of Souppes.

In the case of caprines, adult animals are omnipresent, with mature animals occurring on most of the Iron Age sites, while the Late Bronze Age site is characterized by a wider spread of slaughtering ages. For the Iron Age, the site of Braine stands apart by virtue of the high proportion of animals aged between 18 and 24 months. Evidence for the killing of lambs is lacking on most sites.

In the face of such large numbers of animals, which significantly surpass those found on simple farms, the issue of the mass of meat generated has to be considered. Estimations of meat weight are based on the works of several authors (Auxiette *et al.*, 2020; Hachem, 2011a; Vigne, 1988) in which they are calculated for broad age groups. We can therefore easily get an idea of the quantities of meat involved and in general they are impressive, ranging from several hundreds of kilos to tens of thousands of kilos. Not surprisingly, the meat mass estimates obtained for cattle are the highest; if we add the meat masses obtained for the horses identified on the Iron Age sites, then it emerges that large domesticated animals were the principal source of meat even though they were numerically (MNI) inferior to the smaller farm animals.

Next in order of importance to beef is pig, which was also consumed in considerable quantities, ranging from several tens of kilos for the two earliest sites, to several thousands of kilos for the others. The Late Bronze Age site of Villiers-sur-Seine produced an exceptional estimated pig meat mass of more than 13,000 kg; it should be remembered that cattle were marginal on this site and that red deer were the second most important source of meat.

The corollary to the masses of meat yielded by cattle and pigs is the small quantity generated by sheep, which ranges from a few tens of kilos (or even less in the case of the two earlier Bronze Age sites) to a few hundreds of kilos for the Iron Age sites.

Let us not forget that in absolute terms, the quantities of meat consumed must be related to the duration of occupation of the site. Despite everything, the maintenance of the ditches shows that the bones come from one of the last operating states of the sites.

Looked at as a whole, this data arising from the estimation of meat weights from the MNI reveals a number of differences between the various sites: cattle are predominant in the early sites at Mondeville, Blainville, Abbeville and Etaples while pig was preferentially consumed at the Late Bronze Age site of Villiers-sur-Seine (along with venison. During the La Tène C2 and La Tène D1b, cattle and pigs balance each other out and horse, which declines slightly relative to the other two species, nevertheless accounts for 20% of the total meat weight. During La Tène D2, cattle supersede all other species.

It is probable that these "banquets" had multiple significances, not only social and civil but also religious, for the societies involved (cf. citation Goody *supra*). A number of our results find fascinating echoes in the writings of Posidonius of Apameia (135-51 BC) (translation by C. D. Yonge 1854)[1]:

"The Celts place food before their guests, putting grass for their seats, and they serve it up on wooden tables raised a very little above the ground; and their food consists of a few loaves, and a good deal of meat brought up floating in water, and roasted on the coals or on spits. And they eat their meat in a cleanly manner enough, but like lions, taking up whole joints in both their hands and gnawing them; and if there is any which they cannot easily tear away, they cut it off with a small sword which they have in a sheath in a special box...".

"And Poseidonius continuing, and relating the riches of Luernius the father of Bityis, who was subdued by the Romans, says that "he, aiming at becoming a leader of the populace, used to drive in a chariot over the plains, and scatter gold and silver among the myriads of Celts who followed him; and that he enclosed a fenced space of twelve furlongs in length every way, square, in which he erected wine-presses, and filled them

with expensive liquors; and that he prepared so vast a quantity of eatables that for very many days any one who chose was at liberty to go and enjoy what was there prepared, being waited on without interruption or cessation."

On all of the farms where this feasting took place, certain unusual practices have been systematically observed: the best documented include the deposition of pieces of meat at the entrances to the sites, as well as the display of animal heads (skulls) and the presence of human remains including foetuses and dried bones (relics), the outer surfaces of which bear traces of weathering and sometimes cut marks (Pinard, 2016). At Souppes-sur-Loing, for example, deposits of piglets and human foetuses near the site entrance raise questions regarding the original intention: while it is easy to imagine that the piglets

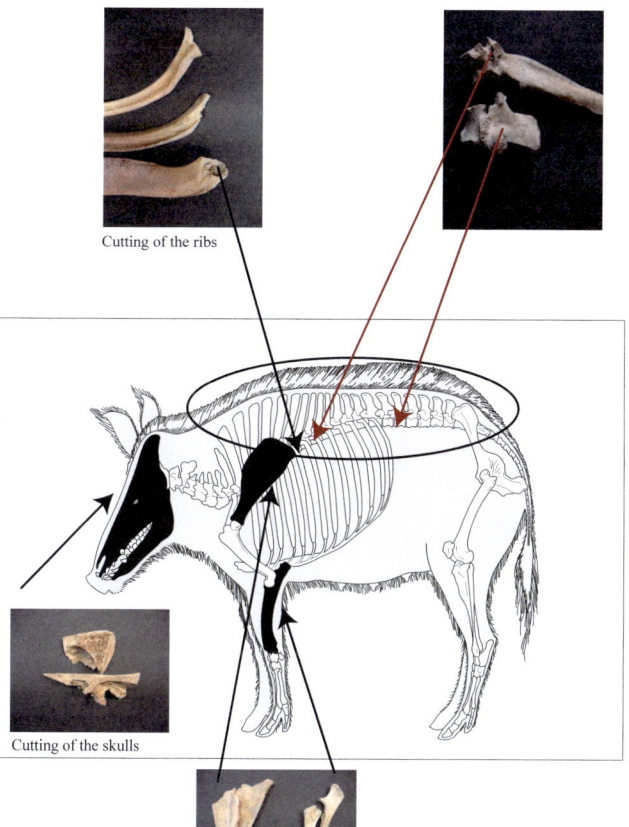

Cutting of the ribs

Cutting of the skulls

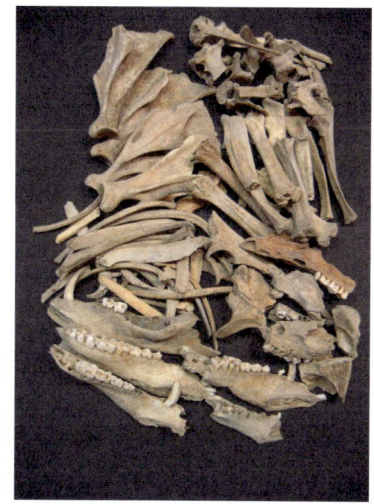

Assemblage of bones

Figure 104: Villeneuve-Saint-Germain "les Etomelles" (Aisne), Late La Tène, particular assemblage in a well (photo: G. Auxiette)

were sacrificed, it is much more difficult to imagine why the Gaulish population might have deposited human foetuses. Among the indigenous Andean tribes of Bolivia, the use of animal (pig, sheep and camelid) foetuses is attested in certain ritual practices *known as* "mesas".

> *"These are complex offerings made up of substances of various origins- mineral, vegetal, animal and even dietary- all of which have precise symbolic values...These mesas have two fundamental aims: one is beneficial in nature, the other evil".* (Girault L., 1975, 220).

Those of a beneficial nature are generally directed towards the earth goddess, the spirits of the ancestors or, sometimes, the god of thunder. They are offered to "please" the gods/ancestors or to "pay" them back; they may also express requests.

Although this example is geographically far removed, the presence of human foetuses in the deposits from Souppes-sur-Loing propels us into the infinitely complex realm of human thought, gesture and practices. The intentional nature of the deposits implies forethought and preparation within the perspective of an event that was fixed in time. The foetuses would thus be the object of particular attention and preservation.

Evidence for collective meals in a well

Excavation at Villeneuve-Saint-Germain "les Etomelles" uncovered a Late La Tène farm, comprising buildings, ditches and two wells, the latters situated on the edge of the site. Well 378 is the most recent feature. This well produced an unusual finds assemblage, including animal remains, pottery, amphorae and metal objects (Hénon *et al.*, 2012). The bulk of the finds come from a layer at a depth of between 1.4 and 1.8 metres.

With pig making up 73 % of the remains, and selection of certain meat pieces, there is a clear predominance of fore quarters (scapulae and radii/ulnae) and elements of the axial skeleton, namely ribs and vertebrae (fig. 104).

These complete bones are well preserved. Cut marks are evident on the vertebrae and ribs (ablation of transverse apophyses and sectioning of proximals) suggesting the preparation of racks of

ribs and fillets. Part of the shoulder (humerus) and the hind part of the skeleton are completely absent. The selection of certain parts of the skeleton and the number of individuals involved (four pigs of around 2 years) are clear evidence of consumption over and beyond that normally encountered in a domestic context. The presence of half a horse pelvis mixed up with the rest of the bones reminds us that this piece (perhaps a choice cut) is also associated with deposits of clearly identified cuts of meat in the Late La Tène of the Aisne valley (as at Braine, for example).

Apart from the fauna, the well yielded selected elements from five to six large amphorae (Dressel 1); principally parts of the necks, handles and rims. The fragments are very well preserved and although there are no signs of wear or re-use and numerous refits were possible, certain fragments bear traces of impacts. Amongst the ceramic remains we also note the presence of decorated carinated vessels with a biconical profile. The corpus of finds also included a "Laconian key".

The finds had all been disposed of together, or at least within a very short time span. The discarded material does not correspond to the category of everyday domestic waste, but instead probably relates to a single event, which was marked by a collective meal.

Endnote

1   http://www.attalus.org/old/athenaeus4.html#1

4. CULTURAL MANIFESTATIONS

At some time in their history, some animal parts are singled out for treatment which diverges from their "normal" trajectory (which is to end up in a rubbish tip or to lie on the ground as consumption waste). This singularity is expressed in the creation of structured deposits (Hill, 1995; Morris, 2008).

According to the logistic approach developed by Gardin: *"the depositional situation, for the objects concerned, corresponds to a particular combination of intrinsic properties (physical, geometric, semiotic), which make them likely candidates for deposition, and of extrinsic properties (location, time, 'function'), which squarely place the deposit in a socially-, culturally- and 'ideally' significant context."* (Gardin, 1979).

'Deposits' constitute a purely archaeological category (...) created first and foremost by the archaeologist as an outside observer. It allows us to group the products of various distinct practices purely on the basis of the resemblances exhibited by these products, or more exactly their material remains; to which can also be added criteria of dissimilarity according to which these products differ from what is regarded as 'normal'." (Auxiette and Ruby, 2009, 118).

We will refer to concepts such as the "social life of things" and the "biography of things" taken from Appadurai (1988a, 1988b) and "biography" from Kopytoff (1988), through which "things" can be envisaged in the perspective of a dynamic "social trajectory".

In our opinion, certain faunal assemblages found in domestic contexts can be seen as the deposition of animal parts for symbolic reasons in

the Neolithic (Méniel, 1987a; Farruggia *et al.*, 1996; Ghesquière and Hachem, 2018) and in Bronze and Iron Ages (Auxiette, 2000a, 2013d, 2013d; Auxiette and Ruby, 2009; Delattre and Auxiette, 2018).

Fauna – as depositional material – is sometimes found in a particular state related to a socially connoted "situation". *"An object may be regarded as a depositional object at certain times but not at others, after having been something else in between."* (Auxiette and Ruby, 2009).

By transposing Appadurai's propositions based on "mercantile" situations, we can treat this "depositional" situation as a biographical sequence in the social life of the object, as its capacity to become depositional material in the social context where an object becomes a depositional object.

Deposits containing entire animals, or animal parts, in "domestic" contexts

The entity under consideration is therefore a "structured deposit" of animal parts. These have been recorded on nearly every site and take various forms which can be organised into categories. They are attested over the complete chronological sequence. Several configurations are observed. Theoretically, the deposits might have been composed of bones alone, bones and meat, bones and skin, bones and skin and meat, or simply of meat. Based on archaeological remains, five empirical categories have been established ranging from the simplest – a single, isolated bone – to the most complex and complete – the entire animal skeleton: Category 1 contains heads and isolated selected bones which do not display the characteristic fragmentation resulting from consumption (for instance scapulae and coxae); Category 2 groups together bone assemblages forming a meaningful anatomical unit, for instance a limb (front or hind); the anatomical connections were preserved in the flesh at the time of deposition, which distinguishes these remains from those arising from "normal" consumption; Category 3 and Category 4 differ by the absence (3) or the presence (4) of anatomical connections; they correspond for the most part to bone assemblages from different parts of the skeleton, equivalent to several selected anatomical groupings, collected from one or several animals, sometimes belonging to different species; Category 5 corresponds to situations where the archaeological remains consist of one or several articulated skeletons.

The interest of faunal remains lies in the fact that they refer inevitably to the initial "whole" that was the live animal. Thus, the physical properties that need to be considered are not only those of the "*pars*" that formed all or part of the object in the deposit, but also those of the "*totus*" from which the part came.

The species involved are preferentially domestic species, *i.e.* cattle, horse, pig, sheep and dog. Birds are rare. Wild animals feature in small numbers (aurochs, red deer and roe deer). Some species tend to be better represented during certain chronological periods than during others.

The large-scale compilation of data, both chronological and geographical, allows different lights to be shed on the evolution of religious practices: the place of the various species across time, deposit configurations, preferential combinations, and the place of these practices in the context of the various categories of sites.

The Neolithic

The Early Neolithic

A ceremonial enclosure
The LBK site of Menneville "Derrière le Village" (Aisne) includes a settlement, graves and a ditched enclosure that extends over an area of 6.4 ha. (fig. 49). A little less than one third of the total surface area of the site has been excavated. The first excavation campaigns revealed complex deposits in several segments of the enclosure ditch: eleven individuals, adults and children, were discovered inhumed in various ways; isolated human bones were also uncovered. Certain individuals were inhumed in a manner that was conventional for the LBK, while others were simply thrown into the ditch. In addition, certain individuals showed clear traces of violence. All of the human remains were associated with animal remains, principally domesticated species, which included cattle skulls that were deliberately arranged in a complex manner (Farruggia *et al.*, 1996; Hachem, 2001).

This data was re-assessed by C. Thevenet in her PhD thesis (Thevenet, 2010) notably outlining an organisational scheme for the deposits in the ditches, as well as proposing an interpretation of the enclosure's function (Thevenet, 2016a, 2017). Four depositional associations, of varying degrees of complexity, can be identified on the basis of the ways in which the deceased were treated, the categories of human bone present, the animal species present, and the stratigraphic distribution of the remains within the ditches (refer to table in Thevenet 2016).

All this evidence can best be interpreted in terms of episodically repeated events of a ceremonial nature. Questions can also be asked about the positive or negative nature of the various treatments of the dead, in particular about the possibility of human sacrifice in some cases.

The stratigraphy of the deposits and probable re-cuts imply that the deposits are spread out over time within each segment of the ditch, with each segment functioning as an independent structure. The development of the segment deposits, like that of the enclosure, was probably progressive but apparently followed a plan (Thevenet *et al.*, forthcoming). This is suggested, on the one hand, by the recurrent organisation of the deposits within the different segments and, on the other, by the fact that their depth is correlated to the type of deposit (the most complex deposits are found in the deepest segments (fig. 105).

Finally, certain features discovered within the enclosed area have a direct link with the events recorded in the ditches. This is the case for a conventional burial (burial 272), in which the body had been manipulated and several anatomical parts removed. It is also the case for a large pit containing several bodies and animal bones (feature 93) where the depositional association is identical in every way to that observed in the most complex deposits in the ditch segments.

Since 2013, new excavations have been conducted the results of which support the patterns already observed and also provide new elements (Thevenet, 2013, 2016b, 2018, 2020).

The faunal material retrieved from the ceremonial enclosure at Menneville is very different to that found in the lateral pits of contemporary LBK houses. First of all, the taphonomy is different. The quality of the preservation is much better if the bones were rapidly, and in some cases deeply, buried, as was not the case with the everyday domestic refuse discarded in the pits alongside houses; the surfaces of bones from the latter contexts tend to show traces of alteration. Secondly, the composition of the samples also differs. On settlements, consumption was varied;

it was based on three domesticated species and also on a certain number of wild species (see chapter 2). In the enclosure ditch, the animals represented are almost exclusively domesticated species. Lastly, the state of the bones is different. Sometimes found in large numbers, the bones in the pits alongside houses are often broken and accompanied by small flakes that result from the breaking of the bones to extract marrow. In the enclosure ditch, bones are not very abundant and are often articulated or the epiphyses are intact in cases where the bone is fractured.

The segments of the ditch have yielded bones from the four main domestic species, namely cattle, sheep, pigs and dogs; these do not occur in the same proportions and were not deposited in the same manner. The first two species predominate. The cattle bones are isolated and often articulated; they are found in the form of joints or large fragments of diaphyses that had been discarded or deposited once the meat had been removed

Figure 105:
Menneville "Derrière le Village" (Aisne), LBK enclosure; the most complex deposits were found in the deepest ditch segments (photo: UMR 8215, Trajectoires).

Figure 106: Menneville "Derrière le Village" (Aisne), LBK enclosure, feature 602, deposit of articulated cattle vertebrae (photo: UMR 8215 Trajectoires).

and eaten. We are not looking at the disposal of whole carcasses (with one exception, a small veal); certain parts were clearly selected. Elements of the vertebral column occur in greater numbers than other parts of the skeleton, and three or four connected vertebrae are sometimes found together (fig. 106). Next come elements of the skull (horn

Figure 109: Menneville "Derrière le Village" (Aisne), LBK enclosure, feature 602 voir fig 23 rapport 2017, a complete bovine scapula and cranial elements, arranged alternately with sandstone blocks (photo: UMR 8215 Trajectoires).

Figure 107: Menneville "Derrière le Village" (Aisne), LBK enclosure, feature 576, cattle horn core (photo: UMR 8215 Trajectoires).

Figure 108: Menneville "Derrière le Village" (Aisne), LBK enclosure, feature 571, bull bucranium deposited at the feet of burial VII. It was placed in symmetry to a female aurochs bucranium (photo: UMR 8215 Trajectoires).

Figure 110: Menneville "Derrière le Village" (Aisne), LBK enclosure, feature 617, a bovine radius-ulna disconnected from the skeleton, deposited in the lower layers of the ditch segment (photo: UMR 8215 Trajectoires).

cores and bucrania), (fig. 107 and fig. 108), scapula (complete) (fig. 109), and front limbs articulated proximal radius and ulna (fig. 110). Hind limbs are represented, but in smaller quantities (tibia, coxal).

Sheep are the second most important species deposited in the enclosure ditch. Here, particular attention has been paid to distinguishing between sheep and goats on the basis of morphological criteria; despite this, no goat remains have been identified. Unlike cattle, sheep are more often present in the form of whole carcasses or articulated carcass parts, and never as isolated bones (fig. 111).

The third domestic species recorded from the ditch at Menneville is pig. Certain segments of the ditch are practically devoid of pig bone, particularly those excavated during the first few excavation seasons; in these cases they are found only occasionally in the upper layers of fill. However, the segments excavated since 2013 have yielded quite large quantities, including from the bottoms of the ditches. In our opinion, this difference could be the result of a change in ritual at the end of the LBK but this remains to be confirmed. Like the sheep, the pigs were deposited as whole carcasses or as articulated pieces (fig. 112). The individuals are often young or very young, so the possibility arises that these are wild boar piglets; we cannot distinguish between the young of the wild and domestic species using morphological criteria. In any case, the animals whose bones are epiphysised are always pigs.

Dog was also implicated in the ceremonial activities at Menneville, but much more sporadically than the other three domesticated species (fig. 113). It was deposited as entire carcasses (very young and adult individuals) or as isolated bones (particularly mandibles).

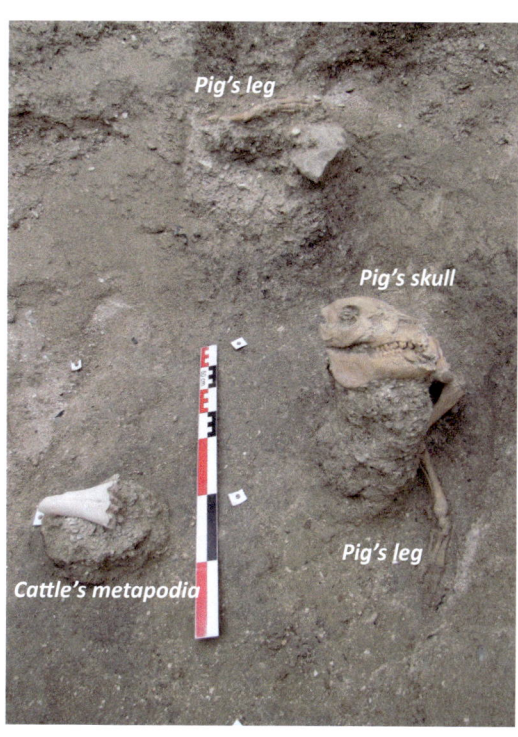

Figure 112: Menneville "Derrière le Village" (Aisne), LBK enclosure, feature 573, skull and articulated fore limb of a pig and the metacarpal of a young bovine (after Thevenet *et al.* 2014, fig. 41 modified; photo: UMR 8215 Trajectoires).

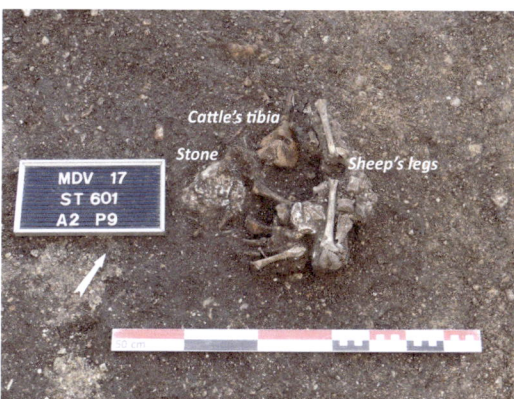

Figure 111: Menneville "Derrière le Village" (Aisne), LBK enclosure, feature 602, articulated fore and hind limb of a sheep, associated with a burnt, perforated limestone macehead and a bovine proximal tibia (photo: UMR 8215 Trajectoires).

Figure 113: Menneville "Derrière le Village" (Aisne), LBK enclosure, feature 600, dog deposited with pottery, a block of limestone and bovine thoracic vertebrae (photo: UMR 8215 Trajectoires).

The deposition ritual was complex. Animal deposition is not always associated with human remains in the ditch. To illustrate the phenomenon using a representative example, we will look in detail at the discovery in one of the ditch segments of a lamb, deposited at the bottom of the feature (Thevenet, 2016b), some 1.8 m below the stripped back ground surface (fig. 114). The animal was laid on its left side, the head to the west, and had been twisted so that the hind quarters were bent over the back, a position that could not be achieved unless the animal was tied or had had its belly cut open. On the basis of the skeletal epiphysation, the lamb had been slaughtered between 3 and 5 months of age, in other words between the months of June and August. Other bones had been buried with the animal:

- a pig's foot was placed parallel to the lamb's back, to the south of the vertebral column;
- a complete lumbar vertebra (the spinous processes of which were fractured) from a young bovine was placed to the north of the lamb's head;
- a fragment of radius/ulna diaphysis from a bovine was placed in front of the right foreleg (right patella) of the lamb; this diaphysis was fractured in a way that does not match the fractures normally encountered in settlement refuse;
- a bovine horn core was placed under the right ribs of the lamb.

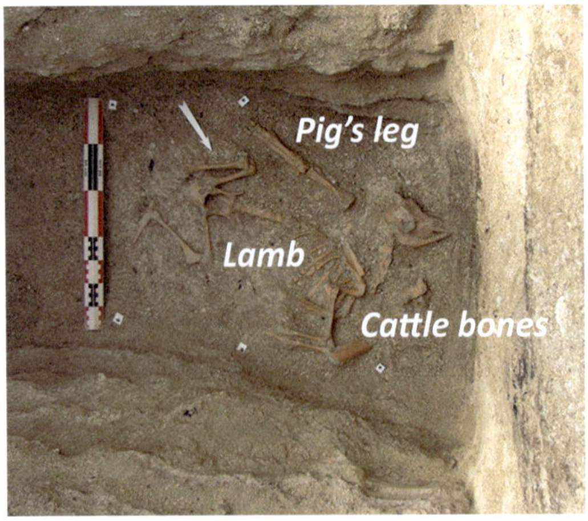

Figure 114: Menneville "Derrière le Village" (Aisne), LBK enclosure, feature 572, a lamb placed on its side in a restrained position, accompanied by a pig's leg and various cattle bones, namely a lumbar vertebra, a radius/ulna and a horn core (after Thevenet *et al.* 2014, fig. 40 and Hachem 2017, fig. 7; photo: UMR 8215 Trajectoires).

But faunal remains also occur in association with human inhumations. During the preparation of an article on the first excavation campaigns, we analysed the stratigraphy of the ditch segments and came to the conclusion that three broad groupings could be identified. The deepest layer (Group 1) included most of the complex inhumations, including those of children. Segment 273, for example, produced the inhumations of two children which were directly associated with animal remains and the deposit extends beyond the burial (Hachem, 2001). In general within the group, the bones were placed close to the deceased and there was a certain degree of separation between the cattle and the caprine remains.

The intermediate layer (Group 2) contained disarticulated human remains. The associated fauna was composed of cattle. Animal remains, particularly connected ribs bearing cut marks, horn cores and bucrania, can also occur in isolation.

The upper layer (Group 3) contained just a few human and animal bones. The rest of the fauna is composed of domestic animals and the presence of pig is a little more pronounced.

The most exceptional deposit was discovered in 2014 (Thevenet, 2016b) (fig. 115). It took the form of an accumulation of animal remains directly associated with the inhumation of an adult female at the extremity of a ditch segment (segment 571) in the lower part of the fill. Next to the corpse, 43 animal remains were deposited; these remains correspond to three incomplete domestic cattle- two adults and a young animal (in MNI).

The bones were disarticulated and partial, apart from a long section of vertebral column that appears to have been laid out to mirror the position of the corpse. Traces of cut marks caused by flint tools were evident on a lower jaw bone and attest to the skinning of the animal. The upper teeth show signs of heat shattering (fig. 116). The spines of three thoracic vertebrae also show traces of burning in the form of light brown discolouration and traces of ashy sediment. This is the first time that we have observed this kind of slight burning, which is quite different in aspect to that observed on burnt bone discarded in the lateral pits of houses. It leads us to envisage the slow roasting of meat over a hearth.

Two bucrania occur within the deposit (fig. 108). A fragment of red-deer antler palm was found to the right side of the woman's head (fig. 117). The charred remains of what seems to have been a small wooden container (tomodensitometry performed by T. Nicolas, Inrap) were found not far from the woman's left hand, near the side of the ditch. The hand itself was positioned on a small sandstone block, while fragments of a decorated pot lay scattered on the eastern edge of the burial deposit.

In addition to the bovine bones, the remains of caprines and pigs were deposited as entire carcasses or as articulated legs; four lambs and two piglets have been identified.

Other animals were placed directly beneath the corpse; these included at least four complete lambs and two piglets aged between 3 and 5 months, which suggests that they were slaughtered between June and August (fig. 118).

Figure 115: Menneville "Derrière le Village" (Aisne), LBK enclosure, feature 571, female burial (VII), surrounded by the incomplete remains of at least three cattle (bucranium, scapula, vertebrae, upper jaw, limb bones), the bucranium of a female aurochs, four lambs and two piglets, a worked red-deer antler, a broken ceramic vessel, a block of limestone and a burnt wooden container (after Thevenet *et al.* 2014, fig. 23; photo: UMR 8215 Trajectoires).

Figure 116: Menneville "Derrière le Village" (Aisne), LBK enclosure, feature 571, shattered teeth due to slow burning, in the upper jaw of a bovine deposited with burial VII (photo: L. Hachem).

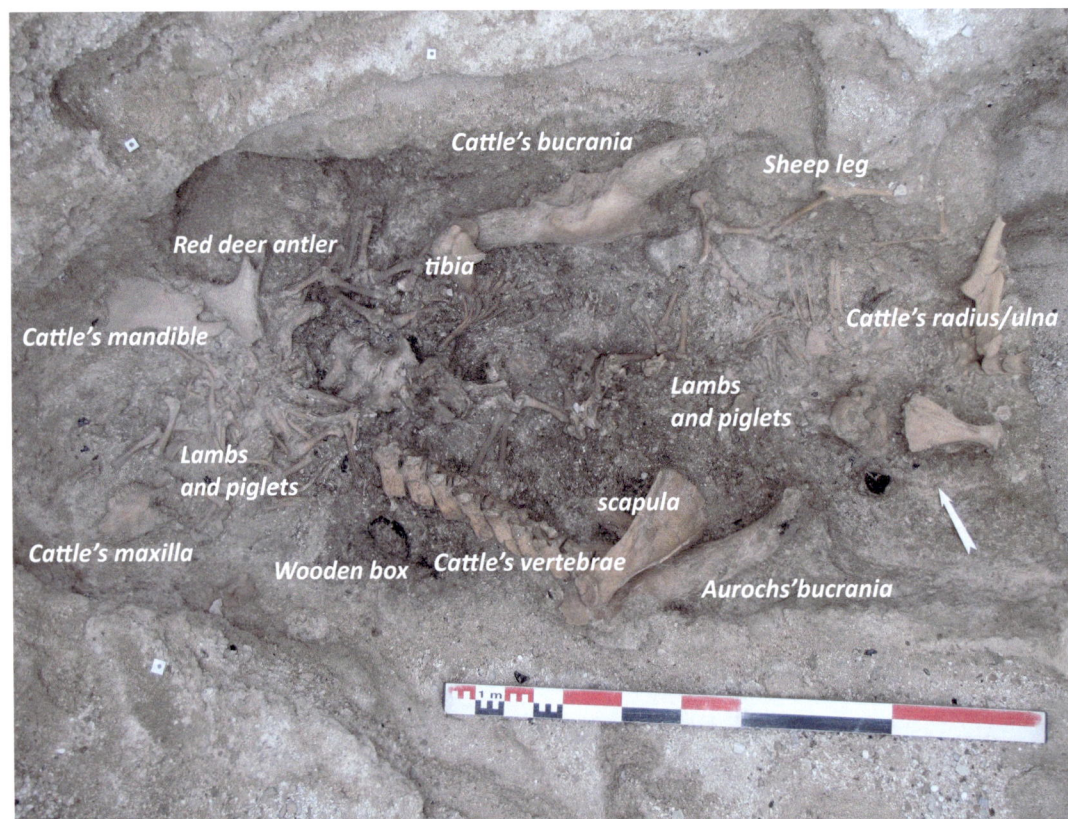

Figure 117: Menneville "Derrière le Village" (Aisne), LBK enclosure, feature 571, wild fauna deposited with burial VII, antler (located under the abdomen of subject VII); half-bucrania of aurochs deposited upside down (after Thevenet *et al.* 2014, fig.34 modified; photo UMR 8215 Trajectoires).

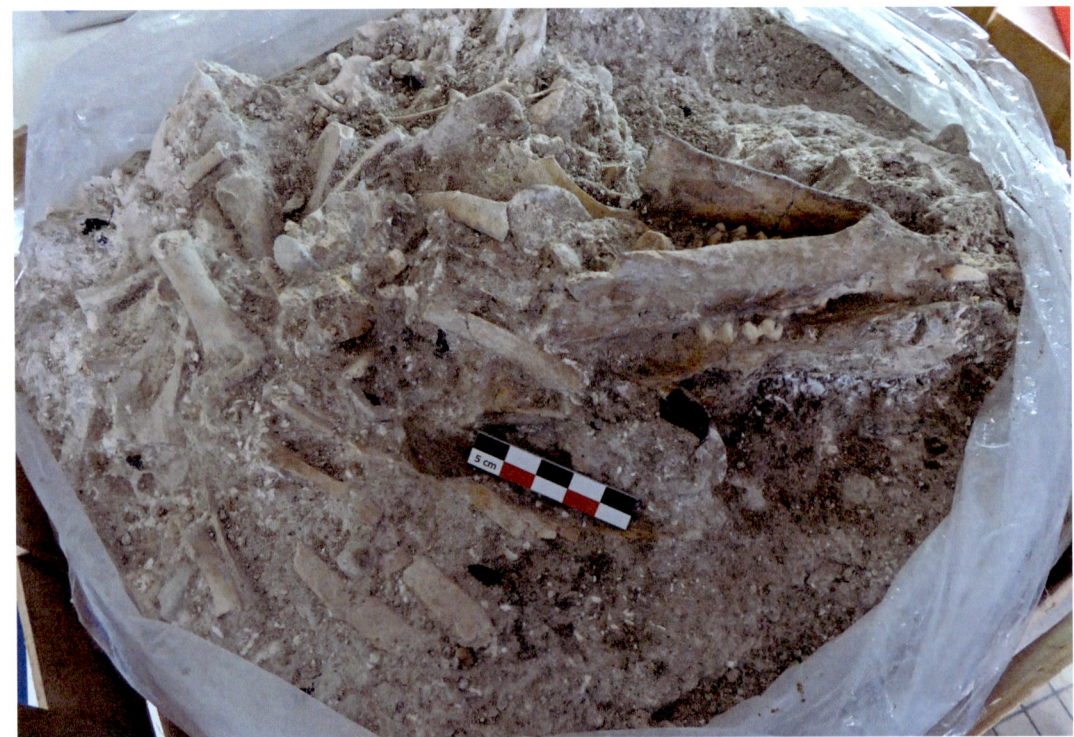

Figure 118: Menneville "Derrière le Village" (Aisne), LBK enclosure, feature 571, fauna deposited under burial VII, comprising at least four lambs and two piglets aged 3 to 5 months (after Thevenet *et al.* 2014, fig. 35 modified; photo: UMR 8215 Trajectoires).

Object-Signs

The sociologist J. Baudrillard (Baudrillard, 1972) developed the concept of the object-sign in modern society: a consumable resource not only has a material function, it can also have a social function, which allows differentiation between individuals.

In the field of Neolithic archaeology, P. Pétrequin uses this term to qualify the large jade axes that were items of exceptional symbolic and prestige value for Neolithic societies and which were the object of exchange (Petrequin *et al.*, 2017) . There are other object-signs, such as personal ornaments (LBK spondylus beads, BVSG schist bracelets, etc.), items such as pottery whose decoration is significant, and, as a more specific example, a bovine scapula decorated with punched motifs that was found on the site of Tinqueux (fig. 119). These types of objects are relatively easy to spot because of the particular care taken in their manufacture.

However, the ethnologist of techniques P. Lemonnier believes that it is practically impossible for an archaeologist to detect certain objects, which appear banal, but which were originally invested with an important role; this is not the case for ethnologists who have the advantage of being able to study objects within their original context. Divorced from this societal context, it is difficult to perceive the central role of such objects in the production and reproduction of the social relationships of the groups concerned due to their insignificant external appearance. Nonetheless, Lemonnier believes that some of these objects can be discerned by an attentive observer, because their form, mechanical characteristics and raw material indicate that they are something more than simply functional (Lemonnier, 2015). We can add a further element to these distinctive characteristics: archaeological context. It is context that provides archaeologists with clues regarding the special nature of these objects.

In what follows we are going to look at a number of cases that, in our opinion, fall within this category of object-signs; namely horns, tibiae, phalanges and teeth. These are not beautiful objects, nor are they rare or the result of hours of painstaking work; these are bones, some of which we might even fail to notice at first sight. It is our opinion, however, that in Early Neolithic society, they may have carried meaning that contributed to the construction of a world of shared ideas and practices (Lemonnier, 2010).

In the LBK and BVSG settlements of the Paris Basin, we sometimes uncover, in the lateral pits of houses (fig. 120) the bucrania or horn cores of domestic cattle or aurochs; These cranial elements were charged with meaning in the Neolithic (Chaix,

1981; Marciniak, 2008), which is why we believe their deposition to have been intentional and not a simple act of disposal.

At Cuiry-lès-Chaudardes, complete *bovinae* horn cores and bucrania are only known from certain houses and do not appear before the third phase of occupation of the site (fig. 121). In the majority of cases these houses can be qualified as unusual, two of them beacause of their large size (n°380 and 225), which led them to be interpreted as meeting houses (Gomart *et al.*, 2015), and a third (n°425), which in this case was small but which yielded the

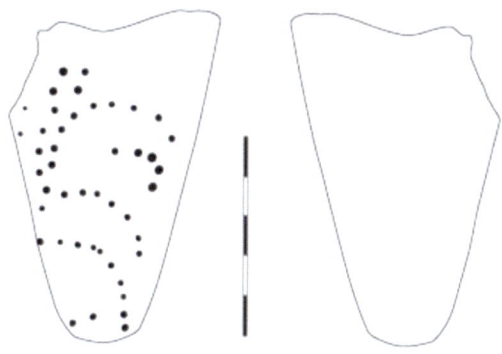

Figure 119: Tinqueux "la Haubette" (Marne), Early Neolithic BVSG settlement, feature 5, decorated tool made from the scapula of a large ruminant (cattle or red deer) (after Hachem *et al.* 2007, fig. 22. Photo: S. Oboukhoff; drawing: P. Allard).

largest assemblage of wild fauna from the whole site (fig. 122, see chapter 3).

Furthermore, the only aurochs bucranium from the site was found in the largest house which also happens to be the house with the largest assemblage of domesticated animal remains (house 225).

In the ceremonial enclosure at Menneville, horn cores and bovinae bucrania were directly associated with inhumation burials at the bottom of the ditch or were deposited in isolation (fig. 123).

In the intermediate layer of the enclosing ditch (Group 2), these *bovinae* cranial elements occur in isolation; however, a number of arguments can be made for a direct link between the skulls and the funerary ritual applied to the inhumations situated in the lower level. In fact, these elements are principally present in those segments of the ditch that also feature burials.

Moreover, several complete bucrania were placed over the exact locations of complex child burials -segments 188 and 189- (Farruggia *et al.*, 1996) which might lead us to interpret them as a sign of commemoration.

Another example is the one mentioned above from Menneville (Thevenet, 2016b), concerning

the deposition of two bucrania, one complete bucranium of a bull, and the other demi-bucranium of a female aurochs, which were placed face-to-face (the aurochs skull was upside-down) at the feet of a woman (fig. 118). The "mirrored" conjunction of the domestic and wild forms of *bovinae* is clearly evident in this example but it is also apparent in settlement contexts. At Dachstein (Alsace), for example, two complete bucrania, one of a domesticated bovine and the other of an aurochs, were buried together in an isolated pit (Schneider, 1980).

A particular element attracted our attention when examining the Early Neolithic faunal remains: a number of caprine and roe deer tibias bore an unusual perforation close to their lower end. The perforation could occur on one face of the bone or on both faces. The perforation on the front face is generally oblique and bean-shaped, while that on the opposite side is larger and less regular, almost as if a chip had been broken away (fig. 124). Since first noticing this feature we have established that it is recurrent and appears to carry some significance.

These perforated bones have been found on several RFBS and BSVG sites in the departments of

Figure 121: Cuiry-lès-Chaudardes "les Fontinettes" (Aisne). Inventory of complete cattle and aurochs horn cores and bucrania, pierced roe deer tibias and antlers found in the lateral pits of twenty-six LBK houses.

House n°	Rear unit	Phase	Dom./Wild of AHC Groups	House category	Pierced tibia	Roe deer antler	Cattle	Aurochs	NISP horn core
90	1	Aisne 1	Caprines/Aurochs	Herding	Sheep/Goat	Antler			
440	1	Aisne 2	Cattle/Wild boar/Aurochs	Mixed herding	Sheep/Goat	Hunted animal			
360	2	Aisne 3	Caprines/Aurochs	Herding	Sheep/Goat	Hunted animal	Horn core		16
280	2	Aisne 3	Caprines/Roe deer/Aurochs	Herding	Sheep/Goat		Horn core (male)	Bucrane	2
225	3	Aisne 3	Caprines/Roe deer/Aurochs	Herding	Sheep/Goat*	Hunted/ worked	Horn core	Bucrane	6
245	3	Aisne 3	Caprines/Roe deer/Aurochs	Herding	Sheep/Goat	Hunted animal	Horn core		2
126	1	Aisne 1	Pig/Wild boar	Hunting +	Roe deer	Hunted animal			
640	?	Aisne 1	Pig/Roe deer	Mixed hunting +	Roe deer	Hunted/ Shed	Horn core		8
400	1	Aisne 2	Cattle/Wild boar	Mixed hunting +	Roe deer	Hunted animal			
425	1	Aisne 3	Pig/Wild boar	Hunting +	Roe deer	Hunted animal	Horn core (female)		1
570	1	Aisne 2	Pig/Roe deer	Mixed hunting +	Roe deer				
320	1	?	Caprines/Aurochs	Mixed hunting +		Hunted/ worked			
635	1	?	Pig/Wild boar	Hunting +		Hunted animal			
112	1?	Aisne 1	Pig/Wild boar	Hunting +		Shed antlers			1
380	3	Aisne 2	Cattle/Red deer/Aurochs	Mixed		Shed antlers	Horn core (female)		22
580	1	Aisne 3	Pig/ Red deer	Mixed			Horn core (male)		1
80	1	?	Cattle/Roe deer/Aurochs						2
85	1	?	Pig/ Wild boar	Hunting +					5
89	1	3	Pig/ Wild boar	Mixed					1
126	1	3	Pig/ Wild boar	Hunting +					3
330	1	3	Pig/ Roe deer	Mixed					1
690	1	3	Pig/ Roe deer	Hunting +					3
450	2	?	Cattle/Red deer/Aurochs	Herding +					1
500	3	2	Pig/Red deer/ Aurochs	Herding +					1
11	3	1	Caprines/Roe deer/Aurochs	Herding+					1
650	?	?							1

* radius

Aisne, Seine-et-Marne and Champagne, from the beginning of the chronological sequence (Aisne 1) to the Final BVSG. The bones treated in this manner are mostly the distal tibias of caprines and roe deer, but the treatment has also been observed on proximal radius.

The site of Cuiry-lès-Chaudardes has yielded the largest number of these artefacts; they occur in the faunal assemblages of eleven houses dating to all the occupation phases, with a single example per house (Hachem, 2011a). Half of these houses contained the tibias of caprines, the other half those of roe deer. It turns out from our investigations that there appears to be a link between the proportions of species consumed in each house and the presence of these perforated tibias. In fact such a link appears to exist: in houses where there is a preponderance of domestic species, especially caprines, the tibias are of caprines, while in houses where hunting was important, they are roe deer tibias.

It is also worth noting that the perforated tibias come from houses that have also yielded roe deer antler, particularly antler from slaughtered animals. Yet, only half of the houses of the village were found to contain roe deer antler (fig. 125 and fig. 126).

We therefore think that there is a link between the roe-deer antler and the presence of perforated tibias (fig. 126).

Unlike red deer antler, roe deer antler is rarely worked because of its poor resistance (Y. Maigrot pers. com.). However, certain unshed roe deer antlers discovered in the lateral pits of LBK houses display deliberate transformation (fig. 121). This transformation consists of traces of polish on the tips of the tines which does not appear to have a practical function. A single example of one of these pointed tines is recorded from the enclosure ditch

Figure 120: Venizel "le Creulet" (Aisne), excavation of an Early Neolithic LBK pit, feature 136 (after Ilett *et al.* 2015, fig. 3-3; photo: M. Ilett).

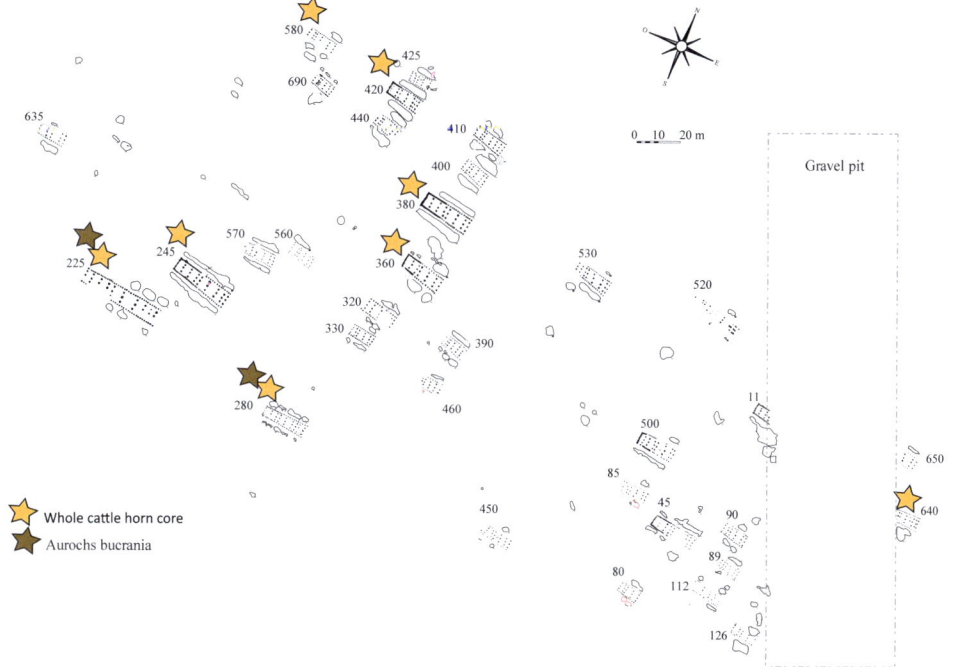

Figure 122: Cuiry-lès-Chaudardes "les Fontinettes" (Aisne). Distribution of complete cattle horn cores and aurochs bucrania.

Figure 123: Menneville "Derrière le Village" (Aisne), LBK enclosure, feature 600, a female aurochs bucranium accompanied by a dog skull, pottery sherds and blocks of sandstone and lime-stone (photo: UMR 8215 Trajectoires).

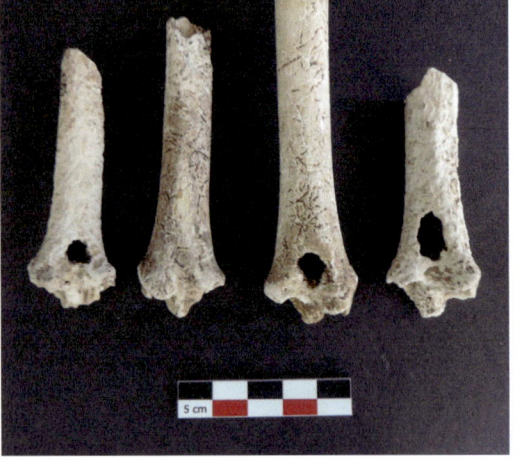
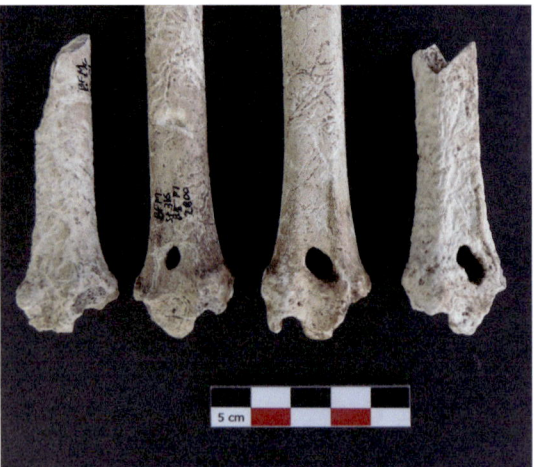

Figure 124: Pierced caprine tibias dating to the Early Neolithic (BVSG) in the Aisne and Aube, faces A and B. Bottom: detail of keyhole-shaped perforation, faces A and B, Pont-sur-Seine "le Haut de Launoy", feature 7205, phase 3 (after Hachem 2015, fig. 176, modified). Top, from left to right: 1- Bucy-le-Long "le Fond du Petit Marais", feature 376; 2- Bucy-le-Long "le Fond du Petit Marais", feature 316; 3- Villeneuve-Saint-Germain "les Grandes Grèves", feature 112; 4-Buchères "Parc Logistique de l'Aube", feature 1563.

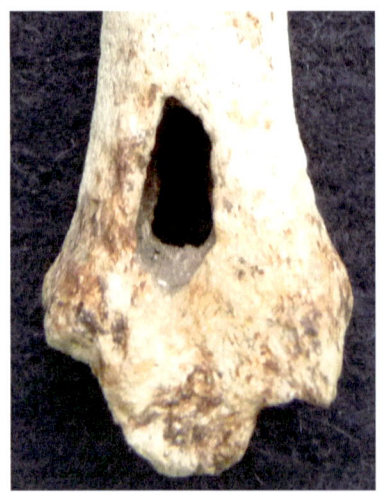

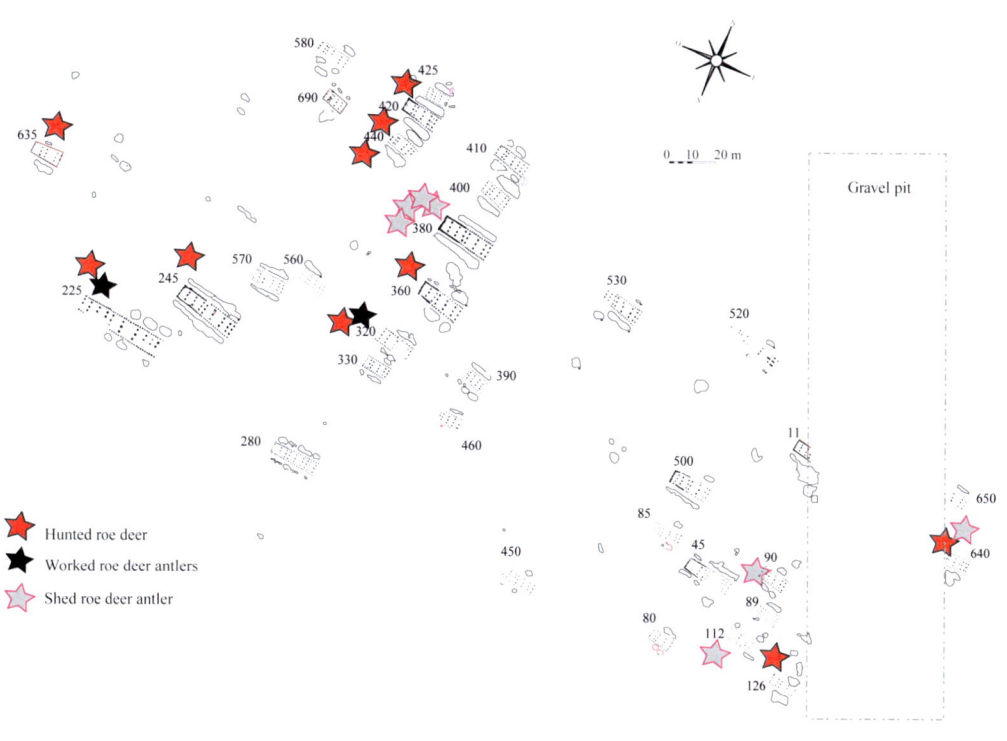

Figure 125: Cuiry-lès-Chaudardes "les Fontinettes" (Aisne). Distribution of roe deer antler.

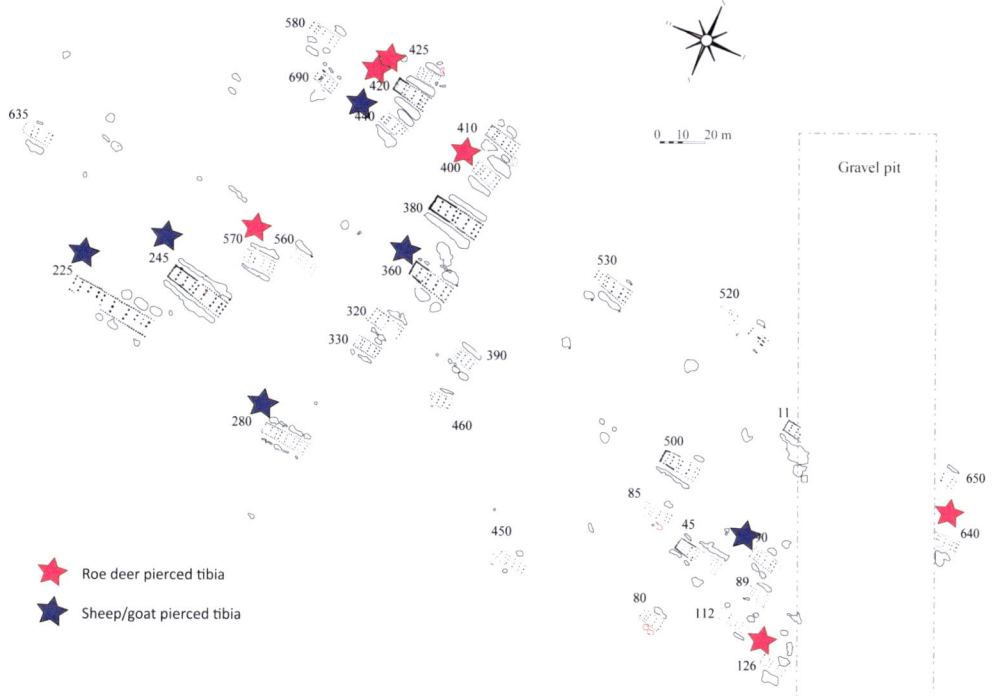

Figure 126: Cuiry-lès-Chaudardes "les Fontinettes" (Aisne). Distribution of pierced caprine and roe deer tibias.

at Menneville (fig. 127). It was discovered inserted between the ribs of the left thorax of a woman whose skeleton bears traces of violent trauma (Thevenet, 2013, fig. 65; Thevenet *et al.*, forthcoming). It seems plausible to suggest that these transformed roe deer antlers may also have had a role as object-signs.

Perforated tibias are also found on BVSG sites (fig. 124). On the site of Villeneuve-Saint-Germain (Aisne), an example from the lateral pit of an Early BVSG house, was an almost complete caprine tibia which sets it apart from other examples where only the distal epiphysis generally survives. The proximal epiphysis is broken, but a perforation is discernible, which raises the possibility that these objects, when they were complete, were originally perforated at both ends. The perforation of the distal epiphysis is regular and only visible on one side of the bone. At Bucy-le-Long "le Fond du Petit Marais" (Aisne), two distal caprine tibias were found in the lateral pits of two different houses dating to the Middle or Late BVSG. One of the tibias displays a worked surface above the perforation. At Balloy "les Hauts de Borne" (Seine-et-Marne), a caprine distal tibia was again discovered in the lateral pit of an Early BVSG house. In this case the perforation is visible on both sides of the bone. Lastly, at Buchères "Parc Logistique de l'Aube" (Aube), an isolated pit dating to the Final BVSG produced a roe deer distal tibia, perforated on both sides. This pit (feature 1563) contained an exceptionally large quantity of game (150 remains

representing at least three wild boars, one red deer and one roe deer).

In all of these cases, a relationship exists between the predominant species consumed in the house and the species from which the tibia originates. Caprine tibia come from houses with a very high percentage of caprines, while the roe deer tibia from Buchères came from a pit containing nothing but game (principally wild boar and red deer). An interpretation of these objects, which do not fall into the categories of personal ornament or worked bone, might be found through anthropology, and particularly the branch that studies technical know-how and highlights the role of objects and material actions (Revolon *et al.*, 2012). Clearly, many possibilities present themselves, nonetheless, the range of possibilities is not infinite, because, as pointed out by the ethnologist Maurice Godelier (1999, 19):

> *"Societies are always particular, their histories are always singular, and yet we observe that similar processes are reproduced. The process of the emergence of chiefdoms is reproduced in various eras and in societies that have had no contact with each other. There is, therefore, something in social logic that always leads to this result (…). There is not an infinity of sociological forms, there exist multiple variations of a few types and there exist relationships and transformations between these types."*

Taking inspiration from this notion, we have explored what comparisons with other societies might offer in terms of object-signs that are technically and aesthetically simple but which are deemed to be "irreplaceable". This leads us to two hypotheses. The first is that the perforated tibiae could be objects similar to the *illas* of the highlands of Peru and Bolivia. In these areas, where animist beliefs attribute active social identities to places and things, *illas* are believed to be living entities representing a powerful source of fertility for animals and crops (Silar, 2012). *Illas* are passed down from generation to generation and can take the form of pebbles, other strangely shaped objects found by members of the household or even small carved stones.

> *"Illas are not objects destined to be exhibited; they play no role in social hierarchies, they are solely intended for personal and domestic use. Nevertheless, they play a role in maintaining value systems and commitments of family members. The raw material and the appearance*

Figure 127: Menneville "Derrière le Village" (Aisne), LBK enclosure, feature 563, pointed object made from a roe-deer antler and found in the left side of the thorax of burial IV (after Maigrot 2013, fig. 80; photo: Y. Maigrot).

of the illa are unimportant, what matters is the relationship that each illa has with its household." (Silar, 2012, 7).

The second hypothesis, which has something in common with the first, is that the perforated tibiae are similar to the *"lekan"* of certain animist hunting communities in Siberia (Lot-Falk, 1953). The *"lekan"* is a crafted object, often of animal origin, which acts as a receptacle or prison for a wandering spirit known as the *"Ongone"*. In return, the spirit will no longer cause illness or drive away game; instead it will aid the hunters. The *Ongone* is prayed to and fed, but if hunting is unsuccessful over a prolonged period, the spirit is beaten and destroyed. The *Ongones* of the ancestors are transmitted from generation to generation but the lifespan of a *lekan* can vary. If the spirit deserts its container, then the latter is discarded without ceremony or abandoned on the roadway.

A third example are the "fetish objects" found in certain African societies. These are roughly fashioned objects, often made using a combination of raw materials, which are common, for example, among the Bwaba of Burkina Faso: this is a society of farmers, blacksmiths and weavers who share a community religion (Coquet, 1987). Generally portable, these objects are supposed to be kept close to the body or to remain in the house and one of their properties is that they are inhabited by a "force" or "power" which can be either good (protection) or evil (witchcraft).

In our opinion, these three ethnological cases, which share a common base, provide plausible explanations for the Neolithic perforated tibias: the evolution of these objects as domestic amulets for the protection of livestock and the assurance of plentiful game. Their connection to the household is, we believe, underlined by the fact that they occur singly in the pits of LBK and BVSG houses. Thus, the presence of these perforated tibias in the refuse pits could be the result of their intentional breaking and the abandonment of the distal portion. This could have occurred without the objects being put completely out of use, since the breaking through of the surface opposite the original perforation is not always present. An area of wear is visible on the shaft of two of the bones, indicating that the objects had been manipulated. It is therefore probable that the perforation originally held something made from perishable material, perhaps wood. As the object was used this piece of wood gradually bored through the bone before emerging on the opposite side.

Certain rare wild animals, such as wolf (fig. 128), bear and horse, occur in exceptionally small proportions in the faunal remains from settlement

Figure 128: Wolf remains are rare in Neolithic sites and include cranial elements (upper and lower jaws, etc.), fibulas and phalanges which probably had a symbolic significance (photo:Wikilmages, Pixabay).

Figure 129: Buchères "Parc logistique de l'Aube" (Aube), feature 1563 dating to the Final BVSG; a complete male aurochs phalange bearing traces of slow burning, discovered in a pit containing only wild animal remains, including a red deer antler hammer and a pierced roe deer tibia (photo: L. Hachem).

sites, a fact that qualifies them as "rare species". While they could have been hunted for their skins, which happen to be quite difficult to work, the bones themselves point us to an another hypothesis.

The skeletal elements found on each site are unusual and show no traces of transformation. In each case the elements are recurrent and consist of teeth, autopods – such as phalanges and taluses – and sometimes metapodials.

In a way, the treatment of the remains of these rare species can be likened to that of certain aurochs bones. We have seen how aurochs phalanges and metapodials might attest to the sharing out of an animal during LBK communal feasts (see chapter 3). It is interesting to note that in the isolated BVSG pit at Buchères "Parc logistique de l'Aube" (feature 1563), which as we have seen contained only game remains and a perforated roe deer tibia, the only element of aurochs present was a phalange. The bone came from a male and bore visible, brown traces of slow burning (fig. 129).

Furthermore, even though they relate to different domains, the indications of the singularity of these species are worth noting.

Thus, in addition to the fact that bear teeth can be used as pendants, a statuette of a bear's head was found on the LBK site of Merzenich-Valdersweg, in Germany (Husmann and Cziesla, 2014).

Moreover, on the settlement at Menneville, the large burial pit mentioned previously (feature 93) contained a horse mandible, along with a caprine foot and the upper jaw of a bovine (fig. 130).

Finally, while wolf was present in only two of the thirty-three houses at Cuiry-lès-Chaudardes, one came from what is interpreted as a meeting house (n° 380) and the other from the house with the largest proportion of game remains on the entire site (n°425). The same is true for horse and bear.

Given these elusive but recurrent elements, it appears that the bones of these rare species were initially vested with a symbolic value. If we consider the danger involved in tracking and killing these animals, particularly the adult males, and given that at least two of them – the bear and bull – have symbolized brute force for centuries (Pastoureau, 2007), the role of their remains as hunting trophies must be considered (Hell, 2012). However, these artefacts could equally have been amulets or objects used in divination, just as bear paws are used in certain Siberian shamanic rituals today (Stépanoff, 2009).

While the European pond turtle (*Emys orbicularis* L. 1758) is relatively well represented in the South of France, it is rarely found on northern sites.

In settlement contexts, turtle bones have been recorded in the lateral pits of houses dating to the LBK, as, for example, at Berry-au-Bac "le Vieux Tordoir" and Bucy-le-Long "La Fosselle" (Aisne) (Hachem, 2018a); they have also been found in pits of BVSG date, as at Longueil-Sainte-Marie "la Butte-de-Rhuis II", Oise (Bostyn *et al.*, 2015) and Fresnes-sur-Marne "les-Sablons" (Brunet, 1992).

The Middle Neolithic

Turtle shells

Turtle shells have been found in funerary contexts and at the bottom of an isolated pit dating either to the Early/Middle Neolithic transition or to Middle Neolithic I.

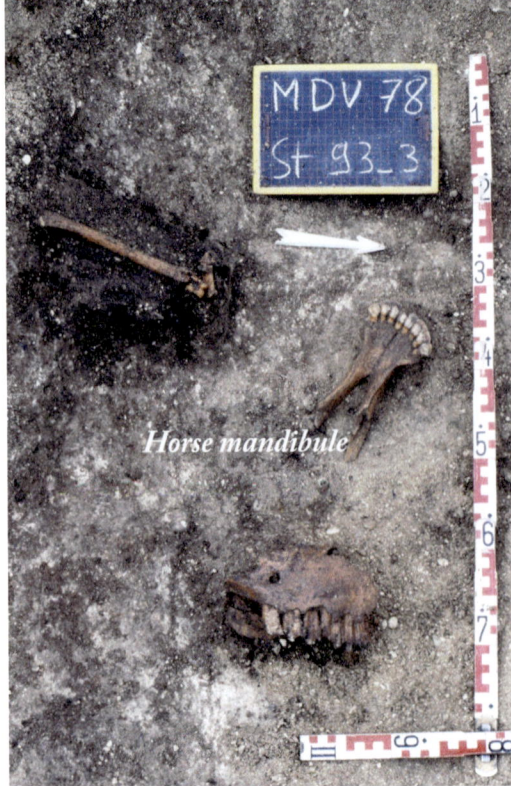

Figure 130: Menneville "Derrière le Village" (Aisne), LBK settlement, feature 93; a horse mandible discovered at the base of a pit containing three burials, accompanied by the upper jaw of a bovine (after Thevenet 2014, fig. 11; photo: UMR 8215 Trajectoires).

Thus, turtle has been recorded in two female burials. Tomb 269 at Buthiers-Boulancourt (Seine-et-Marne), attributed to the very end of the BVSG, or the early beginning of Cerny Culture, contained a tortoise shell (archaeozoological analysis C. Bemilli, Inrap) along with several lithic artefacts (Gosselin and Samzun, 2008). The grave goods were deposited in a pit to the south-east of the head of the deceased who had been inhumed in a flexed position on a layer of ochre.

The other example is Tomb 47 at Balloy "les Réaudins" (Seine-et-Marne), which probably dates to the Middle Neolithic I, Cerny Culture (Mordant, 1997). In this case the body of a woman was laid out in a supine position and a turtle shell was placed against her right shoulder, open side up, mirroring the rotated position of the skull.

In the context of isolated pits, a turtle shell was discovered at the base of a pit at Villiers-sur-Seine "le Défendable", Seine-et-Marne (Quenez, 2020) (fig. 131)[1]. The upper shell was complete and turned with its internal face facing upwards. The lower shell had been broken in two. Two bones (right and left) from the scapular girdle were also present; none of the remains bore cut marks (Hachem, 2020a).

Finally, mention must be made of an isolated deep pit discovered at Neuville-sur-Oise (Val d'Oise),

Figure 131:
Villiers-sur-Seine "le Défendable", pit 1241, the turtle shell *in situ* (after Quenez 2020, photo: J.-Ph. Quenez).

which contained four turtle shells deposited at its base (Marti, 2013). While the dating of this pit appears to be later (Final Neolithic), the incongruous presence of these turtles is interesting. A slightly later human cranial cap was found in the upper fill of this pit.

The use of turtles for medicinal and symbolic (as well as alimentary) purposes is well attested in Mesopotamia, Anatolia and in Arab countries of the Middle East where turtle shells have been discovered in tombs dating to various phases of the Neolithic (Berthon *et al.*, 2016). Turtle remains are relatively common in the evolved Neolithic (5th millennium) of Hungary and Poland (especially the Brzesc Kujawski group) (Bogucki, 1988) .

Tortoises are vested with numerous symbolic values and are frequently associated with longevity and strength in traditional societies. The archaeological evidence indicates that they played the role of an animal *psychopomp* (soul guide) who escorted the deceased to the afterlife; their presence might, therefore, be associated with shamanic activity.

Even though they are located in different areas and belong to different cultures, it is tempting to envisage such an explanation for the presence of turtles in the two female burials in the Paris Basin.

A monumental building

An occupation layer associated with a monumental building, both dating to the Middle Neolithic I, Cerny Culture, were discovered in the Aisne valley at Beaurieux "la Plaine" (Colas, 2008; Colas *et al.*, 2018). The building, measuring 80m in length and 20 m in width, is clearly of Danubian inspiration in terms of its shape, but its sheer size represents an architectural feat and a clear break with buildings constructed in the Early Neolithic (fig. 132).

The fauna from the occupation layer near the building amount to 349 poorly preserved remains. The assemblage is largely composed of domestic animals (78 %), with bovines predominant (68 %) (Hachem, 2018b). The second species represented is pig (around 9 %), followed by red deer and aurochs (both at 6 %). The presence of caprines as well as wild boar is very marginal, with only one bone in each case.

Figure 132: Beaurieux "la Plaine" (Aisne), Middle Neolithic 1 (Cerny), view of the monumental building after excavation (after Colas *et al.* 2018, fig. 9, modified; photo: G. Naze).

The composition of this assemblage is unusual. There are only adult animals and most of these are males, whether cattle or aurochs. Furthermore, the absence of caprines, roe deer and small wild animals is surprising.

In addition, very many of the bones are burnt. This is perhaps linked to the particular function of the site, which does not appear to have been a settlement and has been tentatively interpreted as a "sanctuary" (Colas *et al.*, 2018).

Deposits in enclosures

The occurrence of deposits of particular animal bones is a frequently observed phenomenon on enclosure sites of the Cerny Culture (Bedault, 2007; Mordant, 1997), the Michelsberg Culture, in both France and Germany (Guthmann, 2010; Höltkemeier, 2013) and the northern Chasséen (Méniel, 1984; Hachem *et al.*, 2016). These deposits involve elements that differ from normal every-day bone refuse in many respects: the species selected, the presence of intact bones, an over-abundance of certain carcass parts, the presence of articulated bones and isolated bones

(fig. 133). A round table discussion was held on the subject of deposition in various periods and a data base has been created which contains the sites mentioned in this work (Auxiette and Méniel, 2013).

Five broad categories of faunal deposits have been identified with the aim of facilitating archaeozoological analyses (Auxiette and Ruby, 2009).

The application of this system of categories has facilitated comparison of faunal deposits from Michelsberg enclosures in France and Germany, revealing their similarities as well as their differences (Höltkemeier, 2013).

Even though there is considerable variation between sites, a common pattern emerges in which domestic species are in the majority. Cattle are the species most often encountered in these deposits, particularly in the form of bucrania and horn cores. Caprines, pigs and dog are also represented to varying degrees depending on the site.

Wild species present are generally represented by cranial elements: antlers in the case of deer, horn cores in the case of aurochs and mandibles in the case of wild boar.

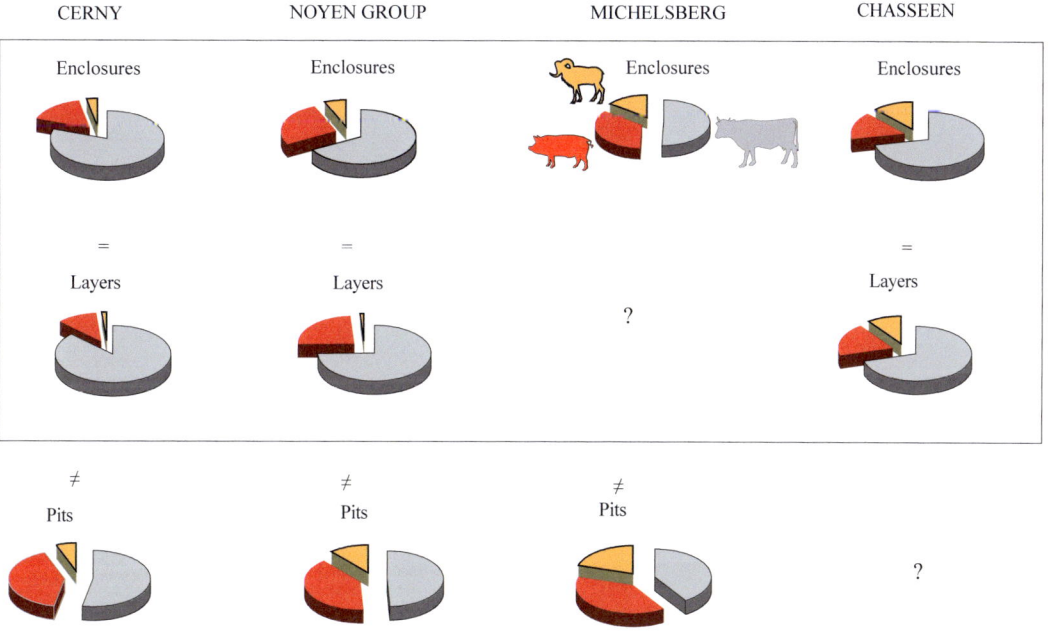

DOMESTIC ANIMALS

Figure 133: Proportion of domestic animals according to archaeological context in the Middle Neolithic 2 in northern France (33 sites, NISP = 54 308). Sites composed of pits yield less cattle and more pigs than enclosures and occupation layers (after Hachem 2011b, table 1 and fig. 5, modified).

Figure 134: Passel "le Vivier" (Oise), cattle bucrania *in situ*. 1- Feature 227, general view and detail of a cattle bucranium. 2- Feature 205, isolated horn core. 3- Complete bucranium associated with pottery. 4- Feature 20, bucrania and horn cores associated with pottery and macrolithic tools (after Hachem *et al.* 2016, fig. 13 modified; photos: N. Cayol).

1.

2.

3.

4.

Figure 135: Bazoches-sur-Vesle "le Muisemont" (Aisne), Middle Neolithic 2 enclosure (Michelsberg), faunal deposit with pottery (photo: UMR 8215 Trajectoires).

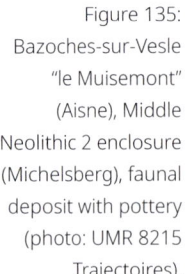

The Chasséen enclosure at Passel in the Oise has produced numerous examples of such deposits (Hachem *et al.*, 2016, 2017; Cayol, forthcoming) and this is why it has been chosen here as a case study. It should be noted, however, that all Middle Neolithic enclosures display some degree of evidence for this kind of cult activity.

Cattle and aurochs horn cores and bucrania have been found on several enclosure sites belonging to the Middle Neolithic I Cerny Culture.

For example, on the site of Gurgy three aurochs horn cores were placed at the junction between two palisades (Bedault, 2007). Another example occurs in the Balloy enclosure where two adult cattle half-mandibles were arranged along with an upturned calf mandible to form the outline of a bucranium; these were placed on top of several cattle bones and four complete ceramic vessels that had been crushed in situ (Augereau and Mordant, 1993; Tresset, 1996,

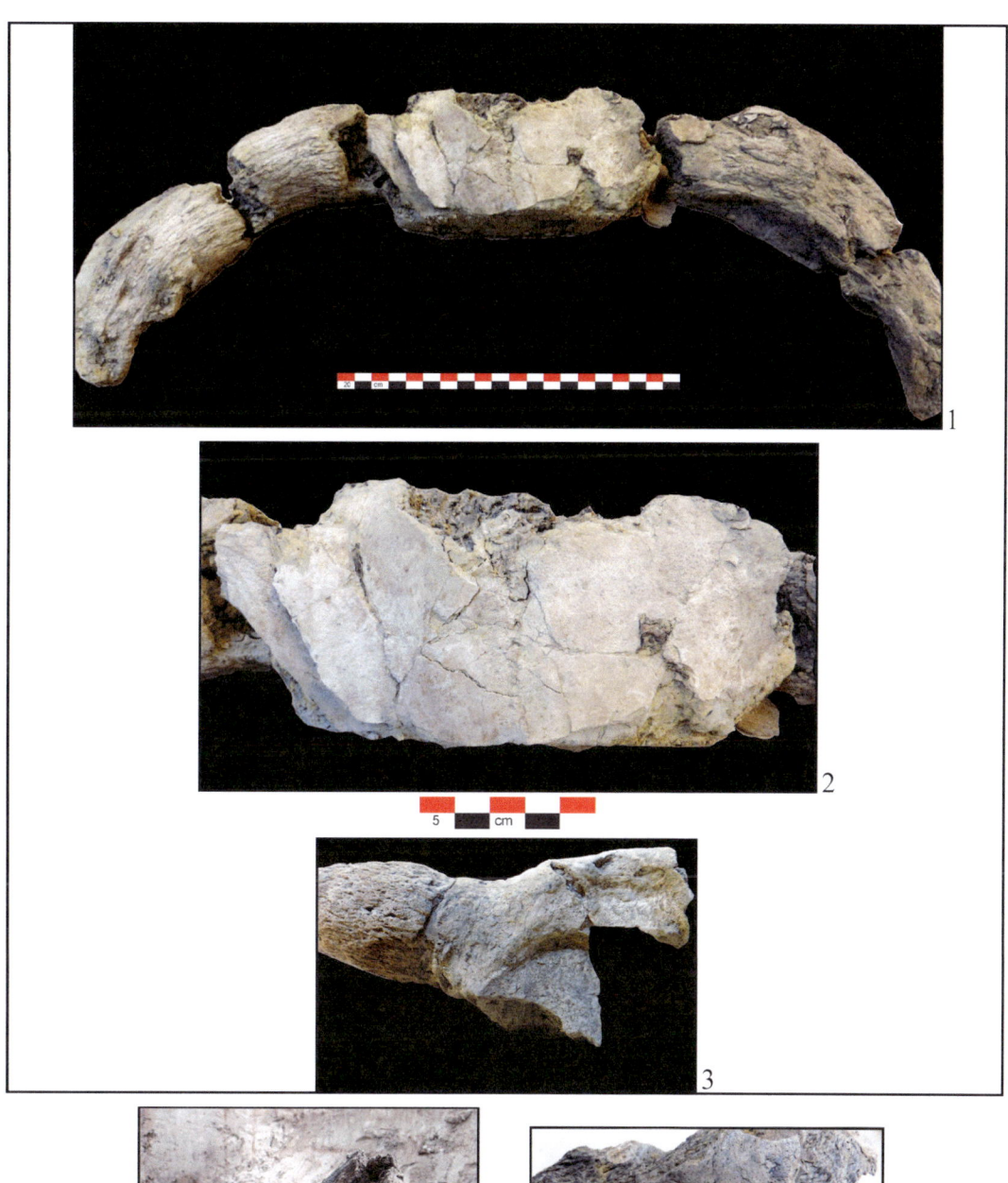

Figure 136: Passel "le Vivier" (Oise), Middle Neolithic 2 enclosure (Northern Chasséen), procedures used for cutting up bovine bucrania: 1- Example from feature 20. 2- Detail of frontal view. 3- Detail of nuchal view. 4 Complete bucranium associated with pottery and exhibiting cut marks made with a flint edge. 5- Detail of cut marks (after Hachem *et al.* 2016, fig. 14, modified; photos: L. Hachem).

fig. 26). A short distance away, a bucranium was found in association with two vessels. A further cattle bucranium was found in another nearby ditch segment.

The practice of depositing cranial elements became even more common in the Middle Neolithic II.

A remarkable number of bucrania and horn cores were found at Passel: thirty bucrania and thirty horn cores. No other site in northern France has produced such a large quantity of deposits of this kind. They are often associated with complete (inverted) pottery vessels and sandstone grinding tools (fig. 134); this association of fauna with pottery and lithic artefacts (fig. 135) has also been observed in the the Michelsberg enclosure at Bazoches-sur-Vesle (Dubouloz *et al.*, 1997).

The poor state of preservation of these bone remains, which were often crushed *in situ* and, in the case of Passel, had suffered as a result of fluctuations in the water table, often makes it difficult to distinguish between wild and domestic animals. The distinction cannot be made in the case of young animals, which are also present in these deposits. Where there is uncertainty the term *bovinae* is used here. However, when gender and sex could be identified, the analysis revealed a predominance of domestic cattle, with many more bulls than cows.

Aurochs bucrania and horn cores are much rarer. At Passel, a majority of these cranial elements are from female aurochs.

The way the skulls were cut up to form the bucrania is standardised, with a rectilinear strip of frontal bone (c. 7 cm long and 5 cm high at the back of the skull) retained between the two horn cores (fig. 136). Traces left by flint tools are sometimes visible on the frontal bone and attest to the skinning of the animals. In contrast, the bases of the horn cores display no traces that might attest to the removal of the horn sheath. In

numerous cases, indentations have been observed on either side of the frontal bone and their origin, whether taphonomic or human, is still unclear.

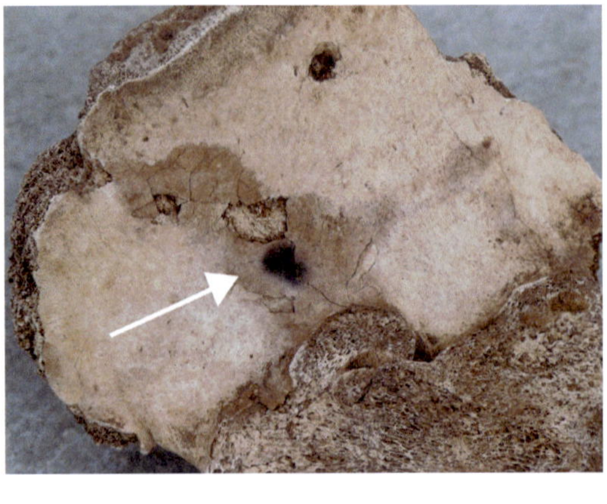

Preliminary spatial analysis for the Passel site indicates that the bucrania are concentrated near the ditch interruptions, one of which was probably a major passage-way. In the other enclosures, these bucrania are again deposited in specific areas; at Vignely "la Noue Fenard", for example, a male aurochs bucranium and two horn cores were discovered in the middle of a segment of the outer ditch (st 1014) which also contained human bones, indicative of secondary burial.

Unlike the bucrania, which are particularly concentrated at certain interruptions, the single *bovinae* horn cores at Passel are regularly spaced within the ditch segments. This phenomenon is particularly marked at the interruption situated on the north side of the enclosure (Hachem *et al.*, 2016, fig. 17). This interruption also yielded a concentration of *suinae* skulls and mandibles as well as a large quantity of waste material. This was almost certainly a principal entrance to the site.

The isolated cattle and aurochs horn cores sometimes display enigmatic circular patches of burning. In the enclosure at Crécy-sur-Serre, an isolated horn core was completely burnt (Hachem, 2015c). In the same enclosure, burnt spots were also recorded on several bones (fig. 137).

The presence of complete scapulae, which are generally rare in refuse assemblages from the Middle Neolithic I and II, is worth highlighting.

For the Middle Neolithic I, Cerny Culture, in Calvados, scapulae have been found in the ditches and graves of long funerary monuments ("STP") and within the tombs themselves (fig. 138).

In the Middle Neolithic II enclosure at Passel, complete scapulae occur relatively frequently. In certain cases they were found in association with fragments of broken pottery and cattle horn cores.

Occasionally, semi-articulated skeletal material has been found. For the Middle Neolithic I, complete calf bones have been found in association with deposits of pottery and mandibles in the Balloy enclosure (see section above dealing with bucrania).

In the Middle Neolithic II, we also find the limbs of very young calves, as for example at Maizy (Hachem, 1989) and Carvin (Hachem, 2014). But pieces of vertebral column are more common in enclosures belonging to the Michelsberg, Chasséen and Spiere cultures.

Figure 137: Crécy-sur-Serre "le Bois de Sort/la Croix Saint-Jacques" (Aisne), Middle Neolithic 2 enclosure (Michelsberg), example of burnt spots observed on certain bones, in this case a bovine sacrum (photo: L. Hachem).

To the best of our knowledge, the only complete calf carcasses found to date are from the Chasséen enclosure at Boury-en-Vexin, where a deposit level yielded a total of 15 individuals (Méniel, 1987a).

Perforated phalanges are not "worked" bones in the classic sense of the term because the perforation is generally roughly executed; it takes the form of a hole in the middle of the diaphysis that either passes right through the bone or pierces just one face. Perforations on the first phalanges of domestic cattle and aurochs (and red deer) have been recorded from several Middle Neolithic II enclosure ditches belonging to the Michelsberg Culture in Germany (Höltkemeier, 2013); the site of Bruchsal "Aue", for instance, yielded 69 examples (Steppan, 2003).

There is some evidence to suggest that these objects first appeared at an earlier period in the

Cattle scapula

Sheep deposit

Deceased inhumated

Figure 138: Fleury-sur-Orne "les Hauts de l'Orne" (Calvados), Middle Neolithic 1 (Cerny), grave 26-5, an example of animal offerings deposited in a burial monument; cattle scapulae and complete sheep (after Ghesquière and Hachem 2018, fig. 9, modified; photo: L. Hachem).

North of France but this cannot be verified at present because only three sites have produced evidence for them: one example was found in the LBK site of Cuiry-lès-Chaudardes (Hachem, 2011a), one example was found in a BVSG pit at Buchères "Parc Logistique de l'Aube" in Champagne (st 16, D 39, (Hachem, 2019) and the

third was discovered in a Cerny layer adjacent to the monumental building at Beaurieux "la Plaine" (Colas *et al.*, 2018).

Less frequent than in the Middle Neolithic II of Germany, a few perforated phalanges are nonetheless known from the Michelsberg and Chasséen enclosures of the North of France, as illustrated by an example from Villers-Carbonnel (fig. 139).

The enclosure at Saint Martin-de-Fontenay (Calvados), dated to the end of Middle Neolithic I and the beginning of Middle Neolithic II, has yielded a semi-complete skull of a female wild boar (Hachem, 2013).

The skulls of pigs and wild boars have been found in pits within and outside the Chasséen enclosure at Passel (Hachem *et al.*, 2016). These are either complete skulls or semi-complete skulls, essentially the occipital and parietal portions. Of particular note is a complete wild boar skull which was found under intentionally broken sherds of pottery, in association with five tranverse arrowheads and cattle bucrania (fig. 140).

The skull of an adult sow was discovered at the centre of the deposit of numerous caprine and bovine remains at the Chasséen enclosure at Boury-en-Vexin (Méniel, 1987a).

In general the Michelsberg enclosures do not feature many pig skulls, a rare example of which

Figure 139: Villers-Carbonnel "la Sole d'Happlincourt" (Somme), Middle Neolithic 2 enclosure (Northern Chasséen), pierced bovine phalange, probably a symbolic object (after Hachem *et al.* 2016, fig. 11; photo: L. Bedault).

Figure 140: Passel "le Vivier" (Oise), Middle Neolithic 2 enclosure (Northern Chasséen), wild boar skull accompanied by intentionally broken pieces of sandstone, five transverse arrow heads and cattle bucrania (after Cayol *et al.* 2017; photo: L. Hachem).

was discovered in a ditch segment at Bazoches-sur-Vesle (Hachem, 1987).

Complete pig or wild boar mandibles have been discovered on enclosure sites. They are quite rare but appear to occur from as early as the Middle Neolithic I as evidenced at the site of Saint Martin-de-Fontenay in Calvados. They are distinguishable from discarded refuse by their completeness: a complete mandible consists of the two halves of the lower jaw, which are joined together at the symphysis; the canines are generally missing.

For the Middle Neolithic II, the Michelsberg site at Crécy-sur-Serre has yielded an intact wild boar mandible (fig. 141) and two isolated domesticated pig mandibles which were found in two different segments of ditch (Hachem, 2015c).

The Chasséen enclosures at Villers-Carbonnel and Passel have also yielded examples. In the case of Passel, it is not always possible to identify the sex of the animal. However, when the canines are preserved and where measurements are possible, it seems that for pigs males and females are represented in equal proportions whereas for wild boars female animals predominate.

Similarly, the complete, or almost complete, mandibles of *suinae* were discovered at regular intervals within segments of the Middle Neolithic

ditches at Duntzenheim in Alsace (Guthmann and Arbogast, 2011).

Pig skeletons rarely occur among deposits of animal bone. The complete skeletons of five very young piglets and a sow were found at Boury-en Vexin, along with a few piglet bones (Méniel, 1987a).

A certain number of pigs, both complete and incomplete, have been discovered in the Passel enclosure, where they seem to have been preferentially deposited in the outer ditch (fig. 142). Finally, at Maizy, the front and hind limbs of a piglet were discovered in the inner ditch.

Caprines played an important role in the rites practised during gatherings held in the enclosures (fig. 143).

A particularly impressive example is the site of Boury-en-Vexin "le Cul Froid" (Méniel, 1987a), where an excavation of a 12 m segment of ditch revealed a sequence in which a number of sterile layers were overlain by deposits of faunal remains, ranging from isolated limbs to complete skeletons, which in turn were overlain by human remains and finally by layers of refuse (see chapter 3). All of the animals are domesticated species and caprines predominate: sheep (MNI =24), goats (MNI = 3), cattle (MNI = 6), pig (MNI =3) and dog (MNI =1).

Figure 141: Crécy-sur-Serre "le Bois de Sort/la Croix Saint-Jacques" (Aisne), Middle Neolithic 2 enclosure (Michelsberg), wild boar mandible deposited in feature T29 (photo: L. Hachem).

Figure 142: Pig increases in importance over the course of the Neolithic (photo: R. Owen-Wahl, FreeImages).

Figure 143: Goats were sacrificed in Middle Neolithic 2 enclosures (photo: Sasin Tipchai, Pixabay).

Figure 144: Escalles "le Mont d'Hubert" (Nord Pas-de-Calais), Middle Neolithic 2 enclosure (Spiere Group), feature 445; articulated upper part of a goat (after Praud *et al.* 2014, fig. 69; photo: Inrap-CG 62).

The majority of the animals are juveniles; the adult individuals are mostly female while the young animals belong to both sexes. The sheep mandibles clearly indicate seasonal slaughtering; they were killed in spring time, either simultaneously or during the same season. Moreover, the slaughter age as a function of the sex of the individuals corresponds to the reality of a living herd. The displaced position of certain anatomical parts indicates that the carcasses were probably exposed before being buried and there is visible evidence that some carcasses were cut up into pieces.

Caprine skulls are rare in the deposits. When they occur they are complete and, unlike cattle, sheep bucrania have not been found in the ditches of enclosures.

There are two recorded goat bucrania; one from the palisade of the Middle Neolithic I enclosure at Osly-Courtil; the other from the Middle Neolithic II enclosure at Escalles.

An adult caprine skull is recorded from Saint Martin-de-Fontenay.

Other individual specimens are known from Bazoches-sur-Vesle and Crécy-sur-Serre.

Deposits of articulated skeletal parts of caprine have been found. These are often parts of the vertebral column, mostly thoracic vertebrae or lumbar vertebrae that in some cases are still attached to the sacrum. Several examples were observed in the enclosures at Escalles and Carvin (fig. 144).

There appear to be concentrations of disarticulated and articulated caprine remains in certain parts of the sites; in contrast, cattle and pig remains tend to be more dispersed among the ditch segments. The enclosure at Carvin is a good example.

Dogs were also deposited as complete or semi-complete skeletons.

All of the enclosures feature deposits of dog remains, but the numbers of individuals represented varies considerably. Nine individuals were identified at Passel, for example, and the remains take the form of complete skeletons or isolated mandibles, skulls and long bones.

Unlike the other domestic species, dog is rarely found outside this type of deposit, indicating that it probably was not consumed on a regular basis (fig. 145).

However, the fact that certain skeletons found in the enclosures are incomplete suggests that

Figure 145: Dog was not commonly eaten in the Neolithic but it does appear in the deposits found in enclosures (photo: M. Ilett).

dogs may have been eaten occasionally during gatherings. Moreover, butchery marks have been identified on a coxal bone found at Escalles (fig. 146) and localised burning has been identified on the proximal end of a radius from the site of Villers-Carbonnel (fig. 147).

The dog remains are sometimes associated with other finds; at Carvin, the remains of a dog were found in association with a pottery vessel and at Crécy-sur-Serre a dog mandible was found close to a human skull cap.

At Boury-en-Vexin, the remains of three dogs – one skeleton and two complete skulls – were uncovered within the layer of animal deposits.

Red deer remains are rare, apart from items related to the bone-working industry. It appears, therefore, that the meat of this species was only consumed sporadically on enclosure sites.

An almost complete red deer was discovered in a pit which appears to have been dug to block a gap in the enclosing ditch at the Michelsberg enclosure at Vignely "la Noue Fenard"; this feature, which is slightly later than the enclosure, was interpreted as a pit dug to mark the formal abandonment of the site (Lanchon *et al.*, 2001). At present there are no other examples of red deer remains in the enclosures of the Paris Basin but the deposition of young red deer in circular pits dating to the Münzingen period is recorded in Alsace (Lefranc *et al.*, 2012).

However, complete red deer antlers do occur, and sometimes in very large numbers.

A number of these are shed antlers, which would have been collected in winter and are clearly not connected with diet. However, antlers removed from hunted animals also occur in the enclosures.

All of these antlers are large and would have belonged to large adults of reproductive age (fig. 148).

They constitute an excellent raw material, perfect for producing substantial heavy implements, but none of them were actually worked. All of the antlers come from the bottom of the enclosure ditches, and in contrast to worked bone (abandoned tools and manufacturing debris) were generally found more to the middle of ditch segments, away from the main passage-ways into and out of the enclosures. In addition, the antlers do not bear any traces of gnawing by rodents, who have a taste for this material because of its mineral salt content, nor were they chewed by carnivores. This suggests that were deliberately deposited and

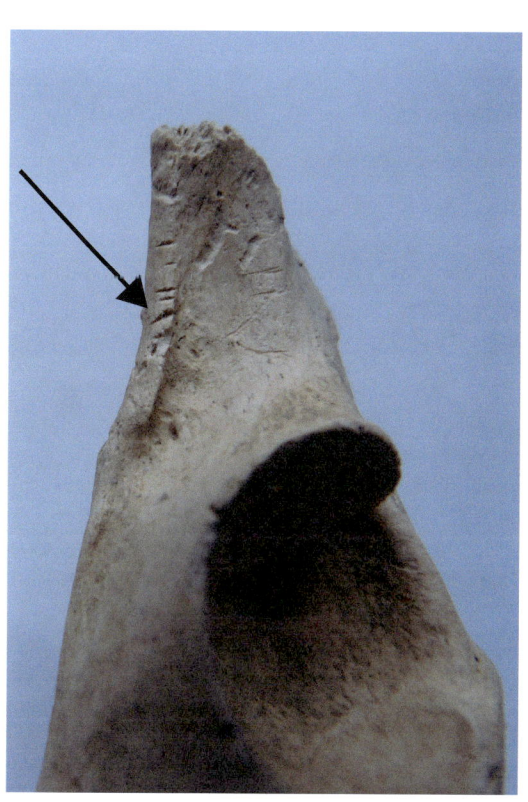

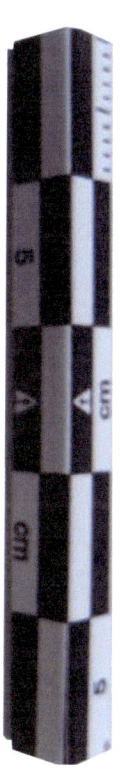

Figure 146: Escalles "le Mont d'Hubert" (Nord Pas-de-Calais), Middle Neolithic 2 enclosure (Spiere Group), feature 445; dog coxal bone with incisions made by a flint tool (photo: L. Hachem, J. Chombart).

Figure 147: Villers-Carbonnel "la Sole d'Happlincourt" (Somme), Middle Neolithic 2 enclosure (Northern Chasséen), feature 325; dog radius with evidence for burning at proximal end (photo: L. Hachem).

Figure 148: Bazoches-sur-Vesle "le Bois de Muisemont" (Aisne), Middle Neolithic 2 enclosure (Michelsberg), shed red deer antler deposited at the bottom of ditch 2 (after Hachem and Maigrot 2019, fig. 21, modified; photo J. Dubouloz, CNRS).

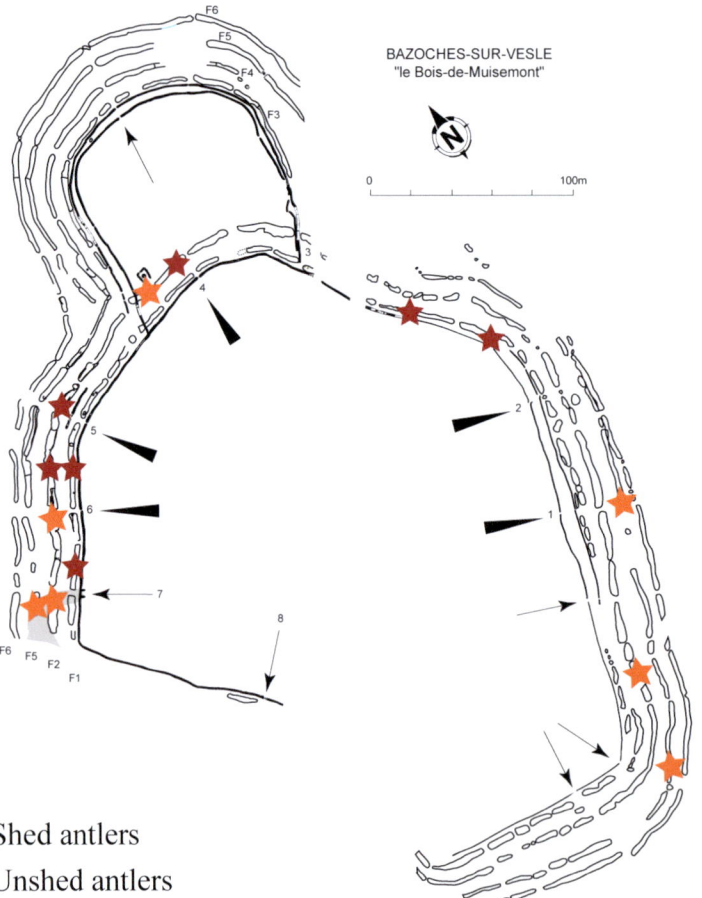

Figure 149: Bazoches-sur-Vesles "le Bois de Muisemont" (Aisne), Middle Neolithic 2 enclosure (Michelsberg), distribution of red deer antler; unshed antler (from hunted deer) was deposited in the inner ditch while shed antler was deposited in the outer ditch (after Hachem and Maigrot 2019, fig. 22).

Red deer bucrania

Wooden pole

Figure 150: Passel "le Vivier" (Oise), Middle Neolithic 2 enclosure (Northern Chasséen); hunted deer antlers mounted on a wooden post, feature 927 (after Cayol *et al.* 2017, fig. 227, modified; photo: N. Cayol, Inrap).

then rapidly covered over to protect them from damage.

It is also noteworthy that at Bazoches-sur-Vesles, antlers from killed deer were only deposited in the inner ditch while shed antlers mainly occur in the outer ditch (fig. 149).

Furthermore, while no complete antlers were found in the bottom of the ditches at Passel, a hunting trophy was erected on a post which seems to have marked the principal thoroughfare (fig. 150). This underlines the ostentation associated with the hunting of red deer stags. A comparable example was discovered in Lake Ilay (Jura); here the fore part of a red deer, along with the antlers,

were found in association with an ash post and were interpreted as a Middle Neolithic ritual deposit (Chaix *et al.*, 1989).

Apart from antlers, a red deer tine was found at the base of a ditch segment at Crécy-sur-Serre (Naze, 2014, fig. 66); interestingly, tines are considered as a separate category of deposit on Michelsberg sites in Germany: Calden, Heilbronn-Klingenberg, Salzkotten, Soest (Höltkemeier, 2010).

Wild species are rare on enclosure sites. We have previously examined the case of certain animal parts that probably fall into the category of ritual deposits. These are shed and unshed red deer antlers, wild boar skulls and mandibles, and aurochs bucrania and horn cores.

Two other wild species were deposited: roe deer and hare. Roe deer was found at Maizy where certain limbs and vertebrae were selected while parts of the front and back legs of a hare were deposited at Crécy-sur-Serre. The deposition of these two species, while rare, is also recorded in the Michelsberg enclosure at "Aue" in Bruchsal, Germany (Steppan, 2003).

These two species were also deposited in tombs: roe deer in the Cerny cemetery at Vignely "la Porte aux Bergers", and hare in the cemetery of Pont-sur-Seine "Ferme de l'Isle", which dates to the end of the Middle Neolithic II.

Figure 151: Passel "le Vivier" (Oise), Middle Neolithic 2 enclosure (Northern Chasséen), feature 1, canine from a very large male bear (photo: L. Bedault).

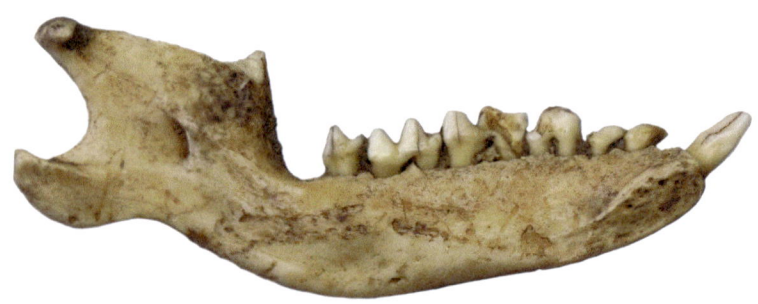

Figure 152: Villers-Carbonnel "la Sole d'Happlincourt" (Somme), Middle Neolithic 2 enclosure (Northern Chasséen), polished hedgehog half-mandible, interpreted as an amulet (after Hachem *et al.* 2016, fig. 12; photo: Y. Maigrot).

Brown bear was identified at Passel where an impressively large male canine (fig. 151) and very large radius and ulna were found. It is also worth noting that a bear skull was found in the Calden Michelsberg enclosure in Germany (Weinstock and Pasda, 2000).

The remains of red deer, wild boar and wolf found at Passel were also large in size suggesting that high-risk hunting of adult males was practiced.

Incidentally, a pendant made from a wolf mandible was discovered in the enclosure at Crécy-sur-Serre (Naze, 2014, fig. 14, 65).

Horse remains in the form of teeth occur on several enclosure sites.

Amulets

The half-mandible of a hedgehog, bearing extensive polishing in both sides, was found at Villers-Carbonnel (fig. 152). It is likely that this polish was the result of its being kept in a supple container, perhaps a leather bag, which prevented the loss of the teeth.

Mandibles bearing similar polish have been found in a Chasséen grave at Saint-Michel-du-Touch in Toulouse (Méroc *et al.*, 1979) and on Middle Neolithic II settlement sites in the Jura and Switzerland (Chaix, 1989). These have been interpreted as amulets and we suggest the same role for the Villers-Carbonnel example.

Burnt bones

Traces of burning, often in the form of concentrations of burnt bone fragments are frequently found in Michelsberg enclosure sites such as Maizy, Crécy-sur-Serre, Bazoches-sur-Vesle, with a particularly notable example being the Spiere Group enclosure at Carvin.

Moreover, the presence of charcoal-rich layers in the ditch segments has been observed in all of these enclosures.

At Crécy-sur-Serre, the burnt bones are black in colour indicating that combustion was not complete. The few that are identifiable are the remains of cattle, pigs and red deer, but most of the material is unidentifiable. In our opinion, this burnt bone refuse relates to the collective activities (purification through fire?) carried out on the site, and is not simply the remains of daily consumption.

On this site, for example, the accumulations of burnt bone are not found in the same areas as the four ovens and hearths that were discovered in the ditch segments.

Excavations at Carvin revealed the presence of small piles of burnt bone (white or grey in colour) throughout the various segments of the inner ditch (fig. 153). They occur in very large numbers in deep spreads in certain gaps in the ditch segments. While burnt bones also occur in the external ditches 2 and 3, the concentrations are not identical.

The site displays a striking difference in the treatment of the animal bones between the various species: only the bones of caprines and pigs were subjected to fire, while the bones of the other species such as cattle, red deer, and dog did not undergo the same alteration. Even when the bones cannot be identified, it is clear that the fragments are from smaller mammals. Do these deposits represent cremations? Or were they the waste material from slow burning hearths? For the moment there are no clear answers to these questions.

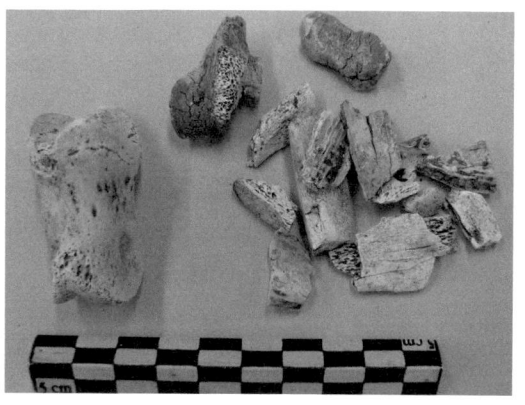

Figure 153: Carvin "la Gare d'Eau" (Pas de Calais), Middle Neolithic 2 enclosure (Spiere Group), feature 1.67: burnt suinae remains (photo: L. Hachem).

The Late Neolithic

Wild animals in deep pits

Narrow, deep pits (referred to as "Y", "V" and "W" shaped pits) are observed from the Mesolithic right up until the Historic Period; the majority date to the Late Neolithic and Final Neolithic (Achard-Corompt *et al.*, 2013). A very small number have revealed the presence of a skeleton or of skeletal parts.

Because of the specific way in which the carcass parts were manipulated and reassembled, we believe that some of these pits were involved in particular ritual practices (Hachem and Durand, 2018). The site of Ramerupt "Cour Première" in the Aube, is an example. Two pits dating to the Late Neolithic yielded aurochs remains in one case (st 1154) and horse remains in the other (st 1144) (Hachem and Bandelli, 2015). The first contained the articulated lumbar vertebrae and mandible of a female aurochs. The second revealed the coaxal bone of a mare along with articulated vertebrae and ribs (fig. 154). It should be remembered that in the Neolithic horse played no part whatsoever in the daily diet.

We refer the reader to the end of the chapter 4 for a more detailed discussion of these pits.

Wild horse

It seems that there was a discernible interest in the horse in the Late Neolithic of the Paris basin which manifests itself in a greater frequency of remains (fig. 155).

At Pont-sur-Seine "le Haut de Launoy" (phase 3) five bones were recovered from a ditch (Peltier and Fournand, 2015). While clearly the remains of a horse, it was not possible to determine if the animal was wild or domestic. Two complete coxal bones, a left and right from the same individual, were found along with an almost complete metatarsal and the core of a vertebra. The intact nature of the bones suggests that the material was

Figure 154: Ramerupt "Cour première" (Aube) feature 1144, deep W-shaped pit dating to the Late Neolithic, coxal bone of a mare (photo: S. Thiol, Inrap).

disposed of in a particular way or may have been a deliberate deposit.

It is worth mentioning that the complete skull of a 7-8 year old mare was deposited in the internal ditch of an enclosure at Salzmünde "Schiepzig" (3400-3000 cal. BC) in Central Germany (Döhle, 2009; Döhle and Schunke, 2014). The remains of three horses (teeth, fragments of radius, and especially metapodials and phalanges) were found in the same in the same part of the ditch (Höltkemeier, 2016). On the basis of genetic analyses and cranial measurements, these horses are believed to have been domestic and very certainly imported from southern or south-eastern Europe. Also of relevance here is the finding of a carefully made ash wood bowl in Layer VIII at the site of Chalain III (3040-3170 B.C.) (Pétrequin, 1997). The bowl features a zoomorphic handle which the authors of the report believe represents a horse's head (Arbogast *et al.*, 1997).

The Bronze Age and the Iron Age

Simple deposits

For the Bronze Age and the Hallstatt periods, deposits in pits are rare. Animals deposits are never complete: for example a limb of a crane found with a vase in a pit at Changis-sur-Marne (fig. 156) (Lafage, forthcoming), a part of a cattle at Beaurieux (fig. 157) (Auxiette, 2008).

Sometimes complete red deer antlers have been placed in pits.

During the Late Hallstatt and early La Tène periods, cattle, pig and dog are involved in ritual practices, albeit on a very small scale.

Animals deposits are complete in pits and silos. For example, a pig at the bottom of a silo at Bailly "le Merisier ouest, le Crapaud" (fig. 158) (Auxiette in Granchon dir., 2012), a dog at the bottom of a silo at Bucy-le-Long "le Fond du Petit Marais" (fig. 159) (Auxiette *et al.*, 1996) and a cattle on the top of a silo at Milly-la-Forêt "le Bois Rond" (fig. 160) (Viand

Figure 155: In the Late Neolithic we observe a growing interest in horses (horses at Gy, Haute-Saône; photo: P. Pétrequin).

et al., 2008). Sometimes we can observe some disturbances (as the dog).

For the middle La Tène period, ditches are the main type of feature used and the deposition of body parts, already observed in the 5th century BC, becomes more firmly established even though the deposition of heads tends to predominate.

During Late La Tène period, the practice of deposition increases significantly, mainly in ditches (fig. 161).

Deposition of cattle remains is significant and sustained throughout the Bronze and Iron Ages. Heads are preferentially deposited, particularly during the Middle and Late La Tène periods. One notable finding is the absence of whole cattle carcasses in the ditches of domestic enclosures.

While the deposition of horse remains is attested as early as the Bronze Age, it reaches a climax in the Late La Tène period.

During the early La Tène, deposition of horse remains becomes more frequent, most notably in silos where skeleton parts and entire skeletons can be found. During the middle La Tène, the deposition of heads in enclosure ditches is seen for the first time and more

Figure 156: Changis-sur-Marne "les Pétreaux" (Seine-et-Marne), Late Bronze Age, limb of a crane, pit 4934 (photo: F. Lafage, Afan).

diverse assemblage categories are observed, although they remain quite rare. From the Late La Tène onwards most of the remains are found in enclosure ditches: heads constitute the majority of these finds, but other body parts and associated bones also occur.

On the farm site at Braine "la Grange aux Moines", the remains of an incomplete, partially-connected horse thorax were unearthed in the upper fill of the enclosure ditch, while other elements were spread around the periphery of the ditch (fig. 162). These bones do not exhibit any cut marks.

Horse and cattle carcasses have also been recorded at "l'Orméon" and "Viviers au Grès" in Longueil-Sainte-Marie. At Chambly "la Marnière", the assemblage is also composed of vertebrae, ribs and coxae. Most of these remains exhibit cut marks indicative of the removal of meat (Méniel, 1994; 2006,197).

At Varennes-sur-Seine "la Justice" (Auxiette, 2013b) two remarkable deposits have been recorded. These contained part of the axial skeleton of a bovine, comprising cervical vertebrae, thoracic vertebrae and ribs.

Let us now return to the skulls. This category must be considered separately from the others as it corresponds to a very particular phenomenon; such deposits were clearly intended to attract attention to and to demonstrate the status of certain places. Dating to La Tène D1b, the high ranking Braine "la Grange aux Moines" farm is one of the most iconic sites in this respect (fig. 163 and fig. 164) (Auxiette and Desenne, 2017). Close to the monumental main entrance to the east, and more generally all along its eastern façade, specific skeletal remains were conspicuously displayed for all to see: the skulls of cattle, horses and pigs, red deer calvariae and antlers, and fragments of human calvariae whose presence in this very location is certainly not accidental.

Most animal heads still retain their maxilla and the teeth are still in their sockets. In the case of fragmentary skulls that can be reconstructed, the teeth are more often than not missing. The differences in preservation may be attributed to distinct episodes during which the heads were displayed; the older, broken ones tend to be more fragmentary with degraded bone surfaces. Several elements of red deer skulls present are possibly

Figure 157: Beaurieux "la Plaine" (Aisne), Early Bronze Age, part of cattle, pit 48 (photo: C. Colas, Inrap).

Figure 158: Bailly "le Merisier ouest, le Crapaud" (Yvelines), Late Hallstatt/Early La Tène, pig skeleton, storage pit 62 (photo: Ph. Granchon, Afan).

Figure 159: Bucy-le-Long "le Fond du Petit Marais" (Aisne), Late Hallstatt/Early La Tène, dog skeleton, pit 612 (photo: UMR 8215 Trajectoires).

Figure 160: Milly-la-Forêt "le Bois Rond" (Essonne), Late Hallstatt, cattle skeleton, pit 2094 (photo: A. Viand, Inrap).

the remains of trophies. Part of a calvaria, with sawn-off antlers, found close to the entrance is quite remarkable (fig. 165).

The meaning behind the removal of the antlers remains a mystery. Also within this category are the numerous pig skulls, which, although their significance is less easily deciphered, contribute to the scene being created on the site. Indeed they are preferentially positioned in the areas where the skulls of large mammals are found in large concentrations. In Souppes-sur-Loing, at "l'Est de Beaumoulin" (Seine-et-Marne) (Séguier and Auxiette, 2006), cattle and horse skulls and calvariae are concentrated along the eastern façade of the enclosures on each side of the entrance, just as in Braine (fig. 166).

They are reported on a regular basis by researchers in contexts ranging from simple farms (Gransar *et al.*, 1997) to sanctuaries such as in Gournay-sur-Arounde (Brunaux *et al.*, 1985) where dozens of them have been recorded (Brunaux et al., 1985, 85). In Montmartin, a high-ranking site

which is also a religious site, cattle skulls – some still retaining their mandibles – have been discovered in the corner of a ditch (Brunaux and Méniel, 1997, 87).

Dog is rarely associated with deposits in the Bronze Age and the middle Hallstatt period. During the Late Hallstatt/Early La Tène period, the first deposits of dog in ditches and pits date to Late Hallstatt/Early La Tène. These practices are recorded throughout the Middle La Tène but become more frequent during the Late La Tène period.

In the Bronze Age, deposition of pigs is rarely recorded. During the Late Hallstatt/Early La Tène period, whole animals, sometimes disarticulated, are found in pits or silos dating to Late Hallstatt/ Early La Tène (Bailly, fig. 45). The first evidence for deposition of whole animals or body parts in ditches dates to the Middle La Tène period; deposition in pits and silos continues fig. 167 and fig. 168).

During the Late La Tène period, animal depositions in ditch contexts increase significantly

Figure 161: Bazoches-sur-Vesle "la Foulerie" (Aisne), Late La Tène, part of a chicken in a ceramic vessel, ditch 121 (photo: G. Auxiette, Afan).

(particularly in Picardy and central France). In most cases, the skeletons are articulated. The individuals can be juveniles, aged from a few days to a few months. Perinatal pigs and piglets were selected on the farms where "collective" consumption has been identified, for instance in Souppes-sur-Loing in Seine-et-Marne (Séguier and Auxiette dir. 2006 Auxiette) and at Cergy in the Val d'Oise, (Pariat *et al.* 2011).

Regarding caprines, only a few instances of deposits containing carcasses or parts of carcasses are recorded for the Hallstatt and the middle La Tène periods. However, during Late La Tène, deposits become more numerous and are found in enclosure ditches and pits.

Even though they were rarely selected for deposition, wild species do occasionally occur, particularly red deer and roe deer. Shed and unshed

Figure 162: Braine "la Grange des Moines" (Aisne), Late La Tène, horse thorax, ditch 83 (photo G. Auxiette, Afan).

Figure 163: Braine "la Grange des Moines" (Aisne), Late La Tène, cattle skulls, ditch 83 (photo G. Auxiette, Afan).

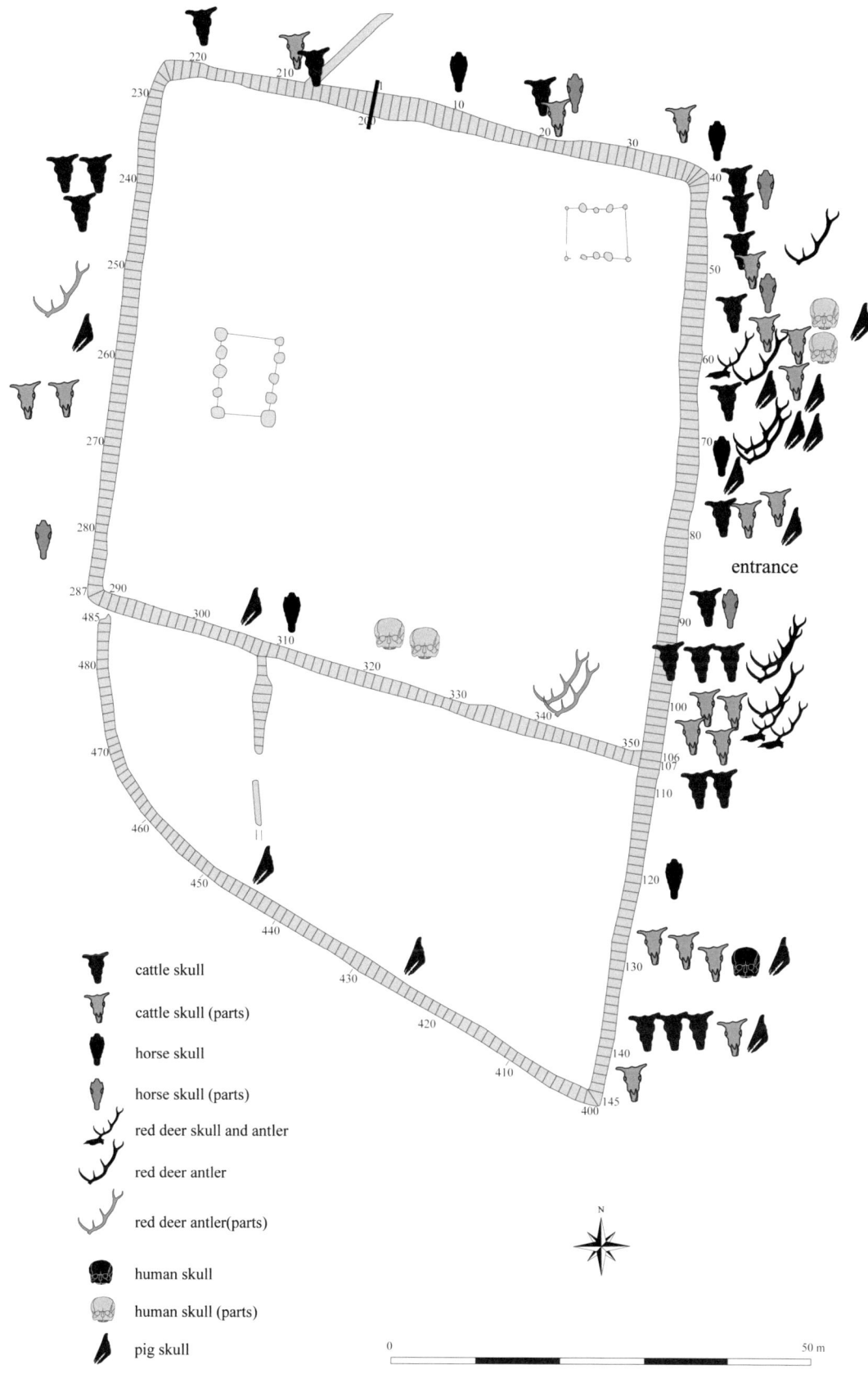

entrance

cattle skull

cattle skull (parts)

horse skull

horse skull (parts)

red deer skull and antler

red deer antler

red deer antler(parts)

human skull

human skull (parts)

pig skull

N

0   50 m

Figure 164: Braine "la Grange des Moines" (Aisne), Late La Tène, distribution of skulls.

Ditch 208 m. 61Bp4-n°5

Red deer skull with sawn antlers (back)

Red deer skull with sawn antlers (front)

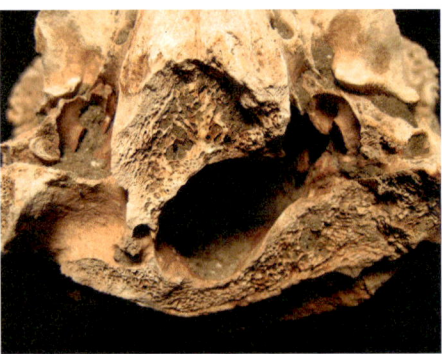

Red deer skull with removal of the occipital condyles

Figure 165: Braine "la Grange des Moines" (Aisne), Late La Tène, skulls of red deer with antlers removed, ditch 208 (photo: G. Auxiette).

sawn left merrain

sawn right merrain

Ditch 208 m. 103Bp3-n°7670

Red deer skull with sawn antlers (back)

Sawn antlers, detail

Red deer skull with opening of the cranium

Str. 208 m. 61Bp4-n°5
Red deer skull with sawn antlers in situ

Figure 166: Souppes-sur-Loing "À l'Est de Beaumoulin" (Seine-et-Marne), Late La Tène, distribution of skulls.

Skull of cattle
Skull of horse
Skull of pig
Skull of sheep
Skull of goat
Skull of dog

Figure 167: Changis-sur-Marne "les Pétreaux" (Seine-et-Marne), Late La Tène, portions of a horse, pit 7337 (photo: F. Lafage, Inrap).

antlers, whole carcasses and carcass parts have all been recorded (fig. 169).

In conclusion, a variety of depositional practices are observed: we see different types of assemblages (skeleton, carcass parts, individual bones) depending on the species, and sometimes even within the same feature.

Analysis of this data has highlighted differences in the choice of species over the centuries, but also a certain consistency from the 5th century BC onward.

The evidence for the practice of deposition in pits is tenuous for the Bronze Age, probably on account of the small number of known sites; an increase in the occurrence of deposits in silos and pits becomes evident in the ensuing Late Hallstatt/Early La Tène period, and a veritable explosion in these practices occurs during Late La Tène when the deposits were mainly located in ditches. Among the domestic species, cattle and horses occupy a privileged place (although horses are absent from the earliest contexts), and dog also becomes prominent in the 1st century BC. Regarding the categories of assemblages, entire carcasses are preferentially deposited in the case of small livestock, whereas body parts and heads are characteristic of deposits involving large mammals. The category of deposit in which heads are the major element increases from the middle La Tène onwards with the development of enclosure ditches. Some farms, particularly those of higher

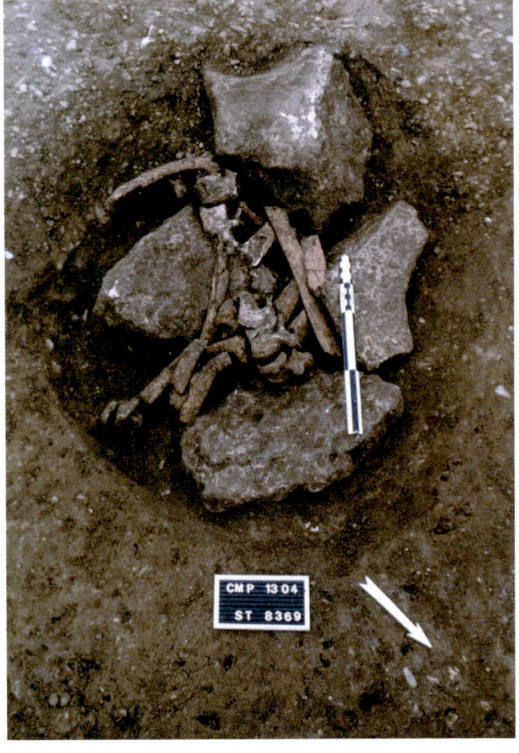

Figure 168: Changis-sur-Marne "les Pétreaux" (Seine-et-Marne), Late La Tène, parts of a horse, pit 8369 (photo: F. Lafage, Inrap).

Figure 169: Romain "la Cense Sauvage" (Marne), Early La Tène, part of a red deer, silo 15 (photo: Y. Rabasté, Inrap).

status, are characterized by large concentrations of deposited bones. While the age of animals selected for deposition varies, pigs, juveniles for the most part, clearly stand apart from the other species in this respect. The recorded deposits of wild animals belong almost exclusively to the Neolithic and Bronze Age. Birds are rare and are the subject of very unique deposits.

The deposition of a dog at Ifs

On Late La Tène farm of Ifs "AR67" (Calvados), the deposition of portions of a dog is unique. The detailed study of dog remains from two features (106 and 107), a pit and a post-hole, revealed that they belonged to a single animal (Auxiette *et al.*, 2010). The remains are essentially made up of the legs and pelvic/scapular girdles, along with a few cervical and lumbar vertebrae (fig. 170). The skeleton of the animal had been cut up; blows were delivered to the sacrum and had also hit the seventh lumbar vertebrae, the head and vertebral column were split up and the lower part of a rear leg had been disarticulated. The division of the leg

elements was particularly complex: the right radius and left ulna, and vice versa, were deposited in each feature.

How can these deposits be interpreted? Do they represent a foundation offering in the case of the bones found in the post-hole, and the remains of pieces consumed on the same occasion in the case of the remains in the pit?

Complex deposits

As well as these so-called "simple" deposits, "complex" deposits have also been recorded. Rarer than their simple counterparts, these complex deposits are particularly concentrated in a geographical area centred on Picardy, Nord-Pas-de-Calais, Ile-de-France and the Champagne-Ardenne region. They seem to have been slightly more frequent in the last two centuries BC. In most cases, they consist of composite bone assemblages (full carcasses or carcass parts and bones) from several species. In order of frequency, we find cattle, horse, dog, and to a lesser extent pig, sheep and red deer. The MNI can be high and corresponds either to

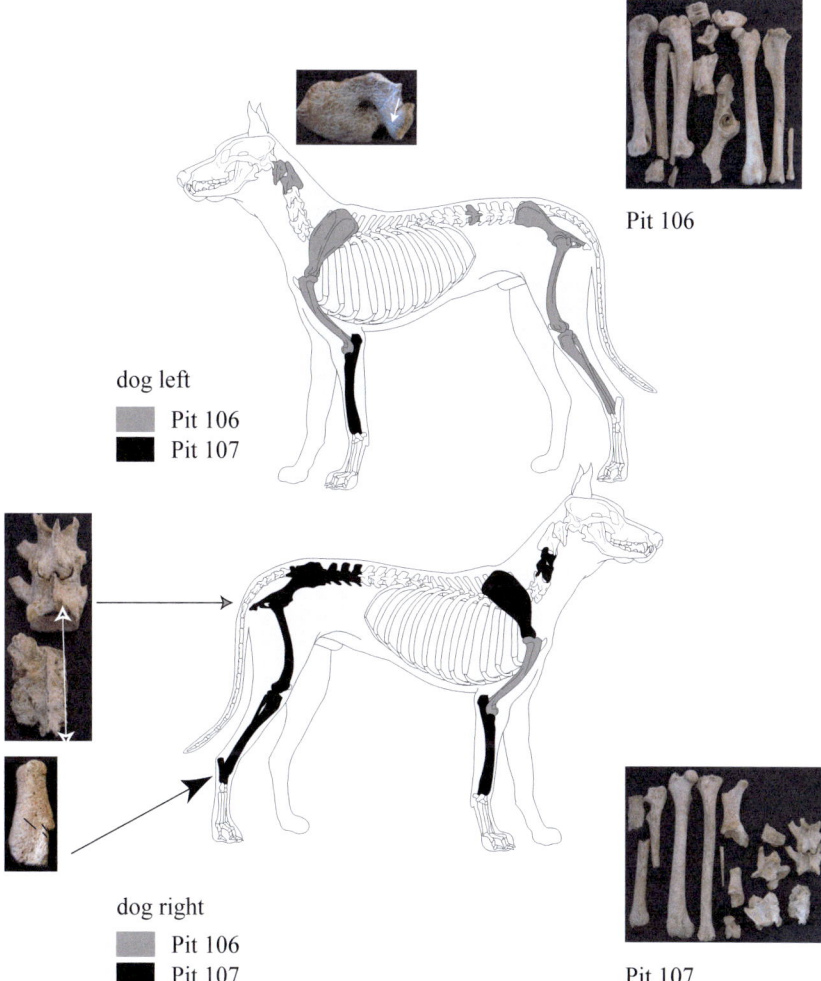

Pit 106

dog left
■ Pit 106
■ Pit 107

dog right
■ Pit 106
■ Pit 107

Pit 107

Figure 170: Ifs "Zac Object'Ifs Sud – AR 67" (Calvados), Late La Tène, part of a dog, pits 106 and 107 (photo: G. Auxiette).

the body parts of several individuals or to several complete animals, or a combination of both. At Bucy-le-Long "le Grand Marais " (Hallstatt final, Aisne), a very important bone deposit associates wild and domestic animals (fig. 171) (Auxiette, 2000a).

Combined deposition of humans and animals in settlement contexts

The occurrence of animal deposits – entire and/ or partial skeletons – and human remains within a settlement context (*i.e.* people deliberately excluded from the communal cemetery) constitutes a depositional category in its own right. Such assemblages are principally recorded in silos dating to the Late Hallstatt/Early La Tène period. These singular assemblages have provoked much discussion regarding the processes of body deposition, decomposition, specific treatments and even planning (Delattre *et al.*, 2018; Delattre and Auxiette, 2018; Duplessis *et al.*, 2013).

Of relevance here is a more general reflection on the deposition of human remains in silos (independent of animals), initiated in the 1980s (Villes, 1986) and which has been significantly developed and substantiated since (Delattre, 2013, 2010; Delattre *et al.*, 2000; Delattre and Séguier, 2007).

For some time, social relegation was the only hypothesis proposed to explain this practice, but in recent years the growing number of recorded sites and silos featuring combined human and animal deposits has led to new interpretations. The presence of deceased individuals in the fill of grain storage pits (silos) appears to be intimately linked to the practice of subterranean deposition. Some of the individuals retained their personal ornaments (torcs, bracelets, rings), and others were accompanied by a large number of animals that, in some cases, had been treated in a particular way, as indeed had some of the human remains. As a whole, this type of deposition brings together codified actions based on practices specific to agricultural communities and assimilated within a ritualized burial rite as a propitiatory and/ or expiatory action within the domestic context. The deposit combinations seem to be infinite. Interpreting the role of animals in domestic rituals during the Iron Age remains a challenge, while this role is well attested in a funerary context (see chapter 5). Pig is the favoured species in deposits, followed in order of importance by sheep and cattle; however, dog and horse, in the form of meat deposits, are very rare in funerary (Auxiette *et al.*, 2002; Bonnabel *et al.*, 2010; Desenne *et al.*, 2009a; Pinard *et al.*, 2010). Conversely,

Figure 171: Bucy-le-Long "le Grand Marais" (Aisne), Late Hallstatt/Early La Tène, parts of cattle, red deer and horse, silo 484 (photo: UMR 8215 Trajectoires).

small proportions of horse and dog are recorded in silos, either as entire carcasses or as carcass parts.

More generally it has been observed that, in silos where humans and animals were deliberately deposited together, horse is preferentially included. It is therefore necessary to look for an alternative logic when attempting to explain the origin of deposits in which humans and animals are intermingled within very specific contexts that are distinct from funerary contexts. The choice of species offers new perspectives on our ancestors' beliefs and practices.

The animal carcasses show signs of weathering from exposure to the elements, and there is sometimes evidence that they were subject to treatment that modified the initial deposits. Taken together, these alterations and manipulations provide us with an indirect measure of time without allowing us to propose an exact scenario.

A fundamental question is who accompanies who? Both humans and animals were manipulated and handled, sometimes even to the point of undergoing removals following similar protocols, the intention clearly being to transform the corpses and remains. The time that elapsed between the various stages is still a mystery but it can be measured by the deliberately introduced layers of soil that separate the deposits and that appear to have played an important role in the ritual.

One of the most emblematic examples is the silo at Chilly-Mazarin "la Butte aux Bergers" (Essonne), which contains a composite deposit including the remains of a woman along with several horses (entire animals and parts of carcasses that have been manipulated), a wild cat and three hares. In succession from the bottom up we find, in the order in which they were placed in the deposit, a wild cat, a five month old foal lying on its left flank, two hares, a five-and-a-half year old stallion lying on its left flank, a second stallion aged more than nine years which was laid out on its back. The two stallions are separated from one another by a layer of fill and are overlain by several carcass parts from a four-and-a-half year old horse, which may have been buried in some kind of box and which had probably been exposed prior to burial; these carcass parts were therefore selected specifically for the deposit. Finally, a hare and various parts from another horse were added prior to the deposition of a human corpse, which is strictly separated from the animal remains (fig. 172) (Duplessis *et al.*, 2013).

Another silo at Puiseaux "Le Chemin de Paris" (Loiret) contained three human bodies and one complete horse and parts of another, deposited in four successive stages. A mare, aged approximately

Chilly-Mazarin «la Butte aux Bergers»: horse number 4

Chilly-Mazarin «la Butte aux Bergers»: human and parts of horses

Chilly-Mazarin «la Butte aux Bergers»: horse number 2

Figure 172: Chilly-Mazarin "la Butte aux Bergers" (Essonne), Late Hallstatt, human skeleton and animals, silo 27 (photo: M. Duplessis, Inrap).

24 months, was deposited first and left exposed in order to initiate its decomposition and to allow the recovery of its skull. Then, it was joined by two human adults (one woman and one man) before the deposit was sealed with a layer of soil. On top of this protective fill the body of a mature woman was then laid out and covered over. Parts of a thirteen year old mare were laid out to close this unusual deposit; the horse remains consisted of parts of the skull (occipital bone and mandibles), probably the entire spine, the right scapula and the complete pelvis (Devilliers, 2006).

Some skeletons appear to be incomplete; the handling of bodies and retrieval of bones from decayed corpses were clearly processes that were applied to humans and animals alike. These corpses were the object of quite significant retrieval procedures, some of which were very precise and point to a high degree of anatomical knowledge on the part of the persons who carried them out (Delattre *et al.*, 2000). In the case of the animals, isolated bones taken as *pars pro toto* might appear to be random but may in fact hold a special meaning when associated with human remains.

In the more complex instances, we observe a succession of depositional acts, sometimes alternating, with several burials. These inhumations within storage pits, singled-out through the treatment they received, appear to be intimately linked with chthonic cults and evoke the practice of subterranean propitiatory and/or expiatory offerings.

Animal deposits in Y- and W-shaped pits: The Neolithic and Bronze Age

Generally located away from settlements, certain large pits displaying characteristic morphologies have been the subject of particular attention in recent years. They are mostly oval and narrow in plan, with a transverse profile characterized by a bottleneck in the lower part which gives them a funnel-like shape. They are commonly referred to as V-, W- and Y-shaped pits. These large pits are almost exclusively found outside settlements and the pits themselves tend to be dispersed over vast areas (Achard-Corompt *et al.*, 2010; Achard-Corompt *et al.*, 2013). They have been found in hundreds in some regions (Champagne-Ardenne and Alsace, for instance), where they can either

form clusters. Numerous isotopic analyses have been carried out in order to try to identify the time period during which these structures were created and used. It was found that this practice appears to have persisted for at least four millennia in Champagne-Ardennes (and also in Alsace), but at higher frequencies during the Neolithic.

These large pits have parallels with Scandinavian and Japanese discoveries (Nespoulous, 2013; Olsen, 2013). In Jämtland in particular, where thousands of these pits are discovered in the 1970s, the question of the function of such structures has been investigated. An initial hypothesis according to which they were used as traps for the capture of large wild herbivores has since been validated (Jordhøy P., 2008). Large, funnel-shaped arrangements of pits have been revealed which would have enabled the capture and killing of large herds of wild mammals. Similar large-scale installations have also been identified in Siberia (Lot-Falck, 1953), in the Near-East and in the American Arctic.

The number of pits with faunal remains is low (less than 10 %). Among these occurrences, we can distinguish between pits where the faunal evidence corresponds to fragments of residual refuse trapped in the upper fills, and those in which the faunal composition and layout suggest that these were intentional deposits. Thus, though these pits were initially used as traps, some have revealed evidence that they were subsequently put to a different use.

These faunal assemblages are located in the lower parts of the pit fill, sometimes at the very bottom. All of the species present are wild. Red deer and aurochs predominate, but there are also roe deer and the wild boar. Parts of skeletons and bucrania are observed in the case of aurochs, skeletons or parts of skeletons for roe deer and wild boar, antlers and parts of skeletons for red deer.

Among the instances of partial skeletons, which are the most frequent occurrence, none were found to be articulated. Overall, this category groups together disorganised skeletons in which some articulated elements may have survived.

Some assemblages suggest that the remains were manipulated (*i.e.* moved from a primary location). There are several instances of deposits where the presence of articulated small bones from the lower leg- carpals in Neuflize and phalanges in Bazancourt – mirrors examples of human manipulation of corpses after their decay, such as those revealed during the La Tène period, notably in the Gaulish sanctuary in Gournay-sur-Aronde (Brunaux and Meniel, 1983).

A - Saint-Léger-Près-Troye "Preslin-PLA-décapage 19", pit 884, aurochs

B - Bazancourt "le Montant de la Sorcière", pit 1, aurochs

C - Palaiseau "Quartier Ouest Polytechnique, Les Trois Mares, avenue de la Vauve", pit 2945, aurochs

Figure 173: Three examples of aurochs deposits : Saint-Léger-Près-Troye "Preslin-PLA-décapage 19", pit 884 ; Late Neolithic, Aube, (photo: V. Riquier, Inrap), Bazancourt "le Montant de la Sorcière" (pit 1 ; Late Hallstatt/Early La Tène, Marne, (photo: S. Degobertière, Inrap), Palaiseau "Quartier Ouest Polytechnique, Les Trois Mares, avenue de la Vauve" (pit 2945 ; Late Bronze Age, Essonne, (photo: J. Durand, Inrap).

The state of preservation of the bones found in these assemblages suggests that some carcass parts were buried rapidly (Neuflize, Palaiseau) while others exhibit alterations of the bone surface that would support the hypothesis that they were exposed (at Buchères and Palaiseau for instance). Among the recorded instances of human manipulation, there are traces of burning on some of the bones in the Buchères assemblage which might indicate preparation and cooking of meat *in situ*.

The Bazancourt, Saint-Léger-Près-Troye, and Neuflize sites in Champagne-Ardenne are emblematic (Achard-Corompt *et al.*, 2013, 22-30), as is the site at Palaiseau in the Essonne (Hachem and Durand, 2018).

At Saint-Léger-Près-Troye "Preslin-PLA-décapage 19" (Aube) the bones come from a large, oval, Y-shaped pit (Aube, pit 884, 4765±40 BP). Again, the bones here came from an aurochs and included skull fragments, vertebrae, left and right ribs, and also the relatively fragmented elements of a radius and a metapodial (fig. 173 A). The lower legs were represented by elements of the right and left tarsi and three phalanges. No anatomical connections were observed and the bones were found to be tightly packed and confined within the constrained space of the pit. Some thoracic and lumbar vertebrae including the sacrum were recorded along with the fourth cervical vertebra, a fragment from a radius and some elements belonging to a lower hind leg. Overall the bones are very well preserved; all are complete, apart from the skull, the radius and a metapodial which was fragmented due to exposure to fire.

In fact, clear traces of exposure to fire can be seen on several bones : the occipital condyle, vertebrae, calcaneus and metapodial. These traces are located on the articulations of the vertebral elements and indicate that the bones were exposed to fire after the various parts had been separated from each other. In contrast to the rest of the bones, the right talus bone is very weathered indicating that it had been exposed outside for a prolonged length of time.

At Bazancourt "Montant de la Sorcière" (Marne, pit 1, 2635±35 BP), the bones of an aurochs (*Bos primigenius*) were retrieved from a large, oval, Y-shaped pit (st. 1), which was devoid of other archaeological material. The bones were generally well preserved.

The bulk of the bones came from a small area in the pit where they lay in a disorganised fashion (fig. 173 B). The assemblage consists of some thoracic elements – a selection of ribs, left and right, and some caudal vertebrae – and of the lower left legs, front and hind, from an animal over 48 months of age. Careful excavation revealed that most of the heap consisted of disconnected ribs, metacarpals and phalanges; the bones were criss-crossed and there was limited sediment infiltration. All three phalanges were disconnected and lay at some distance from the distal end of the metacarpal. Similarly, phalanges from the hind leg, and indeed the tarsal bones, were separated from the metatarsus; in addition, the relatively loose anatomical connection between the phalanges and the sesamoid bones might be explained by a slight slope that could have triggered slippage causing the partial disconnection of the bones. A few isolated carpals and lower teeth lay some tens of centimetres away from the main concentration of bones. The bones do not exhibit any cut marks.

At Neuflize "le Clos" (Ardennes, pit 42), the deposit is composed of a number of carpals and elements from the thorax of a young bovinae (*Bos sp.*, probably an aurochs). The bones are in a state of preservation that is very close to that of fresh bone, with no vermiculation or cut marks (Achard-Corompt N., Riquier V., 2013, p. 29).

A pit discovered at Palaiseau "les Trois Mares" (Essonne, pit 2945, 2905±35 BP) yielded an assemblage composed of several bones from an aurochs: several thoracic elements (thoracic, lumbar and coccygeal vertebrae, some ribs), a humerus, the complete pelvis with the sacrum, and a metatarsal. The fact that all of the bones are complete and in an excellent state of preservation suggests that they were buried quickly (fig. 173 C). Two scenarios can been envisaged: either the animal was trapped and butchered in the pit, and some cuts of meat were abandoned there, or the animal was butchered outside the pit and certain cuts were selected and deposited in the pit.

In these four pits we note similarities in the choice of species (aurochs), the selection of certain pieces of meat, notably ribs, the anatomical disorder of the concentrations with the exception of rare articulated elements, and the manipulation of decayed anatomical units (?). They mainly differ in terms of the states of preservation of the bone surfaces – surface alteration or, on the contrary, very smooth surfaces – and in terms of their treatment, for example their exposure to fire. Burial of the remains may have been immediate (Neuflize, Palaiseau) or deferred, as shown by the weathered bone at Buchères.

What actions might have produced these results? Are we dealing with the (selected?) remains left over after the butchering of an animal that had been trapped, with the waste being discarded following the *in situ* consumption of certain meat pieces? Perhaps we are dealing with entire pieces of meat being left behind? We favour the hypothesis that the carcass was deliberately exposed once most of the meat had been removed. The evidence in support of this comes from Bazancourt, Buchères and Palaiseau, where weathered bone, fallen teeth and the presence of certain articulated carcass parts clearly point to this practice.

One cannot exclude the hypothesis of an immediate burial of fresh pieces of meat, either rejected or carefully selected and set apart during treatment of the carcass *in situ*. However, the ribs are free of any cut marks that might indicate defleshing; certain methods of cooking might explain this absence of cut marks, for example the use of a Polynesian-type oven that causes the cooked meat to detach without impacting the bone surface. The grouping together of joints of meat, be they raw or cooked, can only be explained in terms of codified practices, gestures and ideas (offerings for instance). Such practices have been observed among recent hunter societies in Asia (Siberia), Africa and North-America. Reciprocal relationships are established between Humans and Animals, which incite human hunters to express their gratitude to their animal prey after a successful hunt (Lot-Falck, 1953).

Some of these deposits are composed of entire skeletons. It is impossible to tell how much time elapsed between the moment when an animal became trapped and the moment when the trap was visited by the hunters. Indeed, once an extensive system of pitfall traps had been created, hunters had to walk relatively long distances to retrieve game and the rhythm of these visits remains unknown. It is easy to imagine, however, that the intervals between visits might occasionally be long and that any animals trapped might have begun to decompose; the depth of the pits may have protected the remains from exposure to the elements and predators. In deposits containing a complete animal, the remains are invariably found in a position dictated by the bottleneck profile of the pit. These skeletons can be interpreted as trapped animals that were overlooked by the hunters who had initially built the traps (Sélestat and Achicourt, fig. 174) (Auxiette and Guthman, 2016; Lorin *et al.*, 2013).

Therefore, in most cases the observed disorder in the deposits is probably due to taphonomic processes and reflects the position of the animal after it fell – sometimes on its back, sometimes crouching – and the subsequent collapse of the carcass. Detailed examination of certain skeletons has revealed the absence of some anatomical elements from carcasses that were otherwise apparently complete. This raises the possibility of subsequent human interference with partly decayed carcasses and the retrieval of certain carcass parts. We can therefore distinguish between two broad categories of deposit: traps in which an animal was captured and forgotten, and visited traps, which sometimes became the locus of specific actions.

Endnote

1   The turtle remains are currently undergoing ^{14}C dating, but the lithic artefacts from the upper levels appear to date the pit to the Early Neolithic.

Figure 174: Achicourt "le Fort" (Pas-de-Calais), Late Bronze Age, red deer skeleton, pit 177 (photo: H. Trawka, Inrap).

5. THE ANIMAL IN THE FUNERARY REALM

"Everyone knows that nothing varies as much as a function of person's group, age, sex and social position as funerary rites." (Van Gennep 1981, 210).

"It is most notably at the level of the ritual that death introduces us right into the heart of the symbolic realm." (Thomas, 1975, 438).

Everywhere in the world, and at almost every period from the Upper Palaeolithic onwards, human beings have integrated their dead within rites of varying degrees of elaboration. In most cases these rites are organised around three stages: separation rites, liminal rites and incorporation rites (Thomas, 1975, 439; , 210).

This complexity involves three main aspects : the treatment of the bodies, the construction of the tombs, and lastly the nature and arrangement of items deposited with the burial, including food offerings.

Without adhering to strict rules, the organisation of the inner space of the tomb appears nonetheless to have been rather stereotyped; in many instances "staging" was used to give the illusion that an entire animal had been deposited when, in fact, only certain pieces of meat had been deposited. In many cases, the effect of the show of wealth behind these deposits is lessened by their perishable nature. The offerings themselves have disappeared for ever, leaving only large empty voids inside the tomb.

We also note that funerary practices – inhumation and cremation – varied over time, sometimes replacing each other and sometimes being practiced concurrently.

Graves may be isolated, grouped together or concentrated in clusters, which can be randomly laid out or highly structured. Empty spaces possibly attest to the presence of places of worship. During La Tène, so-called aristocratic graves appear to have acted as foci, attracting other graves to their immediate vicinity.

The Neolithic

The Early Neolithic

Burials
Apart from Menneville "Derrière le Village", other LBK funerary contexts in the Paris basin rarely yield faunal remains.

Some are genuine deposits of specific bones, such as at Bucy-le-Long "La Fosselle" (Aisne), where the proximal metatarsal of a young *bovinae* was discovered near the pelvis of a child (burial n° 54) (Hachem *et al.*, 1997). At Cys-la-Commune "les Longues-Raies", a crane bone was discovered in a burial in the settlement (Labriffe, 1986). Other cases of deposition are less certain such as burial n° 70 at Bucy-le-Long "la Fosselle", where fragments of animal bone were found; however, because the burial had been cut by an LBK refuse pit, we cannot be sure that the bone was actually a grave good.

Outside of the Paris Basin, a number of cases of animal offerings have been recorded. These consist of bones of young caprines and *suinae* found in LBK cemeteries at Aiterhofen and Dillingen in Bavaria (Nieszery, 1995), as well as at Vendenheim "les Hauts du Coteau", a cemetery in Alsace (Böes *et al.*, 2007).

Figure 175: Berry-au-Bac "le Vieux Tordoir" (Aisne), LBK settlement. Left: burial 586, an adult woman buried with the radius of a young bovine and a decorated roe-deer tine (photos:UMR 8215 Trajectoires). Right: burial 607, a child burial accompanied by two anthropomorphic figurines made from a caprine metacarpal and phalange (after Allard *et al.* 1995, fig. 5 modified; drawing: J. Dubouloz, photos: UMR 8215 Trajectoires).

There are also a number of worked bone objects. At Berry-au-Bac "le Vieux Tordoir" (Aisne), a child burial (n° 607) contained an anthropomorphic figurine made from the metacarpal of a very young caprine, as well as second figurine fashioned from the phalange of an older caprine (Allard *et al.*, 1997, fig. 109). Two eyes made from mother of pearl discs were attached to the proximal ends of the two bones. This discovery is echoed in a contemporary burial (n°13) at Ensisheim "les Octrois", Alsace, where a similar figurine was discovered; in addition, the same site yielded a caprine metacarpal which bears traces of intense manipulation (Jeunesse, 1993; Mathieu, 1992). Another grave at Berry-au-Bac (n° 586, female) contained a "container" carefully made from the diaphysis of the radius from a young *bovinae* and a polished "spatula" probably made from a roe deer antler (fig. 175). Both objects bear deliberate incisions that may be decorative or figurative motifs.

It should be noted that these three burials date to the end of the LBK sequence.

At Bucy-le-Long "la Fosselle" (Hachem *et al.*, 1998a) the body of a woman was buried with a set of ornaments made from perforated red deer canines which had probably been sewn onto a hood (fig. 176). The ornaments represent the canines of 41 red deer stags and does and were clearly precious; use wear analysis of the surfaces and areas surrounding the perforations indicate that the teeth had been worn in different ways which might suggest that they had been passed down over several generations (Bonnardin, 2009).

At present, we know of only one example of a faunal burial offering in the Paris basin during the BVSG period. This was a burial discovered at

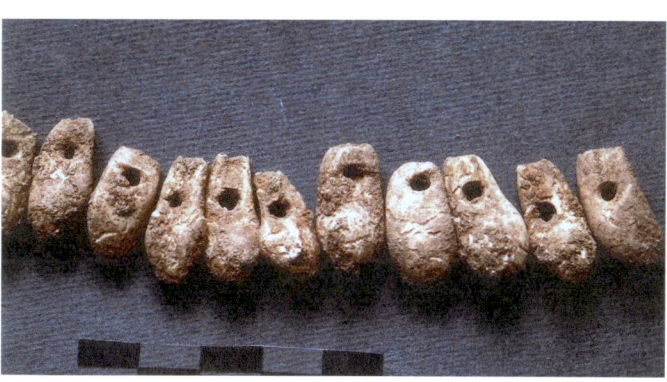

Figure 176: Bucy-le-Long "la Fosselle" (Aisne), LBK settlement, grave 70 of an adult woman; detail of red-deer canines sewn onto a hood (after Bonnardin 2009, drawing: S. Bonnardin).

Buthiers-Boulancourt (Seine-et-Marne) dating to the very end of the Blicquy/Villeneuve-Saint-Germain (Samzun *et al.*, 2012). An adult male (whose arm had been amputated) was inhumed with a very young whole caprine placed at his feet. The other finds from the burial included an exceptionally large flint pick and a long stone axe blade which imitates polished axes of the Alpine region. The particular nature of these lithic objects suggests that the person buried was someone of high social status.

The Middle Neolithic

While evidence for burials in the post-Rössen period is currently lacking in the North of France (apart from Alsace), important cemeteries have been discovered for the Cerny period in the valleys of the Seine and Yonne and, most recently, in Normandy. In what follows we will focus in particular on the cemetery at Fleury-sur-Orne "les Hauts de l'Orne" (Calvados), where we have personally carried out work and where the earliest phases correspond to the Middle Neolithic I (Early Cerny), between 4500 and 4700 cal. BC.

Moreover, the funerary monuments of the Middle Neolithic I and burials of the Middle Neolithic II have yielded animal remains.

Sheep, cattle and pigs

The monumental cemetery at Fleury-sur-Orne (Normandy) was discovered during an aerial survey in the early 1990s by J. Desloges who also discovered the cemetery of Rots a few kilometres away (Desloges, 1997). The site underwent an initial campaign of excavations in the 1990s, and was excavated again in 2014 (Ghesquière *et al.*, 2019a). While other cemeteries have been recorded in the region, the excavations carried out at Fleury-sur-Orne, covering more than 35 hectares, have made it the reference site for an important phenomenon that took place in the northern half of France on the 5th millennium BCE, the funerary monumentalism.

In total, thirty-five funerary monuments of Passy type (STP), comparable to long barrows and twenty burials have been revealed. In half of the monuments, individual burials were found in the central axis of the structures. Grave goods are rare

and comprised mainly of lithic artefacts, most of which are arrowheads.

Animal remains were discovered in five graves. Two of these, excavated in 2014, are particularly spectacular in terms of animal offerings and their layout in the graves (Ghesquière and Hachem, 2018).

Burial pit 19-5, which was very large, contained a minimum of eight sheep. Three rams and four ewes were deposited as an offering on the left side and at the feet of the burial, a man aged over 50; the rams were positioned to face the man while the ewes faced away from him (fig. 177).

Two of the rams were slaughtered at the same age, between 3.5 and 4.5 years, while the third was older, *i.e.* 4.5 to 8 years old. The ewes were aged 3.5 to 4 years, 4.5 to 5 years, and 1.5 to 2 years respectively, while the fourth was a lamb aged less than 6 months.

Apart from the sheep and the arrowheads, the grave also contained four domestic cattle bones: a bull's scapula was placed at the north-west of the pit, next to the man's head, and three other bones – a fragment from the base of a horn core, a fragment of tooth and a fragment of metacarpal – were also recorded.

Grave 26-5 contained the remains of 12 sheep, rams and ewes, mostly deposited in the northern half of the feature (fig. 178).

Six of the sheep had been killed between the ages of 1.5 and 3 years, with a strong possibility that they were killed around the age of 18-20 months. Two others were killed at a younger age, between 1 and 3 months and 8 to 12 months respectively. One individual was older, about 3.5 to 4.5 years.

The positioning of the animals within the tomb is still under investigation but we note that the caprine carcasses were much more numerous in the northern half of the structure than in the southern half. The carcasses appear have to have slipped down from the northern edge of the pit, causing some displacement of sediment from the pit side, but were stopped by an obstacle before reaching the man's corpse. This obstacle has long since disappeared but we were able to observe the effect of this barrier within the fill of the pit. The faunal remains in the southern part of the pit were less well preserved and to a large extent disturbed by a smaller pit that had been dug at a later stage.

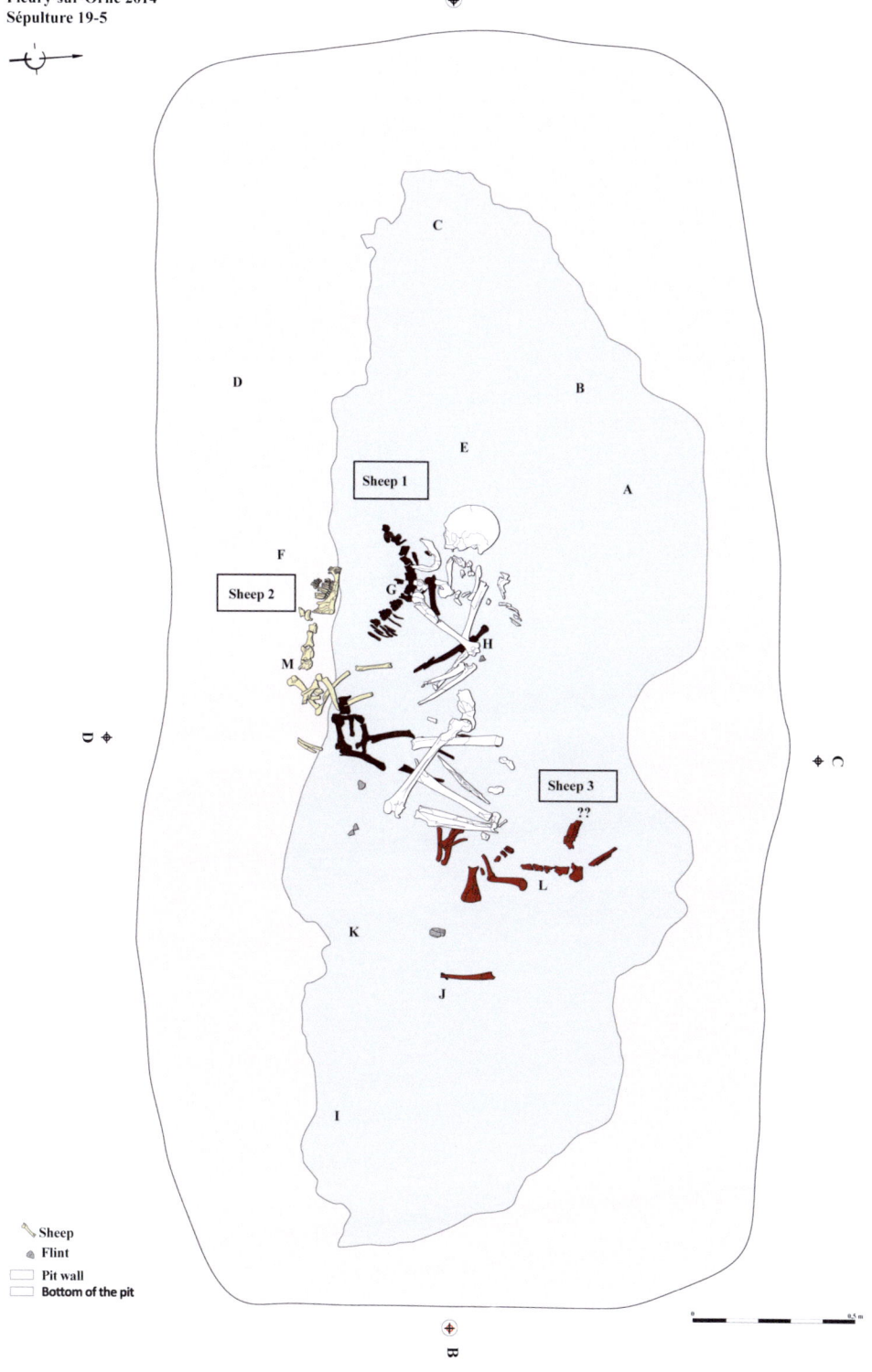

Fleury-sur-Orne 2014
Sépulture 19-5

Sheep 1

Sheep 2

Sheep 3

Sheep
Flint
Pit wall
Bottom of the pit

Figure 177: Fleury-sur-Orne "les Hauts de l'Orne" (Calvados), Middle Neolithic 1 (Cerny), grave 19-5 of an adult man; position of rams amongst the eight sheep deposited in the grave (after Ghesquière and Hachem 2018, fig. 4, modified; CAD: C. Thevenet).

Three domestic cattle bones were also found in tomb 26-5: the scapula of a cow, a spine from a cervical vertebra and a fragment of horn core. The scapula had been placed next to the man's head, in an isolated position, away from the bulk of the sheep remains. The piece of horn core and vertebra were probably not in their original positions but the former was placed towards the northern side of the pit, near the man's head, while the latter was in the north-eastern part of the grave.

In the case of these two burials, the poor state of preservation of the bones did not allow us to determine if all of the sheep skeletons were largely complete but it seems likely that this was the case. However, two types of anatomical part were absent and clearly deliberately removed: the skulls, including horns, and phalanges. All of the individuals were sheep; no goats were present.

Figure 178: Fleury-sur-Orne "les Hauts de l'Orne" (Calvados), Middle Neolithic 1 (Cerny), grave 26-5 of an adult man; twelve sheep were deposited in the grave, the majority on the north side, and three cattle bones were placed near the man's head (after Ghesquière and Hachem 2018, fig. 4, modified; photo: E. Ghesquière, Inrap).

Tombs 19-5 and 26-5 can be compared to three other burials containing animal offerings excavated by J. Desloges in the cemeteries of Fleury-sur-Orne (first excavation campaign) and Rots. No zooarchaeological studies were carried out on these burials but a written description of the finds exists (Desloges, 1997; Arbogast *et al.*, 2002b). The report mentions that complete adult caprine skeletons were found but that the extremities of the limbs (phalanges 3) were absent. After carefully examining the original report and published illustrations (Desloges, 1997, figs. 4-5-6-17) it appears that the horn cores were also absent. It seems, therefore, that as was the case in graves 19-5 and 26-5, the bucrania and hooves were removed prior to deposition of the carcasses.

Other details are also worth highlighting. For example, mention is made that the burial in Monument 1 at Fleury-sur-Orne (M1-F1) contained the remains of five caprines, piled up on the left side of the human skeleton and that they were separated from the latter by a wall-like effect; apart from the fore quarters of one animal which lay partly beneath the skeleton. The similarities with tomb 26-5, with an accumulation of animals on one side of the pit, revealing a deliberate arrangement next to the corpse, are striking.

The grave in Monument 1 (M1-F1) at Rots contained caprine long bones, showing that at least two animals had been placed in the south side of the pit, alongside the corpse. Furthermore, the report mentions the presence of "fragments of a shoulder blade from a large ruminant" in the south-east of the pit, close to the head of the corpse; this is similar to the deposition of scapulae in tombs 19-5 and 26-5 at Fleury-sur-Orne. Also at Rots, the burial in Monument 2 (M2-F2), contained two caprines, positioned at the feet of the corpse, on the south-east side. The position of the offerings in graves M1-F1 and M2-F2 at Rots is thus similar to that observed in burial 19-5 at Fleury-sur-Orne.

Figure 179: Buchères "Parc Logistique de l'Aube" D39 (Aube), Middle Neolithic 1 (Cerny), feature 203, view of the cattle and pig cranial elements in one of two oblong pits in the burial chamber (photo: C. Paresys, Inrap).

Even though these comparisons are limited by the fact that we have to work from published documents, and in the absence of an archaeozoological study, the resemblances observed nonetheless appear to be significant and lead us to believe that we are dealing with identical funerary practices. On the basis of available records, the only difference is the absence at Fleury-sur-Orne of wild boar tusks and bone awls, which occur among the grave goods at Rots. Wild boar tusks are also present in certain Passy-type burials in the Yonne.

It is important to highlight the fact that the ditches that make up the twelve funerary monuments at Fleury-sur-Orne were completely excavated, which is a great advantage; the other Cerny cemeteries have only been partially excavated. This fact has allowed us to retrieve bones that have proven to be quite specific. In fact, the ditches of the monuments generally yield small numbers of bones (one to three bones in the majority of cases, with only a small number yielding ten to twenty pieces), but these are exclusively domestic cattle bones and, notably, the majority are scapulae and cranial elements such as maxillary bones and mandibles (Ghesquière and Hachem, 2018). Comparisons with other Cerny sites reveal similarities. Thus, at Cuverville "le Clos du Houx" (Calvados) we note the presence of seven bones from domestic cattle found at the bottoms of the ditches; these include two mandibles and two scapulae (Hachem, 2016). A cattle mandible was retrieved from the fill of the ditch of Monument 2 in the Rots cemetery (Desloges, 1997).

At the site of Buchères "Parc Logistique de l'Aube", Champagne, (Riquier et al., 2015), one of two ditches belonging to a Passy-type monument (D39) dating to the Cerny also yielded faunal deposits (fig. 179). Here about forty selected bones were discovered: they are almost exclusively mandibles and maxillaries apart from two cattle metatarsals (Hachem, 2019). Altogether, these bones come from six domestic cattle, as well as three pigs and a caprine.

Numerous cranial elements are also mentioned in reports on the sites of Ernes, Orville and Saint-Sylvain in Normandy (Fromont, 2009). At the site of Balloy in Seine-et-Marne, a fragment of cattle skull was discovered at the north-eastern end of Monument V (Mordant, 1997).

In another context, the Cerny period burials discovered at the site of Vignely "la Porte aux Bergers", Seine-et-Marne, (Bostyn et al., 2018b) have yielded a rich assemblage of bone tools (principally made from red deer bone) which are generally associated with other categories of finds: lithic tools, personal ornaments, ceramics and animal offerings.

Faunal remains, often cranial elements from cattle, have been retrieved from the grave fills but other bones have also been found in direct association with the human remains (Chambon et al., 2018). Thus, pig phalanges, without signs of working, have been found in some male graves and are generally positioned at thorax level.

Fox, wolf and birds

Several sites have yielded fox, wolf or bird remains in the context of Middle Neolithic I (Cerny) funerary monuments and Middle Neolithic II burials.

At Beaurieux "la Plaine" (Aisne), a pit belonging to a Cerny funerary monument (st. 29, Monument I) contained an articulated red fox skeleton (Colas et al., 2018). The monument, which is composed of two parallel rows of pits, encloses an axial grave. The adult fox was lying flat on its right side at the base of one of the pits at the south of the monument. There are no parallels for this deposit on known Cerny sites in the North of France. However, a fox skeleton was discovered in a pit on the site of Entzheim in Alsace, of later date than the Münzingen period, and was interpreted as the possible burial of a pet fox (Guthmann et al., 2016); this might also be the explanation for the Beaurieux example.

The presence of fox, and more generally of canids (dogs, foxes, wolves), is recorded in a number of Cerny funerary contexts, most often in the form of personal ornaments, as for example at Balloy (Mordant, 1997) and Passy (Duhamel, 1994).

Somewhat later, at the end of the Michelsberg, fox teeth are found as grave goods buried with children on three sites:

- Beaurieux/Cuiry-lès-Chaudardes "les Gravelines", Grave 21 (Thevenet, 2019),
- Beaurieux "la Plaine", Grave 35 (Thevenet, 2008),

- Beaurieux "les Grèves", Grave 247, (Hachem, 2017b). The latter burial is dated to the very end of the Middle Neolithic II.

In these instances, the fox teeth were perforated at the root to allow them to be threaded on a string (fig. 180). These ornaments do not occur in adult burials.

While this study does not deal with bone or antler tools and ornaments made from hard animal materials we feel that it is important to mention certain objects made from the bones of rare wild animals found in Cerny graves at Vignely "la Porte aux Bergers" (Bostyn *et al.*, 2018b). Here, a perforated bear molar was found in a grave containing the remains of three children (Grave 148).

The exceptional grave of a very young infant (n° 245) was found to contain fifty perforated red deer canines and a wolf vertebra that had been shaped to resemble the head of a bird of prey. The traces of wear observed on this animal figurine indicate that it was both worn and manipulated (fig. 181).

Figure 180: Top: Beaurieux "les Grèves" (Aisne), Late Neolithic, feature 247, set of personal ornaments made up of cattle and fox teeth found in the region of the neck and thorax of a child aged about 7 years (photo: L. Hachem). Bottom: Beaurieux "la Plaine" (Aisne), Middle Neolithic, a pit belonging to a Cerny funerary monument (st. 29, Monument I) contained an articulated red fox skeleton (photo: C. Colas, Inrap).

Figure 181: Vignely "la Porte aux Bergers" (Seine-et-Marne), Middle Neolithic 1, Cerny, grave 257 (child burial), reconstruction of the set of personal ornaments composed of forty-eight red deer canines and a pendant made from a wolf vertebra (after Bostyn *et al.* 2018, fig. 269; photo: V. Brunet).

Wild boar tusks and hares' feet

Suinae tusks have been found in several tombs, three of which are particularly interesting. Again at Vignely "la Porte aux Bergers", a man (Grave 197, Cerny) aged over 30, was found to be wearing two intact wild boar's tusk bracelets on his left arm; four other tusks, which appear to have been used as scrapers, were arranged, like bracelets, on his right arm (Bostyn *et al.*, 2018b, fig. 189) . Similar tusk tools and tusk ornaments are also recorded from the cemeteries of Balloy, Passy and Orville (Thomas *et al.*, 2018).

The three individuals buried in Grave 209 on the same site, two adult women and an adolescent, were accompanied by a boar's tusk that bears traces of sharpening, as well as a complete roe deer leg and two fragments of cattle humerus.

Lastly, Grave 274 at Vignely contained an adult, probably a woman, whose grave goods included two canines from female *suinae*, an unperforated bird of prey talon, and a white limestone bead.

The cemetery dating to the end of Middle Neolithic II at Pont-sur-Seine "Ferme de l'Ile" produced quite a large assemblage, including 74 bone and antler artefacts (Dugois and Loiseau, 2019). The principal species represented in the graves is red deer, the antlers of which were used to make tools and the metatarsals to make points; roe deer bone was also used to make points (Maigrot, 2019).

A few graves contained unworked bones (Auxiette and Hachem, 2019). While certain bones (predominantly cattle) were found in the grave fill, and are thus potentially residual elements, others are directly associated with the deceased.

These include dog bones in two burials, one of which is a child, two burials with bird bones, five with wild boar and one with hare.

Two somewhat unusual deposits deserve special mention.

The first is a deposit in Grave 3072 which associates hares' feet with a boar's tusk and a worked red deer metatarsal. In the case of the hare, the remains represent at least two individuals, of which only the metatarsals (bones of the hind feet) were deposited: an incomplete right hind foot (metatarsals 2 and 5), an incomplete left hind foot (metatarsals 4 and 5), a complete left hind foot (metatarsals 2, 3, 4, and 5). Intriguingly, it was observed during excavation that the position of the tarsals was anatomically incorrect. Metatarsals of the complete left foot (by "complete" we mean that the number of tarsals present is the correct number for a foot) had been rearranged to give the illusion of a "normal" foot.

The second deposit of particular interest was found in Grave 3206 and included boar tusks associated with a point made from a metapodial bone, probably from a roe deer. The four wild boar tusks, which were placed close to the head of the burial, were from two individuals: there were two lower right and two lower left canines (fig. 182).

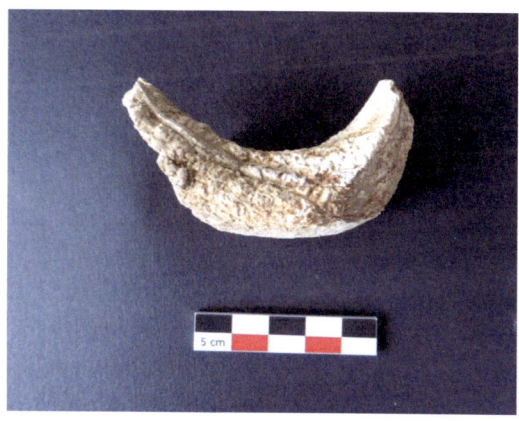

Figure 182: Top: Blignicourt "les Voies de Brienne" (Aube), feature 530, Bell Beaker period; upper tusk from a wild boar found in an adult burial (photo: L. Hachem). Bottom: Pont-sur-Seine "Ferme de l'Ile"(Aube), feature 3206, end of the Middle Neolithic; one of four wild boar tusks discovered in an adult burial (photo: S. Oboukoff, CNRS).

The Late Neolithic and Final Neolithic

Carnivores and suinae

The collective burials of the Late Neolithic and Final Neolithic have yielded very few unworked bone remains (Sohn, 2008) and these have not been systematically recorded. Nonetheless, we can mention the site of Chamigny in Seine-et-Marne, where the faunal assemblage found in the grave consisted of articulated carnivore bones: dog and fox were identified (Hachem, 2020b).

We will briefly review the species present in funerary contexts, either as offerings deposited in the final sealing of monuments or as personal ornaments made from animal materials.

In the North of France and in Germany, faunal remains have been found in the layers sealing certain monuments (Sohn, 2008).

These can be complete animals, including canids (Chaussée-Tirancourt in the Somme) and *suinae* (Warburg 1 in Hesse), but bones of consumed animals also occur (Grave MXI at Sion "Petit Chasseur") (Gallay and Chaix, 1984).

Ornaments made from teeth are generally perforated at the root and are usually canid canines (dog, fox, fig. 180) and pig incisors; more rarely, *bovinae* and equid incisors and cervid canines and incisors were used (Polloni, 2008). There is also a small number of brown bear teeth.

Other objects found in tombs occur in their original form; these are mainly unperforated teeth, present in large numbers (over 100) in certain tombs in Germany and which are frequently found mixed with groups of perforated teeth, as for example Marly-le-Roi "Mississipi" (Yvelines) or in Crécy-en-Brie in Seine-et-Marne (Sohn, 2008).

Sometimes complete half mandibles of marten and hedgehog are recorded. This suggests the survival of the tradition of keeping amulets that we also observed in the Middle Neolithic (see chapter 4).

Lastly, tools made from hard animal material are found frequently; these are mainly made from red deer bone and antler, for example points made from metatarsals and antler sleeves (Maingaud, 2003).

The Bronze Age and the Iron Age

For the Bronze Age, faunal deposits have been recorded in both cremation and inhumation burials, and even in the fill of the ring ditches associated with certain monumental tombs. In the period spanning from the Late Hallstatt to La Tène D2, animal deposits, either fresh or cremated, were frequently placed in graves; this practice was very common in some areas.

The funerary practices of the Bronze Age are well documented in the form of cemeteries, generally composed of small pits containing cremated bones, or, more rarely, simple inhumation burials. Some tombs, originally covered by a tumulus, are now visible in the landscape as a ring ditch. In the Bronze Age, burial mounds were invariably circular and, in most cases, arranged within organised groups. During the Late Bronze Age-Hallstatt period, their shape becomes more variable and they are generally smaller than their predecessors. They tend to be clustered together in a nucleated configuration. Some cemeteries simply consist of circular pits.

We have a wealth of information regarding funerary practices from the end of early Iron Age to Late Iron Age. Small clusters of burials and large cemeteries, some of which included chariot burials, are recorded in south-eastern Picardy and in Champagne-Ardennes. Both inhumation and cremation were practiced, and they may even have coexisted.

Animal deposits in cremation and inhumation burials, in Bronze Age ring ditches

Into the graves, the presence of animal bones is not common. In addition to cuts of meat deposited to accompany the deceased we also find remains in certain levels of the fills of the ring ditches.

In the documented cemeteries where we have definite evidence for the deposition of fresh meat, the main domesticated herd species, namely pig and sheep, predominate and are found in more or less equal proportions. Cattle are less commonly included and are sometimes found in association with sheep and dog, for instance in several cremation and inhumation burials at the cemetery site of Marolles-sur-Seine "les Gours aux Lions" (Mordant *et al.*, 1970). At "la Croix Saint-Jacques", also in Marolles-sur-Seine, there is evidence for pig and cattle cremated bones in some cremation burials (Peake and Delattre, 2007; Roscio, 2011).

In Late Bronze Age cemetery of Marolles-sur-Seine "la Croix de la Mission", one particular grave was found to contain the non-epiphysis distal metapodial of a juvenile suinae which had not been cremated (Peake *et al.*, 1999).

In the cemetery at Jaulnes "le Bas des Hauts Champs-Ouest", a cremation burial was found with fresh bones of a sheep (Auxiette, 2011d).

In the cemetery site of Barbuise "les Grèves de Frécul/la Saulsotte"- "Le Bois Pot de Vin" (Aube) horse and cattle are the main species recorded, particularly for burials where the deceased was placed in a seated position, followed by pig and, in a very small number of graves, by dog and caprines; these species sometimes occur together. The study of this cemetery has revealed certain unique characteristics. Indeed, examination of the bones found in certain graves has led researchers to suggest that these deposits did not in fact consist of cuts of meat but rather were simply deposits of bones (Rottier *et al.*, 2012, 196). Patrice Méniel highlights the importance of the minimum number of individuals and the deliberate selection of certain bones, such as the metapodials and ulnae of horses, and also of dogs, which suggests that bones were retrieved from skeletons that had first been allowed to decay (Méniel analysis in Rottier *et al.* 2012, table 25).

These traditions are very different from those identified in the cemeteries of Picardy, Ile-de-France, Champagne-Ardenne and Normandy, where faunal remains are rare in deposits of cremated bone. For example, at Ciry-Salsogne "la Cour Maçonneuse" (Aisne), only one in over 33 cremation burials contained animal bone, in this case a coracoid of a goose (*Anser sp.),* with evidence of exposure to fire at both ends (proximal and distal) (Auxiette, 2012). In the cemetery at Beaurieux "les Grèves" (Aisne), part of a pig was found together with several vessels (Baillieu *et al.*, 1999). At Vignacourt (Somme), the excavators describe the presence of animal bones mixed with the pyre residue (Buchez, 2011, 179).

Faunal remains have sometimes been found in the fill of ring ditches. Although it is difficult to ascribe them to formal deposits, a certain number nonetheless exhibit the necessary character-istics allowing them to be categorised as such: for example at Bucy-le-Long "le Fond du Petit Marais"(Aisne) one special cattle deposit with head and mandibles, forelimb, and tibiae at the entrance level (Auxiette, 2020). Others cattle deposits are mentioned at Marolles-sur-Seine "les Gours des Lions" (Seine-et-Marne, Late Bronze Age) (Gouge *et al.*, 1991) and at Coquelles "R.N.1" (Pas-de-Calais, Late Middle Bronze Age/Early Late Bronze Age) (Bostyn *et al.*, 1992, 423) and perhaps at Fresnes-lès-Montauban "Motel" (Middle Bronze Age) (Desfossés *et al.*, 1992b, 326-327).

In the cemetery of Jaulnes "le Bas des Hauts Champs-Ouest" (Seine-et-Marne), a cattle skull found on the boundary between two graves can be interpreted as having been part of an open air display rather than a buried deposit, because the bone is weathered. At the cemetery site of Marolles-sur-Seine "les Gours aux Lions" (Seine-et-Marne), cattle bones are recorded as having been found in the fill of an enclosure ditch. These include pairs of mandibles, limbs and part of a thorax, with the peculiarity that the mandible pairs were laid out in a regular manner (Gouge *et al.*, 1991, 93-98).

Late Bronze Age cemeteries located in the northern part of the Escaut Basin have yielded faunal remains in cremations. In a publication reviewing thirteen cemeteries located in the North of France (Nord and Pas-de-Calais administrative regions), Belgium and the Netherlands, sheep and goats are identified as the most common species (De Mulder, 2014, 44) among other recorded species such as dog and roe deer, and also birds and fish.

Bones recently identified in four cremation burials at Rouvignies "Parc d'activités de l'aéro-drome ouest, phase 16" (Nord) conform to these northern traditions. The only species present was sheep and the remains were clearly cremated on the funerary pyre together with the human remains (Auxiette, 2018). The lower leg was preferentially selected for both mature and juvenile animals.

However, deposition of joints of meat to accompany the deceased remains a marginal practice in relation to the thousands of cremation burials recorded to date in the northern half of France.

Faunal deposits in La Tène graves

"Next to the deceased, food offerings were deposited in clay vessels, along with joints of meat, which suggests that a funerary banquet, associating the living and the dead, probably took place during the burial." (Demoule, 2011, 49).

There is a large number of La Tène period ceme-teries and they can be very rich in faunal deposits, depending on the region: the Somme, the Aisne and the Marne are among the best documented areas.

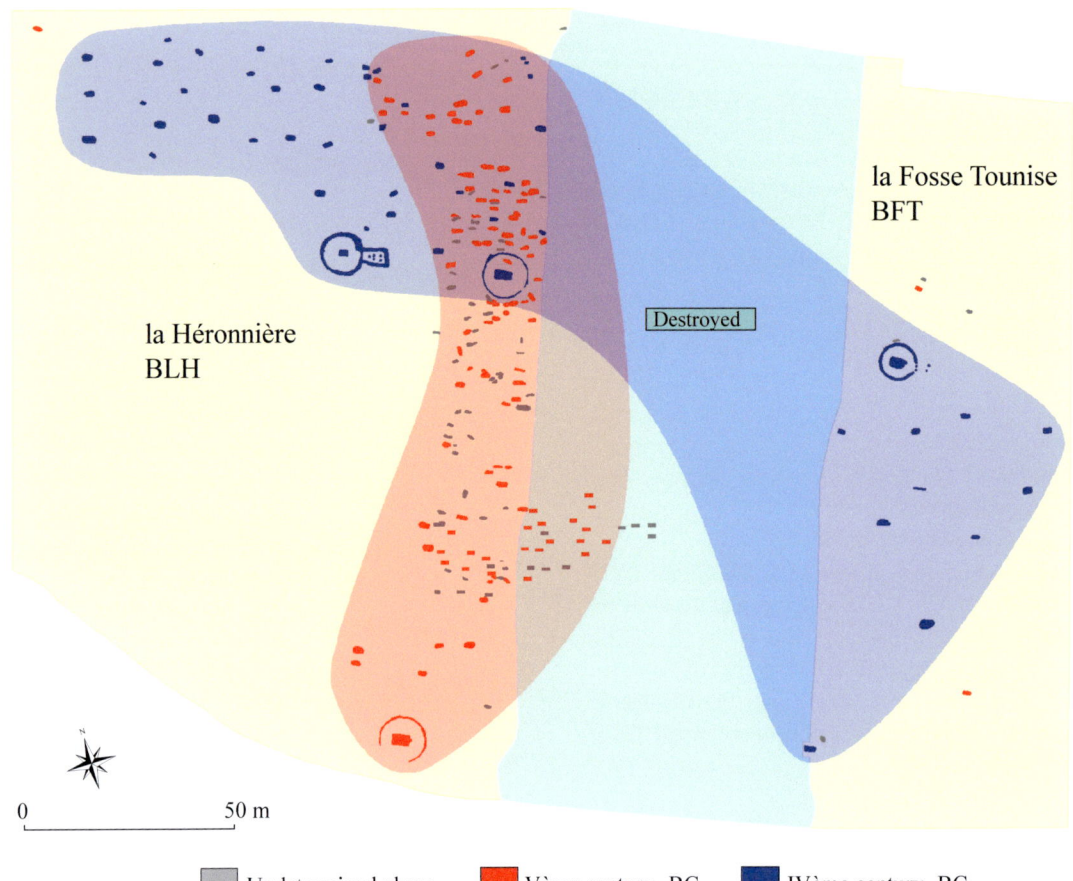

la Fosse Tounise
BFT

la Héronnière
BLH

Destroyed

| | Undetermined phase | | Vème century- BC | | IVème century- BC |

Figure 183: Cemetery of Bucy-le-Long "la Héronnière" (Aisne), Early La Tène (La Tène A and B) (CAD: S. Desenne, Inrap).

Figure 184: Cemetery of Bucy-le-Long "le Fond du Petit Marais" (Aisne), Middle and Late La Tène, aerial view (photo: UMR 8215 Trajectoires).

Rescue archaeology has provided an ever-increasing body of information about the place of animals in cemeteries during the last five centuries BC (Auxiette, 1995; Auxiette *et al.*, 2002; Bonnabel *et al.*, 2010; Desenne *et al.*, 2009a, 2007; Méniel, 1998; Méniel *et al.*, 1994; Méniel and Lambot, 2002; Pinard *et al.*, 2010; Pinard and Desenne, 2009; Pommepuy *et al.*, 2000, 1998).

In the Somme, the Oise and the Aisne, the detailed study of over 70 cemeteries has identified the presence of faunal deposits on more than twenty sites, corresponding to approximately 200 graves (fig. 183, fig. 184, fig. 185).

The data show that, over a five century period (475 BC – 50 CE), on average half of the graves contained an animal deposit.

The La Tène A and B1 periods are characterized by abundant animal deposits. Following a noticeable decline in offerings during La Tène C1 (Auxiette, 1995; Desenne *et al.*, 2009a; Pinard and Desenne, 2009; Pommepuy *et al.*, 1998) the practice becomes common once more in the later period.

Throughout the La Tène period, the cemetery evidence indicates a high degree of social ranking. The differences are particularly identifiable in the deposits that accompany the burials: ceramic vessels, cuts of meat, personal ornaments, weapons and tools. Unfortunately, the disappearance of perishable materials has deprived us of a significant component of these deposits (boneless meat, plants, textiles, wooden items, etc.). Their original presence is often reflected by empty spaces in the burial pits, however. All of these elements attest to the social rank of the deceased and in some cases indicate gender.

The deposits were composed of joints of meat deliberately placed beside the deceased (figs. 186 to 190). Was this carried out to provide food for the deceased during the journey to "the sacred world beyond" (*viaticum*?), or are we looking at ritualised practices expressing the desire on the part of the living to share a last meal – in form of a funerary banquet – with the deceased? Was this perhaps an eschatological rite?

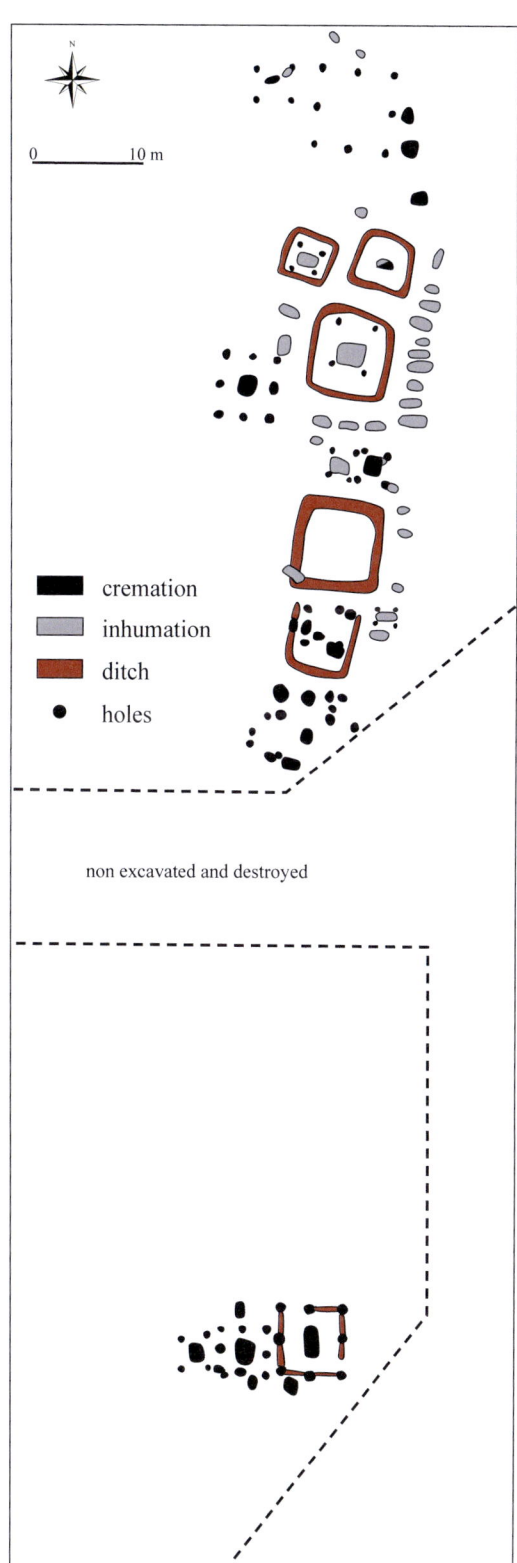

cremation
inhumation
ditch
holes

non excavated and destroyed

Figure 185: Cemetery of Bucy-le-Long "le Fond du Petit Marais" (Aisne), Middle and Late La Tène (La Tène C1/C2/D1a) (CAD: UMR 8215 Trajectoires).

Figure 186: Bucy-le-Long "la Héronnière" (Aisne), Early La Tène, cremation, grave 364 with many portions of pig (photo: UMR 8215 Trajectoires).

Figure 187: Bucy-le-Long "la Héronnière" (Aisne), Early La Tène, woman, grave 368 with portions of pig (photo: UMR 8215 Trajectoires).

Figure 188: Bucy-le-Long "la Héronnière" (Aisne), Early La Tène, man, grave 408 with portions of sheep (photo: UMR 8215 Trajectoires).

Figure 189: Bucy-le-Long "la Héronnière" (Aisne), Early La Tène, woman, grave 441 with portions of sheep, pig and cattle (photo: UMR 8215 Trajectoires).

Figure 190: Bucy-le-Long "le Fond du Petit Marais" (Aisne), Late La Tène, grave 299 with portions of pig (photo: UMR 8215 Trajectoires).

several decades (fig. 191) (Auxiette, 1995; Collectif, 2011; Desenne *et al.*, 2009a).

In inhumation burials the pieces of fresh meat were deposited alongside the body or at the head or feet; in cremation burials they were placed next to the ceramic vessels or mixed with the bones from the pyre. The deposits can be composed of one or more parts from one or more animals from one or more species. The number of pieces of meat, which is sometimes high, is more or less linked to the status of the deceased: some graves contain no meat deposits while some others contain several. In this case, there is also ceramic vessels and/or personal ornaments and/or weapons and/or tools.

Pig is by far the most common type of meat deposited, but sheep are also well represented. Cattle remains occur rarely and it seems that deposition of beef was reserved for high-ranking individuals. Even though rare, chicken is also one of the species selected for deposition (figs. 192 and 193).

The different anatomical parts are carefully selected – *e.g.* pigs' heads, front and/or hind legs from pigs and sheep – and change over time. For example, pig feet and birds make a gradual appearance, either in complete or partial state. Certain anatomical parts, such as sheep heads and feet and the tails of pig, sheep and cattle, are never encountered in deposits (with the exception of the cattle feet found in the chariot

"As regards the funerary aspect, observed among the Celts and the Germans, it also occurs in Rome: according to Cicero, a tomb is not completed until a pig has been sacrificed within it. It is therefore not a question of providing funerary food, but rather a rite of an eschatological nature: was the pig regarded by the Romans, and indeed the Celts, as a chosen intermediary between this world and the next?"
(Sergent, 1999, 21).

The fact that other species, besides pig, are present means that this hypothesis cannot be fully validated, but the predominant place occupied by pig means that its status in these practices is worthy of further consideration.

The composition of the deposits informs us about the choice of species, the age of the animals sacrificed, and the cuts of meat selected. The cuts of meat can vary, depending on the species and the period. The study of large cemeteries enables us to observe variations in depositional practices over

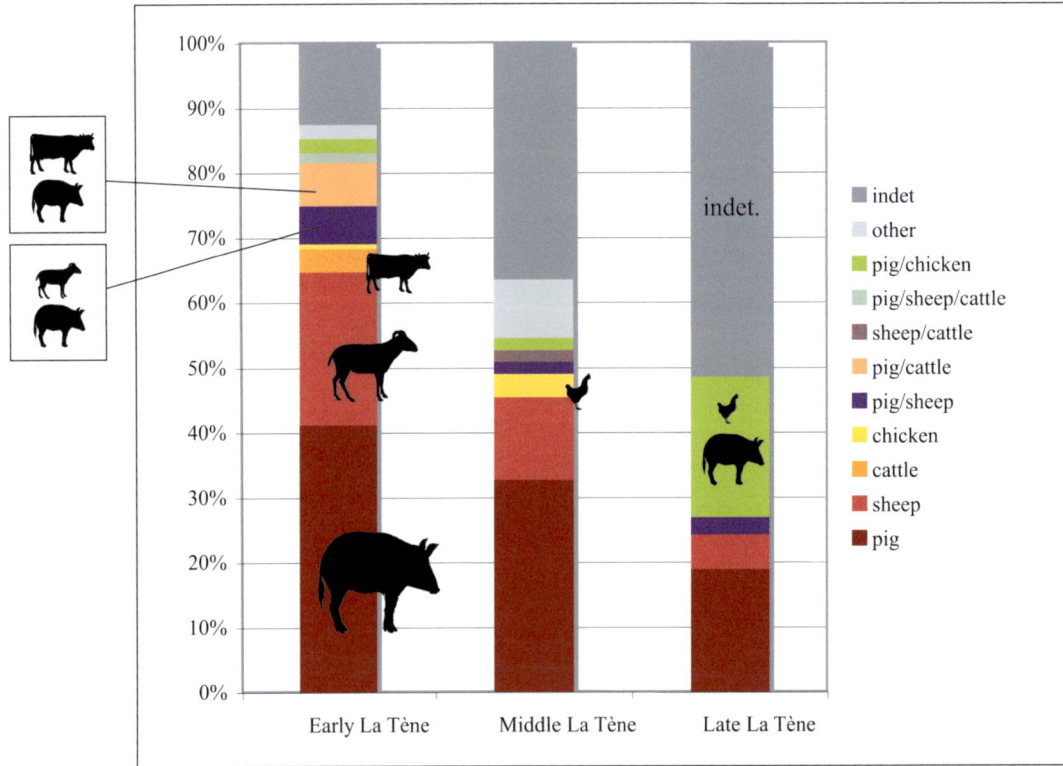

Figure 191: Evolution of the main species in Iron Age cemeteries of northern France.

burial at Bucy-le-long "la Héronnière" (Auxiette, 2009b; Desenne *et al.*, 2009a, 182-183). Elements from the axial skeleton – rib cages, portions of the rachis – occur less frequently than limbs. The organisation of these deposits appears complex, possibly on account of the fact that a large proportion of the deposits may have been entirely perishable in nature (boneless meat). We do not know if the joints of meat received any preliminary treatment, *e.g.* were they raw, cooked, boiled or roasted? The lack of significant traces on the surfaces of the bones greatly limits our understanding of this aspect of the deposition.

The association of two or even three species is sometimes observed, but certain combinations never occur.

For example, the combination of pig and chicken is recorded but the combination of sheep and chicken is never encountered. From the first half of the 4th century BC onwards Late La Tène, the association of joints of pig with chicken is recorded, and also deposits of birds alone. Known species associations in graves, also called *multiple* deposits, are pig and cattle, pig and sheep, and sometimes cattle and sheep. In rare instances, inhumed individuals were honoured with triple offerings: pig/sheep/cattle, pig/cattle/chicken, and pig/sheep/chicken. The association of several species in the same grave is a feature of some of the richer burials. Throughout most of the La Tène period, simple deposits far outnumber multiple deposits.

In the majority of cases, the deposited bones (pieces of meat) are in more or less strict

Figure 192: Bucy-le-Long "la Héronnière" (Aisne), Late La Tène, grave 341, child with part of a chicken (photo: UMR 8215 Trajectoires).

Figure 193: Chicken appears in faunal assemblages around 500 BCE (photo: S. Gaudefroy).

Animal deposits in graves: around -500/-350

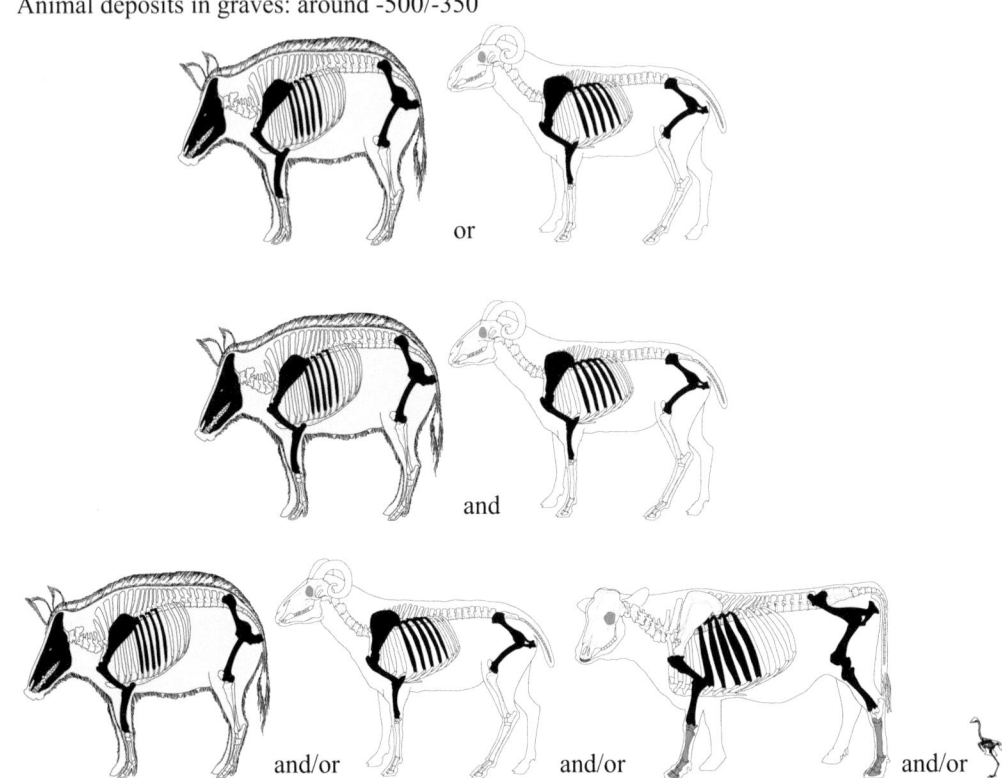

Animal deposits in graves: around -350/200

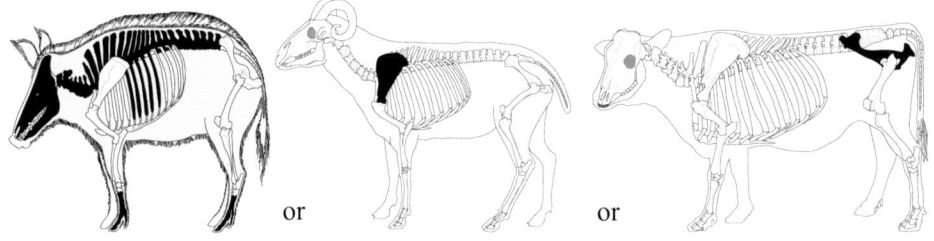

Animal deposits in graves: around -200/-50

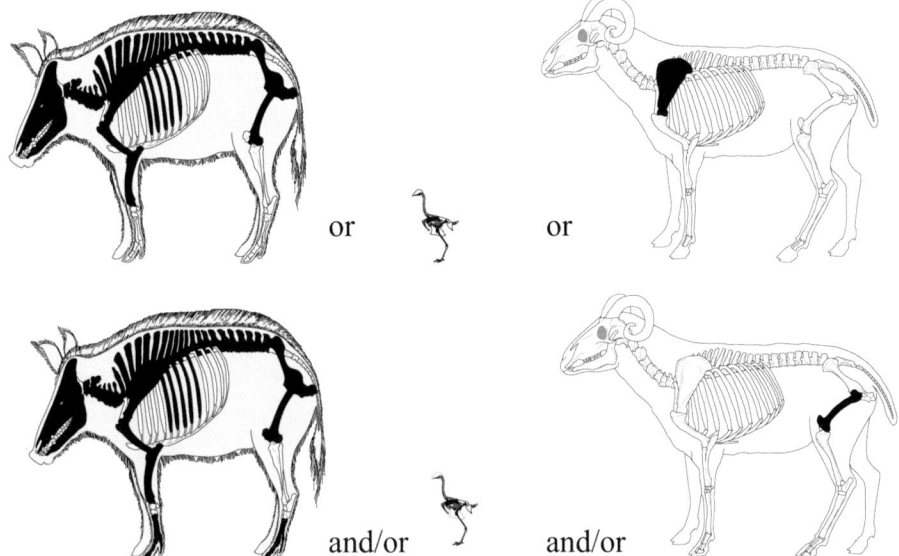

Figure 194: Evolution of portions of meat by species in Iron Age cemeteries of northern France. a: -500/-350; b: -350/-200; c: -200/-50.

anatomical connection. Sometimes they were disarticulated prior to deposition and their layout was staged as a sort of *trompe-l'œil*. Disarticulation with a knife did not leave any marks on the bone surface. However, very clear cut marks appear during middle La Tène: for example, the splitting into two halves of the rachis and the pelvis, through the use of a sharp tool. Pigs' heads are almost systematically split along the sagittal plane (half-heads), regardless of the period. Broadly speaking, in the absence of fractures and cut marks, it is reasonable to envisage that the pieces of meat were cooked prior to deposition.

The lateralization of anatomical parts enables us to determine whether the remains belong to one or more individuals

If we look in greater detail at the various phases of the La Tène, we find that during La Tène A and B1 the shoulder is the most frequently deposited part of the pig. At the start of the La Tène D period, split skulls, shoulders and rear legs (hams) occur in relatively similar proportions. The rachis, which is almost always absent in the earlier phases, becomes gradually more common. For all three of the main

species, the feet are almost never deposited, with the exception of pigs' feet which occur during the La Tène C period (fig. 194) (Auxiette, 2011e).

We will now take a closer look at the two well-documented cemeteries located on the alluvial plain of the Aisne at Bucy-le-Long: "la Héronnière" and "le Fond du Petit Marais", dating respectively to Early/Middle and Late La Tène.

Both provide detailed information on the evolution of faunal deposition practices over a period of three hundred years, between approximately 475 and 130 BC (Desenne *et al.*, 2009a, 182-183; Pommepuy *et al.*, 1998).

During the early- and middle La Tène periods, inhumation was the dominant form of burial but by the Late La Tène, apart from rare exceptions, it was replaced by cremation. Significant differences were observed between both cemeteries. For La Tène A and B, half of the inhumation burials are accompanied by faunal deposits, while for La Tène C1 the proportion drops to a little over a quarter. This practice appears to have been quite limited in the original core of the second cemetery, while, conversely, the last phase of the first cemetery has

Figure 195: Bucy-le-Long "la Fosse Tounise" (Aisne), Early La Tène, chariot burial 150, aerial view (photo: UMR 8215 Trajectoires).

yielded the richest deposits. The episode of deposit impoverishment recorded for the La Tène C1 – which affects all deposit categories alike – is clearly expressed in the dearth of meat deposits.

For La Tène A and B, among the species found in association within the same grave – *i.e. multiple* associations -pig is found together with sheep but cattle rarely occur with sheep. Two female chariot burials at "la Héronnière" were accompanied by triple faunal deposits that included relatively large amounts of beef (figs. 195 to 197).

Simple deposits are far more numerous than multiple deposits (a ratio of 4 to 1) and there is no correlation with the gender of the deceased. As the practice of deposition declined during La Tène C1, no single species was predominant. Several decades later, at the start of the Late La Tène period (La Tène C2), deposits became rich once more and were principally composed of portions of pig meat and chicken. In this period, bone assemblages composed of mixed human and pig bones start to appear, indicating that both had been cremated together.

In most cases, animal offerings are represented by unburnt bones. In both cemeteries, the predominant species is pig, with most individuals being juveniles. Sheep, which occupy second position in the oldest cemetery, are superseded by chicken in the later cemetery. Cattle rank third in both cemeteries. Dog, horse, and all other mammals are totally absent from faunal deposits. Simple deposits of pig or sheep are found equally in male and female graves. The same is true for double species associations. A clear evolution is observed between La Tène B1-B2 and La Tène C1-C2/La Tène D1. In fact, we observe a shift from a ritual where the association of several bones from the same anatomical assemblage or from several anatomical assemblages from the same species co-exist, to a situation where we see the association of several anatomical assemblages from one or more species almost exclusively.

Regarding the treatment of the pieces of meat, we note the almost total absence of cut marks on the bones, except for some portions of pig rachis found in the latest graves from the LTC and LTD periods.

The selection of pieces of meat generally follows the same rules irrespective of the species. For pigs, we find heads that have been split in two, portions of backbone, rib cages, shoulders, hams and, in some rare cases, feet. For sheep, we find the same cuts of meat, with the exception of heads and feet. As regards cattle, the scarcity of occurrences means that no conclusions can be drawn

Figure 196: Bucy-le-Long "la Héronnière" (Aisne), Early La Tène, chariot burial 114 (photo: UMR 8215 Trajectoires).

regarding preferential selection. The butchering of the animals is clearly standardised. Overall, the offerings consist most often of shoulders and hams. Poultry, almost exclusively chicken (sometimes goose), is often in a very poor state of preservation which makes it impossible to identify a specific selection process.

A further study was conducted on the faunal remains in graves from a dozen cemeteries located in the Marne department. Dating to the La Tène A and B periods, these belong to the same cultural entity as the Bucy-le-Long cemeteries. There are approximately 500 graves, eighty of which contain faunal deposits. This study once again highlights the importance of pig, albeit in lesser proportions since sheep occupy a non-negligeable position in this eastern region of the Aisne-Marne Culture. The occurrence of beef in deposits remains marginal; it can occur as pieces veal.

Again in this case, chicken begins to appear over time. However, it is never deposited on its own but instead occurs in association with one of the three other species. Among the graves that yielded associations of species (approximately 10%), pig is almost always the main component of the deposits. The most commonly deposited pieces of pig and sheep meat are the front and hind legs as well as pigs' heads; rib cages are rarer than limbs, and rarer again in the case of sheep. As observed at Bucy-le-Long, certain anatomical parts are never represented, namely the heads and feet of cattle and sheep, and the tails of all three species. Amongst the oldest graves, the proportion of burials accompanied by faunal deposits is below 15%; this proportion increases during the 5th century BC (to ca. 30%) and reaches a maximum during the 4th century BC when almost 50% of graves contain a deposit of meat. Sometime in the first half of the 3rd century BC meat (on the bone) all but disappears from the deposits. Over a period of decades, a qualitative evolution takes place in the number and variety of the pieces of meat deposited.

Figure 197: Bucy-le-Long "la Héronnière" (Aisne), Early La Tène, chariot burial 114, 3D reconstruction (CAD: S. Thouvenot).

From all the excavated cemeteries in the Marne, one sees a trend emerging that apparently associates sheep (particularly the hind limbs) with women, and pigs (particularly the front limbs) with men (fig. 198) (Bonnabel *et al.*, 2010, 150).

Similar observations regarding preferential animal-human associations as a function of gender have already been made by Patrice Méniel on the basis of the deposits found in the cemetery of Aure "les Rouliers" (Ardennes) (Méniel, 1987b). We also observe that female inhumation burials are never accompanied by deposits with beef alone, or by pig/cattle and pig/sheep associations.

The deposition of meat from horses, dogs and wild species is very rare: dog occurs at the site of Barenton-Bugny "site P" (Aisne) (fig. 199) (Auxiette, 2019c), Tartigny (Oise) (Méniel, 1986) and La Croisette (Ardennes) (Méniel *et al.*, 1994), horse at Tartigny, hare at La Croisette, fox at Bucy-le-Long "le Fond du Petit Marais" (Auxiette unpublised).

Some local singularities are observed such as the presence of calf parts in several cemeteries in the Marne; veal is absent from other cemeteries. Differences are obvious in the quantity and the quality of the meat pieces depending on the graves; this more than likely reflects variations in wealth between burials in the same cemetery. The association of several species is quite rare. The pieces of meat most often deposited are shoulders and hams, regardless of species. The fact that certain parts are systematically excluded may reflect dietary taboos. The heads of sheep and cattle are never deposited in burials, while pigs' heads are very common. Pigs' feet appear during the later phase of the cemeteries in the Aisne-Marne Culture, whereas cattle and sheep feet are never encountered (with the one exception already mentioned).

These funerary practices are typical of cemeteries and were inherited from Hallstatt practices which unfortunately are poorly documented in the territory we are principally concerned with here.

Whether they were inhumed or cremated during Iron Age, the dead were frequently buried with portions of meat, although the degree to which this happened varied according to the period. The faunal remains are often found together with a group of ceramic vessels that would originally have contained other food products. These meat deposits, which can be regarded as offerings and/or *viatica*, are mainly made up of cuts of meat from animals selected from the domesticated herd and destined to accompany the deceased. In decreasing order of importance we find pig, sheep, chicken and cattle. They can be deposited on their own or in combinations. Certain meat pieces are preferred to others: for the early La Tène period, pig shoulder is the piece that occurs most abundantly in deposits; for the middle La Tène and the start of the late La Tène periods, split skulls, shoulders and hams are associated in more or less similar proportions. Similarly, the rachis, almost totally absent from deposits in the early phases, gradually becomes more common. The feet of all three main species are almost never deposited; pigs' feet first appear in funerary offerings some time during the middle La Tène period. A change

Figure 198: Early La Tène, distribution of animal species in Marne cemeteries between men and women.
W: women; M: men

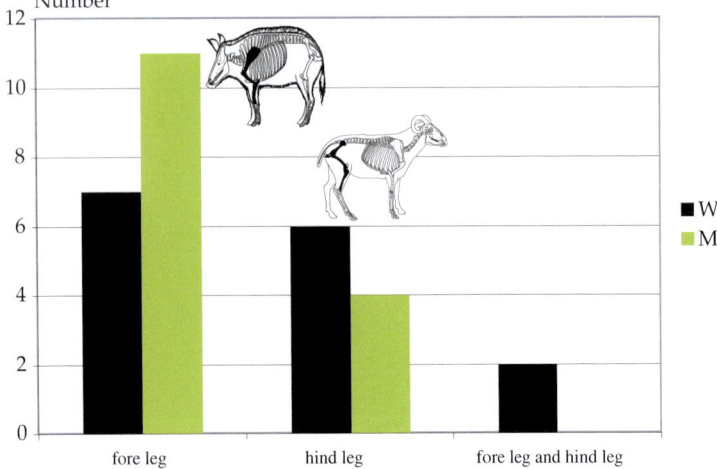

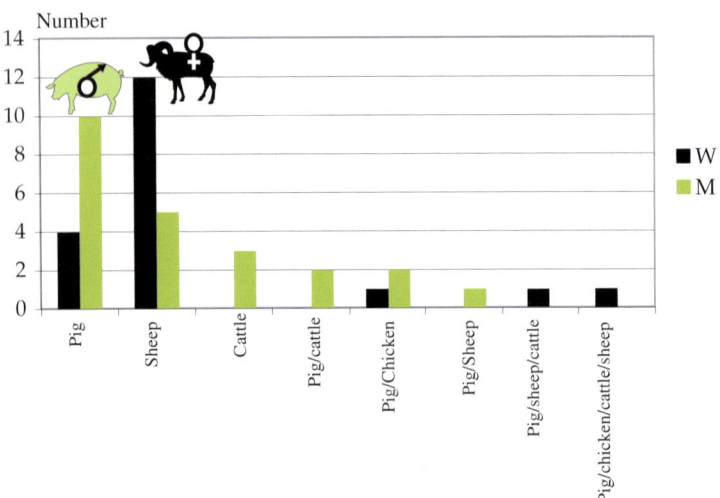

occurred in the way the carcass was cut up compared to the early La Tène period: the carcass was split in two on either side of the back bone and the pelvis was also cut in half.

Cremated animal offerings are far more difficult to characterise; nonetheless, available evidence does not suggest that they differed from deposits of fresh meat in terms of the species or meat cuts selected.

Regardless of whether we are dealing with male or female burials, be they of adults, adolescents or juveniles, no clear rules that might reflect liturgical-like modalities seem to govern the choice of species or the choice of meat cuts (Auxiette *et al.*, 2002; Bonnabel *et al.*, 2010; Desenne *et al.*, 2009a; Pinard *et al.*, 2010).

On the occasion of a funerary banquet, part of the meat from "sacrificed" animals was selected and deposited next to the deceased. This issue of the shares attributed to the dead and the living has been examined on the basis of one hundred or so graves from the largest cemetery at Bucy-le-Long "la Héronnière". In half of the graves, the deposits consist of one to three pieces of meat; for instance one scapula or a grouped scapula and humerus, or a front leg without the foot. Graves that provide evidence for more than fifteen joints of meat include chariot burials and are therefore clearly the burials of high-ranking individuals. In most of the graves these pieces probably correspond to very low proportion of the total amount potentially consumed by the guests at the funerary banquet (Auxiette *et al.*, 2002).

Figure 199: Barenton-Bugny "Pôle d'activités du Griffon"- Secteur P (Aisne), Late La Tène, grave 16000 with portions of dog and pig (photo: E. Pinard, Inrap).

6. THROWING LIGHT ON SOCIAL MECHANISMS

An overview of Neolithic socio-economic systems

The Early Neolithic

In chapter 2. we described the results of studies investigating the relationships in house units at Cuiry-lès-Chaudardes between the faunal assemblages, the macrolithic tools and pottery manufacturing techniques. This research was designed as a contribution to the debate on how domestic life was organised in LBK settlements (Hachem and Hamon, 2014; Gomart *et al.*, 2015). The faunal remains provide insights into livestock and hunting, as well as meat consumption. Macrolithic tools reflect various tasks, in particular grinding of cereal grain and the crafting of implements. Pottery technology provides insights into learning networks. As we saw in chapter 2, comparative analysis of these different lines of evidence reveals a striking pattern, dividing house units into two groups. Thus "Type A" house units are characterised by relatively uniform pottery making techniques, as well as a greater emphasis on livestock and cereal grinding. "Type A" units generally correspond to the larger houses, Type "B" house units, on the other hand, have more varied pottery techniques, a larger contribution to diet from hunting, and more evidence for crafting activities. Furthermore, "Type B" units generally correspond to the smaller houses.

So how can this patterning be plausibly interpreted in social and economic terms?

We suggest that Type A house units constituted households with a greater capacity for food production and storage. They accommodated extended families, established in the village for several generations (up to three or four generations). This hypothesis would explain the ""conservatism"" of their ceramic traditions, which attest to the transmission of technical know-how over the long term within the village. It would also explain the large size of the houses and the intensity of the pastoral and agricultural production associated with them, which was made possible by the ready availability of manpower. These characteristics allow us to classify these households as "economically mature", in contrast to Type B households.

Type B house units were households that were in the process of becoming integrated within the village. They accommodated small-sized family units (perhaps a couple and their children), newly installed in the village. This hypothesis would explain the appearance of new methods for making pots in certain houses, suggesting an input of people from other LBK villages. This ties in well with the results of bioarchaeological studies that suggest that LBK communities functioned according to a patrilocal system, with high mobility among women (Rasteiro and Chikhi, 2013). We can thus envisage the arrival of women from other villages into these houses, probably through marriage alliances.

Our analysis did not reveal major subsistence differences between house units that could be interpreted in terms of "farming households" and "hunting households", and we found no evidence for an input of Mesolithic hunter-gatherer populations (for a more detailed explanation see Gomart *et al.*, 2015). In fact all house units are characterized by the consumption of domestic species and cereals and the only differences relate to over-representation of certain domestic or wild species, over-representation of grinding or craft activities involving macrolithic equipment and the application of different pottery making traditions.

Our hypothesis is that the relationships between houses were based on the production of an excess with the aim of engaging in exchange: those houses characterised by a high capacity for production (Large Type A houses) would have provided the households in the process of integration with products from their agricultural activities (cereal growing and stock rearing) and their technical activities (pottery). In return, households in the process of integration (small Type B houses) provided the long houses with the products of hunting and craft activity (fig. 200).

This model has the advantage of explaining the size differences between houses in a given settlement by their degree of economic maturity and their particular functional status. It also allows us to gain an understanding of the rules governing the establishment, assimilation and integration of family units within a village community and throws light on the enduring rules that underpinned marriage alliances and mobility in the Neolithic (Hofmann, 2012).

We can try to go even further by seeking to identify potential markers of identity and gender through the bone remains of both domesticated and wild species.

Livestock at the centre of wealth, clans and tradition

Livestock takes on several dimensions. From a utilitarian point of view, an animal can be used alive for its physical strength and for certain products (portage, blood, milk, hair); others require the killing of the animal (meat, leather, horn, antler, bone). But the means involved in developing animal husbandry – protection of livestock, procurement of fodder, promotion of

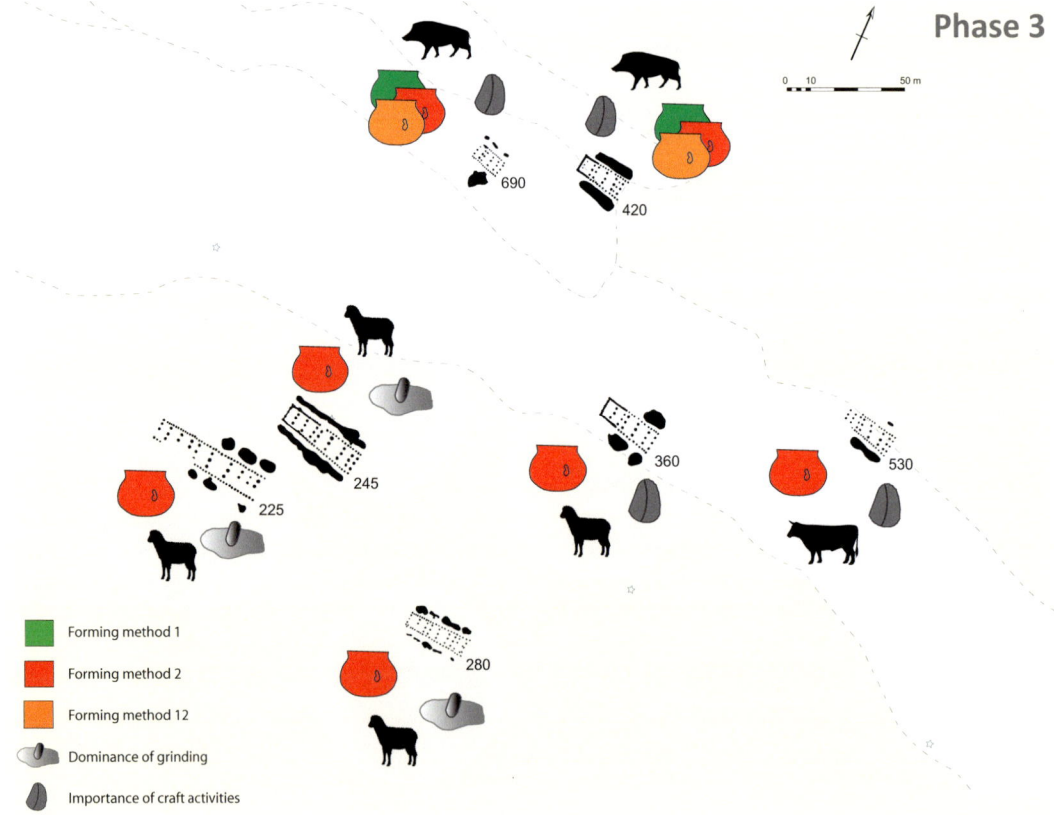

Figure 200: Cuiry-lès-Chaudardes "les Fontinettes" (Aisne), LBK, latest settlement phase, contemporary households with over-representation of sheep, cattle or wild boar, over-representation of grinding or craft activities involving macrolithic equipment and use of different pottery forming traditions (after Hachem 2018; Gomart *et al.* 2015).

Phase 3

Forming method 1
Forming method 2
Forming method 12
Dominance of grinding
Importance of craft activities

reproduction – also serve to make the animals dependent on people (Digard, 1988).

These general assertions can be used as a basis for interpreting our archaeozoological data.

Our research has shown that livestock raising by the LBK communities of the Paris basin mainly involved two species, cattle and caprines (sheep and goats). This can be described as sedentary animal husbandry within a mixed farming system.

The third livestock animal, the pig, is a non-negligible component but seems to correspond to a different system from the two other. Furthermore, pig increases in importance in the Blicquy-Villeneuve-Saint-Germain culture.

This shows us that the domestic animals were not all regarded in the same way and did not have the same status in symbolic representation systems (Bánffy, 2001).

The value of livestock in this agro-pastoral society was considerable, one can perceive it indirectly through the animal figurines which are numerous in the Carpathian Basin (Bánffy, 2001).

These animals are present in all levels of society: the household, the village and in burial ritual.

In illustrating these three situations, we should remember that the large houses can be divided into two groups: those with higher proportions of cattle, and those with higher proportions of caprines. Furthermore, among the large houses in Cuiry-lès-Chaudardes, there are a small number that can be interpreted as meeting houses that may have hosted communal feasts. One of these (n° 380) has higher proportions of cattle, while the other, (n° 225), probably slightly later in date, featured higher proportions of caprines.

This division is also reflected in village layout: at Cuiry-lès-Chaudardes the houses with higher proportions of cattle are situated in the east of the village and the those with higher proportions of caprines are in the south.

In the following hypothesis we attempt to provide a social framework for this zooarchaeological data. Assuming that the house operates according to the rules of society, that is to say was built and used according to the social and ideal norms underlying the community's system of representations (Coudart, 2015, 2009), we can

postulate that the social group that lived in the large houses associated with higher proportions of cattle, or caprines, belonged to a clan. The clan "is defined in a minimal way as a group of unilineal descent whose members cannot establish real genealogical links to a common ancestor, who is often mythical" (Copet-Rougié, 1991, 152). The development of a group of distinct households, each carrying the same name and, in particular, each attached to the same totem and its associated rituals, constitutes a clan (Ghasarian, 1996). A given clan generally has its own totem; this often takes the form of an animal, as is the case, for example, among the Haudenosaune (Iroquois) of North America (Haudenosaunee Confederacy, 2016). The totem may be associated with dietary taboos or, on the contrary, with preferential consumption (Lot-Falk, 1953).

The higher proportions of cattle, or caprines, in the large, economically mature houses – houses that accommodated extended families, long-established in the village, whose ceramic traditions exhibit a "conservatism" that suggests that technical know-how was transferred over a long time period, who were capable of producing agricultural surpluses due to the availability of labour – seems to reinforce the hypothesis of two enduring clans, one with cattle as a totem, the other with sheep, who perpetuated the characteristics that made up the social structure.

In parallel, the two animals essential to village identity, cattle (fig. 201) and sheep, participated in a very fundamental way in the sacred domain, in its various forms. In the ceremonial enclosure at Menneville, cattle predominate and were probably consumed during feasts held in the context of large gatherings. Certain carcass parts (especially horn cores, vertebrae and scapulae) were then selected for deposition close to burials, particularly adults, in what were probably strategic spots. Sheep were deposited either in the form of complete carcasses, which were evidently not eaten, or as partial carcasses, placed close to burials or in isolation. Moreover, unlike cattle, no sheep bucrania have been found. While the link between bucrania and Neolithic burial ritual has long been attested (Chaix, 2012; Marciniak, 2008), we can also confirm, for the LBK, a link between the dominant animal in a society (namely cattle) and its presence in the deposits.

Based on these elements, certain avenues can be envisaged for characterising this Neolithic society with regard to the animals that it shares its space with, that it exploits or that it associates with its religious rituals.

The village was the basic unit of the community, in which the inhabitants all occupied a relatively equal position in terms of consumption, since the same range of animal species was available to each household: but were they all equal in terms of livestock ownership and access to grazing?

Ethnology can provide us with some tentative answers. The issue of ownership of animals is complex, however, and is tied up with marriage alliances and filiation, as illustrated by the example of the Fula people of Benin.

"On marriage, the woman brings oxen into her husband's herd, but she does not have the right to dispose of them without her husband's consent. The husband, however, can sell his wife's animals without her permission. But there are exceptions by which the woman can dispose of her animals as she sees fit. In any case, her concern will be to increase the number of her animals so as to ensure that her children have an inheritance." (Chabi Toko, 2016).

If we return once more to the archaeozoological data, the presence of a common minimum number of species consumed per house suggests that we are looking at a family-owned herd of several

Figure 201: Cattle were at the heart of LBK identity (Alpine pasture at Alpetto, Piedmont; photo: P. Pétrequin).

head of cattle, sheep and pigs, similar to the herd composition in western peasant societies in the 19th century. Nevertheless, the situation was probably more complex, because it is often the case in traditional societies that the herds are owned by families (or individuals) while the grazing land remains communal (Brisebarre, 2013).

These deposits (funerary and ceremonial) of domestic animals could be considered as a marker of high social status, such as the status enjoyed by a large herd owner within pastoral societies. A similar hypothesis has been proposed for interpreting the faunal deposits in the Hinkelstein and Grossgartach burials in Trebur, where the deceased are believed to have enjoyed high status as livestock rearers/crop growers within the social hierarchy (Spatz and Von Den Driesch, 2001; Von Den Driesch, 1992). While this hypothesis cannot be excluded, in the specific case of LBK groups settled in the Paris Basin, it appears to us to be very unlikely. In fact the uniformity of the architecture, the homogeneity of discarded refuse, the parity in the consumption of meat, the rarity of "prestige" artefacts on the settlements and the absence of wealth production (judging from the nature of the settlements) all point to an equilibrium and structural equivalence of the fundamental socio-economic entities (Coudart, 2009; Hamon and Allard, 2010). This is not strict equality between individuals, but rather equality between the segments of the society in terms of production and access to resources and sources of information (Coudart, 2015). The possibility that livestock rearers might have had a higher social status does not, in our opinion, accord with the data at our disposal.

However, it is quite certain that the value of livestock, as has been demonstrated by ethnologists, was economic, but was not always a major factor in the production of material assets.

Livestock played a central role in social, cultural and religious life (Bonte, 2007, 1981; Leroi-Gourhan, 1964). These functions attributed to animals could come into play in certain important domains. An example here concerns the maintenance of the active exchange networks that form the basis of relationships between people, enabling society to continue to reproduce itself (Sahlins, 1965).

These relationships established between partners can involve co-operation, competition or even domination. They can take many forms and in what follows we will cite just a few relevant examples from contemporary societies.

In Africa, among the Fula livestock rearers of the Sahel and Guinea, cattle are central to social exchanges; the cattle herd can be used as a means of saving and hording, kept for marriage payments or as an inheritance (Sonko, 1986). Among the Nuer of Sudan, cattle *"allow them to carry out sacrifices essential for the harmony of human groups and of the cosmos. They are closely associated with lineages, within which these rituals are practiced"* (Bonte, 2007, 132).

In the highlands of New Guinea, the Dani are specialised in the rearing of pigs, which serve as a powerful means of exchange and as compensation for spilt blood; in the latter case, they are seen as a substitute for human life (Pétrequin and Pétrequin, 2006). The personal power of a leader is not measured so much by the accumulation of these "sign" animals, but more so by the control over their redistribution.

Keeping these two elements in mind – a Neolithic society without an obvious hierarchy, and livestock which represents intrinsic wealth – we can forward a hypothesis for the ceremonial acts observed at the Menneville enclosure that is identical to the hypothesis formulated on the basis of the study of households. Thus, the division between cattle and sheep is a sign that society was made up of a cattle rearing clan and a sheep (and goat?) rearing clan. The status of livestock rearer would effectively be higher, but its clan dimension would render this status more collective and less individual than we might expect if, for example, a "Big man" was present (Godelier, 1982). This collective identity would explain the apparent absence of wealth production in the material culture. It is possible that pig was the emblem of a third clan; this possibility is suggested by the presence of pigs, sacrificed and deposited in a manner similar to sheep, in the enclosure at Menneville. But we could also postulate that pigs were associated with male status; in the Hinkelstein cemetery at Trebur, pig and wild boar were found in male graves (Spatz, 1997; Spatz and Von Den Driesch, 2001, 1999). Even though the Hinkelstein period is a little later than the LBK, it is possible that this gender-based division of grave goods had its roots in an earlier tradition.

The realm of burial provides other clues, linked to the ages of the animals and the people buried. The cattle, and even more so the caprines, pigs and dogs, involved in the funerary rituals at Menneville were mostly young or very young animals. This pattern is also evident at the Bavarian LBK cemeteries at Aiterhofen and Dillingen (Nieszery, 1995) and at the site of Vendenheim "les Hauts du Coteau" (Boës *et al.* 2007), Alsace, where several deposits of young lambs and piglets were discovered in tombs. The rare examples of grave goods made from animal remains in the Paris basin indicate that these objects were made from the bones of young animals. As regards the caprine discovered in the Final BVSG burial at Buthiers-Boulancourt (Samzun *et al.*, 2012), here again the animal was very young. Furthermore, there seems to be a link between children and caprines: in addition to the close proximity of lambs to child inhumations at Menneville, the anthropomorphic figurines made from caprine bone, from Berry-au-Bac (Aisne) and Ensisheim (Alsace), were found in child burials (see chapter 5) (fig. 175). For the moment, it is more difficult to draw conclusions regarding the other two species but we can make one pertinent observation: pigs never occur near the graves of children, while cattle and sheep do. This raises the question of a link between pigs and adult status, as we will shortly see.

Grave goods, which can include personal ornaments, are social indicators that involve complex identification processes that are both individual and collective; they might mark the social affiliation and position of the wearer within a group (Callender, 1978; Hodder, 1979) or they might indicate a particular passage during the individual's life (Clifton, 1978). They can also throw light on the social and symbolic world of the group to which the deceased belonged (Bonnardin, 2009; Rigaud, 2014). LBK grave goods from burials in the Aisne and Alsace regions, made from the bones of significant species, and which bear specific decorative motifs (fig. 175 and 176), suggest a link between these objects and the biological status (maturity or gender) and/or or identity (affiliation to a lineage or clan) of the wearer. The deposition of a single unmodified bone, such as the example from grave 54 at Bucy-le-Long "la Fosselle" (see chapter 5) is more difficult to interpret but may be just as significant.

Thus, the choice of animals (as transformed or untransformed remains) deposited with the dead in graves and in ceremonial enclosures, appears to lend support to the hypothesis of affiliation to one of two clans, one associated with cattle and the other with sheep.

Hunting, a marker of gender

Hunting was also an important component in Neolithic society. Neolithic people were not hunter-gatherers in the strict sense of the word because their survival did not depend on wild animals and plants. Nevertheless, evidence for the consumption of wild animals on settlement sites and for their deposition in funerary contexts, indicates that hunting was systematic and was probably central to the functioning of society. We must ask ourselves why this activity continued into the Neolithic even though it was no longer indispensable for survival.

It seems important to once again turn to the ethnographic literature to discover how hunter-gatherers perceive the act of hunting and to see if certain fundamental, psychological elements might have survived within Neolithic society.

As shown by a major study of the hunting peoples of Siberia (Lot-Falk, 1953), traditional hunters consider wild animals as partners and forge reciprocal relationships with them; such relationships follow the model of social relationships but with an added symbolic layer. Moreover, the art of hunting is tied up with specific rituals; it requires a deep knowledge of several areas including techniques, magic, religion and laws.

> *"It is not a simple duel between man and beast in which man has achieved an enormous technical superiority.* [...] *In isolation, technique is useless; it remains ineffective unless it is coupled with the appropriate rituals. Technique and magic are inextricably linked, but the latter renders the former effective.* [...] *It is almost a business transaction: the right to kill has to be paid for, like a hunting license issued by the higher powers. In exchange for sacrifices, the gods provide abundant game. Without sacrifice, luck will desert the hunter."* (Lot-Falk, 1953, 7-9).

When it co-exists with agriculture and livestock rearing, hunting is often entrusted to specialists. It also tends to be an important part of social events, for example large gatherings and certain ritual activities. The hunt is a time of consecration, marked by a ceremonial departure and return, comparable to a time of sacrifice or of war, and the method and frequency of the capture of game maintains the proper order of Nature.

This activity also regulates collaboration between people, exchange and sharing between male partners or between men and women (Héritier, 1984). This way of thinking creates relationships based on social otherness, on the acknowledgement of "the other" as different[1].

Anthropologists have observed that in contemporary hunter-gatherer societies, hunting makes only a limited contribution to the diet, often secondary to the gathering of plant foods. The value attributed to it is a function of the meanings it has for the populations concerned, with a particular stress on a number of elements. Thus, hunting, almost exclusively the preserve of men, is a highly ritualised activity which plays a central role in defining social and symbolic order, particularly through the sharing of the kill. This means that hunting is a fundamental activity, not only because it is central to the material reproduction of a society, but also because

> "it is imbued with values that correspond, from a symbolic point of view, to the superiority of hunting over gathering, and, from a social point of view, to the superiority of the masculine over the feminine." (Bonte, 2007, 132).

Nevertheless, there are examples of women participating in the hunt. According to A. Testart (2014), what differentiates men and women is not the activity itself, but the way in which women practice it, by using weapons that do not spill blood. The level of physical proximity to the game, the use of weapons and traps that kill by stabbing or bludgeoning, involve a gender-based division of tasks in which only men are generally permitted to spill blood. In Testart's opinion, blood is in fact the determining factor in the gender-based division of labour. In this instance, the issue is not the biological substance but rather menstrual blood, which is shrouded in taboos that often forbid "afflicted" women from approaching anything that is symbolically associated with it (Héritier, 1984). Testart used these principles to formulate a general rule.

> "Sexual division of labour arises from the fact that Woman was excluded from tasks that too closely evoked the secret and worrying wound that she harboured within her." (Testart, 2014, 133).

Prohibiting women from engaging in certain activities such as hunting and warfare meant that there was no danger of "*like meeting like*" (*ibid,* 140). In other words, societies have always sought to avoid contact between entities that they consider to be symbolically similar because "*the meeting of two beings that are similarly affected by blood could trigger catastrophic consequences.*" (*ibid,* 143).

These universal principles lead us to believe that in LBK culture the wild boar was a mark of male status and may even have been associated with male initiation rites (fig. 202). There are several reasons why we have come to this conclusion. According to the model that we have developed for LBK settlement, small houses represent households that were still in the process of integrating within the village (Gomart *et al.,* 2015). They were inhabited by small family units who had recently arrived in the village. In this model hunting would have compensated for a scarcity of livestock products during the first few years of the household's establishment; this was the period required to build up a herd and to prepare land for crop cultivation. During this period, the newcomers may have engaged in exchanges with the inhabitants of the large, economically mature houses. In this context, the production of articles made from the by-products of hunting, as shown by macrolithic evidence for working bone (Hamon, 2006), could be seen as a supplementary activity carried out during the integration phase, the aim of which was to produce potentially exchangeable goods.

Nevertheless, wild boar, unlike red deer, was rarely if ever exchanged with the large houses; instead, the food refuse from the latter tends to contain red deer. Hence there appears to have been a desire to limit the consumption of wild boar to the small houses. Furthermore, at Cuiry-lès-Chaudardes, some small houses form a separate cluster in the north-western part of the settlement. Similarly, at the site of Bucy-le-Long "la Fosse Tounise/ la Héronière" (Constantin *et al.,* 1995), an independent group of small houses was identified to the north-west of the settlement (Hachem, 2009). Comparing the plans of the two sites (fig. 50) we notice a similar settlement layout which suggests that the establishment of LBK villages followed some kind of standard template. We envisage a situation where these small houses might have been inhabited by young couples, of which the male partner had to kill one or more wild boars before becoming fully integrated into the village community.

Anthropological literature mentions several types of hunting, including solitary initiatory hunting, such as that practiced in the case of wild boar and bear in European western forests (Pastoureau, 2007; Hell, 2012). The hunting of wild boar is reputed to be a very dangerous activity, even more so than the hunting of red deer. This is why we suggest that the killing of a wild boar, and the bringing back of its remains to the village, may have been a rite of passage for young men in the Neolithic.

Furthermore, since domestic pigs tend to be associated with the small houses where wild boar is also

over-represented, we are also tempted to suggest that these young couples practiced pig rearing; pigs have the advantage of being a more immediate source of meat than cattle or sheep. Nonetheless, we can also suggest that pigs were associated with male status as the two hypotheses are not mutually exclusive.

In the settlements, two species are in opposition to their original taxonomic group: pig, which is preferentially associated with hunting (of wild boar) and aurochs with cattle rearing. As regards roe deer, it is associated with either the rearing of sheep or the hunting of red deer. Red deer is better represented in houses where livestock rearing predominates.

These pairings are reminiscent of the "*mirroring of the domesticated and the wild, indissociable, like the two poles of a mag*net" which is part of the principle of the "*non-exclusion of opposites in profound thought.*" (Poplin, 1993, 531 and 538).

Aurochs and red deer appear to have been the only wild animals whose presence was allowed in the ceremonial enclosure at Menneville: their presence is not marked by isolated bones or carcasses, indicators of consumption or of funerary *viatica*, but rather by ostentatious elements such as bucrania

Figure 202: Wild boar, numerous bones of which are found in the lateral pits of small houses, may have been a marker of masculine gender in the LBK (photo: P. Martorana, FreeImages).

or in items requiring technical investment such as worked red deer antler.

We can add a third wild animal, the roe deer. Only two pieces of roe deer antler were found in the Menneville enclosure ditch, but one of these was from a very unusual context that is worthy of mention. The object in question was a roe deer antler point, created from a left roe deer antler; the distal portion is made from the median point of the antler. Longitudinal lines, which do not correspond to natural wear, are visible on the object. The posterior and anterior points have been removed and the detachment surface has been flattened (Maigrot, 2013). The proximal part of the object was broken *in situ*, discovered between the ribs of a human skeleton (Thevenet, 2013; Thevenet *et al.*, forthcoming) (fig. 127). Because roe deer antler was very rarely used for the manufacture of tools, we believe that this object, which may have been used to to stab the man in the back, falls into the category of object-signs that we defined earlier (see chapter 4).

Given the pronounced emphasis on domestic species in the ceremonial enclosure, we might ask why certain wild elements were allowed to be deposited there. It seems to us that the answer lies in the fact that aurochs and red deer (and to a lesser extent, roe deer) are the two wild species linked in the settlements to the large houses with higher proportions of livestock.

Aurochs, the mirror animal of cattle, is abundant in the faunal assemblages of the large meeting houses and it seems that parts of the animals were distributed among the other houses (see chapter 3).

In an attempt to explain this pattern we can postulate that the hunting of aurochs was a communal activity, much like the hunting of bison by the Plains Indians of North America in the 19th century (Ours Debout, 2014); if this was also the case in the Neolithic, then the presence of aurochs in collective ceremonial events and the redistribution of its remains would make sense.

At this point it is worth mentioning an exceptional zoomorphic vessel found on the BVSG site of Aubevoye in Normandy (Riche, 2004); incidentally, one of the houses also yielded a complete aurochs bucranium (pers. comm. L. Bedault). The vessel is a representation of a bull or aurochs, or perhaps it symbolizes both at the same time (fig. 203). We have already seen how aurochs and domestic cattle are closely linked to each other in LBK culture and it is likely that the BVSG followed the same tradition. Ritual practices involving bovinae are part of a much older tradition, perpetuated in central Europe (Marciniak, 2008; Bánffy, 2001) and very probably originally inherited from the Near East, where representations of these animals are common from as early as the Pre-Pottery Neolithic A – 9500-8700 cal. BC. (Stordeur *et al.*, 2004).

Figure 203: Aubevoye "la Chartreuse" (Eure), decorated zoomorphic ceramic vessel representing a bovine, Early Neolithic BVSG (after Riche 2004, fig. 3; photo: H. Paitier, Inrap).

As regards red deer we can state that they were a key animal in that they are present to some extent in the faunal remains of all the houses (fig. 204). They made an important contribution to the diet, to clothing requirements and to craft activity, and they were also included, in the form of personal ornaments, in LBK funerary rites.

Thus, for example, at Bucy-le-Long "La Fosselle", the exceptional set of ornaments in the grave of an adult woman (see chapter 5) had been attached to an item of clothing, and the extent to which clothes function as a sign of both individual and collective identity is well attested (Cassagnes-Brouquet and Dousset-Seiden, 2012). An essential element of material culture, clothing is a marker of identity common to all human society; loaded with meaning, and not simply a means of protecting and ornamenting the body, clothing is in fact a reflection of social norms (fig. 205). In traditional societies it is an instrument for controlling and hierarchizing the sexes, age groups and social statuses (Cassagnes-Brouquet and Dousset-Seiden, 2012). The deer

tooth beads therefore appear to be a strong identifier of the deceased's female status.

We tentatively suggest that red deer was linked to the female world because, not only have other LBK female burials in the Paris Basin yielded deer tooth beads (Bonnardin, 2009), but they are also characteristic of female personal ornaments in the Hinkelstein group (Spatz and Von Den Driesch, 1999). However, since there is also evidence for red deer teeth ornaments from from two male LBK burials in Austria and the Czech Republic: Rutzing, tomb 13 (Kloiber, von and Kneidinger, 1970); Vedrovice, tomb 15/75 (Podborsky, 2002), this hypothesis needs to be checked against other corpora.

Another surprising aspect of the red deer is its close link to domesticity. In fact, red deer is the game animal of choice of the large houses, which are otherwise closely associated with the keeping of livestock and agriculture. Furthermore, it is worth noting that the majority of deer hunted are does. Taken together, these elements suggest that we can place red deer in a category within which the wild aspect has, in some ways, been deliberately erased.

Figure 204: Hunting is a unifying activity essential to the functioning of Neolithic societies and red deer was the preferred game (photo: Free Photos, Pixabay).

A dual society

In conclusion, it is important to highlight the parallels that we have observed between these animals associated with LBK symbolism and the faunal evidence from the Hinkelstein cemetery at Trebur in Germany, where 79 tombs have been identified (Spatz and Von Den Driesch, 1999). In short, at Trebur, three types of faunal deposits have been recorded in the burials of men, women and children; for the most part, these deposits are made up of parts from domestic species.

In some cases, articulated cattle ribs were placed on the head and thorax of the bodies; these cattle carcass parts were recorded in 36% of the male burials and in 23% of the female burials.

Figure 205: Clothing is a language destined to convey meaning as much as to protect or adorn; Syrian musicians in fur-lined coats (photo: C. Lallemand, 1863-1864).

Furthermore, thirty-four burials contained haunches of beef, mutton, goat and pork. Ten burials contained the skeletons of one or two sheep, the heads and ends of limbs of which had been removed. Pig meat was only included in male burials while there seems to be a preference for sheep in female burials.

Wild fauna is very poorly represented. The evidence mainly consists of haunches of wild boar, but some tusks have also been found. Like domestic pig, wild boar is associated with male graves. Red deer is even rarer, with only a single haunch and two antlers recorded. However, it occurs more frequently in the form of personal ornaments made from teeth. A particularly rich grave (n°63) contained a set of 230 red deer canines (equivalent to the killing of 115 animals) that originally decorated a belt. Such belts decorated with red deer canines are a particularity of female personal ornament. Versions made with shell beads are generally associated with males.

Finally, in the same cemetery, but in graves associated with the Grossgartach culture, two representations of caprines were found on pottery accompanying offerings of cattle and pigs (Spatz and Von Den Driesch, 2001, figs. 4-5).

These similarities between the integration of animals in the LBK rituals of the Paris basin and the Hinkelstein examples in Germany, suggest that these practices were deeply rooted in the Early Neolithic.

We have used the model of settlement organisation and the analysis of funerary structures in order to identify one of the essential dimensions of LBK society: a system founded on a real duality between stock-raising and hunting. The society was structured around cattle, caprines, wild boar, aurochs and red deer, which are systematically found in the houses, settlement pattern and funerary structures.

The interpretation that we forward to define a social framework on the basis of this archaeozoological data from the profane and sacred domains, is that the animal remains are intended as markers (Hachem, 2018a). We believe that these markers signal the identity of certain units such as clans and that they relate to groups of cattle rearers, sheep rearers, and possibly pig rearers. It is our belief that these markers could also incarnate the gender of an individual; thus, the male gender is associated with the wild boar, and possibly the domestic pig, while red deer may be associated with the female gender.

It is also possible to envisage markers of age, but this question is more difficult to deal with. If we take the example of certain present-day societies, such as the Nuers of southern Sudan, cattle, which are closely linked to lineages that engage in sacrifices, are also linked to male individuals who, when they are young, receive a "name cow" with whom they will identify for the duration of the animal's life.

This example tempts us to postulate that, in the Early Neolithic, there may have been a link between animals and children and that if the child died, the animal was sacrificed and part of the carcass selected to accompany the deceased or to mark his/her belonging to a particular clan.

The notion of clans or lineages in LBK society has already been discussed elsewhere (Van de Velde and Amkreutz, 2018), without an association being made with animals. Hence, at Trebur, the clan hypothesis was forwarded in an attempt to interpret two concentrations of tombs containing both male and female burials (Spatz and Von Den Driesch, 1999). Clans have also been mentioned to explain the ranking of households and access to better land at the site of Vaihingen in Baden-Württemberg (Strien, 2005; Bogaard et al., 2017). Finally, given that only a small proportion of the population was formally buried, the question arises whether these individuals represent certain groups within the community, such as lineages (Bickel and Whittle, 2013). Moreover, possible similarities have been observed between LBK (Whittle, 2009), and indeed BVSG (Hachem and Price, forthcoming) houses and graves located near these houses, which might reinforce this hypothesis. The hypothesis of patrilocality discussed in the context of the cemetery at Nitra in Slovakia and supported by the results of strontium analyses (Bentley et al., 2012) is not the only one forwarded for LBK society. The older hypothesis of matrilineal descent may also be valid.

To conclude this discussion on the duality of Neolithic societies and the social ties that it creates, we will look at an example from more recent societies which illustrates what we have been trying to highlight. The example comes from a study, carried out by the ethnologist Roberte Hamayon (1990), which deals with Siberian shamanism and the hunting lifestyle with which it is associated. The study focuses on the Buryat ethnic group which was affected by colonisation in different ways in the 13th century. In the west they remained forest dwelling hunters while in the east they became nomadic herders. This comparative approach between two opposing ecological, sociological and economic realms is instructive in the context of our topic. Within the forest world of the Siberian hunters, where the prey of choice is deer, the kill has to be shared, which precludes any temptation to steal or accumulate goods. In these elementary, undifferentiated societies, alliances exist between humans, nature and the supernatural (collectively known as shamanism); social and supernatural relationships are only envisaged as being horizontal and reciprocal. In contrast, livestock rearing, which is based on the domestication of species for reasons of productivity, may incite its practitioners to develop an investment strategy whereby the animal is no longer a being but a product which can be appropriated, stolen, or passed on as an inheritance. Kinship relations thus supersede alliances, and lineage-based hierarchies replace non-differentiation. There is a clear shift from the horizontal world view of hunters to the vertical world view of herders.

The Middle Neolithic

A tendency towards ostentation

In the Cerny (Middle Neolithic I) important changes occurred in the layout of settlements. Villages as we know them in the LBK and BVSG, *i.e.* composed of clusters of rectangular houses, disappeared. Instead, settlement sites, which are generally less numerous than in the Early Neolithic, take the form of circular or rectangular buildings (Bostyn *et al.*, 2016), but they are no longer grouped together to form villages. To date, the Cerny buildings discovered in northern France are generally isolated and pits only yield small amounts of food refuse.

As regards diet, despite a certain degree of variability, we observe a number of points common to all sites that yield faunal remains. Whether we are looking at the settlement site at Conty, enclosures or the occupation layer adjacent to the monumental building at Beaurieux, we observe the predominance of stock rearing over hunting, a preponderance of cattle over other species, an absence of caprines and roe deer and a significant proportion of pigs. Furthermore, as we have seen in previous chapters, the position occupied by the latter is part of a wider trend, the beginnings of which we see in the middle phase of the Blicquy/Villeneuve-Saint-Germain and which continues throughout the Middle Neolithic II.

As for isolated pits containing skeletons or articulated bones, there is only a narrow range of species, so we can envisage a deliberate selection: the species concerned are roe deer, red deer and domestic cattle.

Cerny funerary contexts clearly reveal an ostentatious dimension. Visually imposing, long earthen barrows, delineated by ditches, were apparently erected over single burials. The construction of such monuments required a considerable collective investment for the benefit of a handful of individual men, women and children, which indicates a highly structured society with significant differences in social status.

Even though there are some specific differences, due either to chronological or regional factors, the funerary monuments of Normandy nevertheless belong to the same phenomenon as the cemeteries of the Seine-Yonne area which formed the basis for current hypotheses about monumental burial (Ghesquière *et al.*, 2019a).

The ditches of these burial monuments contain specific cattle anatomical parts, namely scapulae and cranial remains. At Fleury-sur-Orne, these ditches also yielded massive, red sandstone picks, which were probably used in the construction of these monuments. These bones and picks may have been foundation deposits, a hypothesis that we believe has some merit and which has previously been proposed for certain objects from collective burials dating to the Late or Final Neolithic.

"According to Van Gennep (1981), they form part of rites of passage: sacrifices (of objects, food and human beings) that are described as foundation or construction sacrifices and which initially served to lift a 'taboo' on the structure (whose protection was not guaranteed prior to this). [...] Consecration or foundation deposits are not confined to picks manufactured from stone, but can include all objects that may have been used in the construction of the monuments: picks or hammers made from red deer antler, polished stone axes, etc." (Sohn, 2008, 9).

In certain tombs, the deceased is accompanied by sheep or cattle remains, but the two species are represented differently (see chapter 5.1). Sheep, which are relatively numerous, occur as complete skeletons or as parts, while cattle are represented by isolated bones. Their symbolic dimension is significant as demonstrated by the placing of scapulae, metapodials and horn cores on certain parts of the corpses. The tombs at Fleury-sur-Orne and Rots in Normandy only contain edible parts of domestic animals and never the objects made from wild boar teeth or red deer bone that occur in the burials at Passy in the Yonne valley. Pig, which is associated with every day consumption, is absent, as are wild animals. However, the presence in these burials

of numerous arrowheads, which may have been used for hunting (or warfare), suggests that the person buried was an archer, or at least was represented as such in death (Chambon and Pétillon, 2009). The close link between arrowheads and the masculine identity of the person buried is not only seen in the burials of the Seine-Yonne area (Thomas and Chenal, 2014), but also at Fleury-sur-Orne despite the small number of burials of identified sex.

The fact that several sheep could be killed, and not eaten, in order to accompany the burial, reveals a degree of extravagance that approaches ostentation (fig. 206).

In the following period, the Middle Neolithic II, sedentary village settlements are just as elusive as in the preceding period. However, enclosures become increasingly common, their numbers growing in pace with the population, which we can estimate on the basis of funerary evidence (Bocquet-Appel and Dubouloz, 2004). We thus observe economic, social, and cultural intensification, as well as an increased level of control over territory.

Two interpretations can be forwarded for the development of these enclosures. On the one hand, there are needs that revolve around defence, marking out of territory and social reproduction that led farming communities to

Figure 206: Sheep were sacrificed, sometimes in great numbers, throughout the Neolithic; marked sheep in summer pasture, Mont Viso, Italie (photo: P. Pétrequin).

construct, in certain locations in the landscape, physical features that express their group identity. On the other, we can envisage the existence of a political power that developed through a process of social stratification, within a framework of competition between groups.

"In both cases, the building of enclosed sites illustrates moments when particular social and political regulations emerged; regulations that were linked to greater interaction between human groups and /or more marked territoriality" (Dubouloz, 2018, 198).

Two levels of hierarchy were established, between the less numerous complex enclosures, which probably enjoyed a special or even superior status, and the other enclosures. This hierarchisation was accompanied by a strengthening of territoriality, including the appropriation and control of sources of essential raw materials such as flint (Manolakakis and Giligny, 2011).

The enclosures, due to their monumental nature and their need for maintenance, indicate a capacity on the part of the society to maintain their existence through collective investment over a long period. The construction of such sites required the mobilization of several segments of society, not necessarily in a spirit of competition, but rather in a spirit of cooperation in the undertaking of a common project (Parkinson and Duffy, 2007). It remains to be seen if this cooperation was voluntary or imposed: in the first case, the construction would be an occasion for society to renew its unity and identity, in the second case it would act as a manifestation of the authority of a political entity (Dubouloz, 2018).

We will now try to find evidence in the faunal remains from enclosures in northern France whose characteristics we presented in a previous chapter (see chap. 4) that might throw further light on these issues.

Clues to the activities that took place in enclosures

It is interesting to compare the data from northern France with the evidence from the British Isles, where research into the function of these sites has quite a long history.

Unfortunately, most of these enclosures are late relative to the French Middle Neolithic II, as they appear from around 3700 BC onwards in Britain (Whittle *et al.*, 2011). The hypothesis of feasting is frequently put forward for the sites of Windmill Hill (Whittle *et al.*, 2011) and Hambledon Hill (Mercer and Healy, 2014), and

enclosures are generally considered to have functioned as ceremonial assembly sites (Parker Pearson, 2003). However, a relatively small number of Neolithic enclosures of this date in Britain have had the benefit of archaeozoological studies (Parmenter *et al.*, 2015), and it is therefore difficult to make comparisons. An enclosure of later date, Durrington Walls (2620-2400 BC), built in the period corresponding to the Final Neolithic in northern France, has yielded large quantities of animal bone that are of interest in the context of our study. It includes five buildings and an archaeological layer which has yielded huge quantities of animal bone that are of interest in the context of our study. The majority of these bones are from domestic animals; only a few aurochs, red deer and roe deer bones were found. In the case of pigs, two slaughtering peaks are apparent: one in the middle of winter and the other in summer (Albarella and Payne, 2005). Altogether, the faunal remains, in large quantities suggesting wasteful consumption, support the hypothesis of a site occupied by people (possibly the builders of Stonehenge) who gathered together to feast.

Taking inspiration from this example and the indicators mentioned, we will now see whether or not the enclosures that we have analysed present the same characteristics.

There is a difference, for the moment unexplained, between enclosures that have yielded large quantities of fauna, such as Escalles (Nord-Pas-de-Calais), Villers-Carbonnel (Nord-Pas-de-Calais), and Passel (Oise), and those that have produced small quantities, such as Bazoches-sur-Vesle (Aisne) and Carvin (Nord-Pas-de-Calais). Taphonomy is certainly an important element to be taken into consideration, because the bones are poorly preserved at Bazoches and Carvin, but it is possible that this offers only a partial explanation for the low number of remains, perhaps different activities took place on the two categories of sites (Hachem and Maigrot, 2019).

The other enclosures have not been completely excavated and this limits comparison, but in general the number of bones recovered tends to be quite low, apart from a few exceptions (Hachem, 2011b).

The production of meat necessarily involves the slaughter of an animal (fig. 207). As we saw in the chapter 3.1 the reconstruction of the slaughter profiles of cattle at Passel, Villers-Carbonnel and Escalles, reveals that on each site priority was placed on producing large quantities of meat for consumption on site.

At Escalles, an archaeozoological study was carried out with the specific purpose of determining a potential slaughter season for domestic animals (Hachem and Chombart, forthcoming). The results revealed two specific seasons for the slaughter of cattle and sheep: one at the beginning of autumn and the other at the beginning of spring. A larger number of animals were slaughtered in autumn. While a degree of caution needs to be exercised, it appears that the autumn slaughter involved a larger number of young animals, while predominantly older individuals (over two years of age) were slaughtered in spring. The situation concerning pigs is less clear because

pigs can produce two litters a year, which makes calculations more uncertain; nonetheless, we observe the same two seasonal slaughter peaks that were identified for cattle and caprines.

Among the six attributes used to characterize complex enclosures such as Bazoches-sur-Vesle, two concern animals (Dubouloz, 2018): the sparse evidence for consumption of wild animals and the occurrence of isolated deposits.

A very low rate of game consumption (less than 5 % of remains) has been noted in the following enclosures: Bazoches-sur-Vesle, Boury-en-Vexin, Passel, Villers-Carbonnel and Escalles. But the symbolic importance of wild animals remains, as is evidenced, for example, by the deposition at certain strategic spots within the enclosures of complete red deer antlers at Bazoches-sur-Vesle and of aurochs bucrania at Passel.

The motivation behind such selective hunts may originally have been ritual. In fact, the bulk of the meat supply was ensured, and we can thus state that the search for meat was not the primary motivation behind hunting.

Deposits reflecting ritual acts (fig. 208) are present on all of the enclosure sites as we have seen in detail in the chapter 4.1.2. Nevertheless, manifestations of cult are more ostentatious on certain sites (Boury-en-Vexin, Passel, and Bazoches-sur-Vesle) than on others. In all cases,

Figure 207: Meat necessarily entails slaughter – traces of a heavy frontal blow are visible on certain Neolithic cattle skulls; slaughter of a bovine by a butcher at Quintenas, Ardèche, between 1906 and 1911 (source: familles-de-quintenas.com, Mireille Beile collection).

despite having a common base, these deposits take diverse forms.

We will mention as a reminder that cattle play a primary role; cranial elements, bucrania and horn cores are numerous and occur as isolated deposits or arranged in groups at strategic points within the enclosure, such as principal interruptions in the ditches. While such deposits are particularly numerous, a range of other acts can seen as symbolically important. These acts concern a variety of species, both wild and domesticated, in various forms. Pigs and dogs are often deposited more or less complete (either articulated or disarticulated) while wild boar and red deer are preferentially represented by skulls. At least some of these deposits appear to have been sacrifices.

Furthermore, we can also hypothesise that some of the deposits acted as a kind of show of strength which underlines the monumental character of the enclosure itself, with its massive ditches and imposing palisade as for example at Passel. The deliberate emphasis on dangerous attributes, such as cattle and aurochs horns, is an example, but others include the possible display of a set of red deer antlers on a post (fig. 150), the significant proportion of male animals and the large size of many of the wild animals (wild boar, red deer and bear, fig. 209).

In order to further investigate the faunal evidence for activities, a comparative study of species consumed and species used for making bone tools was carried out on the finds from six enclosures : Bazoches-sur-Vesle, Maizy, Carvin, Passel, Escalles, Villers-Carbonnel (Hachem and Maigrot, 2019).

The results of this study reveal some striking differences between Michelsberg and Chasséen sites in the ways animals were exploited for consumption and tool production. For Spiere's group it is still difficult to say.

As regards the Michelsberg sites, wild fauna provided most of the raw materials for bone working, while consumption was based on domestic animals. This opposition between meat consumption and bone working is less apparent on Chasséen sites because there is rather more use of domestic animals for tool manufacture, even though wild fauna still predominates here.

While the bones of domesticated species destined for the manufacture of tools can be linked to butchering activities, this is not the case with raw materials from red deer. For this species, the amount of meat consumed *in situ* is insufficient to meet the needs of the bone working industry, in terms of both numbers of animals and anatomical parts. This deficit is clear

Figure 208: Maizy "les Grands Aisements" (Aisne), Middle Neolithic 2 (Michelsberg) enclosure, ceramic figurine (photo: UMR 8215 Trajectoires).

Figure 209: Bear teeth testify to the targeted hunting of large individuals during the Neolithic (photo: S. Mitchell, Burst).

on all of the sites. Where, then, did the worked elements come from (use of shed antlers)? Where were the missing joints of meat (limbs, torsos) consumed?

The presence of significant quantities of manufacturing waste indicates that some working of bone and red deer antler occurred on site. Finished tools – at least in the case of the Michelsberg enclosures – are much rarer than manufacturing waste; nevertheless, functional analyses show that they were used before being abandoned (Piliougine, 2015; Maigrot, 2017). Use-wear analyses suggest that they were used in various domestic activities such as the working of ceramics, wood, flint, etc. On the site of Bazoches-sur-Vesle, where there are 10 tools for 100 items of manufacturing waste, it looks as though more tools were manufactured than actually used on site. Furthermore, on all six of the enclosure sites, we find waste from the extraction of cortical strips from red deer antler, but tools made from this kind of blank were not found on the sites.

Where have these tools gone?
The spatial distribution of faunal remains and bone or antler artefacts at Bazoches-sur-Vesle shows rather different patterns for domestic refuse and for what might be refered to as deliberate deposits.

The domestic refuse (food waste, discarded tools and manufacturing debris) is distributed throughout the first two inner series of ditches, with significant concentrations around the interruptions corresponding to passages-ways. As we previously mentioned, the relatively uncommon cases of abandonment or deposition are systematically located away from concentrations of domestic waste (see chapter 4). But this is perhaps the only pattern common to all six enclosure sites under consideration here. In fact, the deposits seem to vary in terms of both their composition and siting. Thus, at Bazoches-sur-Vesle (Michelsberg), the deposits – composed of bucrania, whole red deer antlers and even human remains – tend to be concentrated in the western part of the enclosure, away from the concentrations of domestic refuse. At Carvin (Spiere Group), most of the deposits of caprines and dogs occur in the outer ditch, while domestic waste is concentrated in the inner ditches. At Passel (Chasséen group), nine dogs were deposited in the outer ditch while other types of deposits (bucrania and scapulae) are concentrated in the inner ditch.

The recurrent deficit in certain bone or antler tools and joints of meat highlighted by this overview of animal resources for consumption and tool manufacture, suggests that these

sites, as well as being places of assembly, also acted as places where food stuffs, finished and semi-finished tools and even know-how were exchanged.

The Late Neolithic

The collective and the individual

Collective burials start to become more common in the second half of the 4th millennium and predominate throughout the 3rd millennium BC. While some examples only functioned for a few generations, others continued to receive inhumation burials for over a millennium (Chambon *et al.*, 2017).

Amongst the grave goods accompanying the burials, it is possible to distinguish between "individual" goods and "collective" goods. While tools, weapons and personal ornaments, often used and broken, mark the individual status of the dead, ceramic vessels and polished stone axes are true "funerary symbols", particularly in northern France and Europe at the end of the 4th millennium (Sohn, 2008). They are evidence for deposits of symbolic and ritual value.

In light of the published data, it appears that sets of carnivore teeth ornaments, hedgehog half-mandibles and objects made from red deer bone or antler were associated with special individuals, while complete, or semi-complete, dogs and carnivores are undoubtedly linked to the collective sphere (see chapter 4.2 and chapter 5.1). Nevertheless, a more focused study of these artefacts is required in order to be more certain that these conclusions are valid.

In the rare tombs where it has been possible to observe spatially distinct groups of burials, it appears that the grave goods differ from one group to another, both in terms of their quantity and nature. At Schönstedt (Germany) for example, a detailed study has revealed that the two groups present represent two different kinship groups, although they probably both belonged to the same community (Bach and Bach, 1972). These examples prove that, first and foremost, grave goods marked the difference between groups, which may be families, and that they are signs that these groups belonged to a particular social rank or "clan".

Individual grave goods mark the differences between individuals of the same sex or age group, and sometimes the same burial group. As has been observed at the cemetery of Vignely, certain burials are accompanied by exceptional objects that single them out as individuals of particularly special status (see chapter 5).

Status differentiation

The dead are not all equally adorned nor are they all accompanied with the same goods: some have no grave goods and others have many. This difference is not explained by age or sex, because men and women, old and young, can be accompanied by grave goods.

There are important differences between grave goods with children and with adults, and between men and women in western Europe (Sohn, 2007). The grave goods thus act as a sign of one's belonging to a certain age group or gender group.

Quivers, sets of personal ornaments made from teeth, bone or antler, axes, and flint or copper daggers are associated with men; such objects are probably references to hunting, warfare, construction and land clearance. Awls, smoothers, pottery, knives and blades are associated with women and probably evoke activities connected with basket making, pottery and weaving, as well as harvesting and planting. Women are also associated with a greater diversity of personal ornaments than men.

Grave goods associated with children are abundant, particularly in the form of personal ornaments, and present almost the same characteristics as the grave goods associated with women (apart from the absence of certain types of pendants). No weapons (*e.g.* quivers and daggers) have been found with young children, but older children or adolescents are sometimes accompanied by objects associated with adult men. This allows us to postulate that a particular status was acquired through initiation at the moment when an individual officially passed from childhood to adulthood: in the case of boys, this would have involved passing from the feminine realm to the masculine realm during adolescence, a pattern that is observed in many societies today (Van Gennep, 1981).

Also of note is the association of foxes with children in several burials dating to the transition between the Middle Neolithic and Late Neolithic (fig. 210).

The Final Neolithic

In the Final Neolithic we observe a tendency towards gigantism evident in the construction of large buildings, which probably reflect more complex social relationships. The scale of the work involved in building these structures undoubtedly indicates the mobilisation of a large portion of the community and the application of considerable inherited know-how.

Moreover, the combination of palaeoenvironmental data and the results of technological

Figure 210: Fox canines were transformed into personal ornaments, as illustrated by child burials dating to the Middle-Late Neolithic transition (photo: Skeeze, Pixabay).

and functional analyses provide useful insights into the breakdown of activities both at site and territorial level in the North of France (Martial *et al.*, 2011).

The spatial distribution of evidence for textile production suggests that the *chaîne opératoire* was divided up within the territory, which may reflect the organisation of the territory itself (Martial, 2008). The large number of spinning whorls found on the sites accords with the fact that a significant investment of time and labour was required to produce the kilometres of yarn destined for the weaving of fabric. Conversely, the more limited distribution of archaeobotanical evidence for linen and of loom weights suggests that the growing of this plant, and the weaving of linen fibres, two activities at either end of the *chaîne opératoire*, were the preserve of certain sites and, thus, of certain groups (Martial *et al.*, 2011).

Other contrasts are apparent within the Deûle-Escaut group, such as the dichotomy between imported tools and local tools (Martial and Praud, 2011). Thus, the importation of flint blades, and the diversity in their origins, do not respond to a functional need and may therefore be linked to systems of exchange between communities within a shared territory.

The existence of distinct domains of exchange, which define the Neolithic more than preceding periods, and which undergo a perceptible intensification in the Final Neolithic, suggests varied social motivations including the forging of alliances and resolution of conflicts, the affirmation of status through the acquisition of prestige goods, and the exchange of information (Hofmann, 2012; Perlès, 2012).

The fairly regular spatial distribution of funerary monuments probably indicates that they were used by quite localised communities. Their long duration of use also indicates territorial stability and the number of people buried within them suggests that they were the burial places of a particular group, perhaps a dominant lineage, within each community (Dubouloz *et al.*, 2005).

Over the course of the 3rd millennium, there is an evolution in the destination of grave goods. We witness a veritable transfer of "signs": from axe to dagger, between northern Europe and the Mediterranean regions, from collective to individual ideology (Sohn, 2007). Thus, the grave goods in collective burials attest to major social changes in Europe during the 3rd millennium BC, one of the consequences of which was a return to individual burials.

At present, too few archaeozoological studies have been carried out to allow us to attempt to identify patterns in the social structure of the Final Neolithic using this approach alone. We can simply highlight the fact that cattle undoubtedly remained important within society because their use for draught purposes dates to this period (Pétrequin *et al.*, 2006). We must also stress that wild fauna do not disappear from the record, as evidenced by personal ornaments and other objects made from worked bone.

Conclusions regarding Neolithic social systems

Animal remains in settlement sites and in funerary rituals, and the ways in which they were discarded or deposited, provide useful insights into the important role animals played in Neolithic societies.

> *"Every society develops its own mental representations: a collection of ideas and values that is unique to it (…). These cultural facts are not simply juxtaposed; they form systems, collective representations. Classifications of natural species and of social groups, symbolic organisation of space and representations of the body are all conceived as interdependent elements of a cultural representation of the world."* (Boyer, 1991, 657).

These populations all kept domestic animals, but also continued to hunt and it is this duality that is of particular interest here. In our opinion, these populations embody a transitional stage between the "fully wild" of the Mesolithic and the "fully domesticated" of the end of the Iron Age; a stage in which the physical and the ideal borrow from the two intrinsically linked domains with domesticity, perhaps more closely linked to the village domain, and hunting, no doubt more closely tied to the forest.

According to anthropologists, hunter-gatherer societies and livestock-rearing societies do not share the same perception of the place of human beings in the environment and of their position with regard to animals.

For hunter-gatherers, an animal is not simply a prey: like human beings, an animal has a soul and a language. All nature is living and there are no barriers between the animal, vegetal and even mineral kingdoms; there are simply different aspects and changing appearances (Lot-Falk, 1953; Sahlins, 1976). Human beings are no more than a link in the chain and their attitude to all other living beings is founded on mutual agreement and the recognition of rights. Hunting is the culmination of long ritual preparations because the hunter's feeling of guilt in the face of this "murder" drives him to seek a reconciliation with the victim.

With domestication, another frame of mind develops. Mankind sets itself apart, because its culture leads to a perception of self as superior to nature; in this way humans see themselves as entitled to appropriate part of the animal world. Through the adjusting of the volume and composition of fodder, the weaning of young animals, the control of siring, the selection of individuals for reproduction, and by seeking particular crosses, humans condition animals, shape them and improve them for their own ends (Digard, 2009). These animals, which become the property of a human group, are dependent on people (and vice versa), and this relationship of mutual dependence brings about a behavioural change in human beings themselves. Nevertheless, this concept of domination must rid itself of all manichaeism, as is demonstrated by the example of the Amazonian Achuar tribe and their attitude towards animals (Descola, 1993). Members of this group domesticate certain species without the intention of eating them (because they live by hunting) or changing them, and without considering themselves superior.

However, as has been highlighted by the anthropologist P. Descola (Descola, 2015), the opposition between nature and culture, within which these theories concerning the relationships between humans and non-humans developed and on which western anthropology has been founded since the 19th century, is not shared by all societies. A significant proportion of the world does not think in terms of this opposition, but rather follows other "ontologies" (Descola, 2005). The author presents a theoretical model that is based on the conceptions of continuities and discontinuities between humans and non-humans; he thus identifies four broad modes of identification that are found throughout the world[2].

First, animism, which envisages mental continuity but physical discontinuity between humans and non-humans. Thus humans, animals and plants are not differentiated, they share subjectivity, conscience and intentionality. Second, totemism, which holds that there are moral and physical resemblances between humans and non-humans. When certain attributes are shared by the two entities, they are grouped together in families. Third, analogism which supposes that there are total and permanent discontinuities between humans and non-humans. The beings that inhabit the world are perceived as an infinity of unique entities.

Finally, naturalism, which presupposes dissimilarity in interiority but similarity in physicality.

The author emphasises the fact that in reality these broad ontologies never occur in their pure form. Instead, anthropological investigations reveal nuanced situations, progressive gradations and partial integrations.

While it is not possible to know which ontologies inspired Neolithic populations, the archaeological record seems to offer tantalising glimpses of some of the above conceptions of the world. Thus, for the Early Neolithic, we mentioned totemism in relation to settlement organisation and with regard to funerary contexts in which the remains of cattle, caprines and wild boars are found.

Perhaps we could also speak of animism, at least in the case of the Middle Neolithic: this was the period that saw the inclusion of animals in collective rituals within enclosures, which acted as assembly sites and were constructed in particular natural locations (marshes, hills, promontories, etc.). Clearly the expertise of anthropologists would be beneficial for a more thorough investigation of these aspects.

Dependence on food production requires the invention of new socio-cultural systems. Following the work of archaeologist and ethnologist A. Leroi-Gourhan, and of anthropologist P. Bonte, certain links between animals and pastoral or agro-pastoral societies have been brought to light (Leroi-Gourhan, 1964; Bonte, 2004, 2007). These interactions are evident, for example, in pastoral societies – be they nomadic (Touaregs and Nuers) or sedentary (Maasai) – where we observe a parallelism between a herd dominated by one animal and a human group which is governed by rules of filiation and alliances. Livestock, as the functional basis of society, is considered as true wealth and the modification of the herd brings about the modification of the group.

But livestock does not simply act as a link; it can be at the very origin of interactions. Thus, in the myths of a number of East African populations, cattle are at the root of social relationships. They form a weft that structures the relationships between human beings and between human beings and the supernatural, particularly through the vehicle of ritual sacrifices (Evans-Pritchard, 1954). Its effectiveness, which is of course of a symbolic nature, actually materially determines the way in which livestock rearers organise their pastoral activities through a set of specific techniques and knowledge.

In village-based agricultural societies, in which the animal is linked to the sedentary farmer (the most common type of husbandry practised in Europe in recent centuries), interactions are more diverse: the animal can be kept as part of a larger communal herd, or in a smaller family-owned herd, or as a single animal. Consequently, a certain hierarchisation of animals emerges as they become the object of specialised surveillance (shepherds, herders), or otherwise (watched over by the elderly and children), and receive unequal treatment (integrated within the family or kept outside).

These various anthropological considerations can act as a crucible from which we can draw interpretations to make sense of the archaeozoological data.

In the Early Neolithic period, we have seen how animals were an integral part of the social structure at several levels.

The first level is the house, since a certain range of domestic and wild species is affiliated with each house and, as far as we can tell from the evidence for consumption, there is little differentiation between them (fig. 211). The second level is the village, with probable exchange of goods between houses at different stages of economic integration; the larger houses exchange their excess livestock and game with the smaller houses. The third level is the ceremonial enclosure, where domestic animals are preferentially included in ritual sacrifices (cattle, sheep, pigs and dogs), but where two wild species (red deer and aurochs) also feature.

All of these elements lead us to the hypothesis of an LBK society without marked social inequalities,

at least between the different social entities, as far as we can tell from the material culture. A large part of this society was orientated towards exchange, reciprocity and interdependence based on economic and social status. Domestic and wild animals could be involved in maintaining active networks created through exchanges between the large houses and small houses, a division of labour that would have generated ties within the population.

These animals are also at the centre of funerary practices carried out both in ceremonial enclosures and in graves; this, in turn, gives rise to a parallelism between the social structure and the animal species, which can be regarded as markers of identity, gender and age. This community would have been made up of at least two clans (perhaps three): the cattle-rearers and the sheep-rearers. Sheep may also have been associated with children. Hunting, nonetheless, remained an essential activity, not so much as a source of food, but more as an integral part of social and symbolic rituals, evident for example in the communal sharing and redistribution of aurochs remains. The hunting of wild animals may also have acted as an affirmation of male status (hunting of wild boar) and potentially of female status (personal ornaments made from red deer teeth).

In the Middle Neolithic I, animals continued to be part of the social structure to various degrees.

The small number of settlement sites known from this period means that, at the moment, we can say very little about the role of animals in these contexts. It seems that pig was important since phalanges of this species are found on the torsos of inhumed individuals.

Another category of site that is difficult to interpret are the isolated pits discovered in certain locations and which have been found to contain the deposited remains, either whole or in pieces, of roe deer, red deer and calves. We can hypothesise that roe deer had an emblematic status since it was placed in these pits (and in certain graves), while it is absent from everyday domestic refuse. Similarly, red deer, which is poorly represented in ordinary consumption waste, is given prominence by the inclusion of antlers in the upper fills of these pits. Cattle are more widely consumed that the last two species, but we also know that this animal had a symbolic status in funerary contexts.

Signs of social valorisation exist in both the domestic and wild domains.

Thus, cattle appear to have a prestige value in the Cerny that seems to underline social inequalities which are more marked than in the previous period. In the monumental graves of Normandy, the presence of symbolically significant deposits of whole sheep and cattle bones provides ample evidence of this. Such deposits may have been intended to highlight the deceased's status as a cattle rearer or sheep rearer. In the same graves, we also find arrows placed close to the man's body. The position of hunter (or warrior) is not incompatible with that of a livestock rearer or herder, particularly as it has been noted that arrows represented a statement of status rather than being an individual asset.

Similarly, the personal ornaments made from wild boar tusks, red deer teeth or bear teeth, and wolf vertebrae that are found in burials without monuments, including those of children, are indicators of the importance of wild animals in the realm of the imagination and probably also in the construction of myths.

Interpreting this funerary data is not easy; rather than dominant individuals, it could just as well indicate dominant segments of society and this social configuration may have been linked to rituals surrounding the Ancestors (Dubouloz, 2018). In either case, whether they are representatives of a dominant family or of a lineage, these buried individuals bear witness to a well-established ranking system which would eventually lead to the institutionalization of power.

For the Middle Neolithic II period, archaeologists have forwarded various interpretations regarding the function of enclosures (fig. 212). For the moment, the most widely accepted hypothesis is that they served as supraterritorial assembly sites that had an important economic and ideological role (Gronenborn, 2009; Whittle *et al.*, 2011; Dubouloz, 2018).

There are multiple forms of cooperation, authority and leadership that may have underpinned these assemblies. There exist entities that surpass the nuclear family or the extended family, entities that are founded on a common descendancy which might be real or fictitious. These are tribes or chiefdoms composed of lineages and/or clans that cooperate or are in competition with each other (Eggert, 2007). The clan, as a political entity, is therefore susceptible to form federations with other clans in order to form tribes, or to dominate them and thus form a chiefdom.

Figure 211: Faunal remains from an Early Neolithic BVSG house at Jablines "la Pente de Croupeton" (Seine-et-Marne) (photo: L. Hachem).

Funerary data for the Chasséen and Michelsberg shows that distinct social groups can exist within a single cemetery (Augereau and Chambon, 2011) and that monuments were also constructed for just one or two individuals (Colas *et al.*, 2007).

Taken together, the evidence from the valleys of the Aisne, the Marne, and the Yonne suggests a strengthening of the elite system, compared to the Cerny, and the increasing concentration of power around a "chief" (Dubouloz, 2018).

However, the scarcity of information on settlement and burial in this period is an obstacle to a clear interpretation of the data.

Can the archaeozoological analysis perhaps allow us to interpret the assemblies held within the (complex) enclosures as gatherings of different tribes? This term, which is very flexible, is a 19th century concept that was re-used and modified at various times; it indicates the formation of social segments that are autonomous and equivalent in economic and political terms (Bonte and Izard, 1991). The word is later associated with the "Neolithic revolution", a form

of organisation situated between a band and a state (Sahlins, 1968), and even though the concept attracted criticism, it was never abandoned (Godelier, 2013). It has the advantage of giving a name to the form of social organisation that concerns us here.

The tribal model is based on the generalised opposition of groups, their segmented hierarchy and the homogeneity of these segments: the economy, politics and religion are not confined within distinct institutions, but more often depend on affiliation and, to an even greater extent, on the sharing of sovereignty. The fact of belonging to a tribe, for example, entitles its members to access land, favours certain types of marriages and guarantees common protection.

The available archaeozoological evidence shows that there was indeed collective behaviour. Even though the evidence, taken on its own, cannot prove that gatherings of the different segments of society did take place, we believe that it is a strong possibility. We can envisage the sacrifice of dozens of animals, the consumption of a large quantity of meat in the context of

Figure 212: Middle Neolithic enclosures were the sites of communal gatherings and had an important economic and ideological function (photo of a market in the Bigouden area of Brittany, taken by an unknown photographer).

15   MŒURS ET TYPES BRETONS. — *Vaches bretonnes.* — *Coin de Marché en Bretagne.* — LL.

communal feasts, the arrangement of carcasses and the deposition of significant animal parts, or whole animals, as gestures that were shared by several social segments who each contributed a part of its herd. What was the purpose of such actions? Ethnographic examples show that in societies considered to be tribal, there exist pan-tribal integration mechanisms, such as age groups, religious acts, feasts and/or communal work tasks that cut across the constituent lineages, thereby reaffirming a common identity and preventing fragmentation. The tribes are expected to provide stability in the face of a changing environment, and yet they are composed of different groups and individuals (Kienlin, 2012). As we observed previously in this chapter, the Middle Neolithic enclosures of northern France were not only places where meat was consumed and animal bone worked, but also centres where people met to exchange products and probably know-how as well.

We do not believe that it is currently possible to apply the same level of analysis to the Late Neolithic and Final Neolithic periods on account of the scarcity of faunal data. However, we stress the continuing importance of livestock (in particular cattle and caprines) and the persistence of hunting, as manifested by the presence of carnivores (fox, wolf and lynx), in societies at the end of the Neolithic. Dog is also a significant element in funerary rites.

An overview of Bronze and Iron Age socio-economic systems

Throughout the chronological sequence under consideration, crops and livestock formed the backbone of societies.

For the Bronze Age in the study area, the large numbers of sites, as well as the mass of data recovered from excavations, enable us to envisage a complex social organisation centred on crop-growing/animal-rearing communities. These communities are divided into small agricultural units anchored within a local area or regional territory dependent on a local power and can operate at different scales: either a territorial community composed of dispersed, small farms dominated by a local chief, himself the occupant of one of the farms; or a village-like cluster made up of a larger number of agricultural units. In these more densely populated areas, some villages are enclosed.

Throughout the entire sequence, each politically autonomous territory occupies an area 7 to 15 km across.

"This space-occupation model means that social hierarchy is still based at this time on long-distance exchanges, and not on control over the land." (Brun and Pion, 1992, 125-126).

The primary products and in particular the surpluses, circulated across small distances.

The societies that developed during the Early Bronze Age (2300/1700 BC) do not represent a break with the latest Neolithic societies.

The ensuing period, stretching from the end of the early Bronze Age until the end of the middle Bronze Age (1700/1400 BC), corresponds to a period of instability, attested by all available data. During this period, diversified exchange networks were established, which contributed to the emergence of a conspicuous elite. Singular practices sustained over decades are identified in the form of the non-funerary deposition of bronze objects, predominately axes and swords (especially between 1425 and 1350 BC) (Pennors, 2004; Quilliec, 2007). While many hypotheses have been forwarded to explain these practices, today the most commonly accepted interpretation is that they involve offerings to supernatural powers and that they indicate the intensification of religious practices under the auspices of a social elite.

"These religious symbols are often linked to other identity markers, which convey above all a sense of community belonging: they set one apart from the members of other communities. But the stylistic message is often more complex." (Brun and Ruby, 2008, 31-32).

The social elite displayed its status through the possession of goods: weapons as a symbol of war, banqueting ware and musical instruments for ritual practices, harness parts for mobility, personal ornaments and grooming equipment for majesty. Together they form the four fundamental themes behind the "aristocratic style" flaunted by the male social elite.

"Within these aristocratic groups... fraternity was certainly being strengthened through hospitality and relationships of reciprocity involving the exchange of goods, and based on rituals, belief systems and honor codes." (op. cit., 36).

Certain forms of hierarchical stratification are expressed through livestock and the way in which animals were consumed. Meat consumption focused primarily on the meat provided by the livestock herd, mainly composed of cattle, sheep (sometimes goats) and pigs, in various proportions depending on the location. Each farm/agricultural unit kept its own small herd and followed its own trajectory. However, various trends can be seen in the areas studied, depending on whether they are related to the North-Alpine or to the Atlantic complex. These units were autonomous and managed their livestock as they saw fit. For territories in the Seine-et-Marne, Hauts-de-France and Champagne, it seems that caprines and pigs formed the foundation of the herd, while in Calvados, herds were essentially composed of bovids (cattle and caprines). Horses and dogs are very scarce across the entire geographical area we are concerned with. Nonetheless horse is always present in the faunal remains, regardless of the nature of the site, and a strict association with the elite is therefore difficult to assert. The contribution of venison to the meat diet should not be overlooked; even though it cannot be compared to the Neolithic period in terms of variety and quantity, it remains present as food in various forms depending on the status of the farm. In any case, the daily diet was probably not based on daily meat consumption but rather on the preparation of largely vegetable-based dishes.

Some sites clearly stand out very clearly from the small farms, particularly on account of the abundance of meat consumption refuse and the singularity of their faunal deposits.

The paroxysmal expression of the complex relationships maintained by these farming families, dispersed over a large territory under the control of local chieftains, was manifested during community assemblies, which may have been seasonal and which were generally large in scale.

In this instance, the corpora studied allow us to measure the scale of the solidarity required to host such occasions.

During the Early/Middle Bronze Age (1800/1350 BC) ditched enclosure sites in the north (in Pas-de-Calais and Normandy in particular) were used periodically as communal gathering places; the faunal evidence for these gatherings reveals a higher than usual consumption of cuts of beef which clearly stands out from the "ordinary".

During the Late Bronze Age (930/800 BC), the major sites found at Villiers-sur-Seine and Boulancourt in the Seine-et-Marne and Choisy-au-Bac in the Oise provide the most abundant record attesting to episodes of large-scale consumption, involving more people than the actual inhabitants of the site. The slaughter of hundreds of animals, mainly pigs and cattle, taken from several herds, indicates that the wider community was directly involved in these meals (which would also have involved other foodstuffs such as plant foods) and provided a contribution in kind. The slaughter pattern, involving a high proportion of juveniles, contravenes what would be regarded as reasonable practice in herd management.

The preparation of the meat cuts is calibrated to serve standardized portions to the guests; the amounts of meat involved reach into the thousands and even the tens of thousands of kilos. It seems that this meat was shared on certain occasions during the year, depending on the need of the moment. These feasts might have been held, for instance, to celebrate alliances, fraternal banquets, to reinforce social cohesion, or to celebrate the weddings and funerals of local chieftains. They may have marked the rhythm "dictated" by the agricultural calendar according to territorial rules about which we know nothing: or example, were the feats open to all or restricted only to certain categories of the population? We have no idea of the number of guests involved. One thing is certain, however: large quantities of food were eaten on the occasion of seasonal events fixed by well-established conventions. Hunting might have played an essential role in the preparation of these feasts, notably at Villiers-sur-Seine. One can envisage for example initiatory hunts or male-emulation hunts providing the opportunity to measure one's strength.

Between these two extremes – i.e. between the simple farm/agricultural unit and the site with a supra-community function – we find settlements consisting of several houses, each with several head of livestock, and possibly mutually sharing the breeding animals as well as a horse and dog (fig. 213).

Among these farm groups, some seem to have geared their animal husbandry towards a prefered species, in this case pig (Grez-sur-Loing in the Seine-et-Marne, Buchères and Villemaur-sur-Vanne in the Aube, Osly-Courtil in the Aisne). These intermediate farm groups can be seen as places where smaller-scale gatherings could have been held, but nonetheless requiring the provision of sustenance for many guests; the pig would thus come to play a particular role.

The main objective of livestock rearing is the production of meat, except for rare instances where evidence for milk stimulation can be inferred from some of the slaughter patterns for sheep (*i.e.* Grisy-sur-Seine and Changis-sur-Marne). Hunting is a complement to cattle breeding and contributes only marginally to the food supply; red deer is undoubtedly the preferred game species.

Evidence for the use of animals in funerary rituals is relatively rare when we consider the hundreds of graves excavated. Two geographical areas contradict this general rule: the south of the Seine-et-Marne and the Aube on the one hand, and the Nord and the Pas-de-Calais on the other. The deposits, whether they come from inhumation or cremation burials, consist of fresh cuts of meat,

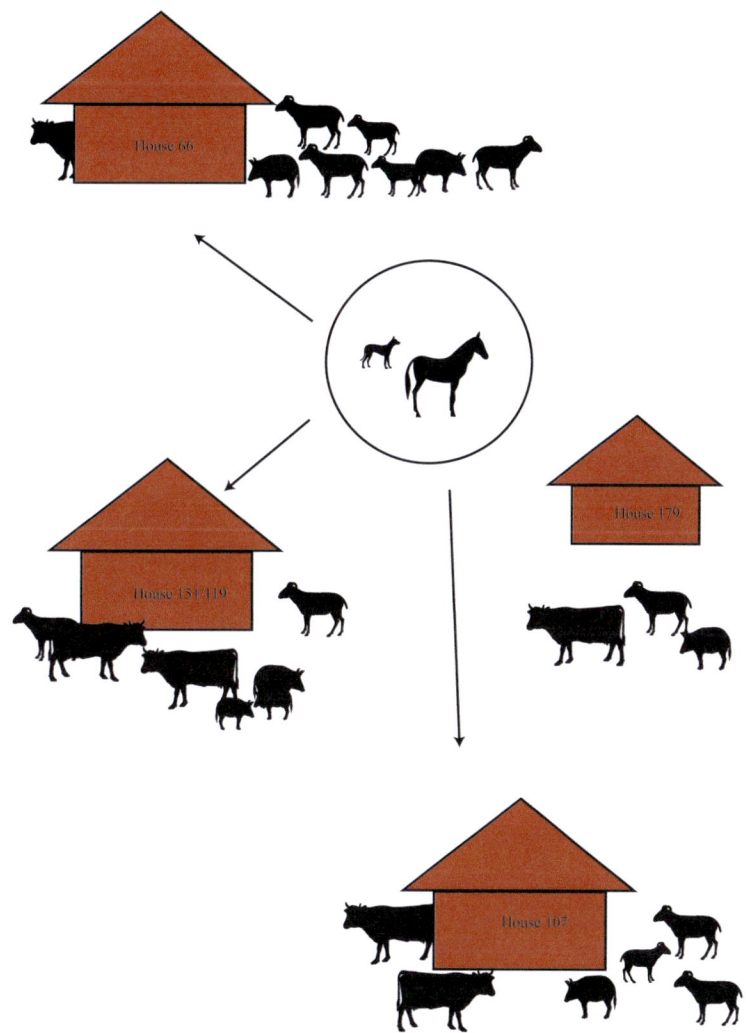

Figure 213: Changis-sur-Marne "les Pétreaux" (Seine-et-Marne), Late Bronze Age, domestic units and their herds.

mainly from pig and sheep and more rarely from cattle or dog. The cemetery at Barbuise "les Grèves de Frécul"/la Saulsotte "le Bois Pot de Vin" (Aube) stands out, mainly because there are some isolated cattle and horse bones, which do not match the deposit categories recorded in the other cemeteries (Rottier *et al.*, 2012). These remains are interpreted as singular acts such as the retrieval of bones from decayed carcasses. In the Nord and the Pas-de-Calais, sheep largely prevail over all other species, even though dog, roe deer as well as some birds and fish have also been recorded (Cahen-Delhaye and De Mulder, 2014).

Plant food is based on a diversified polyculture which also follows particular patterns depending on the location. In Picardy and Ile-de-France, production and food supplies relied heavily on a range of crops including cereals such as hulled barley (*Hordeum vulgare*) and emmer wheat (*Triticum dicoccum*), pulses such as lentils (*Lens culinaris*), peas (*Pisum sativum*) and ervil (*Vicia ervilia*), and oleaginous plants, including the camelina (*Camelina sativa*). Millet (*Panicum miliaceum*) is sometimes used in significant proportions as on the site of Villiers-sur-Seine, which, it should be remembered, is dedicated to collective consumption (Auxiette *et al.*, 2015, 54-61). In Normandy, studies conducted in the Caen Plain have also demonstrated the importance of barley and emmer wheat. This region stands out because of the occurrence of spelt (*Triticum spelta*) and millet and the absence of pulses (Dietsch-Sellami, 2000). In Champagne-Ardenne, recent studies on a number of farms in the Troyes Plain (Matterne, 2009a; Toulemonde, 2013) have revealed the importance of hulled barley, more so than anywhere else, in conjunction with a variety of wheat species, mainly of the hulled type. In certain well-documented regions, evidence appears to point to the existence of a parcel system of land use as early as the first millennium BC (Marcigny *et al.*, 2008). Animals were permitted to roam across the land and to improve it with manure throughout the year. Settlements probably shifted as a function of the crops (Blouet *et al.* 1992). Studies carried out in north-western Europe have revealed the generalized use of mixed-cropping (barley and bearded barley, barley and wheat) which ensures a certain yield level and allows larger surfaces to be cultivated (De Hingh A.-E., 2000; Matterne, 2001). Pollen analyses have confirmed this trend (Leroyer, 1997).

The Hallstatt C/ Late Hallstatt period (800-625 BC) is often difficult to distinguish from the Hallstatt B2/B3-Late Bronze Age IIIb (930-800 BC). The farm settlements are short-lived and very dispersed, yet the best-documented sites do not suggest that social organisation was modified to any extent. The ongoing climate deterioration, observed as early as the 15th century BC (Middle Bronze Age-Bronze Age C1/C2), had a durable impact on the availability of resources and the keeping of the larger cattle (Hachem and Auxiette, 2006, 128), as well as on exchange networks.

Consequently, data regarding farm livestock becomes more scarce compared to the evidence available for the sequence stretching between the Hallstatt A1 (Late Bronze Age IIa) and Hallstatt B2/B3 (Late Bronze Age IIIb) periods, *i.e.* between 1250 and 800 BC. The agricultural units are extremely small in size and only comprise a few head of cattle. While the large sites where people gathered together, such as Boulancourt and Villiers-sur-Seine, are abandoned some time during the Hallstatt C/Late Hallstatt period (800-625 BC), the site at Choisy-au-Bac is apparently still occupied and featuring large-scale slaughtering of juvenile pigs, it represents the continuation of this site category.

From the Hallstatt D-Middle/Late Hallstatt period (640-450 BC), and more specifically during the 6th century BC, thanks to increased agro-pastoral productivity, which benefits from improved climatic conditions, and the amplification of external exchange networks, we observe the creation of more complex chiefdoms. Funerary data demonstrated the strengthening of the social hierarchy.

"This renewed political and economic complexity is clearly validated by the ranking and the diversity of settlements." (Brun, 2015, 51).

In the settlements, often attributed to the transition period between Hallstatt D3 and La Tène A1 (450-400 BC), we observe different configurations of unenclosed farms of variable size, like as shown for example by the site at Villers-en-Prayères "les Mauchamps" (Brun *et al.*, 2005a; Hénon *et al.*, 2002). Enclosed farms are also recorded, for instance at Bucy-le-long "le Grand Marais" (Aisne) or at Grisy-sur-Seine (Seine-et-Marne), and even fortified such as Basly "la Campagne" (Calvados) where the consumption of juvenile animals is significant (Baudry, 2005, 2018).

Among the enclosed farms situated on the valley floors, the site at Bucy-le-Long "le Grand Marais" is surrounded by a palisade with a monumental entrance (fig. 18). Seventeen granaries were found within the enclosed space. This centralization of storage facilities, which ensures control over grain (and their redistribution), and the disproportionately large volumes involved, together with the presence of a significant faunal deposit composed of parts of cattle, deer and horse in a large pit (Auxiette, 2000a), strongly suggest that this was a high-ranking settlement (*i.e.* a "latifundial" residence or *Herrenhof)* (Gransar, 2001, 347).

Analysis of the large number of farms provides a variable picture depending on geographical location. In the Aisne, Oise and Seine-et-Marne, we notice a clear trend in the rearing of small mammals, with caprines and pig occupying a privileged position. However, on the farms in the Marne department, the farmers favoured cattle. The picture revealed by the analysis of the farms located in the western geographical area, on the Caen Plain, is very different: cattle were preferentially reared, followed by pig and, in smaller proportions, caprines. Evidence for cynophagy and hippophagy becomes increasingly common and game disappears completely from the diet.

Palynological analyses show clear evidence for a sharp increase in agricultural activity: cereal pollen gradually becomes more common, with a marked rise in frequency during the La Tène period, matched by ruderal plants, while tree pollen declines (Matterne, 2001, 218). Climate studies have shown that it became colder and wetter during the early Iron Age (Hallstatt D/La Tène A) (Magny R. and Richard H., 1992). The same period also sees more denser arrangements of specialized storage facilities, with clusters of dozens of silos and granaries, as at Bucy-le-Long "le Grand Marais" (*cf. supra*) which centralize the storage of hundreds of cubic metres of grain (Gransar, 2001). These communal reserves were possibly designed as a security in the face of climatic deterioration.

The practice of mixed-cropping is maintained and agricultural management is designed so as to minimize the risk of failure by diversifying animal and vegetable production.

> *"In terms of the landscape, this diversity probably translates into a mosaic of small plots, separated from one another and cultivated in rotation, which further restricts the propagation of diseases and parasites."* (Matterne, 2009b, 2001, 181).

As regards livestock, we note that the marked presence of small mammals, already well established in the preceding decades, is sustained and reaches even higher levels in the Aisne, Oise and Seine-et-Marne. This strategy may have been adopted as an answer to the worsening climatic conditions, since the maintenance of larger mammals is more demanding: it requires greater areas of good quality pasture (one hectare per animal), and far larger amounts of fodder and water than those needed for the rearing of caprines and pigs.

One of the most intensively investigated areas, the Caen Plain displays the best evidence for the establishment of a structured agricultural landscape during Hallstatt D2/D3, in conjunction with the emergence of enclosed settlements. The exploited territory is organised around a network of paths and ditches spread out across a loam-covered limestone plateaus that is particularly favourable for agriculture (Besnard-Vauterin *et al.*, 2016; Le Goff, 2009, 2000, 1207). In the territories of the Caen Plain, the composition of the livestock herd changes, shifting from herds mostly composed of sheep with a complement of cattle, to a situation where cattle become omnipresent. This trajectory clearly sets this area apart from all of the others we have discussed. At the end of the first Iron Age we can observe the appearance of the first ditched enclosures that are sufficiently deep to be preserved. They constitute the organisation of a small territory that would continue to evolve and assert itself in the following centuries (Le Goff, 2000, 1207).

Evidence for communal consumption is discernible in certain assemblages containing preferentially selected species, notably sheep bones that have been treated (slaughtering age, butchering) in a manner that distinguishes them from simple domestic refuse. Most of these assemblages are found in the fill of storage pits and some indicate possible episodes of seasonal consumption. They relate to a sort of "sacralisation" of the refuse inside a feature, the storage pit, which was a symbol of survival for these communities.

Other storage pits, some with large capacities, contained human remains sometimes associated with animals, in particular horses, a species that is almost absent from the meat deposits found in cemeteries. These were long considered as "relegation burials", but are now interpreted in a different manner. In fact, these people were the object of specific treatment that involved exposure of the corpse (male or female), manipulation following a period of decomposition, and then "association" rituals with animals. Can these deposits in silos, which are particularly well documented for the Late Hallstatt and Early La Tène periods, be likened to a "gesture of peace"? Are they the result of a sacrificial practice involving humans, thus associating in the same "cereal-oriented propitiatory intent" humans and animals, and possibly other materials as well (Gransar et al., 2007)? For these sedentary, agricultural populations, silos were the means of cereal storage par excellence (even though raised granaries also existed at the time and became ubiquitous in the Late La Tène period), both for daily consumption and for the preservation of grains. However, it appears that after they had been emptied certain silos were re-used in a different way and became the places of seasonal exchanges between human beings and the invisible world. Within a situation, which we envisage as peaceful and seasonal, the deposition of humans and animals in a state of controlled decay seems rather to be invoking the forces of fertility/fecundity (Delattre, 2013, 2010; Delattre et al., 2018; Delattre and Auxiette, 2018; Duplessis et al., 2013). In such a context, the question of possible human sacrifice remains open.

Faunal deposits in pits become increasingly frequent, either as entire animals or as parts of carcasses; such deposits are sometimes associated with blocks of sandstone or limestone.

In the rare cemeteries attributed to the Hallstatt D2-D3 period (530-475 BC), deposits of pieces of meat on the bone in burials reveal preferential deposition of sheep and calves (Saint-Etienne-au-Temple "Champ Henry" in the Marne) (Paresys et al., 2009). The frequency of such deposits in graves is far lower than in subsequent periods, in particular the early (middle) La Tène – i.e. La Tène A1/A2/B1/B2 periods (475-250 BC) – and the Late La Tène – i.e. La Tène C2/D1 (200-130 BC). Numerous cemeteries of different sizes, varying from a few graves to several hundred, are known for the early La Tène period and provide abundant evidence for meat deposits, notably in Champagne (Marne) and Picardy (Aisne, Somme), where this ritual occurs more frequently than in other regions. Pigs, sheep, and in some rare cases, chicken and cattle, were deposited as meat cuts in variable amounts, sometimes arranged, sometimes combined, and sometimes even staged. In many instances the deposits reflect the social rank of the buried person; indeed, a correlation has been established between the faunal remains and other categories of ostentatious grave goods (Auxiette, 2009b, 1995; Desenne et al., 2009b, 2009a).

The La Tène B2-C1 (325-180 BC) period witnesses the establishment of settlements in areas that until then had been sparsely inhabited, such as the plateau edges and interiors (Malrain et al., 2015, 2002). Technical improvements brought about by the generalization of access to iron, the multiplication of forge sites, and the use of iron ardshares in particular allowed heavier soils to be worked; this represents a major advance in agricultural technology. Farmers gradually abandoned their mixed-cropping practices and opted instead for monospecific agriculture. Hand-held grinding equipment was now replaced by rotating grindstones (Pommepuy, 1999).

The evidence for animal husbandry varies considerably from one region to another. The number of valley floor sites decreases and settlements located on the plateaus or their edges have yielded insufficient evidence to allow us to deal adequately with the issue of livestock management; furthermore, the soils on the plateaus do not favour the preservation of bones. The composition of herds appears to have been very diverse. Cattle prevailed in the Oise, whereas caprines predominated in the Aisne; in Seine-et-Marne, no preference for a given species emerges from analyses and in the Crould Basin (Val d'Oise), herds are based on cattle and pig.

For the most part animal husbandry was geared towards the production of meat.

On the plateaus of the Caen Plain, animal husbandry favoured cattle and livestock management was resolutely oriented towards the production of meat from juvenile and sub-adult individuals. This choice is a sign of the relative prosperity of farmers in this area. In the case of caprines, lambs make up a significant share of the corpus; this might indicate milk production, or might indicate deliberate selection for the production of particularly tender meat.

Four high-ranking farms – three in Picardy, Glisy "les Terres de Ville" and "les Champs Tortus" (Auxiette, 2011a; Gaudefroy, 2000); Braine "la Grange des Moines" (Auxiette and Desenne, 2017) and Fontenay-en-Parisis "la Lampe" in Ile-de-France (Daveau and Yvinec, 2001) display characteristics of assembly sites where collective feasting took place. The number of animals slaughtered and their age (predominantly juveniles) appear to indicate that the people had a desire for high-quality meat.

The previously mentioned climate deterioration gradually recedes at this time (between 500 and 300 BC) and is succeeded by a warmer period (Tegel *et al.*, 2016, 644). Some concomitant population movements take place leading to the establishment of a growing number of farms on the plateaus (Malrain *et al.*, 2013, 105). While the valley bottoms were more favourable to pastoral activities, the conquest of plateaus may have been related to the development of cereal farming (Matterne, 2001, 181). The number of cultivated plants decreases and the archaeobotanical evidence reveals the scale of mixed-cropping, a practice initiated during La Tène B2-C1. There are two systems of production: the first, more intensive, which focuses on mixed seeding and plant diversity, and the second more extensive, based on monospecific cereal farming (Matterne, 2001, 181-182). A concurrent evolution of milling techniques was also taking place. The invention of the rotary quern was truly revolutionary in this respect (Pommepuy, 1999). In the best documented farms, which are mainly situated on the alluvial plains, we notice an increase in the proportion of cattle in the herd, together with an increase in pigs.

In the Caen Plain, this phenomenon of settlement reorganisation is not observed; in fact the farms that existed since the Late Hallstatt/Early La Tène period were maintained and extended. The organisation of the territory and the agriculture rely on the framework originally established. The status of farms – and consequently of their inhabitants – is difficult to infer from the study of finds from the area, even though some elements do indicate a certain degree of wealth (coin hoards and faunal deposits, for instance). Meat consumption focuses on cattle, a significant proportion of which had not attained their mature weight. This is another indication of relative economic prosperity.

A comparative study of the fauna from enclosed farms for the period covering the last four centuries BC was conducted on the large assemblages from a group of farms at Ifs "Object'Ifs Sud" and "AR67", Mondeville "l'Etoile" (Auxiette, 2009c, 2000b) and Hérouvillette "les Pérelles" (Besnard-Vauterin *et al.*, 2015). This study has revealed the undeniable perpetuation of the earlier animal rearing strategies in which cattle prevail, albeit with a non-negligible proportion of caprines and horses. This choice implies very astute herd management. In spring, summer and autumn the animals would have roamed the lush pastures at the bottom of the Orne valley (Lepaumier *et al.*, 2010) with supplementary fodder provided in winter. The rearing of large numbers of cattle was specific to this geographical area and had been part of the local tradition for centuries (Auxiette *et al.*, 2010). Access by farmers to good quality land provided them with the vast expanses needed for the rearing of cattle, which in turn provided the manure necessary to enrich the soil. In this way, the population was able to maintain large herds and to produce significant quantities of meat, mainly from young adults and calves, particularly during the middle La Tène period. A change occurs however in Late La Tène, when more mature animals were slaughtered. This shift might be explained by the desire to optimize meat production by allowing cattle to reach their full mature weight. It might also suggest that "food restrictions" had been put in place in a weakened economic context; conversely, the slaughtering of young animals would have been synonymous with prosperity for a population enjoying a favourable economic situation. Archaeobotanical analyses have revealed the importance of pulses in the western farms. The keeping of cattle beyond their fourth year requires a supply of fodder that may have taken the form of pulses, as these provide an excellent energy complement (Malrain *et al.*, 2015).

An outcome of these prolific breeding practices is the large-scale involvement of cattle, and also horse, in domestic life and in rituals centred on deposition of cuts of meat in the middle and Late La Tène (Auxiette and Desenne, 2017; Gransar *et al.*, 2007; Le Goff *et al.*, 2007).

Cemeteries are well documented and contain both inhumation and cremation burials, as at Bucy-le-Long "le Fond du Petit Marais"(Auxiette, 1995). Grave goods of all kinds – vases, personal ornaments, tools, etc. – are rare (Pommepuy *et al.*, 1998).

Deposition in burials mainly comprises cuts of pig and sheep's meat, and sometimes beef and chicken. The deposition of dog, hare and horse has been observed in few graves in Aisne, Oise and Ardennes.

Cremated faunal remains also begin to appear in the record. Retrieved from the funerary pyre and mixed with the human bones, they attest to a change in practice from the preceding period. While the deposition of pieces of fresh meat can be interpreted as a *viaticum* (provisions for the afterlife), the meaning of cremated cuts of meat is more difficult to explain. Possibly they were an offering to honour a divinity or the dead person, in a rite of passage between the world of the living and the world of the dead (Lepetz and Van Andringa, 2004).

In certain regions at the end of the 4th century BC, sanctuaries such as Gournay-sur-Aronde (Oise) were established, and played an essential role as territorial markers (Brunaux *et al.*, 1985; Brunaux and Meniel, 1983).

From the La Tène D1a/D1b period onward (150 BC), the social organisation becomes even more complex with the emergence of towns and states (Brun, 2007). These very large agglomerations/ oppida meet all the criteria of urbanization and centralize economic (notably coin manufacture), political and religious powers. Core activity on the farms remains agro-pastoralism but a ranking can be established on the basis of land surface analysis, buildings and craft activities, as well as the presence of relatively ostentatious categories of objects (Brun and Ruby, 2008; Malrain, 2000; Menez, 2008). The size and variety of settlements attest to real trans-formations that undoubtedly created a significant

demand on the resources needed to respond to the strong demographic growth (Brun, 2007, 381).

Despite these profound changes, the available data concerning livestock rearing does not indicate any particular specialization in any of the regions.

The relative share of each domestic species is quite variable and does not appear to be correlated with the types of farms as as almost all of them own vast areas of land; the differences observed in some faunal spectra relate to the practice of collective feasting or to the large-scale consumption of meat within oppida. These differences mainly concern the place of pig and/or cattle within the herds and on the diner's plate. In the territory of the Suessiones (Aisne), these particularities are well illustrated in the high-ranking farms of Bazoches-sur-Vesle "les Chantraines", Braine "la Grange des Moines" and Villeneuve-Saint-Germain "les Etomelles", as well as the nearby oppidum of Villeneuve-Saint-Germain. The sites at Mont-Notre-Dame and Juvincourt-et-Damary, even though they are still difficult to qualify in their current state of investigation, could be ascribed to the same farm category as they share the same particularities.

Furthermore, we can also distinguish a category of farms for which the proportions of caprines exceed 40%; equally distributed to the east and to the west of the Aisne and Vesle valleys, and in a more remote sector in the north of the Aisne department, they are all attributed to the sequence spanning from La Tène C2 to La Tène D1b (*i.e.* from 200 to 80 BC). The values obtained for the main domestic species in the later farms and in the Villeneuve-Saint-Germain oppidum are centred on 25 to 40% for cattle, over 40% for pig and from ±20% to ± 30% for caprines. Identical values have been obtained for the different assem-blages from the Reims-*Durocortorum* oppidum ("Rue d'Anjou", "Villa des Capucins", "Rue Chanzy", "Rue Rockfeller") attributed to La Tène D2, and also from the rather earlier oppidum of Condé-sur-Suippe/Variscourt, attributed to La Tène D1a (150-110 BC). These values undoubtedly reflect the animal husbandry strategies characteristic of farms and oppida from the end of the second and the start of the first century BC, which attracted large numbers of consumers. Three sites share a high proportion of cattle. Two of these are farms dating to the La Tène D1, which exhibit a distinctive fauna that is clearly different from spectra found in agglomerations. As for the third site, situated at "Rue Carnot" within the Reims oppidum, the difference in composition of the sample compared with the other investigated sectors could be explained by the size of the sample and/or the function of this zone which appears to have been non-residential.

For the Oise sites, the values obtained for the high-ranking farm at Verberie "la Plaine d'Herneuse II", with an abundance of pig, match the values for farms of the same rank for the period La Tène D1b to D2 in the Aisne. Conversely, no real coherent picture emerges from the farms of Seine-et-Marne. The two settlements – one dating to La Tène C2, the other to La Tène D2 – where communal and even supra-community consumption have been identified, exhibit a high prevalence of cattle and pig relating to these feasting events. In Varennes-sur-Seine, two La Tène D1-D2 sites display almost the same high values for cattle and pig as do farms of similar rank in the territories of the Suessiones and the Remes.

We now turn our attention to the sites situated on the Caen Plain, which corresponds to the westernmost part of the area under consideration. As we saw for the Late Hallstatt and Early La Tène period, the singularity of the local animal husbandry strategies is striking: farmers clearly focused on the breeding of cattle (over 40% and at times over 70%) as the foundation of their livestock herd, with caprines as a secondary species (±20% to ±50%); the frequencies for pig never exceed 20%. None of these farms, which are sited very close to one another, exhibits any specificities and it becomes difficult to establish a hierarchy between them (Lepaumier *et al.*, 2010).

Some herds were bred for meat with cattle slaughtered before their twelfth month or at the end of their fourth year while others are comprised of dairy cows and draft animals.

We also observed an important part of horses; we can correlate their importance to that of oats *in the botanical assemblages.* However, horses are also well represented in Picardy where oats are not very represented (Malrain *et al.*, 2015, 141-142).

For the La Tène D2 (90-20 BC) archaeozoological analyses reveal an ever increasing proportion of cattle and equine livestock (the latter to a lesser extent) in all regions until this kind of animal husbandry attained the level of large-scale specialization. Cattle and horses, together with pigs, generate large amounts of manure when kept in large herds, which is very useful for fertilizing the soils. A corollary to their increasing numbers, however, is the acute need for fodder production and, therefore, for more cultivated land, a fact which is supported by carpological analyses.

Mixed-cropping was definitively abandoned. A single species was cultivated in each field. The spectrum of cultivated plants became more restricted and crop husbandry relied mostly on emmer wheat *(Triticum dicoccum)* and hulled barley *(Hordeum vulgare)* whose yields are higher than that of pulses. We thus witness the emergence of extensive agriculture. These changes were made possible by improved techniques and by the implication of animals for the ploughing and harrowing of fields and the transport of the harvest (Halstead and Jones, 1989; Matterne, 2001, 182).

The question of identifying production and consumption sites is essential for gaining broader understanding of the organisation of countryside and society in the later Iron Age (Malrain *et al.*, 2009). While salt manufacturing workshops and certain iron workshops can be easily identified as production sites, the situation is more difficult with animal husbandry, which is always attached to a farm. It is often hard to distinguish between autarchic production, aimed simply to meet the needs of one or several families, and production intended for export to a wider group of consumers.

Nonetheless, some sites display a considerable deficit in certain anatomical parts that might point to the import/export of pieces of meat, while others are conspicuously lacking in very young animals, the presence of which would have been expected in the case of *in situ* breeding, in particular of cattle.

The relationship between town and countryside is well-documented in the Aisne valley and, more specifically, in the area close to the oppidum of Villeneuve-Saint-Germain. By the end of the Late La Tène period, the population within the oppidum is estimated to have been about 4,000 (Brun *et al.*, 2000, 85). Impressive amounts of beef were consumed there, whether one counts in numbers of bones or animals (see chapter 2). Yet at the same time there are fewer farms, indicating that some people had deserted the countryside.

> *"A rapid collapse occurs during the first century BC and the number of sites has been halved by the time of the Conquest."* (Gaudefroy *et al.*, 2013, 106).

Can we identify among the few "surviving" large farms the cattle rearing centres that would eventually feed the hundreds of mouths within the oppidum, and perhaps elsewhere? Live cattle were probably moved from the place where they were reared and provisionally kept in pens awaiting slaughter and dismemberment, which took place in a specially-dedicated butchering zone (see chapter 2) (Auxiette, 1996; Auxiette and Paris, 2017). The faunal corpora of these contemporary farms of the oppidum are rather poor; it possibly indicates the transfer of live cattle to the oppidum. Furthermore, the absence of juvenile mortality observed in the cattle processed in the oppidum, despite the hundreds of animals involved, strongly points to an external source for the animals.

As regards the salting of certain meat products with a view to their export, despite the well known remarks of Strabo traces of salting are difficult to identify. At the very least we should observe a systematic absence of certain anatomical parts in the faunal assemblages.

> *"...of all the Gallic people, the bravest are the Belgae, who are divided into fifteen tribes (...) Their flock of sheep and herds of swine are so very large (...) The wool from which they weave the coarse "sagi" is rough and flocky and, through trade, they supply an abundance of these clothes and of salted pig meat not only to Rome, but to most parts of Italy as well."*
> (Strabo, Geographica, Book IV).

Certainly, one cannot disregard the presence of dolia (large ceramic vessels), sometimes in great numbers, which bear traces of salt attack on their inner walls. Salted products were undoubtedly consumed on farms and in the oppida, but we could be dealing with the domestic salting of products for deferred consumption, a practice that still occurs today. Sometimes we also found some fragments of "salt moulds", for example at the oppidum of Villeneuve-Saint-Germain (Robert and Weller, 1995). Experiments have been carried out in an attempt to characterise the alterations caused by salting on the bones of meat joints, but without success to date (Baudry, 2018).

Faunal deposits in funerary contexts are well-documented, in particular in Picardy (Aisne and Somme) and Champagne-Ardenne. Following an episode when such deposits become rare during the middle La Tène period, they become increasingly frequent in the form of fresh and/or cremated bones. Pig and chicken, sometimes combined, are the two species preferentially deposited usually, in parts. Caprines are also involved. Rare instances of cattle, fox, hare, and dog have been recorded. The carcass parts most often deposited were the heads, shoulders, hams and parts of the rachis.

We also see an increase in the deposition of carcass parts or full carcasses – either exposed or buried – mainly in the ditches of enclosed farms, with a predominance of horses and cattle. Dog, which was originally almost absent, occurs more frequently in the context of these practices. The combination of parts belonging to several species and forming composite assemblages is more frequently encountered. Heads (skulls and bucrania) constitute a category in their own right as part of ostentatious deposition/display practices.

Even if the deposit of some dead in silos continues after the 3rd century BC, it becomes scarce and gives way to new practices, both in domestic contexts -isolated bones in pits and ditches- and in separate places like sanctuaries, which mark a new concept of worship, with large collective installations (Delattre *et al.*, 2018).

Evolution of cultual expressions through the ages

A wide variety of species has been recorded including small and large mammals, both domestic and wild, as well as some avifauna which are more often wild than domestic.

The place occupied by each of the six principal species varies depending on the chronological period. Among domestic animals, the highest frequencies are observed for cattle followed by caprines and pigs. Equids are notably absent during the Neolithic.

For the Neolithic, among wild mammals, red deer is preferentially selected, far ahead of roe deer, aurochs and wild boar. In fact the latter two species are generally rather poorly represented in deposits. There follows a series of small fur-bearing mammals, of which the hare is the most frequent.

In the Early Neolithic deposits are rare, unlike in the Middle Neolithic where they are concentrated in the enclosures. There are very few Bronze Age deposits in pits, but numbers are recorded during Hallstatt D.

During La Tène, cattle and equids are the two species preferentially selected for deposition, with cattle being in the majority (fig. 160, fig. 163, fig. 167, fig. 168, fig. 172). The inclusion of dog in deposits increases steadily from the early La Tène onwards, through the last five centuries BC, and peaks during the Late La Tène period. Caprines are the least well represented species.

For more details concerning deposits and species between the Neolithic and the Iron Age we refer the reader to a conference publication on the subject (Auxiette, 2013d).

In the case of animal deposits associated with human remains, distinct and complex configurations have been observed during the Neolithic and Iron Age (see chapter 4).

Bovines and caprines are the main species deposited during Early and Middle Neolithic (Farruggia *et al.*, 1996; Hachem *et al.*, 2016; Thevenet, 2017; Ghesquière and Hachem, 2018). A marked preference for horse is noticed in the Early Iron Age (Delattre *et al.*, 2018; Delattre and Auxiette, 2018). These deposits associated with human bodies clearly in storage pits differ from instances of joints of meat being deposited within graves, where horse is absent (Auxiette, 2009b, 1995; Auxiette *et al.*, 2002; Desenne *et al.*, 2009a, 2009b; Méniel, 2004, 1998; Méniel *et al.*, 1994; Méniel and Lambot, 2002). In the case of deposits accompanying human remains, the animal remains are part of complex processes which suggest that, for certain animals and human bodies, their deposition had been planned in advanced (Delattre *et al.*, 2018).

The question of animal sacrifice through the ages

There is ample evidence, both ethnographical and textual, for strict regulations governing animal sacrifices in past and present societies; in ancient Greece, for example, the sacrificial animal had to signal its consent to the priest before being killed (Detienne and Vernant, 1979). All such evidence highlights the complexity of the practices, which are difficult to discern through the examination of

the faunal remains from prehistoric archaeological sites simply because they leave few visible traces. Furthermore, the implements used for the sacrifices are generally unknown, apart perhaps from the knives that were sometimes placed on top of the carcass parts accompanying the deceased in La Tène graves.

In terms of traces on the bones themselves, clear evidence for blows to the heads of cattle and for throat slitting, which leaves cut marks on the cervical vertebrae, cannot necessarily be taken as signs of sacrifice. Nonetheless, the practice of sacrifice can be deduced from the fact that we observe singular practices, as attested notably by portions of meat on the bone in graves, by certain bone assemblages in enclosure ditches and in all cases by assemblages containing several animals or portions of animals. We can also envisage sacrifice in cases of animal skulls (heads) were displayed on palisade posts and at the entrances of enclosures.

In parallel, we must envisage animal sacrifice in domestic contexts since, in traditional societies, rituals practices are still recorded. Unfortunately such practices leave no trace in the archaeological record.

Sacrificial rites and the sharing of meat evolved over time. Thanks to archaeozoological studies we can reveal the most obvious manifestations, such as in graves and sanctuaries, but also those that are more difficult to detect in enclosure ditches or enclosed sites.

In our study, we have shown that cattle, pigs, sheep, horses and dogs were selected for sacrifice throughout the period from the Neolithic to the Iron Age. Neolithic and Bronze Age hunters, and occasionally those of the Iron Age, killed wild animals in the forest, brought them back to their settlements and put certain carcass parts (aurochs bucrania, red deer antlers, etc.) on display.

Other indications of ritualised hunting are perceptible from the Middle Neolithic to the Late Bronze Age through the deposition of carcass parts (often articulated) of aurochs, roe deer, wild boar and wild horse in certain deep pits that are interpreted as traps.

Sacrifices may also have been collective, forming part of important ceremonies carried out for the well-being of the group and which followed a certain temporal rhythm, which could have been related for example to the changing of the seasons,

a - Bucy-le-Long «la Héronnière» ear rings in gold, chariot burial n°196

b - Bucy-le-Long «la Héronnière» ring in gold, chariot burial n°196

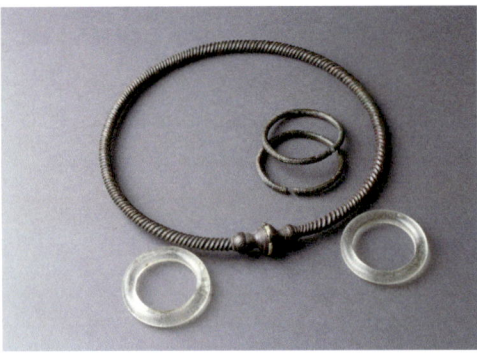

c - Bucy-le-Long «la Fosse Tounise» ornaments in bronze and glass, chariot burial n°150

d - Orainville «la Croyère» torque in bronze and glass, grave n°3

Figure 214: Examples of rich ornaments from Early La Tène graves. a to d- Bucy-le-Long "la Héronnière/la Fosse Tounise" and Orainville "la Croyère" (a to c: photos: Michel Minetto; d: photo: Sylvain Thouvenot). e to f- Orainville "la Croyère" and Vasseny "le Dessus des Grouins" (photos: Sylvain Thouvenot).

e - Orainville «la Croyère» fibule in bronze and coral, grave n°3

f - Vasseny «le Dessus des Grouins» pendant in bronze, grave n°528

the opening of the hunting season, rites of passage, or propitiatory rituals.

The seasonality of such events has been highlighted on the Middle Neolithic site of Escalles (Pas-de-Calais) and on the Bronze Age site of Villiers-sur-Seine (Seine-et-Marne) and the Iron Age site of Souppes-sur-Loing (Seine-et-Marne) through the slaughter patterns for domestic animals.

The assembly sites in which the sacrifices were carried out may have occupied specific locations in the landscape that were considered sacred, which does not exclude other strategic or defensive functions. These locations are for example at the foot of particular trees, within forest clearings, near marshland, on plateaus promontories, on high sites visible from far away or in subterranean structures on the edges of settlements.

Hence, we observe that certain Michelsberg enclosures were preferentially located near marshy environments, or that Chasséen enclosures were located on the edges of plateaus.

Other examples of later sites show a link with a specific environment: for Late Bronze Age the site of Villiers-sur-Seine was located within a meander of the Seine and for the Late La Tène the site of Souppes-sur-Loing was located on the edge of a plateaus overlooking the Loing valley.

From a modern perspective the fact that these deposits involved whole and often juvenile animals, might seem like an economic loss since the living gained no benefit, assuming that none of the meat from these animals was actually eaten. But animal sacrifices are recurrent in a large number of societies and are associated with rituals and practices that go beyond the simple economic role of the animal. Meals organised by and for the living can include portions of meat that were not included in deposits. In certain high-ranking Early La Tène graves (fig. 214), these masses of remains that are not placed in graves, correspond to large quantities of meat shared by all on the occasion of the funerary feast. It is reasonable to hypothesise that other animals were sacrificed on these occasions, cut up, cooked and eaten during a communal banquet; if this was the case, then the numbers of animals taken from the herd(s) would be considerable (fig. 215).

Endnotes

1   These close ties between hunting and the fundamental elements of social life undoubtedy contribute to the importance of the figure of the hunter as a symbol of "the other" in founding myths (Lot-Falk, 1953).

2   For an overview of discussions on the concept of ontology, (Dianteill, 2015).

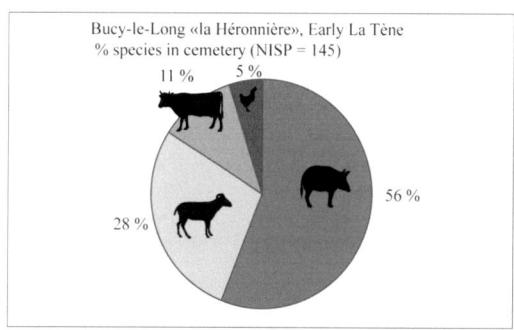

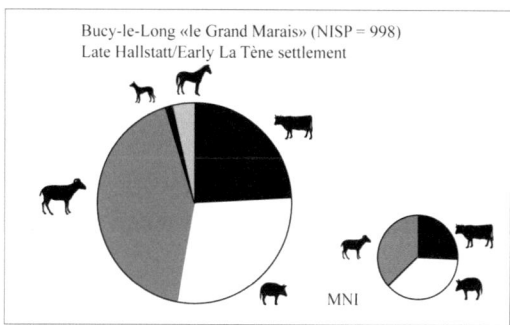

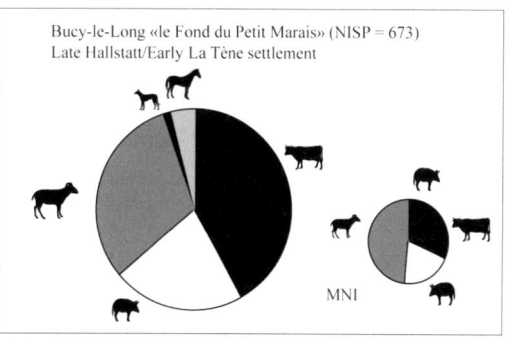

Figure 215: Bucy-le-Long (Aisne), Late Hallstatt/Early La Tène, comparison of frequencies of animal species between settlements and cemeteries. Pigs are more common in graves than in settlements, which probably means they come from several herds.

7. GENERAL CONCLUSION

Culture, defined by the set of representations and values shared by the members of a given society, brings together the ideal and material realms (Godelier, 2010). These components of cultural systems provide an opportunity for archaeozoologists to rediscover, through faunal analysis, many elements of social organisation and its underlying conceptual principles. Foods carry meaning, since as well as having a practical nourishing function they also have a symbolic value. They are shared by members of a social class, group or culture; in this context, meat always occupies a special position due to its social and individual dimensions.

This is how we have been able to retrace on a more global scale, *i.e.* five millennia, the oscillations in the place of species over time in a particularly well-documented sector, the Seine-et-Marne area (fig. 216). For cattle, we note their importance during all the Neolithic in all the regions. There is also a remarkable drop in the Final Bronze Age due to a few major sites that concentrate more pigs than anywhere else. This perception is not shared in the other regions. The place of cattle increases again during the Iron Age but never reaches the level it occupied during the Neolithic period.

Pigs were the third most important resource after cattle and sheep/goats in the Early Neolithic; they became increasingly important from the Middle Neolithic onwards. Pigs are prominent in the herds of Iron Age farms, in varying proportions.

Finally, caprines are important in the Early Neolithic, then more discrete from the Middle Neolithic to the Final Neolithic. The low incidence of caprines is still recorded during the Bronze Age. A strong increase is then observed during the 5th centurey BC.

We have examined the archaeological evidence for this appropriation of animals by looking both at the intrinsic properties of bone remains (whole bones, broken bones, articulated bones, skeletons) and at their extrinsic properties (domestic pits, silos, graves, enclosures, ditches). In addition, we have carried out these analyses at three different spatial scales, namely the domestic unit (houses,

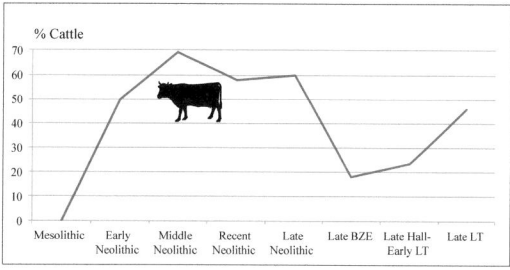

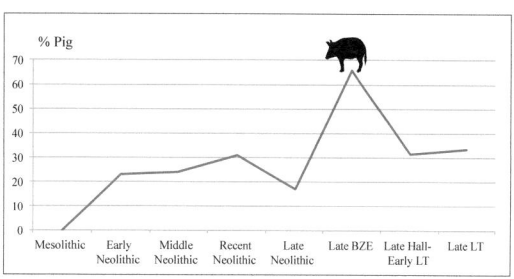

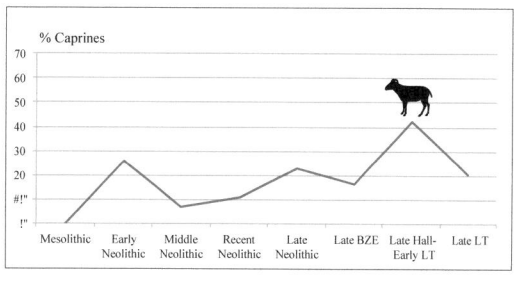

Figure 216: Change of the main domestic species over five millenia in the Seine-et-Marne department (% NISP).

households), the site (villages, farms, oppida) and the region; this allows us to propose a model in which the elements can interlock and interact in a coherent whole (fig. 217).

To quote Barker and Gamble, "*Rigorous contextual analysis and adequate sampling are clearly essential prerequisites for any realistic assessment of the likely relationship between residues from complex sites and the behaviour of the inhabitants...*" (Gumerman, 1997, 111).

In all the models proposed, we have sought to understand the local political and historical trajectories in order to interpret changes in the use of animals over space and time (fig. 217).

Differential consumption and the use of animals for non-nutritional ends, have enabled us to define certain identity /belonging, gender and age statuses relating to. Among the many criteria used to define these status-based differences in consumption are the species represented (wild or domestic, restrictions or excesses), the quantity of meat involved, the pieces of meat involved, the frequency of certain pieces, the age profiles of the animals, a large number of which show a distinct preference for juvenile or sub-adult animals, and the preparation of the pieces (cutting up of legs into large pieces, complete racks of ribs, the degree of fragmentation in settlement contexts). This led us to the hypothesis that in the Neolithic animals acted as markers of clan identity (cattle, caprines), of gender among adults (wild boar for men, red deer for women), as well as to the hypothesis that there were symbolic relationships between certain animals and children (sheep, fox) at this time of Neolithic. These markers are less evident in the Bronze and Iron Ages; nonetheless, in the graves of the Iron Age cemeteries of Champagne-Ardenne, sheep appear to have been preferentially deposited with women.

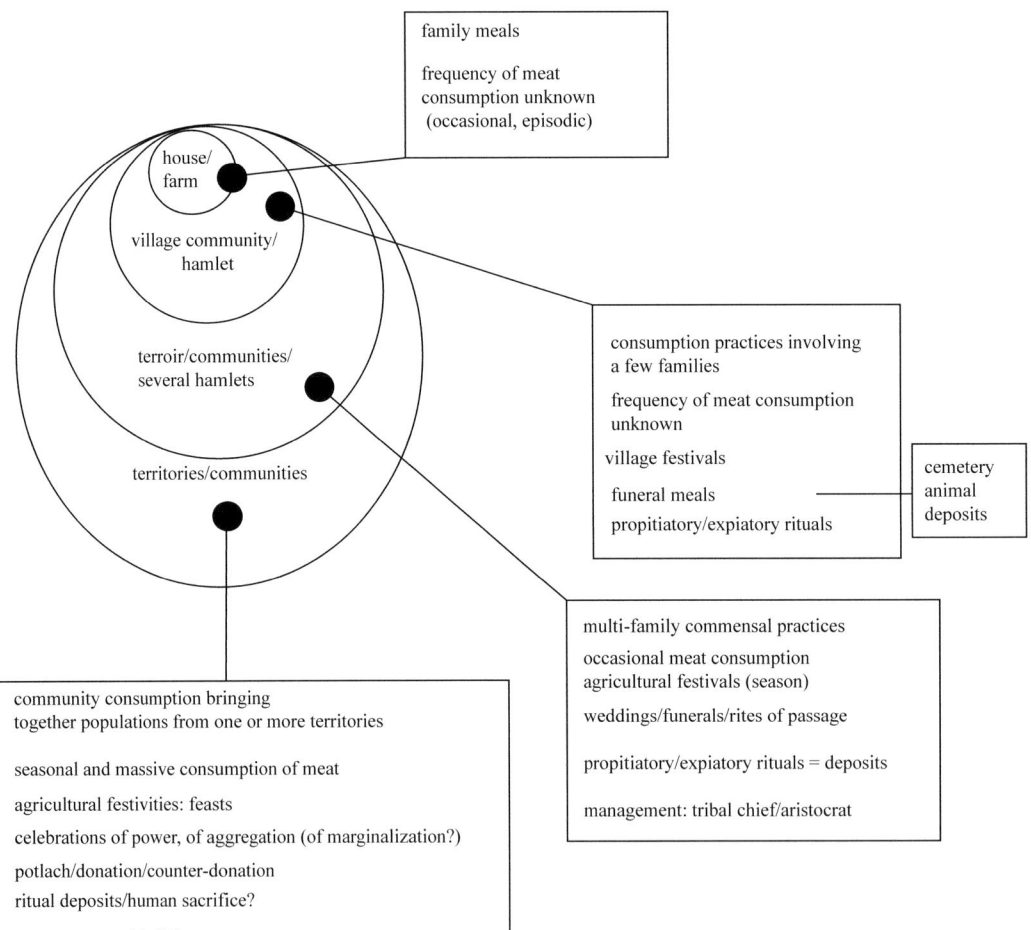

Figure 217: Proposition of a model of meat consumption in relation to the level of economic integration of societies. Areas of consumption meals and ritualized practices involving animals.

family meals

frequency of meat consumption unknown (occasional, episodic)

house/ farm

village community/ hamlet

terroir/communities/ several hamlets

territories/communities

consumption practices involving a few families

frequency of meat consumption unknown

village festivals

funeral meals

propitiatory/expiatory rituals

cemetery animal deposits

multi-family commensal practices

occasional meat consumption
agricultural festivals (season)

weddings/funerals/rites of passage

propitiatory/expiatory rituals = deposits

management: tribal chief/aristocrat

community consumption bringing together populations from one or more territories

seasonal and massive consumption of meat

agricultural festivities: feasts

celebrations of power, of aggregation (of marginalization?)

potlach/donation/counter-donation

ritual deposits/human sacrifice?

management: chief/king

In addition, we have been able to correlate this data with site layouts and house plans.

While there is no doubt that the various cultures that succeeded each other throughout the last five millennia BC regularly consumed meat, it seems that this consumption often took place within a specific context of sharing. The location of these events, characterised on the basis of the nature of features and sites, undoubtedly represents a key to understanding the social organisation of the communities. The meals, which were, to varying degrees, communal and festive in nature, might vary in scale: from a single extended family, or several families, to a whole community, as we have seen in the large houses of the Early Neolithic and in the high ranking farms of the Iron Age. Within the village framework of the first farming communities, the refuse generated by these meals can throw light on the relationships between households; relationships that were cemented through the exchange of products gleaned from livestock rearing, agriculture and hunting.

Finally, at an even larger scale, the sharing of food could take place on certain dedicated sites such as the ditched enclosures of the Middle Neolithic and the enclosed settlements of the Iron Age; here, large populations could assemble to part take in eating and drinking in accordance with strict rules (choice of species, slaughter age, selected cuts) and at specific moments in the year (seasonality). In these cases, the quantities of meat consumed far exceed anything that we see on strictly domestic sites. On each occasion, it is probable that dozens of animals were selected for slaughter from among the herds of the populations involved in the festivities.

The link between the economic sphere and the ideological sphere is unequivocally expressed through the deposition of faunal offerings within enclosures on the occasion of important gatherings (see chapter 3). These rites were specially conceived to unite populations around shared values for the general benefit of the group; we believe that this is particularly well illustrated by the animals sacrificed within the Early Neolithic enclosure at Menneville, which in our opinion convey the particularities

of clan affiliation. This role is also evident on certain aristocratic sites of the Iron Age, for example at Braine, where the faunal remains express the fundamental role of chiefs in the organisation of territories.

In the case of large-scale ceremonies where animals and drink were consumed to a degree that surpassed normal economic rules, the gatherings, whose original function was the cementing of the community or the forging of alliances, also served to reinforce the power and position of those who instigated and controlled them. They led to distinctions in status and to social and political inequalities by underlining the legitimacy of powerful individuals associated with the religious sphere. Through these costly and ostentatious feasts, the elites maintained or reaffirmed the cohesion of the group and reinforced social distinctions. We have been able to observe this in certain Bronze Age enclosure sites and in high ranking sites of Bronze Age and La Tène (see chapter 2).

Collective feasting practices are indissociable from the phenomena of deposition and ostentatious display. The sites mentioned in our study speak volumes in this regard: the deposition of particular carcass parts, the accumulation of portions of meat in ditches, and the display of animal heads (trophies), are practices that endured for five millennia (see chapter 4).

We have been able to show that some of the societies analysed were characterised by significant socio-economic differences based on wealth, prestige and power, and our observations support those made by other authors (deFrance, 2009; Dietler and Hayden, 2001; Hayden, 2014).

In the Middle and Final Neolithic, these inequalities can be perceived in the monumental nature of certain ceremonial and funerary structures and in the animal offerings that they contain. The same is true of certain enclosed sites of the Bronze Age and Iron Age. By highlighting the link between the wealth of certain tombs and the hierarchy of animal species deposited within them (cattle associated with the richest graves) (see chapter 5), we have also been able to demonstrate that the funerary rituals of the Iron Age form an integral part of this system.

On the basis of the faunal evidence from the graves, we have suggested that the parts of the animal that were not placed in the grave were eaten in the feasts or banquets organised for the living. To judge from certain high-status Iron Age graves, large quantities of meat would have been consumed in these funeral meals. During the preparation of the funeral, everyone gets involved. The entire village is in mourning and not just a single family. Among Christians and Druze of Mount Lebanon

"The fraternity symbolised by the shared meal is an essential element of the rituals surrounding death. Thus villagers recount that if they do not share food on these occasions, enmity may follow." (Kanafani-Zahar, 2007, 151).

Animals sacrificed and shared during the funeral bear witness to the complex relationships maintained by these populations with their animals, with certain species and certain parts of the animals being favoured and others excluded. While the principal species associated with funerary rituals are domestic, for example cattle, sheep and pigs, the case of horse and dog is significant because they are rarely involved. Almost always excluded from the funerary realm, in the strict sense of the term, they are nonetheless involved in ritualised practices, sometimes in association with human remains, in specific contexts such as Neolithic enclosures (dog) and in some Early Iron Age silos (horse) (see chapter 4). These two animals also seem to be the subject of dietary taboos from time to time.

With the emergence of clear social inequalities from the Middle Neolithic onwards (the emergence of hierarchies which favour one dominant group or individual) and of political elites in the Bronze Age, the interpretation of faunal remains becomes increasingly complex, in line with the development of different categories of settlements and farms and the emergence of particular orientations that probably reflect the control of production by elites.

The killing of the animal is carried out though ritualised sacrifice. Timeless practices, sacrifices are an integral part of the structuring of societies over the five millennia that we have studied.

Ethnologists, philosophers and sociologists who have studied this issue in ancient and contemporary societies have identified a number of fundamental points, common to all societies.

Through the consecration of a victim, sacrifice is a religious act and a mean by which the profane can communicate with the sacred.

"It changes the state of the individual who carries it out or that of certain objects in which the individual has an interest." (Bonte and Izard, 2013, 643; Mauss and Hubert, 2016).

Even though sacrifice can take different forms, the motives are always the same: sacrifices are essential for the harmony of human groups and the cosmos, for the cohesion of society and for the continuation of the proper functioning of the world. They allow societies to call on the Gods, or on the ancestors, to protect the harvest, or livestock, or to ensure plentiful game.

In a domestic context, animal sacrifice will be a modest affair, carried out by a small group of individuals, as would have been the case for the intact cuts of meat deposited in Iron Age silos, for example.

But sacrifices can also be carried out in a collective context. We can cite important seasonal or cosmogenic ceremonies, which follow a temporal rhythm. In these cases, the role of a specific individual (shaman, druid, etc.) is essential to serve as an intermediary between the world of humans and that of non-humans. The sacrificial offerings, which are repeated at regular intervals, may take place *in situ* at the location where it was hoped that the herd would prosper and be protected from disease.

In conclusion, the following quote summarises what we hope to have highlighted in this work.

"Categories of food, ways of cooking, the place and time of consumption are all elements of a shared system of meaning. We could say that animals, as entities that are cooked and eaten, but also as entities that are storied and loaded with affective and symbolic significance, act as mediators in the relationships between humans." (Armengaud, 1998, 867).

REFERENCES

Achard-Corompt, N., V. Riquier, G. Auxiette, V. Desbrosse, K. Fechner, Achard-Corompt, N., V. Riquier, G. Auxiette, V. Desbrosse, K. Fechner, C. Moreau, J. Van Moerkerke, V. Peltier & G. Achard. 2010. Chasse, culte ou artisanat? Premiers résultats du projet de recherche relatif aux fosses à profil en V-Y-W. *Bulletin de la Société Préhistorique Française,* 107: 588-91.

Achard-Corompt, N., G. Auxiette, K. Fechner, V. Riquier & J. Vanmoerkerke. 2013. Bilan du programme de recherche: fosses à profil en V, W, Y et autres en Champagne-Ardenne, in N. Achard-Corompt & V. Riquier (ed.) *Chasse, culte ou artisanat ? Les fosses « à profil en Y-V-W »: structures énigmatiques et récurrentes du Néolithique aux âges des Métaux en France et alentour,* Supplément 33: 11-82. Société Archéologique de l'Est.

Albarella, U. & S. Payne. 2005. Neolithic pigs from Durrington Walls, Wiltshire, England: a biometrical database. *Journal of Archaeological Science* 32. Elsevier: 589-99.

Allard, P., J. Dubouloz & L. Hachem. 1997. Premiers éléments sur cinq tombes rubanées à Berry-au-Bac (Aisne-France): principaux apports à l'étude du rituel funéraire danubien occidental, in C. Jeunesse (ed.) *Le Néolithique danubien et ses marges entre Rhin et Seine.* Supplément 3: 31-43. Strasbourg: Cahiers de l'Association pour la promotion de la recherche archéologique en Alsace.

Allard P., Hamon C., Bonnardin N., Cayol N., Chartier M., Coudart A., Dubouloz J., Gomart L., Hachem L., Ilett M., Meunier K., Monchablon C., Thevenet C. Linear pottery domestic space: taphonomy, distribution of finds, and economy in the Aisne valley settlements, in C. Hamon, P. Allard & M. Ilett (ed.) *The domestic Space in LBK Settlements*: 9-28 (Internationale Archäologie). Verlag Marie Leidorf.

Andersen, N.H. 1997. *Sarup Vol. 1, The Sarup Enclosures: The Funnel Beaker Culture of the Sarup site, including two causewayed camps compared to the contemporary settlements in the area and other European enclosures.* Jutland Archaeol. Soc. Publ. 33.1. Aarhus University Press, Arhus.

Appadurai A. (ed.). 1988a. *The social life of things: commodities in cultural perspective.* Cambridge: Cambridge University Press.

–. 1988b. Introduction: commodities and the politics of value, in Appadurai A. (ed.) *The social life of things: commodities in cultural perspective*: 3-63. Cambridge: Cambridge University Press.

Arbogast, R.-M. 1989. IV. Les animaux domestiques des fosses-silos. *Gallia Préhistoire,* 31: 139-58. CNRS Editions.

–. 1991. Les restes osseux des grands mammifères du site rubané de Juvigny, Les Grands Traquiers, et leurs indications sur le rôle de la chasse et de l'élevage, *Association régionale pour la protection et l'étude du patrimoine préhistorique*: 15-23.

–. 1994. *Premiers élevages néolithiques du Nord-Est de la France.* Vol. 67. ERAUL.

Arbogast, R.-M., D. Baudais & P. Pétrequin. 1997. Le bol en bois à poignée zoomorphe, in P. Pétrequin (ed.) *Les Sites littoraux néolithiques de Clairvaux-les-Lacs et de Chalain (Jura): 3200-2900 av. J.-C*: 545-50, vol. 3. Les Editions de la MSH.

Arbogast, R.M., C. Jeunesse & J. Schibler. 2001. *Rôle et statut de la chasse dans le Néolithique ancien danubien (5500-4900 av. J.-C.). Table ronde, premières rencontres danubiennes, Strasbourg 20 et 21 novembre 1996.* VML, Rahden/Westf.

Arbogast, R.-M., B. Clavel, S. Lepetz, P. Méniel & J.-H. Yvinec. 2002a. *Archéologie du cheval*. Paris: Errance.

Arbogast, R.-M., J. Desloges & A. Chancerel. 2002b. Sauvages et domestiques: les restes animaux dans les sépultures monumentales normandes du Néolithique. *Anthropozoologica*: 17-27.

Arcelin, P. & J.-L. Brunaux. 2003. Un état des questions sur les sanctuaires et les pratiques cultuelles de la Gaule celtique, in *Cultes et sanctuaires en France à l'âge du Fer*, 60: 5-8. Paris: *Gallia*, CNRS Editions.

Armengaud, F. 1998. Au titre du sacrifice: l'exploitation économique, symbolique et idéologique des animaux, in B. Cyrulnik (ed.) *Si les lions pouvaient parler. Essais sur la condition animale*: 856-87 (Quarto). Paris: Gallimard.

Audebert, A, E. Pinard, G. Auxiette, A. Corsiez, V. Drost, V. Matterne, , A. Morel, S. Normant & J. Siguoirt. 2013. Barenton-Bugny et Chambry (Aisne), Pôle d'activités du Griffon, Tranche 1A. Laon: Rapport de fouille, CG Aisne, Inrap NP.

Audebert, A., G. Auxiette, A. Corsiez, V. Le Quellec, S. Normant, E. Pinard & V. Matterne. 2016. Héritage et évolution des implantations foncières chez les Rèmes dans le nord-laon-nois entre le IIIe s. av. J.C. et le IIIe s. ap. J.C., l'exemple du pôle d'activité du Griffon, à Barenton-Bugny, Chambry et Laon (Aisne), in F. Malrain, G. Blancquaert (ed.) *Evolution des sociétés gauloises du second âge du Fer, entre mutations internes et influences externes*: 113-32. Amiens: *Revue Archéologique de Picardie*, n° spécial 30.

Audebert, A., G. Auxiette, A. Corsiez & V. Le Quellec. 2016. Aux confins de la cité des Rèmes, les secteurs H et K du Pôle d'activités du Griffon, à Barenton-Bugny (Aisne, France), in C. Besson, O. Blin, B. Triboulot B. (ed.) *Franges urbaines, confins territoriaux. La Gaule dans l'Empire*, Mémoire 41: 361-79. Versailles: Ausonius Editions.

Augereau, A. & P. Chambon (ed.). 2011. *Les occupations néolithiques de Macherin à Monéteau (Yonne)*. Mémoire 53. Société Préhistorique Française.

Augereau, A. & D. Mordant. 1993. L'enceinte néolithique Cerny des Réaudins à Balloy (Seine-et-Marne). *Mémoires du Groupement archéologique de Seine-et-Marne*: 97-109.

Auxiette, G. forthcoming. Les consommations carnées au début de La Tène finale sur le site de Villiers-sur-Seine, le Défendable (Seine-et-Marne), in J.-P. Quenez (ed.) *Un établissement aristocratique de La Tène C2 à Villiers-sur-Seine, le Défendable (Seine-et-Marne)*. Rapport de fouille, Inrap CIF.

–. 1994. Mille ans d'occupation humaine. Mille ans d'élevage. L'exploitation des animaux sur les sites du Bronze final à l'Augustéen dans la vallée de l'Aisne. Thèse de doctorat en Archéologie, Ethnologie, Préhistoire, Paris: Paris 1-Panthéon-Sorbonne.

–. 1995. L'évolution du rituel funéraire à travers les offrandes animales des nécropoles gauloises de Bucy-le-Long, in B. Lizet, P. Méniel, F. Audouin-Rouzeau, P. Bonte, Y. Preiswerk & C. Olive (ed.) *L'Homme et l'Animal*: 245-52. Paris: *Anthropozoologica*.

–. 1996. La faune de l'oppidum de Villeneuve-Saint-Germain (Aisne): quartiers résidentiels, quartiers artisanaux. *Revue archéologique de Picardie*: 27-98.

–. 1997. La faune des établissements ruraux du Bronze final au Hallstatt final/La Tène ancienne dans la vallée de l'Aisne. *Revue archéologique de Picardie*: 29-71.

–. 2000a. Les rejets non domestiques des établissements ruraux du Hallstatt final à La Tène finale dans la vallée de l'Aisne et de la Vesle, in S. Marion et G. Blancquaert G (ed.) *Les installations agricoles à l'âge du Fer en France septentrionale*, 6: 169-80. Paris: Editions Rue D'Ulm-Ecole Normale Supérieure. Etudes d'Histoire et d'Archéologie.

–. 2000b. Elevage, chasse et pêche: la faune, in E. Le Goff (ed.) *Les occupations protohistoriques et antiques de la Zac, Object'Ifs Sud, (Calvados)*: 1149-61. Rapport de fouille, Inrap GO.

–. 2002. La faune du site de Bétheny de la Tène C2/D1, in P. Rollet (ed.) *Bétheny, les Équiernolles, Marne. Un établissement agricole gaulois de La Tène C2/D1, une occupation de la période Gallo-Romaine (Ier et IIe s. ap. J.-C.): la faune*. Rapport de fouille, Inrap GEN.

–. 2003. La faune, in A. Balmelle & R. Neiss (ed.) *Les maisons de l'élite à Durocortorum*, 4: 38-39 (Archéologie Urbaine à Reims 96). Reims: Bulletin de la Société Archéologique Champenoise.

–. 2004. La faune des habitats protohistoriques, in A. Koehler (ed.) *Barreau Est de Reims « Contournement de Witry-lès-Reims, Itinéraire de substitution ». Habitats laténiens*: 38-56. Rapport de fouille, AFAN.

–. 2006. Etude de la faune de l'âge du Bronze moyen, in A. Henton (ed.) *Guînes, Jardins du Couvent 2 (Pas-de-Calais)*: 38-40. Rapport de fouille, Inrap NP.

–. 2008. Les dépôts de bœufs de la fosse 48 de l'âge du Bronze, in C. Colas (ed.) *Beaurieux/Cuiry-lès-Chaudardes, la Plaine*: 21-24. Rapport de fouille, Inrap NP.

–. 2009a. Les restes paléo-environnementaux: la faune, in C.-C. Besnard-Vauterin (ed.) *En plaine de Caen. Une campagne gauloise antique. L'occupation du site de l'Etoile à Mondeville (Calvados)*: 133-52 (Archéologie et Culture). Rennes: Presses Universitaires de Rennes.

–. 2009b. Les offrandes animales, in S. Desenne, C. Pommepuy & J.-P. Demoule (ed.) *Bucy-le-Long (Aisne, France), une nécropole de la Tène ancienne: (Ve – IVe siècle avant notre ère)*: 397-406. Revue archéologique de Picardie, n° spécial 26.

–. 2009c. Les restes paléo-environnementaux: la faune, in C.-C. Besnard-Vauterin (ed.) *Ifs, ZAC Object'Ifs Sud – AR 67 (Calvados). Un habitat du second âge du Fer*: 47-53. Rapport de fouille, Inrap GO.

–. 2010. La consommation carnée du Ier au IVe à Reims – Durocortorum, in R. Chossenot, A. Estéban, R. Neiss (ed.) *Académie des inscriptions et des belles lettres, Ministère de l'Education Nationale, Ministère de la Recherche, Ministère de la Culture et de la Communication, Maison des Sciences de l'Homme, Paris,* (Carte Archéologique de La Gaule, Reims, 51/2 Sous La Direction de Michel Provost), 103-4.

–. 2011a. La faune de l'occupation de la fin du Bronze final/Hallstatt C/D1, in C.-C. Besnard-Vauterin (ed.) *Ifs, Zac Object'Ifs Sud – dernière tranche, (Calvados). Habitats et lieux funéraires*

protohistoriques et vestiges antiques: 266-68. Rapport de fouille, Inrap GO.

–. 2011b. La faune du site laténien, in F. Lafage (ed.) *La ferme gauloise de Jossigny, Pré Chêne, Pré du But (IIe s. av. n.è.), Marne-la-Vallée, secteur 3 (Seine-et-Marne)*: 177-80. Paris: Rapport de fouille, Inrap CIF.

–. 2011c. Études et interprétations des assemblages fauniques de l'établissement de l'état 1 (La Tène C2), in A. Gapenne (ed.) *Longueau et Glisy, Zac Jules Verne, le Champ Queutoir et les Champs Tortus. Un établissement rural de La Tène C2-D1*: 99-132. Rapport de fouille, Inrap NP.

–. 2011d. Les offrandes animales, in R. Peake & V. Delattre (ed.) *Jaulnes, Le Bas des Hauts Champs-Ouest. Ensembles funéraires et habitats de l'âge du Bronze et du premier âge du Fer*: 72. Rapport final d'opération archéologique, Inrap CIF.

–. 2011e. Les pièces de viande, in Collectif (ed.) *Celtes et Gaulois. Deux chemins vers l'au-delà*: 119-21. Soissons: Musée de Soissons/INRAP.

Auxiette, G. 2012. Etude de la faune du site 3-Annexe 13, in P. Granchon & A. Mondoloni (ed.) *Gonesse, ZAC des Tulipes Nord (Val d'Oise). Nécropole de l'âge du Bronze final, enclos de La Tène moyenne, habitat de La Tène finale à l'époque gallo-romaine*: 219-29. Rapport de fouille, Inrap CIF.

Auxiette, G. 2012. La faune des sépultures à incinération, in P. Le Guen & E. Pinard (ed.) *Ciry-Salsogne « la Cour Maçonneuse » (Aisne). Carrière Desmarest: nécropole de l'âge du Bronze final*: 35. Rapport de fouille, Inrap NP.

–. 2013a. La faune du site de la Tène moyenne, in A. Audebert & E. Pinard (ed.) *Barenton-Bugny-Bugny et Chambry (Aisne), Pôle d'activités du Griffon, Tranche 1A*. Rapport de fouille, CG Aisne, Inrap.

–. 2013b. Étude et interprétations des assemblages fauniques du Bronze final à la période romaine, in Séguier J.-M. (ed.) *Varennes-sur-Seine, La Justice – Le Marais de la Fontaine du Cœur (Seine-et-Marne)*, 2: 57-148. Rapport de fouille, Inrap CIF.

–. 2013c. La faune du site d'Abbeville, Mont à Cailloux, sud (Bronze ancien/moyen), in P. Le Guen (ed.) *Abbeville (80), Faubourg de Menchecourt, Mont à Cailloux, sud. Fouille 2004, enclos d'habitat de l'âge du Bronze ancien / moyen*: 41-44. Rapport de fouille, Inrap NP.

–. 2013d. Evolution des dépôts du Néolithique à l'Antiquité tardive en contexte non funéraire: un premier état des lieux, in G. Auxiette & P. Méniel (ed.) *Les dépôts d'animaux en France, de la fouille à l'interprétation*, 4: 167-76 (Archéologie Des Plantes et Des Animaux). Montagnac: Editions Monique Mergoil.

–. 2013e. Evolution des dépôts du Néolithique à l'Antiquité tardive en contexte non funéraire : un premier état des lieux., in G. Auxiette & P. Méniel (ed.) *Les dépôts d'animaux en France, de la fouille à l'interprétation*, 4: 167-76 (Archéologie Des Plantes et Des Animaux). Montagnac: Editions Monique Mergoil.

–. 2014a. Elevage, consommation carnée et productions secondaires au Bronze final/Hallstatt ancien sur le site de Buchères/Saint-Léger-Près-Troyes, Parc logistique de l'Aube, in V. Riquier, J. Grisard (ed.) *Buchères, Moussey, Saint-Léger-près-Troyes (Aube). Parc Logistique de l'Aube. L'évolution d'un terroir dans la plaine de Troyes. Campagnes de fouille 2005 et 2006*: 201-31. Rapport de fouille, Inrap GEN.

–. 2014b. L'étude de la faune du site de Méaulte, ZAC du Pays du Coquelicot, in N. Soupart (ed.) *Méaulte et Bécordel-Bécourt, ZAC du Pays du Coquelicot, site 06, Somme. Un habitat rural daté de La fin de La Tène ancienne à La Tène moyenne*: 96-104. Rapport de fouille, Inrap NP.

–. 2014c. La faune du Pôle d'activités du Griffon, tranche 2-2A, secteurs H, K et L, in A. Audebert & E. Pinard (ed.) *Barenton-Bugny (Asine), Pôle d'activités du Griffon, tranche 2-2A, secteurs H, H', I, J, K et L*: 383-418. Rapport de fouille, CG 02, Inrap NP.

–. 2015. La faune du site de Glisy-site C, in S. Gaudefroy (ed.) *ZAC de la Croix de Fer, pôle Jules Verne, site C. Glisy, les Quatorze*: 188-97. Rapport de fouille, Inrap NP.

–. 2016. La faune du Pôle d'activités du Griffon, tranche 2-1, secteurs M-M', M" et N, in A. Audebert (ed.) *Barenton-Bugny (Aisne), Pôle d'activités du Griffon – Tranche 2-1, Secteurs M-M', M" et N*: 191-214. Rapport de fouille, CG 02, Inrap NP.

–. 2017a. Les consommations carnées à l'âge du Bronze: bilan et perspectives, in L. Carozza, C. Marcigny & M. Talon (ed.) *L'habitat et l'occupation des sols à l'âge du Bronze et au début du premier âge du Fer*: 327-36 (Recherches Archéologiques). Paris: INRAP/CNRS éditions.

–. 2017b. Etude du mobilier faunique, in L. Duvette (ed.) *Amiens ZAC de Renancourt, Sites 1, 2, 3, 5a, 5b*: 883-948. Amiens: Rapport de fouille, Inrap NP.

–. 2017c. Autour du mouton: analyse de consommations singulières au cours du premier millénaire avant notre ère, in M.-P. Horard-Herbin & B. Laurioux (ed.) *La Viande: fabrique et représentations*: 117-30 (Tables Des Hommes). Tours: Presses Universitaires François-Rabelais et Presses Universitaires de Rennes.

–. 2017d. Étude archéozoologique du site d'Etaples, ZAC du Chemin des Près, daté du Bronze ancien/moyen, in Y. Lorin (ed.) *Fouille d'un enclos circulaire atypique du Bronze ancien/moyen*: 165-72. Rapport de fouille, Inrap NP.

–. 2018. Les offrandes animales incinérées, in D. Labarre (ed.) *Rouvignies, parc d'activités de l'aérodrome ouest, phase 16: occupations diachroniques sur le versant nord de l'Escaut*: 124-25. Rapport de fouille, Inrap NP.

–. 2019a. Etude archéozoologique – Annexe II, in V. Le Quellec (ed.) *Beaurieux, Les Grèves » (Aisne), tranches 1 et 2*. Rapport de fouille, CG Aisne, Inrap NP.

–. 2019b. La faune et la consommation carnée au Bronze ancien sur le site de Blainville-sur-Orne, Zac Terres d'Avenir, zones 1A et 1B (Calvados), in E. Ghesquière (ed.) *Blainville-sur-Orne, Zac Terres d'Avenir, zones 1A et 1B (Calvados). Un enclos elliptique de l'âge du Bronze ancien*: 130-47. Rapport de fouille, Inrap GO.

–. 2019c. Les offrandes animales, in A. Audebert & E. Pinard (ed.) *Barenton-Bugny (Aisne), Pôle d'activités du Griffon – Tranche 2-2B – Secteurs O, P et Q*: 110-15. Rapport final d'opération, CG Aisne, Inrap NP.

–. 2019d. La faune issue de la double enceinte du Bronze final sur le site de Mondeville, Rue Nicéphore Nièpce (Calvados), in C.C. Besnard-Vauterin (ed.) *Mondeville, Calvados, Rue Nicéphore Nièpce. Une double enceinte du*

Bronze final et un enclos d'habitat du Hallstatt D. Vestiges de la seconde Guerre Mondiale: 118-26. Rapport de fouille, Inrap GO.

–. 2020. La place de l'animal dans les espaces funéraires au Bronze ancien/moyen: les dépôts de faune à Bucy-le-Long (Aisne). *Revue Archéologique de Picardie*: 5-11.

Auxiette, G. & L. Bedault. 2015. La consommation carnée à Choisy-au-Bac, Canal Seine-Nord, au Bronze final IIIb et au Hallstatt ancien, in C. Riche (ed.) *Choisy-au-Bac, à la confluence de l'Aisne et de l'Oise : 10 000 ans d'histoire du Paléolithique à l'âge du Fer*. Rapport de fouille, Canal Seine-Nord Europe, Inrap NP.

Auxiette, G. & S. Desenne dir. 2017. *Une trajectoire singulière. Les enclos de Braine, la Grange des Moines (Aisne) à La Tène finale*. Vol. 3-4. Revue Archéologique de Picardie.

Auxiette, G. & E. Guthman. 2016. La faune des fentes, in Y. Thomas (ed.) *Sélestat, Westrich-Riedwasen Bas-Rhin (67). Un ensemble de dispositifs de Schlitzgruben néolithiques*: 137-54. Rapport de fouille, Inrap GEN.

Auxiette, G. & L. Hachem. 2019. La faune, in F. Dugois & S. Loiseau (ed.) *Pont-sur-Seine, Ferme de l'Ile» (Aube). Exploitation d'un territoire en bord de Seine: de l'enceinte monumentale du Néolithique moyen II à la ferme fossoyée médiévale*: 487-503. Rapport de fouille, Inrap GEN.

Auxiette, G. & G. Jouanin. 2018. Les relations homme/animal durant La Tène dans les sites du bassin du Crould, in C. Laporte-Cassagne (ed.) *Les sites ruraux du second âge du Fer dans le bassin versant du Crould en Val d'Oise et ses marges*, supplément n°5: 297-309. Paris: *Revue Archéologique d'Ile-de-France*.

Auxiette, G. & Méniel. 2005a. Les études de faunes de la Protohistoire récente. *Revue archéologique de Picardie*: 167-76. https://doi.org/10.3406/pica.2005.2464.

Auxiette, G. & P. Méniel. 2005b. Les études de faunes de la Protohistoire ancienne. *Revue archéologique de Picardie*: 121-126. https://doi.org/10.3406/pica.2005.2461.

– (ed.). 2013. *Les dépôts d'ossements animaux en France, de la fouille à l'interprétation: Actes de la table ronde de Bibracte, 15-17 octobre 2012* (Archéologie Des Plantes et Des Animaux (APA). Montagnac: Editions Monique Mergoil.

Auxiette, G. & P.-E. Paris. 2017. L'oppidum de Villeneuve-Saint-Germain: sa place, son rôle dans l'économie de subsistance à la fin de La Tène finale à travers le prisme des études archéozoologiques, in S. Marion, S. Deffressigne, J. Kaurin, G. Bataille (ed.) *Production et proto-industrialisation aux âges du Fer. Perspectives sociales et environnementales*: 101-16. Bordeaux: Ausonius Editions, Mémoires 47.

Auxiette, G. & P. Ruby. 2009. La vie sociale de la viande, in S. Bonnardin, C. Hamon, M. Lauwers, B. Quilliec (ed.) *Du matériel au spirituel. Réalités archéologiques et historiques des « dépôts » de la Préhistoire à nos jours*: 257-66. Antibes: APDCA.

Auxiette, G., S. Desenne & C. Pommepuy. 1995. Bazoches-sur-Vesle, la Foulerie, in *Fouilles Protohistoriques de la Vallée de l'Aisne*: 133-90 (23). Rapport de fouilles, Afan Centre-Nord.

Auxiette, G., B. Hénon, S. Desenne, F. Gransar, C. Pommepuy & M. Boulen. 1996. Le site du début du second âge du Fer de Bucy-le-Long, le Fond du Petit Marais, in *Fouilles Protohistoriques de la Vallée de l'Aisne*: 73-114. Rapport de fouille, Afan Centre-Nord.

Auxiette, G., S. Desenne & C. Pommepuy. 2002. Des viatiques et des banquets: alimentation des défunts, alimentation des vivants sur la nécropole de La Tène ancienne de Bucy-le-Long (Aisne), in P. Méniel & B. Lambot (ed.) *Repas des vivants et nourriture pour les morts en Gaule*, 16: 317-36. Bulletin de la Société Archéologique Champenoise.

Auxiette, G., M. Boulen, S. Desenne, J.-M. Pernaud, P. Ponel, C. Rocq & J.-H. Yvinec. 2003. Un site du Hallstatt à Villeneuve-Saint-Germain, les Etomelles (Aisne). *Revue archéologique de Picardie*: 21-65.

Auxiette, G., M.-P. Horard-Herbin, S. Frère, P. Méniel & J.-H. Yvinec. 2005. Les Gaulois d'Ile-de-France au second âge du Fer et leur cheptel: état de la question, in O. Buchsenschutz, A. Bulard, Th. Lejars (ed.) *Les Gaulois d'Ile-de-France au second âge du Fer et leur cheptel: état de la question*. Supplément 26: 261-72. Revue Archéologique du Centre de la France.

Auxiette, G., A. Baudry & P. Méniel. 2010. Une histoire de l'élevage dans l'ouest de la Normandie: les sites de Mondeville, Ifs, Fleury, Creully (Calvados) et les autres, in P. Barral, B. Dedet, F. Delrieu, P. Giraud, I. Le Goff, S. Marion, A. Villard-Le Tiec (ed.) *L'âge du fer en Basse Normandie*, 1: 185-202. Besançon: Presses universitaires de Franche-Comté, Annales littéraires, Série « Environnement, sociétés et archéologie ».

Auxiette, G., R. Peake & F. Toulemonde. 2015. Food production and diet during the Late Bronze Age in upper Seine valley (France). International Open Workshop 2013, The third food revolution ? Setting the Bronze Age Table: Common trends in economic and subsistence strategies in Bronze Age Europe., in J. Kneisel, M. Dal Corso, W. Kirleis, H. Scholz, N. Taylor, V. Tiedtke (ed.) *The third food revolution? Setting the Bronze Age Table: Common trends in economic and subsistence strategies in Bronze Age Europe*, 6: 47-62. Kiel 2013: Universitätsforschungen zur Prähistorischen Archäologie.

Auxiette, G., L. Hachem & P. Ruby. 2020. La consommation animale : du quotidien au festin collectif saisonnier, in R. Peake (ed.) *Villiers-sur-Seine, un habitat aristocratique du IXe siècle avant notre ère*: 136-82 (Recherches Archéologiques 18). Paris: co-édition INRAP/CNRS éditions.

Bach, A. & H. Bach. 1972. Anthropologische Analyse des Walternienburg/Bernburger Kollektivgrabes von Schönstedt im Thüringer Becken. *Alt-Thüringen* 12. H. Böhlaus Nachf.: 59-107.

Baillieu, M., P. Brun, Y. Guichard & G. Auxiette. 1999. *Beaurieux, la Justice (Aisne)*. Rapport de fouille, Afan Centre-Nord.

Bălăşescu, A., D. Simonin & J.-D. Vigne. 2008. La faune du Bronze final IIIb du site fortifié de Boulancourt, le Châtelet (Seine-et-Marne). *Bulletin de la Société Préhistorique Française* 105: 371-406.

Balasse, M. & A. Tresset. 2002. Early Weaning of Neolithic Domestic Cattle (Bercy, France) revealed by Intra-tooth Variation in Nitrogen Isotope Ratios. *Journal of Archaeological Science* 29: 853-859.

Bánffy, E. 2001. Notes on the connection between human and zoomorphic representations in the Neolithic, in B. Biehl, F. Bertemes & H. Meller (ed.) *The archaeology of cult and religon*, Prehistoric Society Research Paper 8: 53-71. Budapest: Oxbow Books.

Baudrillard, J. 1972. *Pour une critique de l'économie politique du signe*. Gallimard Paris.

Baudry, A. 2005. Approvisionnement et alimentation carnée sur les sites de l'âge du Fer en Bretagne et en Normandie, première approche. L'exemple du site de La Campagne à Basly. *Revue Archéologique de l'Ouest*: 165-80.

–. 2018. *Ressources animales et alimentation carnée à l'âge du Fer: le cas du nord-ouest de la France (Bretagne et Basse Normandie)*. Recherches Archéologiques 13. Paris: INRAP/CNRS éditions.

Bedault, L. 2004. Tinqueux, la Haubette, site du Villeneuve-Saint-Germain du nord-est de la France (Champagne), Étude de la faune issue des maisons 70 et 60. Paris: Mémoire de Maîtrise, Université de Paris 1 Panthéon Sorbonne.

–. 2005. La faune au Néolithique ancien VSG, état de la question. Paris: Mémoire de D.E.A., Université de Paris 1 Panthéon Sorbonne.

–. 2007. La faune, in K. Meunier (ed.) *Gurgy, le Nouzeau (Yonne)*. Rapport de fouille: 71-77. Inrap GEN.

Bedault, L. 2009. First reflexions on the exploitation of animals in Villeneuve-Saint-Germain society at the end of the early Neolithic in the Paris Basin (France), in D. Hofmann & P. Bickle (ed.) *Creating Communities. New advances in Central European Neolithic Research*: 111-131. Oxford: Oxbow Books.

Bedault, L. 2012. L'exploitation des ressources animales dans la société du Néolithique ancien du Villeneuve-Saint-Germain en Bassin parisien: synthèse des données archéozoologiques. Thèse de doctorat Art et archéologie. Préhistoire. Ethnologie. Anthropologie, Paris: Paris I Panthéon-Sorbonne.

Bemilli, C. 2007. La faune du Néolithique final, in P. Brunet & R. Cottiaux (ed.) *Meaux (Seine-et-Marne), Route de Varreddes, Chemin de Flandres: occupations du Néolithique sur les Côteaux de la Justice:* Rapport de fouilles: 100-109. Inrap CIF.

Bentley, R.A. *et al.* 2012. Community differentiation and kinship among Europe's first farmers.

Proceedings of the National Academy of Sciences 109. National Acad. Sciences: 9326-30.

Berthon, R., Y.S. Erdal, M. Mashkour & G. Kozbe. 2016. Buried with turtles: the symbolic role of the Euphrates soft-shelled turtle (Rafetus euphraticus) in Mesopotamia. *Antiquity* 90. Cambridge University Press: 111-25.

Besnard-Vauterin, C.-C., G. Auxiette, M. Besnard, V. Deloze, C. Fiant, S. Giazzon, M. Le Puil-Texier & E. Séhier. 2015. L'occupation d'un micro-terroir de la Protohistoire à l'Antiquité: le site d'Hérouvillette, les Pérelles (Calvados). La faune. *Revue Archéologique de l'Ouest*: 129-76.

Besnard-Vauterin, C.-C., P. Giraud, H. Lepaumier & D. Giazzon. 2016. Genèse d'un réseau de fermes du second âge du Fer en Plaine de Caen, in G. Blancquaert, F. Malrain (ed.) *Evolution des sociétés gauloises du Second âge du Fer, entre mutations internes et influences externes*: 61-81. Numéro spécial 30. Revue Archéologique de Picardie.

Bickle, P. & A. Whittle. 2013. *The First Farmers of Central Europe: Diversity in LBK Lifeways*. Oxbow Books.

Billard, C., F. Bostyn, C. Hamon & K. Meunier. 2014. *L'habitat du Néolithique ancien de Colombelles, le Lazzaro (Calvados)*. Mémoire 58. Société Préhistorique Française.

Blouet, V., Th. Klag, M.-P. Petitdidier & L. Thomashausen. 2013. Le Néolithique ancien en Lorraine. Société Préhistorique Française: 377.

Bocquet-Appel, J.-P. & J. Dubouloz. 2004. Expected paleoanthropological and archaeological signal from a Neolithic demographic transition on a worldwide scale. *Documenta Praehistorica. Neolithic Studies* 31: 25-33.

Bocquet-Appel, J.-P., R. Moussa & J. Dubouloz. 2015. Multi-agent Modelling of the Neolithic LBK, in F. Giligny, F. Djindjan, L. Costa, P. Moscati & S. Robert (ed.) *CAA2014 – 21st Century Archaeology. Concepts, methods and tools*: 611-22. Oxford: Archeopress. https://www. academia.edu/12945923/.

Böes, U., C. Jeunesse, R.-M. Arbogast, P. Lefranc, M. Mauvilly, F. Schneikert & I. Sidéra. 2007. Vendenheim, le Haut du Côteau (Bas-Rhin): remarques sur l'organisation interne d'une nécropole du Néolithique danubien, in

M. Besse (ed.) *Sociétés néolithiques: des faits archéologiques aux fonctionnements socio-économiques*: 279-83. Lausanne: *Cahiers d'Archéologie Romande*, 108.

Bogaard, A., R.-M. Arbogast, R. Ebersbach, R.A. Fraser, C. Knipper, C. Krahn, M. Schäfer, A. Styring & R. Krause. 2017. The Bandkeramik settlement of Vaihingen an der Enz, Kreis Ludwigsburg (Baden-Württemberg): an integrated perspective on land use, economy and diet. *Germania: Anzeiger der Römisch-Germanischen Kommission des Deutschen Archäologischen Instituts* 94. Römisch-Germanische Kommission des Deutschen Archäologischen Instituts: 1-60.

Bogucki, P. 1988. *Forest farmers and stockherders: early agriculture and its consequences in north-central Europe*. Cambridge: Cambridge University Press.

Bökönyi, S. 1974. *History of domestic mammals in Central and Eastern Europe*. Akademiai Kiado.

Bonnabel, L., C. Moreau, M. Saurel, I. Richard, G. Auxiette & E. Vauquelin. 2010. Pratiques funéraires entre le Hallstatt final et La Tène moyenne en Champagne-Ardenne: un genre de point de vue, le point de vue du genre, in P. Barral, B. Dedet, F. Delrieu, P. Giraud, I. Le Goff, S. Marion, A. Villard-Le Tiec (ed.) *Gestuelles funéraires au second âge du Fer*, 2: 129-54. Besançon: Presses universitaires de Franche-Comté, Annales littéraires, Série «Environnement, sociétés et archéologie».

Bonnardin, S. 2009. *La parure funéraire au Néolithique ancien dans les bassins parisiens et rhénan: Rubané, Hinkelstein et Villeneuve-Saint-Germain*. Mémoire 49. *Société Préhistorique Française*.

Bonte, P. 1981. Ecological and economic factors in the determination of pastoral specialisation. *Journal of Asian and African Studies* 16: 33-49.

–. 2004. Des «peuples du bétail». Origines mythiques et pratiques rituelles de l'élevage en Afrique de l'Est. *Techniques et Culture. Revue semestrielle d'anthropologie des techniques*. https://doi. org/10.4000/tc.1116.

–. 2007. *Essai sur les formations tribales du Sahara occidental Approches comparatives, anthropologiques et historiques*. Bruxelles: Editions Luc Pire.

Bonte, P. & M. dir. Izard. 2013. *Dictionnaire de l'anthropologie et de l'ethnologie* (Quadrige). Presses Universitaires de France.

Bonte, P. & M. Izard (ed.). 1991. *Dictionnaire de l'ethnologie et de l'anthropologie*. Paris: Presse Universitaire de France.

Bostyn, F. (ed.). 2002. *Néolithique et protohistoire du site des Antes à Rungis, Val-de-Marne*. Artcom Editions.

– (ed.). 2014. *Des systèmes d'enceintes au Néolithique moyen II. Canal Seine-Nord, fouille 12, Picardie, Somme, St Christ-Briost et Villers-Carbonnel.* Rapport de fouille, Inrap NP.

Bostyn, F. & S. Denis. 2016. Specialised production and distribution networks for flint raw materials during the Blicquy/Villeneuve-Saint-Germain culture (early Neolithic), in T. Kerig, K. Nowak & G. Roth (ed.) *Alles was zählt... Festschrift für Andreas Zimmermann*: 195-208. Bonn: Habelt: Universitätsforschungen zur prähistorischen Archäologie.

Bostyn, F. & Y. Lanchon (ed.). forthcoming. *Le Néolithique ancien de la Basse vallée de la Marne. Evolution des modalités d'occupation et d'exploitation d'un territoire au début du Ve millénaire BC*, Rapport d'Action Collective de Recherche (ACR).

–. 2003. Synthèse générale, in F. Bostyn (ed.) *Néolithique ancien en Haute-Normandie: le village Villeneuve-Saint-Germain de Poses, Sur la Mare, et les sites de la Boucle du Vaudreuil*. Mémoire 4: 27-64. Société Préhistorique Française

Bostyn, F., L. Hachem & Y. Lanchon. 1991. Le site néolithique de la Pente de Croupeton à Jablines (Seine-et-Marne): premiers résultats, in *La région centre: carrefour d'influences? Association Régionale pour la Protection et l'Etude du Patrimoine Préhistorique*, 45-82.

Bostyn, F., G. Blancquaert, Y. Lanchon, G. Auboire & P. Méniel. 1992. Les enclos funéraires de l'âge du Bronze de Coquelles, RN 1(Pas-de-Calais). *Bulletin de la Société Préhistorique Française*, 89, 10-11: 413-28.

Bostyn, F., E. Martial & I. Praud (ed.). 2011a. *Le Néolithique du Nord de la France dans son contexte européen: habitat et économie aux 4e et 3e millénaires avant notre ère*. Numéro spécial 28. Revue Archéologique de Picardie,

Bostyn, F., C. Monchablon, I. Praud & B. Vanmontfort. 2011b. Le Néolithique moyen II dans le sud-ouest du bassin de l'Escaut. Nouveaux éléments dans le Groupe de Spiere, in F. Bostyn, E. Martial & I. Praud (ed.) *Le Néolithique du nord de la France dans son contexte européen. Habitat et économie aux 4ème et 3ème millénaires avant notre ère*, Numéro spécial 28: 55-76, Revue Archéologique de Picardie.

Bostyn, F., R.-M. Arbogast, B. Clavel, C. Hamon, C. Kuhar, D. Maréchal, E. Pinard & I. Praud. 2015. Habitat et sépultures du Blicquy/Villeneuve-Saint-Germain à Longueil-Sainte-Marie, le Barrage (Oise), in F. Bostyn & L. Hachem (ed.) *Hommages à Mariannick Le Bolloch*, 3-4, Revue Archéologique de Picardie: 155-206.

Bostyn, F., L. Hachem, F. Joseph, C. Hamon & Y. Maigrot. 2016. L'apport du site d'habitat de Conty, ZAC Dunant (Somme) à la connaissance de la culture de Cerny. *Bulletin de la Société Préhistorique Française* 113: 291-332.

Bostyn, F., M. Ilett & K. Meunier. 2018a. Tendances évolutives de l'organisation des habitats, au Néolithique ancien (Rubané/BVSG) dans le bassin de la Seine, in O. Lemercier, I. Sénépart, M. Besse & C. Mordant (ed.) *Habitations et Habitat du Néolithique à l'âge du Bronze en France et ses marges*: 27-40. Archives d'Ecologie Préhistorique.

Bostyn, F., Y. Lanchon & P. Chambon (ed.). 2018b. *Habitat du Néolithique ancien et nécropoles du Néolithique moyen I et II à Vignely, la Porte aux Bergers, Seine-et-Marne*. Mémoire 64. *Société Préhistorique Française*.

Boyer, P. 1991. Système de représentation, in P. Bonte & M. Izard (ed.) *Dictionnaire de l'ethnologie et de l'anthropologie*: 626-27 (Quadrige). PUF.

Braguier, S. 2000. Economie alimentaire et gestion des troupeaux au Néolithique récent/final dans le centre-ouest de la France. Thèse de doctorat, Lille: Université de Lille III. http://www.theses.fr/2000TOU20049.

Bréart, B. 1991. Trosly-Breuil, in *Archéologie de la vallée de l'Oise: Compiègne et sa région depuis les origines*: 50-53 (Catalogue d'exposition). Compiègne: CRAVO.

Brisebarre, A.-M. 2013. Leurrer le bétail: techniques d'adoption et de traite (France,

Maghreb, Afrique Subsaharienne). *Cahiers d'anthropologie sociale*. Éditions de l'Herne: 59-71.

Brun, P. 2007. Une période de transition majeure en Europe: de la fin du IVè au début du IIè s. av. J.-C. (La Tène B2 et C), in C. Mennessier-Jouannet, A.-M. Adam, P.-Y. Milcent (ed.) *La Gaule dans son contexte européen aux IVè et IIIè siècles avant notre ère*: 377-84. Association pour le Développement de l'Archéologie en Languedoc-Roussillon.

–. 2015. L'évolution en dent de scie des formes d'expression du pouvoir durant l'âge du Fer en Europe tempérée, in M.-C. Belarte, D. Garcia, J. Sanmarti (ed.) *Les estructura socials protohistoriques a la Gàllia a libèria*: 49-59. Barcelone: Area d'Arqueologia.

Brun, P. & P. Pion. 1992. L'organisation de l'espace dans la vallée de l'Aisne pendant l'âge du Bronze, in C. Mordant, A. Richard (ed.) *L'habitat et l'occupation du sol à l'âge du Bronze en Europe:* 117-27. Paris: Comité des Travaux Historiques et Scientifiques.

Brun, P. & B. Robert. 1988. L'oppidum de Pommiers. Amiens: Rapport de fouilles, Service Régional de l'Archéologie.

Brun, P. & P. Ruby. 2008. *L'âge du fer en France: premières villes, premiers états celtiques*. Paris: la Découverte.

Brun, P., M. Chartier & P. Pion. 2000. Le processus d'urbanisation dans la vallée de l'Aisne, in V. Guichard, S. Sievers, O.-H. Urban (ed.) *Les processus d'urbanisation à l'âge du fer. Eisenzeitliche Urbanisationsprozesse*, 4: 83-96. Glux-en-Glenne: Bibracte – Centre archéologique européen.

Brun, P., C. Cathelinais, S. Chatillon, Y. Guichard, P. Le Guen & E. Nere. 2005a. L'âge du Bronze dans la vallée de l'Aisne., in M. Talon & J. Bourgeois (ed.) *L'âge du Bronze dans le nord de la France dans son contexte européen:* 189-208. Paris: Editions du CTHS/APRAB.

Brun, P., N. Buchez, S. Gaudefroy, M. Talon, I. Le Goff, F. Malrain & V. Matterne. 2005b. Protohistoire ancienne en Picardie. La recherche archéologique en Picardie: Bilans et Perspectives. *Revue Archéologique de Picardie*: 99-120. https://doi.org/10.3406/pica.2005.2460.

Brunaux, J.-L. & P. Meniel. 1983. Le sanctuaire de Gournay-sur-Aronde (Oise): structures et rites, les animaux du sacrifice. *Revue Archéologique de Picardie*, 165-73.

Brunaux, J.-L. & P. Méniel. 1997. *La résidence aristocratique de Montmartin (Oise) du IIIe au IIe s. av. J.-C.* Paris: Document d'Archéologie Française 64.

Brunaux, J.-L., P. Méniel & F. Poplin. 1985. *Gournay I. Les fouilles sur le sanctuaire et l'oppidum (1975-1984).* Numéro spécial 4. Revue Archéologique de Picardie.

Brunet, P. (ed.). 1992. *Le site de Fresnes-sur-Marne, les Sablon*. Rapport de fouilles. AFAN Ile-de-France.

Brunet, P. & R. Irribarria. 2018. Quelles datations pour les bâtiments des Lignères à Mareuil-lès-Meaux (Seine-et-Marne). Néolithique récent-final ou post-Campaniforme-Bronze ancien?, in O. Lemercier, I. Sénépart, M. Besse & C. Mordant (ed.) *Habitations et habitat du Néolithique à l'âge du Bronze en France et ses marges*: 247-60. Archives d'Écologie Préhistorique.

Brunet, P., M.-F. André, C. Bémilli, V. Brunet, R. Cottiaux, J. Durand, R. Gosselin, Y. Le Jeune & C. Renard. 2004. Deux sites de la fin du Néolithique en vallée de Marne: Lesches, les Prés du refuge et Meaux, Route de Varreddes (Seine et Marne): résultats préliminaires, in *Internéo et Société Préhistorique Française*: 101-13..

Brunet, P., M.-F. André, V. Brunet, L. Hachem, R. Irribarria, H. Mathat, C. Monchablon, A. Salavert & A. Samzun. 2014. Le site des Lignères à Mareuil-lès-Meaux (Seine-et-Marne), in *Internéo et Société Préhistorique Française*: 81-87

Buchez, N. 2011. La Protohistoire ancienne. Recherche et fouille de sites de l'âge du Bronze à La Tène ancienne sur les grands tracés linéaires en Picardie occidentale. Questions méthodologiques et résultats scientifiques. *Revue Archéologique de Picardie*: 121-99. https://doi.org/10.3406/pica.2011.3262.

Cahen-Delhaye, A. & G. De Mulder (ed.). 2014. *Des espaces aux esprits, l'organisation de la mort aux âges des Métaux dans le nord-ouest de l'Europe* (Etudes & Documents, Archéologie 32). Wallonie, Belgique: SPW Editions.

Callender, C. 1978. Shawnee, in B. Trigger (ed.) *Handbook of North American Indians*: 622-35. Washington DC: Smithsonian Institution.

Casanova, E. *et al.* 2020. Accurate compound-specific [14] C dating of archaeological pottery vessels. *Nature*. Nature Publishing Group, 1-5.

Cassagnes-Brouquet, S. & C. Dousset-Seiden. 2012. Genre, normes et langages du costume. *Clio. Femmes, Genre, Histoire*: 7-18.

Cayol, N. (ed.). forthcoming. L'enceinte du Néolithique moyen II de Passsel (Oise), in *Statut des objets, des lieux et des Hommes au Néolithique*. Internéo et Société Préhistorique Française, 32, Le Mans, novembre 2017.

Chabi Toko, R. 2016. Place de l'élevage bovin dans l'économie rurale des Peuls du nord Bénin. Thèse de doctorat en sciences agronomiques et ingénierie biologique, Belgique: Communauté Française de Belgique, Université de Liège, Gambloux Agro-Bio Tech.

Chaix, L. 1981. Quelques réflexions sur le bucrane, in *L'animal, l'homme, le dieu dans le Proche-Orient ancien*: 33-37. Université de Genève.

–. 1989. La faune des vertébrés des niveaux V et VIb, in P. Pétrequin (ed.) *Les sites littoraux néolithiques de Clairveaux-les-Lacs (Jura), Le Néolithique moyen*: 369-404. Paris: La Maison des Sciences de l'Homme.

–. 2012. Les bœufs africains à cornes déformées: quelques éléments de réflexion. *Anthropozoologica* 39: 335-42.

Chaix, L. & P. Méniel. 2001. *Archéozoologie: les animaux et l'archéologie* (Hespérides). Paris: Errance.

Chaix, L., A.-M. Pétrequin, P. Pétrequin & H. Richard. 1989. Le cerf néolithique du lac d'Ilay (Jura). *Revue Archéologique de l'Est et du Centre* 40: 105-110.

Chambon, P. & J.-M. Pétillon. 2009. Des chasseurs Cerny? *Bulletin de la Société Préhistorique Française* 106: 761-83.

Chambon, P. & L. Salanova. 1996. Chronologie des sépultures du IIIe millénaire dans le bassin de la Seine. *Bulletin de la Société Préhistorique Française*, 93: 103-18. https://doi.org/10.3406/bspf.1996.10104.

Chambon, P., A. Blin, C. Bronk Ramsey, A. Bayliss, N. Beavan, F. Healy & A. Whittle. 2017. Collecting the dead: temporality and disposal in the Neolithic hypogée of Les Mournouards II (Marne, France). *Germania*: 93-143.

Chambon, P. , S. Bonnardin, F. Bostyn & N. Cayol 2018. La nécropole du Néolithique moyen, in F. Bostyn, Y. Lanchon & P. Chambon (ed.) *Habitat du Néolithique ancien et nécropoles du Néolithique moyen I et II à Vignely, la Porte aux Bergers, Seine-et-Marne*. Mémoire 64: 203-400. Société Préhistorique Française.

Chancerel A., Marcigny C. (ed.). 2006. *Le plateau de Mondeville (Calvados): du Néolithique à l'âge du Bronze*. Documents d'Archéologie Française 99. Paris: Maison des Sciences de l'Homme.

Charier, M.-A. 1986. Missy-sur-Aisne, le Culot: étude d'un village néolithique danubien. Paris: Mémoire de Maîtrise, Université de Paris 1 Panthéon Sorbonne.

Chartier, M. 2010. Choix et évolution de l'implantation des sites d'habitat au Néolithique ancien dans la vallée de l'Aisne (France). *Bulletin de la Société Préhistorique Française* 107: 85-95.

Claßen, E. 2005. Siedlungsstrukturen der Bandkeramik im Rheinland., in J. Lüning, C. Frirdich & A. Zimmerman (ed.) *Die Bandkeramik im 21 Jahrhundert*: 113-124 (Internationale Archäologie, Arbeitsgemeinschaft, Symposium, Tagung, Kongres 7). Rhaden/Westf: VML, Verlag Marie Leidorf.

Constantin, C., P. Allard, L. Hachem & I. Sidera. 2014. Deux fosses Seine-Oise-Marne à Cuiry-lès-Chaudardes, Les Fontinettes (Aisne), in R. Cottiaux & L. Salanova (ed.) *La fin du IVe millénaire dans le Bassin parisien. Le Néolithique récent entre Seine, Oise et Marne (3500-2900 avant notre ère)*. Supplément 34: 13-25, 1, Société Archéologique de l'Est/ Revue Archéologique d'Ile-de-France.

Clifton, J.A. 1978. Potawatomi, in B. Trigger (ed.) *Handbook of North American Indians*: 622-35. Washington DC: Smithsonian Institution.

Colas, C. (ed.). 2008. *Beaurieux (Aisne), la Plaine, zone sud, tranche 2*. Rapport de fouilles, Inrap NP.

–. 2007. Le monument funéraire Michelsberg ancien de Beaurieux, la Plaine (Aisne, France), in *Sociétés néolithiques. Des faits archéologiques aux fonctionnements socio-économiques*: 329-34. Lausanne: Cahiers d'Archéologie Romande.

Colas, C., I. Praud, Françoise Bostyn & N. Cayol. 2016. Chasséen septentrional, qui es-tu? Apports des découvertes récentes dans le nord-ouest de la France, in T. Perrin, P. Chambon, J. Gibaja Bao & G. Goude (ed.) *Le Chasséen, des Chasséens... Retour sur une culture nationale et ses parallèles, Sepulcres de fossa, Cortaillod, Lagozza*: 123-40. Toulouse: Archives d'Écologie Préhistorique.

Colas, C., P. Allard, M. Chartier, C. Constantin, L. Hachem, Y. Maigrot, L. Manolakakis & C. Thevenet. 2018. Les ensembles monumentaux du Néolithique moyen de Beaurieux, la Plaine. *Revue Archéologique de Picardie* 3-4: 11-123.

Collectif. 2011. *Celtes et Gaulois. Deux chemins vers l'au-delà*. Soissons: Musée de Soissons, Inrap.

Constantin, C. 1985. *Fin du Rubané, céramique du Limbourg et post-Rubané. Le Néolithique le plus ancien en Bassin parisien et en Hainaut*. BAR International series S273. Oxford.

Constantin, C. & J. Debord. 1982. Les fouilles de Villeneuve-Saint-Germain (Aisne). *Vallée de l'Aisne: cinq années de fouilles protohistoriques*. Numéro spécial 1: 211-64. Revue Archéologique de Picardie.

Constantin, C. & M. Ilett. 1997. Une étape finale dans le Rubané récent du Bassin parisien, in C. Jeunesse (ed.) *Le Néolithique danubien et ses marges entre Rhin et Seine*, Supplément 3: 281-300. Strasbourg: Cahiers de l'Association pour la promotion de la recherche archéologique en Alsace.

Constantin, C., A. Coudart & J.-P. Demoule. 1982. Villeneuve-Saintt-Germain, les Grands Grèves. Les bâtiments de la Tène III. Numéro spécial 1: 195-205. Revue Archéologique de Picardie,

Constantin, C., J.-P. Farruggia & Y. Guichard. 1995. Deux sites du groupe Villeneuve-Saint-Germain à Bucy-le-Long (Aisne). *Revue Archéologique de Picardie* 1-2: 3-59.

Copet-Rougié, E. 1991. Clan, in P. Bonte & M. Izard (ed.) *Dictionnaire de l'ethnologie et de l'anthropologie*: 152-53 (Quadrige). Paris: Presse Universitaire de France.

Coquet, M. 1987. Une esthétique du fétiche. An Aesthetics of Fetishes (Bwaba, Burkina Faso). *Systèmes de pensée en Afrique noire* 8: 111-40. https://doi.org/10.4000/span.1034.

Cottiaux, R. & L. Salanova (ed.). 2014. *La fin du IVe millénaire dans le Bassin parisien: le Néolithique récent entre Seine, Oise et Marne (3500-2900 avant notre ère)*. Supplément 34, 1, Société archéologique de l'Est/ Revue Archéologique d'Ile-de-France. https://halshs.archives-ouvertes.fr/halshs-01159121.

Cottiaux, R., L. Salanova, P. Brunet, T. Hamon 2014a. Les sites d'habitat du Néolithique récent dans la basse vallée de la Marne, in R. Cottiaux & L. Salanova (ed.) *La fin du IVe millénaire dans le Bassin parisien. Le Néolithique récent entre Seine, Oise et Marne (3500-2900 avant notre ère)*. Supplément 34, 1: 151-87, Société Archéologique de l'Est/ Revue archéologique d'Ile-de-France

–. 2014b. Le Néolithique récent dans le Bassin parisien (3600-2900 avant notre ère): péri-odisation et faciès régionaux, in R. Cottiaux & L. Salanova (ed.) *La fin du IVe millénaire dans le Bassin parisien: le Néolithique récent entre Seine, Oise et Marne (3500-2900 avant notre ère)*. Supplément 34, 1: 455-550. Société Archéologique de l'Est/ Revue Archéologique d'Ile-de-France https://halshs.archives-ouvertes.fr/halshs-01159121.

Cottiaux, R., G. Durbet, L. Hachem & E. Martial. 2008. L'enceinte du Néolithique moyen de Maisons-Alfort, ZAC d'Alfort (Val-de-Marne), in *Internéo 7*: 71-86.

Coudart, A. 1998. *Architecture et société néolithique: l'unité et la variance de la maison danubienne*. Les Editions de la Maison des Sciences de l'Homme.

–. 2009. La maison néolithique: métaphore matérielle, sociale et mentale des petites sociétés sédentaires, in J.-P. Demoule (ed.) *La révolution néolithique dans le monde*: 215-235. Paris: INRAP/CNRS éditions.

–. 2015. The Bandkeramik longhouses. A Material, Social, and Mental Metaphor for Small-Scale Sedentary Societies, in C. Fowler, J. Harding & D. Hofmann (ed.) *The Oxford Handbook of Neolithic Europe*: 309-25. Oxford: Oxford University Press.

Daveau, I. & J.-H. Yvinec. 2001. L'occupation protohistorique du site de Fontenay-en-Parisis, la Lampe (Val-d'Oise): un lieu de consommation collective à La Tène C1/C2. *Revue archéologique du Centre de la France* 40: 69-102. https://doi.org/10.3406/racf.2001.2875.

De Hingh A.-E. 2000. *Food production and food procurement in the Bronze Age and Early Iron Age (2000-500 BC), The organisation of a diversified and intensified agrarian system in the Meuse-Demer-Scheldt region (The Netherlands and Belgium) and the region of the river Moselle (Luxemburg and France).* Vol. 7. Leiden: Archaelogical Studies Leiden University.

De Mulder, G. 2014. Les rites funéraires dans le nord du bassin de l'Escaut à l'âge du Bronze final et au premier âge du Fer, in A. Cahen-Delhaye & G. De Mulder (ed.) *Des espaces aux esprits, l'organisation de la mort aux âges des Métaux dans le nord-ouest de l'Europe*, 32: 29-52 (Etudes & Documents, Archéologie). Wallonie, Belgique: SPW Editions.

Debord, J. 1982. Premier bilan de huit années de fouilles à Villeneuve-St-Germain (1973-1980). *Vallée de l'Aisne: cinq années de fouilles protohistoriques.* Numéro spécial 1: 213-64. Revue Archéologique de Picardie.

–. 1990. Les fouilles du site gaulois tardif de Villeneuve-Saint-Germain (Aisne). *Mémoires de la Fédération des Sociétés d'Histoire et d'Archéologie de l'Aisne* 35: 137-70.

–. 1993. Les artisans gaulois de Villeneuve-Saint-Germain (Aisne). Structures, production, occupation du sol. *Revue Archéologique de Picardie*: 71-110.

–. 1995. A propos de la chronologie des sites de Pommiers et de Villeneuve-Saint-Germain (Aisne). *Revue Archéologique de Picardie*: 205-8.

Debord, J. & S. Desenne. 2005. Bucy-le-Long, la Grande Pièce de la Croix Rouge (Aisne). Découverte d'un ensemble caractéristique du début de La Tène ancienne, in G. Auxiette & F. Malrain (Ed.) *Hommages à Claudine Pommepuy.* Numéro spécial 22: 163-74. Revue Archéologique de Picardie. https://doi.org/10.3406/pica.2005.2728.

deFrance, S.D. 2009. Zooarchaeology in complex societies: political economy, status, and ideology. *Journal of Archaeological Research* 17: 105-68.

Delattre, V. 2010. Les dépôts en silos laténiens: une pratique cultuelle? Dépôts atypiques et manipulations de corps au second âge du Fer: l'exemple de la confluence Seine-Yonne (Seine-et-Marne), in B. Boulestin, L. Baray (ed.) *Morts anormaux et sépultures bizarres, les dépôts humains en fosses circulaires ou en silos du néolithique à l'âge du Fer*: 113-26 (Collections Art, Archéologie et Patrimoine). Editions Universitaires de Dijon.

–. 2013. Sacrifices et dépôts composites au second âge du Fer dans le Bassin parisien: quand le défunt échappe à la nécropole et devient offrande, in S. Krausz, A. Colin, K. Gruel, I. Ralston, T. Dechezleprêtre (ed.) *L'âge du Fer en Europe. Mélanges offerts à Olivier Buchschenschutz*: 481-99. Bordeaux: Ausonius Editions.

Delattre, V. & G. Auxiette. 2018. Homme Vs animal: une même intention cultuelle dans les dépôts domestiques du second âge du Fer dans le Bassin Parisien?, in S. Costamagno, L. Gourichon, C. Dupont, O. Dutour et D. Vialou (ed.) *«Animal symbolisé – Animal exploité. Du Paléolithique à la Protohistoire»*: 291-305. Paris: Editions du Comité des Travaux Historiques et Scientifiques.

Delattre, V. & J.-M. Séguier. 2007. Du cadavre à l'os sec: manipulations de corps à caractère cultuel à l'âge du Fer dans le territoire sénon, in Ph. Barral, A. Daubigney, C. Dunning, G. Kaenel, M.-J. Roulière-Lambert (ed.) *L'âge du Fer dans l'arc jurassien et ses marges. Dépôts, lieux sacrés et territorialité à l'âge du Fer*: 605-20. Besançon: Presses Universitaires de Franche-Comté.

Delattre, V., A. Bulard, P. Gouge & P. Pihuit. 2000. De la relégation sociale à l'hypothèse des offrandes: l'exemple des dépôts en silos protohistoriques au confluent Seine-Yonne (Seine-et-Marne). *Revue Archéologique du Centre de la France*: 5-30.

Delattre, V., G. Auxiette & E. Pinard. 2018. *Quand le défunt échappe à la nécropole. Pratiques rituelles et comportements «dits» déviants au second âge du Fer dans le Bassin parisien.* Dijon: Editions Universitaires de Dijon.

Demoule, J.-P. 1999. *Chronologie et société dans les nécropoles celtiques de la culture Aisne-Marne du VIe au IIIe siècle avant notre ère.* Revue Archéologique de Picardie, n° spécial 15.

–. 2011. Les pratiques funéraires, in Collectif (ed.) *Celtes et Gaulois. Deux chemins vers l'au-delà*: 49-50. Soissons: Musée de Soissons/INRAP.

Demoule, J.-P. & F. Lüth (ed.). 2010. *MK Projekt. Emergence de la complexité socio-politique:*

enceintes, ressources et territorialité au Néolithique. Recherches franco-allemandes sur la culture de Michelsberg. Projet franco-allemand ANR/DFG. http://www.anr-mk-projekt.fr/.

Demoule, J.-P., J. Dubouloz & L. Manolakakis. 2007. L'émergence des sociétés complexes (4500-3500), in J.-P. Demoule (ed.) La révolution néolithique en France: 60-77 (Archéologie de La France). Paris: Éditions La Découverte/ INRAP.

Desbrosse, V. & V. Peltier. 2010. Premiers résultats de la fouille préventive du Haut de Launoy à Pont-sur-Seine, Paris: Internéo et Société Préhistorique Française, 8, 111-115.

Descola, P. 1993. Les lances du crépuscule. Avec les indiens Jivaros de haute Amazonie. Paris: Plon.

–. 2005. On anthropological knowledge. Social Anthropology 13. Cambridge University Press: 65-73.

–. 2015. Par-delà nature et culture (Folio Essais). Gallimard.

Desenne, S. dir. 2017. Trois occupations de pente aux âges des Métaux, à Pasly, les Côteaux de Pasly (Aisne), in F. Bostyn, L. Hachem (ed.) Hommages à Mariannick Le Bolloch, 3-4: 279-303. Revue Archéologique de Picardie.

Desenne, S., G. Auxiette, J.-P. Demoule & S. Thouvenot. 2007. Reflet d'une communauté celtique à travers ses pratiques funéraires: étude d'un cas, la nécropole de Bucy-le-Long, la Héronnière (Aisne), in L. Baray, P. Brun et A. Testart (ed.) Nouvelles approches en archéologie et en anthropologie sociale: 134-55 (Collection Art, Archéologie et Patrimoine). Sens, France: Editions Universitaires de Dijon.

Desenne, S., C. Pommepuy & J.-P. Demoule (ed.). 2009a. Bucy-le-Long (Aisne, France), une nécropole de la Tène ancienne (Ve – IVe siècle avant notre ère). Revue Archéologique de Picardie, n° spécial 26.

Desenne, S., G. Auxiette, J.-P. Demoule, S. Gaudefroy, B. Henon, S. Thouvenot & T. Lejars. 2009b. Dépôts, panoplies et accessoires dans les sépultures du 2ᵉ âge du Fer en Picardie, in E. Pinard, S. Desenne (ed.) Les gestuelles funéraires au second âge du Fer, 3-4: 173-86. Revue Archéologique de Picardie.

Desfossés, Y., E. Martial & L. Vallin. 1992a. Le site d'habitat du Bronze moyen du Château d'Eau à Roeux (Pas-de-Calais). Bulletin de la Société Préhistorique Française 89: 343-92.

Desfossés, Y., B. Masson, P. Barbet, A.-V. Munaut, F. Emontspohl, P. Rodriguez, M.-E. Solari & J.-H. Yvinec. 1992b. Les enclos funéraires du Motel à Fresnes-lès-Montauban (Pas-de-Calais). Bulletin de la Société Préhistorique Française 89: 303-42.

Desloges, J. 1997. Les premières architectures funéraires de Basse-Normandie, in C. Constantin, D. Mordant & D. Simonin (ed.) La culture de Cerny, nouvelle économie, nouvelle société au Néolithique: 515-39. Mémoires du Musée de Préhistoire d'Ile-de-France, 6.

Detienne, M. & J.-P. Vernant. 1979. La cuisine du sacrifice en pays grec (Bibliothèque Des Histoires). Paris: Gallimard.

Devilliers, C. (ed.). 2006. Puiseaux, le Chemin de Paris. Rapport d'opération de sondage, Société Archéologique de Puiseaux, SRA région Centre.

Dianteill, E. 2015. Ontologie et anthropologie. Dix ans de controverse (Brésil, France, États-Unis). Revue Européenne des Sciences Sociales: 119-44. https://doi.org/10.4000/ress.3314.

Dietler, M. 1990. Driven by drink: the role of drinking in the political economy and the case of early Iron Age France. Journal of Anthropological Archaeology 9: 352-406.

–. 2001. Theorizing the feast: rituals of consumption, commensal politics, and power in African contexts, in M. Dietler & B. Hayden (ed.) Feasts: Archaeological and ethnographic perspectives on food, politics, and power: 65-114 (Smithsonian Series in Archaeological Inquiry). Washington: Smithsonian.

Dietler, M. & B. Hayden. 2001. Digesting the feast: Good to eat, good to think, in M. Dietler & B. Hayden (ed.) Feasts: Archaeological and ethnographic perspectives on food, politics, and power: 1-20 (Smithsonian Series in Archaeological Inquiry). Washington: Smithsonian.

Dietsch-Sellami, M.-F. 2000. Environnement, agriculture et consommation des espèces végétales: les données carpologiques, in E. Le Goff (ed.) Les occupations protohistoriques et antiques de la Zac Object'Ifs Sud, Ifs (Calvados): 1167-85. Etudes spécialisées, Rapport de fouille, Inrap GO.

Digard, J.-P. 1988. Jalons pour une anthropologie de la domestication animale. *L'homme*. JSTOR, 27-58.

–. 2009. Les voies de la domestication animale, entre tendances, hasard et nécessité, in J.-P. Demoule *La révolution néolithique dans le monde*: 165-78. Paris: INRAP/Universcience, Editions du CNRS.

Döhle, H.J. 1993. Haustierhaltung und Jagd in der Linienbandkeramik-ein Überblick. *Zeitschrift für Archäologie* 27: 105-24.

Döhle, H.-J. 1994. *Die linienbandkeramischen Tierknochen von Eilsleben, Bördekreis: ein Beitrag zur neolithischen Haustierhaltung und Jagd in Mitteleuropa*. Halle (Saale): Landesmuseum für Vorgeschichte.

Döhle, H.-J. 2009. Ein neolithischer Pferdeschädel von Salzmünde bei Halle (Saale). *Beiträge zur Archäozoologie und Prähistorischen Anthropologie*: 23-29.

Döhle, H.-J. & T. Schunke. 2014. Der erste neolithische Pferdeschädel Mitteldeutschlands – ein frühes Hauspferd?, in H. Meller & S. Friederich (ed.) *Salzmünde-Schiepzig – ein Ort, zwei Kulturen: Ausgrabungen an der Westumfahrung Halle (A 143), Teil I*: 257-61. Halle a. d. Saale: LDA, Archäologie in Sachsen-Anhalt.

Domboroczki, L. 2009. Settlement structures of the Alföld Linear Pottery Culture (ALPC) in Heves County (North-Eastern Hungary): development models and historical reconstructions on micro, meso and macro levels, in J.K. Kozłowski (ed.) *Interaction Between Different Models of Neolithisation North of the Central European Agro-Ecological Barrier*: 75-127. Kraków: Polska Academia Umieje̜ Prace Komisji Prehistorii Karpat.

Driesch von den, A. 1978. *A guide to the measurement of animal bones from archaeological sites* (1). Cambridge: Peabody Museum Press, Harvard University Press.

Dubouloz, J. 1998. Réflexions sur le Michelsberg ancien en Bassin parisien., in *Die Michelsberger Kultur und ihre Randgebiete-Probleme der Entstehung, Chronologie und Siedlungswesens*: 9-20. Stuttgart: Konrad Theiss Verlag.

–. 2003a. Datation absolue du premier Néolithique du Bassin parisien: complément et relecture des données RRBP et VSG. *Bulletin de la Société Préhistorique Française* 100: 671-89.

– (ed.). 2003b. *Osly-Courtil, la Terre-Saint-Mard (Aisne): nouvelle enceinte du Style de Menneville; habitat du Bronze final IIIb*. Rapport de fouille, Inrap NP.

–. 2008. Impacts of the Neolithic Demographic Transition on Linear Pottery culture settlement, in O. Bar Yosef & J.-P. Bocquet-Appel (ed.) *The Neolithic Demographic Transition and its consequences*: 207-235. New-York: Springer.

–. 2012a. Interdépendance et cohésion des différents niveaux de territorialité au Néolithique Rubané en Bassin parisien, in V. Carpentier & C. Marcigny (ed.) *Des hommes aux champs. Pour une archéologie des espaces ruraux du Néolithique au Moyen âge*: 23-34 (Archéologie et Culture). Presses Universitaires de Rennes.

–. 2012b. À propos d'implantation, de démographie et de scission villageoises au Néolithique rubané. *Les Nouvelles de l'Archéologie*: 30-34.

–. 2017. Modélisation et simulation de la colonisation néolithique de l'Europe tempérée par la culture à céramique linéaire, in *Archéologie des migrations*: 111-24 (Recherches/INRAP). Paris: La Découverte. https://doi.org/10.3917/dec.garci.2017.01.0111.

–. 2018. Lecture multiscalaire des enceintes du Néolithique moyen (4500-3800 BC) en France du Nord. Hypothèses sur la structuration économique, sociale et politique au Néolithique, in M. Gandelin, L. Jallot & J. Vaquer (ed.) *Les sites fortifiés de la Préhistoire: nouvelles approches, nouvelles hypothèses*: 197-213. Toulouse: Archives d'Ecologie Préhistorique.

Dubouloz, J. & Y. Lanchon. 1997. Cerny et Rössen en Bassin parisien: une approche par la céramique, in *La culture de Cerny, nouvelle économie, nouvelle société au Néolithique*, 6: 239-65. Mémoires du Musée de Préhistoire d'Ile-de-France, Editions A.P.RA.I.F.

Dubouloz, J., D. Mordant & M. Prestreau. 1991. Les enceintes néolithiques du Bassin parisien. Variabilité structurelle, chronologique et culturelle. Place dans l'évolution socio-économique du Néolithique régional. Modèles interprétatifs préliminaires, in *Identité du Chasséen,*

4: 211. Mémoires du Musée de Préhistoire d'Ile-de-France.

Dubouloz, J., D. Hamard & M. Le Bolloch. 1997. Composantes fonctionnelles et symboliques d'un site exceptionnel: Bazoches-sur-Vesle (Aisne), 4000 ans av. J.-C., in G. Auxiette, L. Hachem & B. Robert (ed.) *Espaces physiques, espaces sociaux, dans l'analyse interne des sites du Néolithique à l'âge du Fer*: 127-44. Paris: Editions du Comité des Travaux Historiques et Scientifiques.

Dubouloz, J., F. Bostyn, M. Chartier, R. Cottiaux & M. Le Bolloch. 2005. La recherche archéologique sur le Néolithique en Picardie. *Revue Archéologique de Picardie* 3-4: 63-98.

Dubouloz, J., D. Gronenborn, L. Manolakakis & O. Weller. forthcoming. MK-Projekt: the emergence of social complexity. Enclosures, resources and territoriality in the Neolithic, in T. Darvill & A. Sheridan (ed.) *Hands across the water: the archaeology of the Cross-channel Neolithic.* Bournemouth: Prehistoric Society and Bournemouth University Archaeology Group.

Dugois, F. & S. Loiseau (ed.). 2019. *Pont-sur-Seine, Ferme de l'Ile (Aube). Exploitation d'un territoire en bord de Seine: de l'enceinte monumentale du Néolithique moyen II à la ferme fossoyée médiévale.* Rapport de fouille, Inrap GEN.

Duhamel, P. 1994. La nécropole monumentale Cerny de Passy (Yonne): description d'ensemble et problèmes d'interprétation, in *La Culture de Cerny: nouvelle économie, nouvelle société au Néolithique*: 397-448. Nemours: Mémoires du Musée de Préhistoire d'Ile-de-France, 6. https://halshs.archives-ouvertes.fr/halshs-01839332.

Duplessis, M., V. Delattre & G. Auxiette. 2013. Un dépôt composite et atypique d'humain et d'animaux dans le silo 27 (Hallstatt final/La Tène ancienne) de la Butte aux Bergers à Chilly-Mazarin (Essonne). *Revue Archéologique d'Ile de France* 6: 31-54.

Duval, C., S. Lepetz & M.-P. Horard-Herbin. 2012. Diversité des cheptels et diversification des morphotypes bovins dans le tiers nord-ouest des Gaules entre la fin de l'âge du Fer et la période romaine. *Gallia* 69: 79-114.

Ebersbach, R. & C. Schade. 2004. Modeling the intensity of linear pottery land use: an example from the Mörlener Bucht in the Wetterau Basin, Hesse, Germany, in *Enter the past: the E-way into the four dimensions of cultural heritage*, 1227: 337-48. BAR International Series.

Eggert, M.K. 2007. Wirtschaft und Gesellschaft im früheisenzeitlichen Mitteleuropa. Überlegungen zum "Fürstenphänomen". *Fundberichte Baden-Württemberg* 29: 255-302.

Evans-Pritchard, E.E. 1954. The Meaning of Sacrifice Among the Nuer. *The Journal of the Royal Anthropological Institute of Great Britain and Ireland* 84: 21-33. JSTOR. https://doi.org/10.2307/2843998.

Farruggia, J.-P. & C. Constantin. 1984. Le site néolithique et des âges des métaux de Missy-sur-Aisne, le Culot, in *Fouilles Protohistoriques de la vallée de l'Aisne, 12*: 61-94. Paris: Université de Paris I, Panthéon-Sorbonne.

Farruggia, J.-P., Y. Guichard & L. Hachem. 1996. Les ensembles funéraires rubanés de Menneville, Derrière le Village (Aisne), in P. Duhamel (ed.) *La Bourgogne entre les bassins rhénan, rhodanien et parisien: carrefour ou frontière?*, 14ème supplément: 119-74. Revue Archéologique de l'Est.

Forest, V. & I. Rodet-Belarbi. 2002. À propos de la corpulence des bovins en France durant les périodes historiques. *Gallia* 59: 273-306.

Fournand, S. (ed.). 2012. *Pont-sur-Seine/Marnay-sur-Seine, la Gravière (Aube). Habitat et Nécropole du Néolithique ancien à la Tène ancienne.* Rapport de fouilles, Inrap GEN.

Fournand, S., P. Allard, E. Bonnaire, K. Fechner, L. Hachem, C. Hamon, Y. Maigrot, K. Meunier & A. Salavert. 2010. Un habitat Rubané à Pont-sur-Seine/Marnay-sur-Seine (Aube). *Internéo* 8: 9-22.

Frangipane, M., E. Andersson Strand, R. Laurito, S. Möller-Wiering, M.-L. Nosch, A. Rast-Eicher & A. Wisti Lassen. 2009. Arslantepe, Malatya (Turkey): Textiles, Tools and Imprints of Fabrics from the 4th to the 2nd Millennium BCE. *Paléorient* 35. Maison René-Ginouvès: 5-29.

Frère, S. 2012. Étude archéozoologique de l'établissement rural laténien, in F. Lafage (ed.) *Pierrefitte-sur-Seine, les Tartres, Rue Emile Zola, Rue Guynemer (Seine-Saint-Denis). La ferme gauloise des Tartres (II-Ier av. n.è.)*: 229-53. Rapport de fouille, Inrap CIF.

Fromont, N. (ed.). 2009. *Saint-Sylvain, Rue Vilaine/ Chemin rural d'Argences (Calvados)*. Rapport de fouilles. Inrap GO.

Frontin, D. 2014. Les restes de poissons, in V. Riquier & J. Grisard (ed.) *Buchères, Moussey, Saint-Léger-près-Troyes (Aube), Parc Logistique de l'Aube. L'évolution d'un terroir dans la plaine de Troyes. I et 2: campagnes de fouille 2005 et 2006*, Rapport de fouille, Etudes spécialisées. Inrap GEN.

–. 2017. Les restes de poissons, in C. Laurelut (ed.) *Bréviandes (Aube) ZAC St Martin 1, les Pointes et les Grèvottes (OA 5115). Un village de la colonisation danubienne initiale à forte composante non rubanée. Occupations mésolithiques, sépultures collectives du Néolithique final, nécropole de l'âge du Bronze ancien-moyen, habitat RSFO-Hallstatt C*. Etudes spécialisées: 206-7. Rapport de fouille, Inrap GEN.

Gallay, A. & L. Chaix. 1984. Le site préhistorique du Petit Chasseur (Sion, Valais), 5. Le Dolmen M XI. Texte et planches. *Cahiers d'Archéologie Romande de la Bibliothèque Historique Vaudoise*: 1-182.

Gardin, J.-C. 1979. *Une archéologie théorique*. Paris: Hachette.

Gasnier, M., J. Wattez, P. Wuscher, J. Durand, S. Durand, L. Hachem, C. Monchablon & C. Verjux. 2014. Une succession d'occupations du Néolithique ancien à la fin du Néolithique sur le site de Limay, Rue Nationale (Yvelines): résultats préliminaires, in C. Louboutin & C. Verjux (ed.) *Colloque interrégional sur le Néolithique*, supplément 51: 241-52. Revue Archéologique du Centre de la France.

Gaudefroy, S. dir. 2000. Glisy, les Terres de Ville, ZAC de la Croix de Fer (Somme). L'occupation du premier âge du Fer et l'établissement agricole de La Tène moyenne. Rapport de fouille, Afan Centre-Nord.

Gaudefroy, S., F. Gransar & F. Malrain. 2013. La Picardie (Chapitre IV), in F. Malrain, G. Blancquaert, Th. Lorho (ed.) *L'habitat rural au second âge du Fer. Rythmes de création et d'abandon au nord de la Loire*. Recherches Archéologiques 7. Paris: INRAP/CNRS Editions.

Ghasarian, C. 1996. *Introduction à l'étude de la parenté*. Editions du Seuil.

Ghesquière, E. & L. Hachem. 2018. Place et rôle de l'animal dans la nécropole de Fleury-sur-Orne (Calvados), in S. Costamagno, L. Gourichon, C. Dupont, O. Dutour & D. Vialou (ed.) *Animal symbolisé – Animal exploité. Du Paléolithique à la Protohistoire*. Paris: Editions du Comité des Travaux Historiques et Scientifiques. Open Editions Books. https://books.openedition.org/ cths/4686.

Ghesquière, E., F. Charraud, L. Hachem, L. Manceau, C. Marcigny & H. Seignac. 2016. Le bâtiment 6 NMII de Saint-André-sur-Orne, la Delle du Poirier (Calvados), in *INTERNEO*, 11: 71-87. Internéo et Société Préhistorique Française,.

Ghesquière, E., P. Chambon, D. Giazzon, L. Hachem, C. Thevenet & A. Thomas. 2019. Monumental cemeteries of the 5th millenium BC: The Fleury-sur-Orne contribution, in H.-H. Müller, M. Hinz & M. Wunderlich (ed.) *Megaliths- Societies-Landscapes. Early Monumentality and Social Differentiation in Neolithic Europe*: 177-90 (Frühe Monumentalität and Soziale Differenzierung 18).

Ghesquière, E., F. Charraud, D. Giazzon, L. Hachem, L. Manceau, C. Marcigny, C. Mougne, C. Nicolas & H. Seignac. 2019b. Grands bâtiments du Néolithique final à Saint-André-sur-Orne (Calvados): 235-38. Paris: Editions du Comité des Travaux Historiques et Scientifiques.

Giligny, F. (ed.). 2006. *Le Néolithique des Yvelines* (Rapport Final de Projet Collectif de Recherches 2002-2006). Service Régional de l'Archéologie d'Ile-de-France, Service Archéologique Départemental des Yvelines, UMR 7041, INRAP.

–. 2005. *Louviers, la Villette. Un site Néolithique moyen en zone humide*. Document Archéologique de l'Ouest, Collection Documents Archéologiques.

Gillis R. E., Kovačiková L., Bréhard S., Guthmann E., Vostrovská I., Nohálová H., Arbogast R.-M., Domboróczki L., Pechtl J., Anders A., 2017 . The evolution of dual meat and milk cattle husbandry in Linearbandkeramik societies. *Proceedings of the Royal Society B: Biological Sciences* 284. The Royal Society: 20170905.

Girault L. 1975. Les fœtus animaux dans le rituel des indiens andins (Bolivie), in R. Pujol (ed.) *L'homme et l'animal*: 217-26. Paris: Institut international d'ethnosciences.

Godelier, M. 1982. *La production des grands hommes: pouvoir et domination masculine chez les Baruya de Nouvelle-Guinée* (L'Espace Du Politique). Paris: Fayard.

–. 1999. Chefferies et États, une approche anthropologique, in *Les princes de la Protohistoire et l'émergence de l'État*: 19-30. Publications de l'École Française de Rome. https://www.persee.fr/doc/efr_0223-5099_1999_act_252_1_6002.

–. 2010. Systèmes de parenté, formes de famille. Quelques problèmes contemporains qui se posent en Europe occidentale et en Euro-Amérique. *La Revue Lacanienne*: 37-48.

–. 2013. En guise de conclusion: Tribus, ethnies et États, in H. Dawod (ed.) *La constante «Tribu»: Variations arabo-musulmanes*: 245-71 (Quaero). Paris: Demopolis. http://books.openedition.org/demopolis/249.

Gomart, L. 2014. *Traditions techniques et production céramique au Néolithique ancien. Etude de huit sites rubanés du nord est de la France et de Belgique*. Sidestone Press.

Gomart, L. & M. Ilett. 2017. From potters' hands to settlement dynamics in the Early Neolithic site of Cuiry-lès-Chaudardes (Picardy, France). *Archeologické Rozhledy* 69: 209-26.

Gomart, L., L. Hachem, C. Hamon, F. Giligny & M. Ilett. 2015. Household integration in Neolithic villages: a new model for the Linear Pottery Culture in west-central Europe. *Journal of Anthropological archaeology* 40: 230-249. http://dx.doi.org/10.1016/j.jaa.2015.08.003.

Gonzalez-Villaescus, R.G. 2010. Problematique archéologique sur la production de laine et de étoffes en Gaule Belgique, in L.-P. Pujol (ed.) *Hispania et Gallia: dos provincias del Occidente romano*, 38: 125-43. Barcelone: Col.leccio Instrumenta.

Gosselin, R. & A. Samzun. 2008. Un dépôt associé à une sépulture de la fin du Néolithique ancien à Buthiers-Boulancourt (Seine-et-Marne, France). Approche tracéologique et techno-fonctionnelle du mobilier lithique. *Préhistoires Méditerranéennes*: 91-104.

Gouge, P. & J.-M. Séguier. 1994. L'habitat rural de l'âge du Fer en Bassée et à la confluence Seine-Yonne (Seine-et-Marne): un état des recherches, in *Les installations agricoles de l'âge du Fer en Ile-de-France*: 45-70. Paris: Rue d'Ulm. Presses de l'École Normale Supérieure.

Gouge, P., P. Chambon, P. Méniel & P. Pihuit. 1991. La nécropole de Marolles-sur-Seine, les Gours des Lions au Bronze final IIIb-Hallstatt ancien. *Bulletin du Groupement Archéologique de Seine-et-Marne*.

Granchon, P. (ed.). 2012. *Bailly, le Merisier ouest, le Crapaud. Bailly, liaison A86/A12, Yvelines*. Rapport de fouille, Inrap CIF.

Gransar, F. 2001. Le stockage alimentaire à l'âge du Fer en Europe tempérée. Paris: Thèse de Doctorat, Université de Paris, Panthéon-Sorbonne.

Gransar, F. & C. Pommepuy. 2005. Bazoches-sur-Vesle, les Chantraines (Aisne): présentation préliminaire de l'établissement rural aristocratique de La Tène D1, in G. Auxiette, F. Malrain (ed.). *Hommages à Claudine Pommepuy*. Numéro spécial 22: 193-216. Revue Archéologique de Picardie.

Gransar, F., S. Desenne, M. Gransar, G. Auxiette, V. Matterne & C. Pommepuy. 1997. *Sermoise, les Prés du Bout de la Ville (Aisne)*. Rapport de fouille, Afan Centre-Nord.

Gransar, F., G. Auxiette, S. Desenne, B. Hénon, P. Le Guen & C. Pommepuy. 1999. Essai de modélisation de l'organisation de l'habitat au cours des cinq derniers siècles avant notre ère dans la vallée de l'Aisne, in F. Braemer, S. Cleuziou, A. Coudart (ed.) *Habitat et Sociétés*: 419-38. Antibes: APDCA.

Gransar, F., G. Auxiette, S. Desenne, B. Hénon, F. Malrain, V. Matterne & E. Pinard. 2007. Expressions symboliques, manifestations rituelles et cultuelles en contexte domestique au Ier millénaire avant notre ère dans le Nord de la France, in Ph. Barral, A. Daubigney, C. Dunning & M.-J. Lambert (ed.) *L'âge du Fer dans l'arc jurassien et ses marges. Dépôts, lieux sacrés et territorialité à l'âge du Fer*: 549-64. Besançon: Presses Universitaires de Franche Comté.

Gronenborn, D. 2003. Ancestors or Chiefs? Comparing Social Archaeologies in Eastern North America and Temperate Europe. *Leadership and Polity in Mississippian Society. Center for archaeological investigations.-Occasional Paper*: 365-97.

–. 2009. Climate fluctuations and trajectories to complexity in the Neolithic: towards a theory. *Documenta praehistorica* 36: 97-110.

Gumerman, G. 1997. Food and complex societies. *Journal of Archaeological Method and Theory* 4: 105-39.

Guthmann, É. 2010. Signification des dépôts animaux dans les structures d'habitat et les fossés d'enceinte au Néolithique récent. Les cultures de Munzingen, Michelsberg et Münchshöfen (4400 – 3500 av. J.-C.). Master 2, Université de Strasbourg.

Guthmann, E. & R.-M. Arbogast. 2011. Des reliefs de banquets au Néolithique moyen? Les vestiges de faune des sites à enceinte cérémonielle de Duntzenheim et de Meistratzheim (Alsace), in A. Denaire, C. Jeunesse & P. Lefranc (Ed.) *Nécropoles et enceintes danubiennes du Ve millénaire dans le Nord-est de la France et le Sud-Ouest de l'Allemagne*, 29: 86-103. Strasbourg: UDS-MISHA.

Guthmann, É., P. Lefranc & R.-M. Arbogast. 2016. Un dépôt de renard roux (*Vulpes vulpes*) du 4e Millénaire av. J.-C. à Entzheim, les Terres de la Chapelle (Bas-Rhin): offrande ou sépulture animale? *Revue archéologique de l'Est*: 257-68.

Habermehl, K.H. 1961. *Die Alterbestimmung bei Haustieren, Pelztieren und beim jagdbaren Wild*. Berlin and Hamburg: Verlag Paul Parey.

Hachem, L. forthcoming. La faune des bâtiments, in F. Joseph (ed.) *Le site d'habitat du IIIe millénaire avant J.-C. de la ZAC Jules Verne à Glisy (Somme)*. Rapport de fouilles, Inrap CIF.

–. 1987. La faune de l'enceinte Michelsberg de Bazoches-sur-Vesle. Etude préliminaire, in *Fouilles Protohistoriques de la vallée de l'Aisne*: 135-46. Paris: Université de Paris I, Panthéon-Sorbonne.

–. 1989. La faune et l'industrie osseuse de l'enceinte Michelsberg de Maizy (Aisne): approche économique, spatiale et régionale. *Revue Archéologique de Picardie* 1: 67-108. https://doi.org/10.3406/pica.1989.1541.

–. 1995. La Faune rubanée de Cuiry-lès-Chaudardes (Aisne-France): essai sur la place de l'animal dans la première société néolithique du Bassin parisien. Thèse de Préhistoire-Ethnologie-Anthropologie, Paris: Paris I, Panthéon-Sorbonne.

–. 1997. Structuration spatiale d'un village du Rubané Récent, Cuiry-lès-Chaudardes (Aisne). Analyse d'une catégorie de rejets domestiques: la faune, in G. Auxiette, L. Hachem & B. Robert (ed.) *Espaces physiques, espaces sociaux, dans l'analyse interne des sites du Néolithique à l'Age du Fer*: 245-61. Editions du Comité des Travaux Historiques et Scientifiques.

–. 1999. Apport de l'archéozoologie à la connaissance de l'organisation villageoise rubanée, in S. Cleuziou, A. Coudart & F. Braemer (ed.) *Habitat et société*: 325-38. Antibes: APDCA.

–. 2000a. New observations on the Bandkeramik house and social organization. *Antiquity* 74: 308-312.

–. 2000b. La faune du site Néolithique moyen, in R. Cottiaux (ed.) *Le site de Maisons-Alfort, Zac d'Alfort (Val-de-Marne)*. Rapport de fouilles, Inrap CIF.

–. 2001. La conception du monde animal sauvage chez les éleveurs du Rubané Récent du Bassin parisien, in R.-M. Arbogast, C. Jeunesse & J. Schibler (ed.) *Rôle et statut de la chasse dans le Néolithique ancien danubien (5500-4900 av. J.C.)*: 91-111. VML, Rahden/Westf.

– (ed.). 2003. *Tinqueux, La Haubette (Marne)*. Rapport de fouilles. Inrap GE.

–. 2009. Elevage, chasse et société dans le Néolithique français: exemples dans le Danubien du Nord de la France, in J.-P. Demoule (ed.) *La révolution néolithique dans le monde*, INRAP/Universcience: 197-213. Editions du CNRS.

–. 2011a. *Le site néolithique de Cuiry-lès-Chaudardes – I: de l'analyse de la faune à la structuration sociale* (Internationale Archäologie 120). Rahden: Marie Leidorf GmbH.

–. 2011b. Les faunes du Néolithique moyen dans le Nord de la France: bilan et pistes de recherches, in F. Bostyn, E. Martial & I. Praud (ed.) *Habitat et économie aux 4e et 3e millénaires avant notre ère*, Numéro spécial 28: 313-329. Revue Archéologique de Picardie.
http://www.persee.fr/doc/pica_1272-6117_2011_hos_28_1_3338.

–. 2011c. La faune, in M. Gasnier (ed.) *Les occu-pations néolithiques du site de Limay, Rue Nationale, Yvelines.* Rapport de fouilles: 209-11. Inrap CIF.

–. 2013. Les restes archéozoologiques, in C. Marcigny & E. Ghesquière (ed.) *Une enceinte palissadée du Néolithique moyen 1, Le Diguet, Saint Martin de Fontenay (Calvados).* Rapport de fouilles: 132-38. Inrap GO.

–. 2014. La faune néolithique, in C. Monchablon (ed.) *Carvin, la Gare d'Eau, une enceinte du Néolithique moyen II,* Rapport de fouille: 101-14. Inrap NP.

–. 2015a. La faune, in V. Peltier (ed.) *Pont-sur-Seine, le Haut de Launoy, phase 3. De nouvelles données sur les enceintes palissadées au Néolithique récent: un système d'entrée complexe,* Rapport de fouilles: 242-47. Inrap GEN.

–. 2015b. La faune et l'industrie osseuse Cerny, in R. Peake (ed.) *La Saulsotte, le Vieux Bouchy (Aube).* Rapport de fouilles. Inédit.

–. 2015c. Consommation et dépôts dans l'enceinte Michelsberg de Crécy-sur-Serre (Aisne), in F. Bostyn & L. Hachem (ed.) *Hommages à Mariannick Le Bolloch,* 3-4: 60-80. Revue Archéologique de Picardie.

–. 2016. La faune, in N. Fromont (ed.) *Le Clos du Houx, Cuverville, (Calvados). Un monument funéraire prémégalithique du Néolithique moyen dans la Plaine de Caen.* Rapport de fouille: 97-98. Inrap GO.

–. 2017a. La faune et l'industrie osseuse du Bâtiment 1, Néolithique final, in E. Ghesquière (ed.) *Saint-André-sur-Orne, la Delle du Poirier (Calvados). Grands bâtiments néolithiques.* Rapport de fouilles: 185-91. Inrap GO.

–. 2017b. La parure en matière dure animale, in B. Hénon (ed.) *Sépulture néolithique et habitats protohistoriques. Beaurieux, les Grèves (Aisne), tranche 1.* Rapport de fouilles: 93-95. Inrap NP.

–. 2018a. Animals in LBK society: Identity and gender markers. *Journal of Archaeological Science: Reports* 20: 910-21. https://doi.org/10.1016/j.jasrep.2017.09.020.

–. 2018b. La faune, in C. Colas (ed.) *Les ensembles monumentaux du Néolithique moyen de Beaurieux, la Plaine,* 3-4: 79-84. Revue Archéologique de Picardie.

–. 2019. La faune néolithique des D39 et D40, in C. Paresys (Ed.) Buchères, les Terriers (Parc Logistique de l'Aube, Aube). L'évolution d'un terroir dans la Plaine de Troyes (V: campagne de fouilles 2012-2013). Vol 2. Etudes spécialisées. Rapport de Fouille, Inrap GEN.

–. 2020a. Faune et objets en os néolithiques de Villiers-sur-Seine, le Défendable, in J.-P. Quenez (ed.) *Villiers-sur-Seine, le Défendable (Seine-et-Marne).* Rapport de fouilles, Inrap CIF.

–. 2020b. Les carnivores de la sépulture collective de Chamigny, in N. Mahé-Hourlier (ed.) *Chamigny (Seine-et-Marne), rue de la Marne – RD 80 – Lieu-dit la Grande Maison, Vol. 1: Les occupations néolithiques.* Rapport de fouilles, Inrap CIF

Hachem, L. & G. Auxiette. 2006. Une histoire des bovinés durant les cinq millénaires précédant notre ère: l'exemple de la vallée de l'Aisne et de la Vesle (France), in *Les bovins: de la domestication à l'élevage,* 79: 127-35. Paris, 2006: *Ethnozootechnie.*

Hachem, L. & A. Bandelli. 2015. La faune des struc-tures 1144 et 1154 de Ramerupt, Cour Première, in S. Thiol (ed.) *Ramerupt, Cour Première (Aube). Occupations néolithiques, enclos funéraires protohistoriques et nécropole tardo-antique.* Rapport de fouilles: 67-70. Inrap GEN.

Hachem, L. & L. Bedault. 2008. Recherches sur les sociétés du Néolithique danubien à partir du Bassin parisien: approche structurelle des données archéozoologiques, in L. Burnez-Lanotte, M. Ilett & P. Allard (ed.) *Fin des traditions danubiennes dans le Néolithique du Bassin parisien et de la Belgique (5100-4700 BC) – Autour des recherches de Claude Constantin:* 222-43. Mémoire 44. Société Préhistorique Française.

–. 2014. La faune néolithique de Villers-Carbonnel, la Sole d'Applincourt, in F. Bostyn (ed.) *Des systèmes d'enceintes au Néolithique moyen II. Canal Seine-Nord, fouille 12, Picardie, Somme, St Christ-Briost et Villers-Carbonnel.* Rapport de fouille: 187-200. Inrap NP.

Hachem, L. & J. Chombart. forthcoming. L'élevage et la chasse, in I. Praud & E. Panloups (ed.) *Escalles, Mont d'Hubert: une enceinte du Néolithique moyen II, des fosses du Néolithique moyen I et du Bronze final sur le littoral de la Mer du Nord.*

–. 2014. Etude de la faune sauvage et domestique du Néolithique, in I. Praud (ed.) *Une enceinte du Néolithique moyen II, des fosses du Néolithique moyen I et du Bronze final sur le littoral de la Mer du Nord. Escalles, Mont d'Hubert, Nord-Pas-de-Calais*. Rapport de fouille: 163-214. Inrap NP.

Hachem, L. & J. Durand. 2018. Un aurochs du Bronze final à Palaiseau (Essonne, France) dans une fosse aménagée. *Revue Archéologique d'Ile de France*: 23-38.

Hachem, L. & C. Hamon. 2014. Linear Pottery Culture Household Organisation. An Economic Model, in A. Whittle & P. Bickel (ed.) *Early farmers, The view from Archaeology and Science*: 159-80 (Proceedings of the British Academy 198). Oxford University Press.

Hachem, L. & Y. Maigrot. 2019. Faune et travail des matières osseuses dans les enceintes du Néolithique Moyen II du Bassin parisien, in F. Bostyn, C. Hamon, A. Salavert & F. Giligny (ed.) *L'exploitation du milieu au Néolithique dans le quart nord-ouest de l'Europe: contraintes environnementales, identités techniques et choix culturels*: 95-117. Société Préhistorique Française.

Hachem, L. & T.D. Price. forthcoming. La mobilité dans la société Blicquy/Villeneuve-Saint-Germain, in F. Bostyn & Y. Lanchon (ed.) *Le Néolithique ancien de la Basse vallée de la Marne. Evolution des modalités d'occupation et d'exploitation d'un territoire au début du Ve millénaire BC*. Bulletin de la Société Préhistorique Française.

Hachem, L., M. Ilett, S. Bonnardin, M. Chartier, C. Constantin, J. Dubouloz, J.-P. Farruggia, C. Monchablon & I. Sidéra. 1997. *Le site néolithique de Bucy-le-Long la Fosselle (Aisne)*. Rapport de fouilles, Inrap NP.

Hachem, L., P. Allard, C. Constantin, J.-P. Farruggia, Y. Guichard & M. Ilett. 1998a. Le site néolithique rubané de Bucy-le-Long, la Fosselle (Aisne), in *Internéo,2*: 17-22.

Hachem, L., Y. Guichard, J.-P. Farruggia, J. Dubouloz & M. Ilett. 1998b. Enclosure and burial in the earliest Neolithic of the Aisne valley, in M. Edmonds & C. Richards (ed.) *Social Life and Social Change: the Neolithic Western Europe*: 127-40. Glasgow: Cruithne Press.

Hachem, L., P. Allard, N. Fromont, C. Hamon, K. Meunier, V. Peltier & J.-M. Pernaud. 2007. Le site Villeneuve-Saint-Germain de Tinqueux, la Haubette (Marne, France) dans son contexte régional, in F. Le Brun-Ricalens, F. Valotteau & A. Hauzeur (ed.) *Relations interrégionales au Néolithique entre Bassin parisien et Bassin rhénan*, 7: 229-73. Luxembourg: Musée National d'Histoire et d'Art.

Hachem, L., L. Bedault & C. Leduc. 2016. L'élevage et la chasse au Chasséen septentrional: renouvellement des connaissances d'après l'étude des enceintes de Villers-Carbonnel (Somme) et de Passel (Oise), in T. Perrin, P. Chambon, J. Gibaja Bao & G. Goude (ed.) *Le Chasséen, des Chasséens… Retour sur une culture nationale et ses parallèles, Sepulcres de fossa, Cortaillod, Lagozza*: 241-58. Toulouse: Archives d'Écologie Préhistorique.

–. 2017. La faune du Néolithique moyen de Passel, in N. Cayol (ed.) *Passel, le Vivier, Oise, Une enceinte du Néolithique moyen II*. Rapport de fouille: 351-87. Inrap NP.

Halstead P. & Jones G. 1989. Agrarian ecology in the Greek Islands: time stress, scale and risk. *The Journal of Hellenic Studies*: 41-56.

Hamayon, R. 1990. *La chasse à l'âme: esquisse d'une théorie du chamanisme sibérien*. Vol. 1. Mémoire de la Société d'Ethnologie.

Hamon, C. 2006. *Broyage et abrasion au Néolithique ancien. Caractérisation technique et fonctionnelle des outillages en grès du Bassin parisien*. British Archaeological Reports. Vol. 1551 (BAR International Series). Oxford.

Hamon, C. 2008. Functional analysis of stone grinding and polishing tools from the earliest Neolithic of north-western Europe. *Journal of Archaeological Science* 35: 1502-1520.

Hamon, C. & P. Allard (ed.). 2010. *Economie et société des populations rubanées de la vallée de l'Aisne*. Rapport final d'activité 2007-2010. Picardie: Service Régional de l'Archéologie.

Haselgrove, C. & P. Lowther. 1992. *Damary, le Ruisseau de Fayau (Aisne)*. Rapport de fouille programmée, Service Régional de l'Archéologie, Amiens.

Haudenosaunee Confederacy. 2016 *http://www.haudenosauneeconfederacy.com/culture.html*.

Hayden, B. 2014. *The Power of Feasts. From History to the Present*. Cambridge: Cambridge Press University.

Hell, B. 2012. *Sang noir. Chasse, forêt et mythe de l'homme sauvage en Europe*. Paris: L'Oeil d'or, Essais et Entretiens.

Hénon, B., G. Auxiette, M. Boulen, S. Desenne, F. Gransar, P. Le Guen, C. Pommepuy & B. Robert. 2002. Trois nouveaux sites d'habitat du Hallstatt final/La Tène ancienne dans la vallée de l'Aisne, in P. Méniel & B. Lambot (ed.) *Repas des vivants et nourriture pour les morts en Gaule*: 49-66. Mémoire 16. Revue Archéologique Champenoise.

Hénon, B., G. Auxiette, S. Bauvais, F. Gransar, V. Legros, C. Monchablon & V. Pissot. 2012. Villeneuve-Saint-Germain, les Etomelles (Aisne): huit siècles d'occupation (IVe s. avant J.-C. – IVe s. après J.-C.). *Revue Archéologique de Picardie* 3-4: 47-205. https://doi.org/10.3406/pica.2012.3281.

Héritier, F. 1984. Le sang du guerrier et le sang des femmes. *Les cahiers du GRIF* 29: 7-21.

https://doi.org/10.3406/grif.1984.1629.

Hermetey, C. 1994. La faune de la structure 126 à Villemaur-sur-Vanne, les Gossements (Aube, âge du Bronze final IIIb). Paris: Mémoire de Maîtrise, Université de Paris 1 Panthéon Sorbonne.

–. 1995. Potentialités informatives des petits assemblages de faune de l'âge du Bronze: l'exemple de la Bassée, Seine-et-Marne (IXème-VIIème siècle avant J.-C.). Paris: Mémoire de DEA, Environnement et Archéologie, Université de Paris 1 Panthéon Sorbonne.

Higgs, E.S. & C. Vita-Finzi. 1972. Prehistoric Communities: A Territorial Approach. *Papers in Economic Geography*, 30.

Hill, J.D. 1995. *Ritual and rubbish in the Iron Age of Wessex: a study of the formation of a specific archaeological record*. Vol. 242. BAR British Series.

Hodder, I. 1979. Economic and social stress and material culture patterning. *American Antiquity*, 446-54.

Hofmann, D. 2012. Bodies, houses and status in the western Linearbandkeramik, in T.L. Kienlin & A. Zimmerman (ed.) *Beyond Elites: Alternatives to Hierarchical Systems in Modelling Social Formations*, 215: 183-96 (Universitätsforschungen Zur Prähistorischen Archäologie). Bonn: Verlag Dr. Rudolph Habelt GMBH.

Höltkemeier, S. 2010. Les dépôts de faune dans les enceintes de la Culture de Michelsberg du Nord de la France et de l'Allemagne. Paris: Mémoire de Master 1, Université de Paris 1 Panthéon Sorbonne.

–. 2013. Les dépôts de faune dans les enceintes néolithiques Michelsberg dans le Nord de la France et en Allemagne, in G. Auxiette & P. Méniel (ed.) *Les dépôts d'ossements d'animaux en France, de la fouille à l'interprétation*: 177-89 (Archéologie Des Plantes et Des Animaux, volume 4). Montagnac: Editions Monique Mergoil.

–. 2016. L'exploitation animale au IVe millénaire avant notre ère en Allemagne centrale: les sites de Wallendorf et Salzmünde dans leur contexte régional. Paris: Thèse de doctorat en Archéologie, Ethnologie, Préhistoire, Université de Paris 1 Panthéon Sorbonne.

Horard-Herbin, M.-P., P. Méniel & J.-M. Séguier. 2000. La faune de dix sites ruraux de la fin de l'âge du Fer de La Bassée (Seine-et-Marne)., in S. Marion, G. Blancquaert (ed.) *Les installations agricoles à l'âge du Fer en France septentrionale*, 6: 181-208 (Etudes d'Histoire et d'Archéologie). Paris: Presses de l'École Normale Supérieure.

Husmann, H. & E. Cziesla. 2014. Bandkeramische Häuser, Brunnen und ein Erdwerk. *Autobahn A4. Fundplatz der Extraklasse-Archäologie unter der neuen Bundesautobahn bei Arnoldsweiler*, 71-118.

Ilett, M. 2010. Le Néolithique ancien dans le nord de la France, in Clottes, J. (ed.) *La France préhistorique, essai d'histoire*: 281-307. Paris: Gallimard.

–. 2012. Linear Pottery and Blicquy/Villeneuve-Saint-Germain settlement in the Aisne valley and its environs: an overview, in S. Wolfram & H. Stäuble (ed.) *Siedlungsstrukturen und Kulturwandel in der Bandkeramik*, 25: 69-79.

Ilett, M. & L. Hachem. 2001. Le village néolithique de Cuiry-lès-Chaudardes (Aisne), in *Communautés villageoises du Proche-Orient à l'Atlantique, 8000-2000 avant notre ère*, Errance: 171-184 (Collection Des Hespérides). Séminaire du Collège de France.

Ilett, M. & K. Meunier. 2013. Avant-propos –
Chronologie du Rubané dans le Bassin parisien.
Bulletin de la Société Préhistorique Française
110: 415-20.

Ilett, M. & M. Plateaux (ed.). 1995. *Le site néolithique
de Berry-au-Bac, le Chemin de la Pêcherie (Aisne)*.
Monographie du CRA 15. CNRS Editions.

Ilett, M., C. Constantin, J.-P. Farruggia & C. Bakels.
1995. Bâtiments voisins du Rubané et du groupe
de Villeneuve-Saint-Germain sur le site de
Bucy-le-Long, la Fosse Tounise (Aisne). *Revue
Archéologique de Picardie*: 17-39.

Ilett, M., L. Hachem & A. Coudart (ed.). 2006.
*L'implantation du Néolithique rubané dans la
vallée de l'Aisne*. Projet d'Action Collective de
Recherche (ACR), rapports 2003, 2005, 2006.
France: Service régional de l'Archéologie de
Nord-Picardie.

Issenmann, R. 2009. Hiérarchie des habitats et
formes de l'occupation du sol à la confluence
Seine-Yonne à la transition entre le premier et
le second âge du Fer, in I. Bertrand, A. Duval,
J. Gomez de Soto, P. Maguer (ed.) *Habitats et
paysages ruraux en Gaule et regards sur d'autres
régions du monde celtique*: 519-27. Chauvigny:
Mémoire 32, 2, Association des Publications
Chauvinoises.

Jeunesse, C. 1993. La nécropole rubanée d'Ensisheim, les Octrois. La parure. *Association pour
la Promotion de la Recherche Archéologique en
Alsace* 9: 59-72.

Jones, G.G. & P. Sadler. 2012. Age at death in cattle:
methods, older cattle and known-age reference
material. *Environmental Archaeology*: 11-28.

Jordhøy P.,. 2008. Ancient wild reindeer pitfall
trapping systems as indicators for former
migration patterns and habitat use in the Dovre
region, southern Norway. *Rangifer* 28: 79-87.

Joseph, F. 2008. Le site d'habitat du IIIe millénaire
avant J.-C. de la ZAC Jules Vernes à Glisy
(Somme): présentation préliminaire, *Internéo et
Société Préhistorique Française, 22*: 163-72.

Joseph, F., M. Julien, E. Leroy-Langelin, Y. Lorin & I.
Praud. 2011. L'architecture domestique des sites
du IIIe millénaire avant notre ère dans le Nord
de la France, in F. Bostyn, E. Martial & I. Praud
(ed.) *Le Néolithique du Nord de la France dans
son contexte européen: habitat et économie aux
4e et 3e millénaires avant notre ère*, n° spécial
28: 249-73. Revue Archéologie de Picardie.

Jouanin, G. & O. Robin. 2010. Etude archéozoologique du site de Villiers-le-Bel/Gonesse.
Déviation RD 10-370. Période protohistorique,
La Tène moyenne/La Tène finale. Rapport
d'étude archéozoologique. Compiègne: CRAVO.

Jouanin, G. & C. Touquet Laporte-Cassagne. 2013.
Sacrifices et repas communautaires sur le site
du Mesnil-Aubry, le Bois Bouchard IV (Val-
d'Oise), in G. Auxiette, P. Méniel (ed.) *Les dépôts
d'animaux en France, de la fouille à l'interprétation*: 69-76 (Archéologie Des Plantes et
Des Animaux, Volume 4). Montagnac: Éditions
Monique Mergoil.

Kanafani-Zahar, A. 2007. Chrétiens et druzes du
Mont Liban: la rupture du "partage du pain
et du sel". Mémoire de l'inimaginable, in A.
Kanafani-Zahar, S. Mathieu, S. Nizard (ed.) *A
croire et à manger. Religions et alimentation*:
143-66. Paris: L'Harmattan.

Kienlin, T.L. 2012. Beyond elites: an introduction, in
T.L. Kienlin & A. Zimmerman (ed.) *International
Conference at the Ruhr-Universität Bochum,
Germany, October 22-24, 2009*, 215: 15-32
(Universitätsforschungen Zur Prähistorischen
Archäologie). Bonn: Verlag Dr. Rudolph Habelt
GMBH.

Kloiber, von, A. & J. Kneidinger. 1970. Die
neolithische Siedlung und die neolithischen
Gräberfundplätze von Rutzing und Haid,
Ortsgemeinde Hörsching, politischer Bezirk
Linz-Land, Oberösterreich. *Jahrbuch des
Oberösterreichischen Musealvereines* 115: 21-36.

Kopytoff I. 1988. The cultural biography of things:
commoditization as process, in Appadurai
A. (ed.) *The social life of things: commodities
in cultural perspective*: 64-99. Cambridge:
Cambridge University Press.

Květina, P. 2010. The spatial analysis of non-ceramic
refuse from the Neolithic site at Bylany, Czech
Republic. *European Journal of Archaeology* 13.
Cambridge University Press: 336-67.

Květina, P. & M. Končelová. 2013. Settlements
patterns as seen in pottery decoration style:
a case study from the early Neolithic site of
Bylany (Czech Republic), in C. Hamon, P. Allard
& M. Ilett (ed.) *The domestic space in LBK*

settlements: 99-110 (Internationale Archäologie). Rahden, Westf: Verlag Marie Leidorf GmbH.

Labriffe, P.-A. 1986. Les sépultures danubiennes dans le Bassin parisien. Paris: Mémoire de Maîtrise, Université de Paris 1 Panthéon Sorbonne.

Lafage, F. (ed.). forthcoming. *Changis-sur-Marne, les Pétreaux (Seine-et-Marne)*. co-édition INRAP/ CNRS.

Lafage, F., G. Auxiette, P. Brunet, E. Martial, V. Matterne, I. Praud & N. Laplantine. 2006. Premières tentatives d'interprétation spatiale d'un site rural du Bronze final à Changis-sur-Marne (Seine-et-Marne). *Bulletin de la Société Préhistorique Française*, 323-77.

Lafage, F., G. Auxiette, P. Brunet, V. Delattre, Y. Le Jeune, E. Martial, V. Matterne & d'Ivan PRAUD. 2007. Changis-sur-Marne, les Pétreaux: trois siècles d'évolution d'établissements ruraux de la fin du Bronze final au début du premier âge du Fer. *Bulletin de la Société Préhistorique Française*, 307-41.

Lanchon, Y. (ed.). 2006. *Action Collective de Recherche: le Néolithique ancien dans la basse vallée de la Marne*. Rapport d'activité, 2e année (état des travaux au 31 décembre 2005). Inrap CIF.

–. 2008. La Culture de Blicquy/Villeneuve-Saint-Germain dans la basse vallée de la Marne: première approche chronologique à partir de la céramique, in L. Burnez-Lanotte, M. Ilett & P. Allard (ed.) *Fin des traditions danubiennes dans le Néolithique du Bassin parisien et de la Belgique (5100-4700 BC) – Autour des recherches de Claude Constantin*: 143-59 (Mémoire XLIV). Société Préhistorique Française.

–. 2012. Le Néolithique ancien dans la basse vallée de la Marne: premières réflexions sur les sites, leur statut et leur organisation territoriale, in V. Carpentier & C. Marcigny (ed.) *Des Hommes aux Champs. Pour une archéologie des espaces ruraux du Néolithique au Moyen Age*: 35-54 (Archéologie et Culture). Presses Universitaires de Rennes.

Lanchon, Y. & P. Marquis. 2000. *Le premier village de Paris il y a 6000 ans: les découvertes archéologiques de Bercy*. Paris: Editions Paris-Musées.

Lanchon, Y., F. Bostyn & L. Hachem. 1997. L'étude d'un niveau archéologique néolithique et ses apports à la compréhension d'un site d'habitat: l'exemple de Jablines, la Pente de Croupeton (Seine-et-Marne), in G. Auxiette, L. Hachem & B. Robert (ed.) *Espaces physiques, espaces sociaux, dans l'analyse interne des sites du Néolithique à l'âge du Fer*: 447-66. Paris: Editions du Comité des Travaux Historiques et Scientifiques.

Lanchon, Y., Brunet Paul, V. Brunet & P. Chambon. 2001. Le site néolithique de Vignely, la Noue Fenard (Seine-et-Marne), in *Journées archéologiques d'Ile-de-France*: 64-77. Service Régional de l'Archéologie d'Ile-de-France.

Lanchon, Y., P. Brunet, V. Brunet & P. Chambon. 2006. Fouille de sauvetage d'un monument funéraire et d'une enceinte néolithiques à Vignely, la Noue Fenard (Seine-et-Marne): premiers résultats, in *Colloque sur le Néolithique 2001*, supplément 25: 335-51. Revue Archéologique de l'Est.

Lanchon, Y., F. Bostyn, L. Hachem, Y. Maigrot & E. Martial. 2008. Le Néolithique ancien dans la basse vallée de la Marne: l'habitat de Changis-sur-Marne, les Pétreaux (Seine-et-Marne). *Revue archéologique d'Ile-de-France* 1: 43-94.

Lanchon, Y. *et al.* 2013. *Un hameau du Néolithique ancien, le Pré des Bateaux à Luzancy (Seine-et-Marne)*. Recherches archéologiques 6. Paris: INRAP/ CNRS Editions. https://hal-inrap. archives-ouvertes.fr/hal-02063514.

Last, J. 1998. The residue of yesterday's existence: settlement space and discard at Miskovice and Bylany. *Bylany Varia*, 17-46.

Latour, B. & P. Lemonnier. 1994. *De la Préhistoire aux missiles balistiques. L'intelligence sociale des techniques*. La Découverte.

Laurelut, C. 2010. Bréviandes (Aube), un site danubien à forte composante "non-rubanée" dans la région de Troyes. Premiers éléments de réflexion, in C. Billard & M. Legris (ed.) *Premiers néolithiques de l'Ouest. Cultures, réseaux, échanges des premières sociétés néolithiques à leur expansion*: 291-304. (Collection Archéologie et Culture). Rennes: Presses Universitaires de Rennes.

Laurelut, C. (ed.). 2017. *Bréviandes (Aube) ZAC St Martin 1, les Pointes & les Grèvottes. Un village de la colonisation danubienne initiale à forte composante non rubanée. Occupations mésolithiques, sépultures collectives Néolithique final, nécropole Bronze ancien-moyen, habitat RSFO-Hallstatt.* Rapport de fouilles. Inrap GEN.

Le Bolloch, M., J. Dubouloz & M. Plateaux. 1986. Sauvetage archéologique à Maizy (Aisne): les sépultures rubanées et l'enceinte de la fin du IVe millénaire. *Revue Archéologique de Picardie*: 3-12.

Le Goff, E. 2000. Les occupations protohistoriques et antiques de la Zac Object'Ifs Sud, Ifs (Calvados). Rapport de fouille, Inrap GO.

–. 2009. Habitat, terroir et paysage rural: aménagements et structuration du territoire et de la campagne gauloise, Ifs, ZAC Object'Ifs Sud (Calvados), in I. Bertrand, A. Duval, J. Gomez de Soto, P. Maguer (ed.) *Habitats et paysans ruraux en Gaule et regards sur d'autres régions du monde celtique*: 93-107 (Mémoire XXXV). Chauvigny: Association des Publications Chauvinoises.

Le Goff, E., G. Auxiette & I. Le Goff. 2007. Manifestations et pratiques cultuelles au sein des habitats et du territoire agraire laténien de la Z.A.C. Object'Ifs Sud à Ifs (Calvados): un exemple de l'ouest de la Gaule., in Ph. Barral, A. Daubigney, C. Dunning, & M.-J. Lambert (ed.) *L'âge du Fer dans l'arc jurassien et ses marges. Dépôts, lieux sacrés et territorialité à l'âge du Fer*: 566-78. Besançon: Presses Universitaires de Franche-Comté.

Le Guen, P., G. Auxiette, P. Brun, J. Dubouloz, F. Gransar & C. Pommepuy. 2005. Apport récent sur la transition âge du Bronze-âge du Fer dans la vallée de l'Aisne, Osly-Courtil, in G. Auxiette & F. Malrain (ed.) *Hommages à Claudine Pommepuy*. Numéro spécial 22: 141-61. Revue Archéologique de Picardie.

Lefranc, P. 2007. *La céramique du Rubané en Alsace: contribution à l'étude des groupes régionaux du Néolithique ancien dans la plaine du Rhin supérieur*. Monographie Arch. Grand Est 2, Université Marc-Bloch.

Lefranc, P., R.-M. Arbogast, F. Chenal, E. HILDBRAND, M. MERKL, C. Strahm, S. Van Willigen & M. WÖRLE. 2012. Inhumations, dépôts d'animaux et perles en cuivre du IV e millénaire sur le site du Néolithique récent de Colmar, Aérodrome (Haut-Rhin). *Bulletin de la Société Préhistorique Française*. JSTOR: 689-730.

Lemonnier, P. 2010. L'Étude des systèmes techniques. Une urgence en technologie culturelle. *Techniques et Culture. Revue semestrielle d'anthropologie des techniques*: 46-67.

–. 2015. Anthropologie des objets ordinaires: faire, faire, faire et faire penser, nouveaux regards sur les techniques, in *Conférences du Musée du Quai Branly*.

Lepaumier, H., C.-C. Vauterin, E. Le Goff & J. Villaregut. 2010. Un réseau de fermes en périphérie caennaise, in Ph. Barral, B. Dedet, F. Delrieu, P. Giraud, I. Le Goff, S. Marion et A. Villard-Le-Tiec (ed.) *L'âge du Fer en Basse-Normandie-Gestes funéraires en Gaule au second âge du Fer*, 20: 139-58. Caen: Presses Universitaires de Franche-Comté.

Lepetz, S. & W. Van Andringa. 2004. Caractériser les rituels alimentaires dans les nécropoles gallo-romaines: l'apport conjoint des os et des textes, in L. Baray (ed.) *Archéologie des pratiques funéraires. Approches critiques*, 9: 161-70. Glux-en-Glenne: Bibracte.

Leroi-Gourhan, A. 1964. *Le Geste et la Parole*. Editions Albin Michel (Tome 1: Technique et Langage).

Leroyer, C. 1997. Homme, climat, végétation au tardi-et postglaciaire dans le Bassin parisien: apports de l'étude palynologique des fonds de vallée. Paris: Thèse de Doctorat, Université de Paris 1 Panthéon Sorbonne.

Lichardus, J. & M. Lichardus-Itten. 1985. *La Protohistoire de l'Europe: le Néolithique et le Chalcolithique entre la Méditerranée et la mer Baltique* (Nouvelle Clio 1 bis). Paris: Presses universitaires de France.

Liétar, C. 2017. *Territoires et ressources des sociétés néolithiques du Bassin parisien: le cas du Néolithique moyen (4500 – 3800 av. n. è.)*. Oxford: Archaeopress Archaeology.

Liétar, C. & F. Giligny. 2016. Territoire des géomatériaux. Occupations, environnement et ressources minérales dans les vallées de l'Aisne et de l'Oise. *Les Nouvelles de l'Archéologie*. Editions de la Maison des Sciences de l'Homme: 14-19.

Lorin, Y., G. Auxiette & S. Loicq. 2013. Les fosses à profil en V et Y dans le Nord-Pas-de-Calais, in N. Achard-Corompt, V. Riquier (ed.) *Chasse, culte ou artisanat? Les fosses "à profil en Y-V-W". Structures énigmatiques et récurrentes du Néolithique aux âges des Métaux en France et alentour*: 175-90. Dijon: Revue Archéologique de l'Est, 33è supplément.

Lot-Falck, E. 1953. *Les rites de chasse chez les peuples sibériens*. Vol. 9 (L'espèce Humaine). NRF Gallimard.

Louboutin, C. & D. Simonin. 1997. Le Cerny-Videlles: un faciès ancien de la Culture de Cerny, in *La Culture de Cerny: Nouvelle économie, nouvelle société au Néolithique*, 6: 135-37. Nemours, France: Mémoires du Musée de Préhistoire d'Ile de France, Editions A.P.RA.I.F.

Lüning, J. 1982. Research into the Bandkeramik settlement of the Aldenhovener Platte in the Rhineland. *Analecta Praehistorica Leidensia* 15: 1-29.

–. 1988. Frühe Bauern in Mitteleuropa im 6. und 5. Jahrtausen v. Ch. *Jahrbuch des Römisch-Germanischen Zentralmuseums Mainz* 35: 27-93.

Lüning, J. 1998. L'organisation régionale des habitats rubanés: sites centraux et sites secondaires (groupements de sites). *Anthropologie et Préhistoire* 109: 163-85.

Lüning, J. 2005. Bandkeramische Hofplätze und absolute Chronologie der Bandkeramik., in J. Lüning, C. Frirdich & A. Zimmerman (ed.) *Die Bandkeramik im 21 Jahrhundert*: 49-74. Rhaden/Westf: VML, Verlag Marie Leidorf.

Madgwick, R. & J. Mulville. 2015. Feasting on forelimbs: conspicuous consumption and identity in later prehistoric Britain. *Antiquity* 89: 629-44.

Magny R. & Richard H. 1992. Essai de synthèse vers une courbe de l'évolution du climat entre 500 BC et 500 AD. *Les Nouvelles de l'Archéologie*: 58-60.

Maigrot, Y. 2013. L'industrie en matière dure animale, in C. Thevenet (ed.) *L'enceinte néolithique de Menneville, Derrière le Village (Aisne)*, Rapport de fouilles programmées: 46. Picardie: Service Régional de l'Archéologie.

–. 2017. L'outillage en matières dures animales de Passel, le Vivier (Oise), in N. Cayol (ed.) *L'enceinte Néolithique moyen de Passel, le Vivier (Oise)*. Rapport de fouilles: 190-204. Inrap NP.

–. 2019. L'outillage en matières dures animales de la nécropole et de l'enceinte du Néolithique moyen de Pont-sur-Seine, Ferme de l'Ile (Aube), in F. Dugois and S. Loiseau (Ed.) *Pont-sur-Seine, Aube, Ferme de l'Ile. Exploitation d'un territoire en bord de Seine: de l'enceinte monumentale du Néolithique moyen II à la ferme fossoyée médiévale*. Rapport de fouille: 501-15. Inrap GEN.

Maingaud, A. 2003. L'industrie en matières dures d'origine animale de la fin du 4e et du 3e millénaires avant J.-C. de la collection de Baye. *Antiquités Nationales (Saint-Germain-en-Laye)*: 55-82.

Malrain, F. 2000. Fonctionnement et hiérarchies des fermes dans la société gauloise du IIIe siècle à la période romaine: l'apport des sites de la moyenne vallée de l'Oise. Paris: Thèse de Doctorat, Université de Paris 1 Panthéon Sorbonne.

Malrain, F. & E. Pinard dir. 2006. *Les sites laténiens de la moyenne vallée de l'Oise du Ve au Ier s. avant notre ère: contributions à l'histoire de la société gauloise*, Numéro spécial 23. Revue Archéologique de Picardie.

Malrain, F., F. Gransar, V. Matterne & I. Le Goff. 1996. Une ferme de La Tène Dl et sa nécropole: Jaux, le Camp du Roi (Oise). *Revue Archéologique de Picardie*: 245-306.

Malrain, F., P. Méniel & V. Matterne. 2002. *Les paysans gaulois (IIIe siècle-52 av. J.-C.)* (Les Hespérides). Errance.

Malrain, F., G. Blanquaert & T. Lorho. 2009. Un enclos=une ferme?, in I. Bertrand, A. Duval, J. Gomez de Soto, P. Maguer (ed.) *Habitats et paysages ruraux en Gaule et regards sur d'autres régions du monde celtique*, 35: 25-43. Chauvigny: Association des Publications Chauvinoises.

Malrain, F., G. Blancquaert & T. Lorho. 2013. *L'habitat rural au second âge du Fer. Rythmes de création et d'abandon au nord de la Loire*. Recherches Archéologiques 7. Paris: I, INRAP/CNRS Editions.

Malrain, F., V. Zech-Matterne & G. et coll. Blancquaert. 2015. Apprehending Continuity an Discontinuity in Iron Age Soil Occupation and Rural Landscape through a Collective Database, in A. Danielisova & M. Fernandez-Götz (ed.) *Persistent economic ways* of living. *Production, Distribution, Consumption in Late Prehistory and Early History*: 137-44. Budapest: Archaeolingua Alapitvany.

Manolakakis, L. & F. Giligny. 2011. Territories and lithic resources in the Paris Basin during the Middle Neolithic (4200-3600 BC), in *Proceedings of the 2nd International Conference of the UISPP Commission on Flint Mining in Pre-and Protohistoric Times*: 45-50. Hadrian Books Ltd, Archeopress, British Archaeological Reports.

Marcigny, C. & M. Talon. 2009. Sur les rives de la Manche. Qu'en est-il du passage de l'âge du Bronze à l'âge du Fer à partir des découvertes récentes, in M.-J. Lambert, A. Daubigney, P.-Y. Milcent, M. Talon, J. Vital (ed.) *De l'âge du Bronze à l'âge du Fer en France et en Europe occidentale*: 385-404. 27è supplément, Revue Archéologique de l'Est.

Marcigny, C., E. Ghesquière & L. Lespez. 2008. Espace rural et systèmes agraires dans l'Ouest de la France à l'âge du Bronze: quelques exemples normands, in J. Guilaine (ed.) *Villes, villages, campagnes de l'âge du Bronze*: 256-79 (Séminaire Du Collège de France). Paris: Errance.

Marcigny, C., E. Ghesquiere, L. Juhel & F. Charraud. 2010. Entre Néolithique ancien et Néolithique moyen en Normandie et dans les îles anglo-normandes. Parcours chronologique, in *Premiers Néolithiques de l'Ouest. Cultures, réseaux, échanges des premières sociétés néolithiques à leur expansion*: 117-62. Presse Universitaire de Rennes.

Marciniak, A. 2005. *Placing Animals in the Neolithic: Social Zooarchaeology of Prehistoric Farming Communities*. London: UCL Press.

–. 2008. Communities, households and animals. Convergent developments in central Anatolian and central European Neolithic. *Documenta Praehistorica* 35: 93-109.

Marti, F. 2013. Les fosses profondes du Néolithique, in B. Souffi (ed.) *Neuville-sur-Oise, les occupations mésolithiques, les occupations néolithiques*. Rapport de fouilles: 266-76. Inrap CIF.

Martial, E. 2008. Exploitation des végétaux et artisanat textile au Néolithique final sur les sites de la vallée de la Deûle (Nord-Pas-de-Calais). *Les Nouvelles de l'Archéologie*. Editions de la Maison des Sciences de l'Homme: 33-41.

Martial, E. & I. Praud (ed.). 2007. Un site palissadé du Néolithique final à Houplin-Ancoisne (Nord), in *Relations interrégionales au Néolithique entre Bassin parisien et Bassin rhénan*, 7: 403-43. Luxembourg: Archaeologia Mosellana, Musée National d'Histoire et d'Art.

Martial, E. & I. Praud. 2011. Une approche pluridisciplinaire des sites du Néolithique final entre Deûle et Escaut: premiers résultats et perspectives, in F. Bostyn, E. Martial & I. Praud (ed.) *Le Néolithique du Nord de la France dans son contexte européen: habitat et économie aux 4e et 3e millénaires avant notre ère,* Numéro spécial 28: 575-83. Revue Archéologique de Picardie.

Martial, E., N. Cayol, C. Hamon, Y. Maigrot, F. Médard & C. Monchablon. 2011. Production et fonction des outillages au Néolithique final dans la vallée de la Deûle (Nord-Pas-de-Calais, France), in F. Bostyn, E. Martial & I. Praud (ed.) *Le Néolithique du Nord de la France dans son contexte européen: habitat et économie aux 4e et 3e millénaires avant notre ère,* Numéro spécial 28: 365-90. Revue Archéologique de Picardie.

Mathieu, G. 1992. Une figurine stylisée dans une tombe d'enfant de la nécropole rubanée d'Ensisheim (Haut-Rhin), in *Internéo et Société Préhistorique Française, 11*: 27-29.

Matterne, V. 2001. *Agriculture et alimentation végétale durant l'âge de fer et l'époque gallo-romaine en France septentrionale.* (Achéologie Des Plantes et Des Animaux, Volume 1). Montagnac: Editions Monique Mergoil.

–. 2009a. Premier aperçu des activités agricoles en plaine champenoise à partir des études carpologiques, in J. Van Moerkerke (ed.) *Le Bassin de la Vesle du Bronze final au Moyen Age, à travers les fouilles du TGV-est*, 102-2: 45-56. Bulletin de la Société Archéologique Champenoise.

–. 2009b. L'agriculture du VIe au Ier siècle avant J.-C. en France: état des recherches carpologiques sur les établissements ruraux, in I. Bertrand, A. Duval, J. Gomez de Soto et P. Maguer (ed.) *Habitats et paysages ruraux en Gaule et regards sur d'autres régions du monde celtique*, II: 383-416. Mémoire 35. Chauvigny: Association des Publications Chauvinoises.

Mauss, M. & H. Hubert. 2016. *Essai sur la nature et la fonction du sacrifice* (Quadrige). Paris: PUF.

Menez, Y. 2008. Le Camp de Saint-Symphorien à Paule (Côtes d'Armor) et les résidences de l'aristocratie du second âge du Fer en France septentrionale. Paris: Thèse de Doctorat, Université de Paris 1 Panthéon Sorbonne.

Méniel, P. 1984. *Contribution à l'histoire de l'élevage en Picardie. Du Néolithique à la fin de l'âge du Fer*, Numéro spécial 3. Revue Archéologique de Picardie.

–. 1985. Les vestiges animaux du site chalcolithique du «Gord» à Compiègne (Oise). *Revue Archéologique de Picardie* 3-4: 119-20. https://doi.org/10.3406/pica.1985.1475.

–. 1986. La nécropole gauloise de Tartigny (Oise): étude des offrandes animales. *Revue Archéologique de Picardie*: 37-39.

–. 1987a. Les dépôts animaux du fossé chasséen de Boury-en-Vexin (Oise). *Revue Archéologique de Picardie*: 3-26.

–. 1987b. Les restes animaux, in J.-G. Rozoy (ed.) *Les Celtes en Champagne. Les Ardennes au second âge du Fer, le Mont Troté, Les Rouliers*, 4: 357-61. Mémoires de la Société Archéologique Champenoise.

–. 1990. Les restes animaux du fossé gaulois de Beauvais, les Aulnes du Canada (Oise). *Revue Archéologique de Picardie*: 97-107.

–. 1991. Les animaux dans les sanctuaires gaulois du nord de la France, in J.-L. Brunaux (ed.) *Les sanctuaires celtiques et leurs rapports avec le monde méditerranéen*, 3: 257-67. Paris: Errance, Dossiers de Protohistoire.

–. 1994. Les restes d'animaux des établissements ruraux de l'âge du Fer en Picardie: l'exemple de Chambly, la Marnière (Oise), in 0. Buchsenschutz, P. Méniel (ed.) *Les installations agricoles de l'âge du Fer en Ile-de-France*, 4: 205-26 (Études d'Histoire et d'Archéologie). Paris: Presses de l'École Normale Supérieure.

–. 1997. L'apport des restes animaux à l'analyse spatiale des sites fossoyés du second âge du Fer, in G. Auxiette, L. Hachem, B. Robert (ed.) *Espaces physiques, espaces sociaux dans l'analyse interne des sites du Néolithique à l'âge du Fer*: 89-100. Paris: Editions du Comité des Travaux Historiques et Scientifiques.

–. 1998. La question du sacrifice animal dans les rites funéraires en Gaule Belgique. *Revue Archéologique de Picardie*: 245-51.

–. 1999. Les restes animaux du sanctuaire celtique, in L. Bourgeois (ed.) *Le sanctuaire rural de Bennecourt (Yvelines), du temple celtique au temple gallo-romain*, 77: 151-68. Paris: Documents d'Archéologie Française.

–. 2000a. Analyses archéozoologiques des restes osseux, in S. Gaudefroy (ed.) *Glisy, les Terres de Ville, ZAC de la Croix de Fer (Somme). L'occupation du premier âge du Fer et l'établissement agricole de La Tène moyenne*. Rapport de fouille, Afan Centre-Nord.

–. 2000b. Des os dans les fossés et des animaux dans les enclos: diversité des fonctions et limites des interprétations, in *Les enclos celtiques*, 1: 267-70. Revue Archéologique de Picardie. https://doi.org/10.3406/pica.2000.2241.

–. 2004. Les animaux dans les rites funéraires au deuxième âge du Fer, in L. Baray (ed.) *Archéologie des pratiques funéraires. Approches critiques*, 9: 189-96. Glux-en-Glenne: Bibracte.

–. 2006. La faune, in F. Malrain & E. Pinard (ed.) *Les sites laténiens de la moyenne vallée de l'Oise du Ve au Ier siècle avant notre ère: contribution à l'histoire de la société gauloise*, Numéro spécial 23: 181-201. Revue Archéologique de Picardie.

Méniel, P. & M.-P. Horard-Herbin. 1996. La faune de Varennes-sur-Seine, le Marais du Pont, in J.-M. Séguier (ed.) *Varennes-sur- Seine, le Marais du Pont (Seine-et-Marne): occupations du Paléolithique supérieur et du Néolithique, nécropole de l'âge du Bronze et habitat groupé de La Tène finale*. Rapport de fouille, Afan IdF.

Méniel, P. & B. Lambot (ed.). 2002. *Repas des vivants et nourriture pour les morts en Gaule*. Supplément 1, Société Archéologique Champenoise.

Méniel, P., R.-M. Arbogast & J.-H. Yvinec. 1987. *Une histoire de l'élevage: les animaux et l'archéologie*. Paris: Errance.

Méniel, P., B. Lambot & M. Friboulet. 1994. *Le site protohistorique d'Acy-Romance (Ardennes), 2: Les nécropoles dans leur contexte régional*. Société Archéologique Champenoise.

Méniel, P., G. Auxiette, D. Germinet, A. Baudry & M.-P. Horard-Herbin. 2009. Une base de données sur les études de faunes des établissements ruraux en Gaule, in I. Bertrand, A. Duval, J. Gomez de Soto, P. Maguer (ed.) *Habitats et paysans ruraux en Gaule et regards sur d'autres régions du monde celtique*, Mémoire 33: 417-46. Chauvigny: Association des Publications Chauvinoises.

Mercer, R. & F. Healy. 2014. *Hambledon Hill, Dorset, England: excavation and survey of a Neolithic monument complex and its surrounding landscape*. English Heritage Publishing.

Méroc, L., G. Simonnet & H. Duday. 1979. Les sépultures chasséennes de Saint-Michel-du-Touch, à Toulouse (Haute-Garonne). *Bulletin de la Société Préhistorique Française*. JSTOR, 379-407.

Meunier, K. 2012. *Styles céramiques et néolithisation dans le sud-est du Bassin parisien*. Recherches Archéologiques 5. Paris: INRAP/CNRS Editions.

–. 2013. La céramique de Juvigny, les Grands Traquiers (Marne) et le Rubané récent champenois. *Bulletin de la Société Préhistorique Française* 110: 421-46.

Meunier, K., L. Bedault, S. Cary, P. Chambon, F. Convertini, C. Croutsch, C. Hamon & J.-G. Pariat. 2012. Deux enceintes du Néolithique moyen 1 à Gurgy, le Nouzeau (Yonne), in *Internéo et Société Préhistorique Française*, 9: 61-72.

Modderman, P.J.R. 1970. *Linearbandkeramik aus Elsloo und Stein,*. R.O.B., Nederlandse Oudheden. Vol. 3. Amerfoort.

Monchablon, C. 2014. *Carvin, la Gare d'Eau (Pas-de-Calais). Une enceinte du Néolithique moyen II*. Rapport de fouille. Inrap NP.

Monchablon, C., M. Baillieu, M. Bouchet, A. Goutellard & I. Praud. 2011. L'enceinte néolithique moyen II de Carvin, la Gare d'Eau (Pas-de-Calais), présentation préliminaire, in F. Bostyn, E. Martial & I. Praud (ed.) *Le Néolithique du Nord de la France dans son contexte européen: habitat et économie au 4e et 3e millénaires avant notre ère*: 407-19. Revue Archéologique de Picardie.

Mordant, C., D. Mordant & T. Poulain-Josien. 1970. *Le site protohistorique des Gours-aux-Lions à Marolles-sur-Seine, Seine-et-Marne*. Mémoire 8. Société Préhistorique Française.

Mordant, D. (ed.). 1992. *La Bassée avant l'Histoire, Archéologie et Gravières en Petite-Seine*.

Nemours, France: A.P.R.A.I.F., Mémoire du Musée de Préhistoire d'Ile-de-France.

–. 1997. Le complexe des Réaudins à Balloy: enceinte et nécropole monumentale, in C. Constantin, D. Mordant, D. Simonin (ed.) *La Culture de Cerny: nouvelle économie, nouvelle société au Néolithique*: 449-79 (6). A.P.R.A.I.F., Mémoire du Musée de Préhistoire d'Ile-de-France.

– (ed.). 2006. *Dynamique d'occupation humaine de la Bassée et du confluent Seine-Yonne dans le contexte local et régional, à partir de 7000 avant notre ère*. PCR, Rapport final, Programme 2002-2004. Bazoches-lès-Bray: CDA- Bassée; Saint-Denis: SRA Ile-de-France.

Morris, J. 2008. Associated bone groups; one archaeologist's rubbish is another's ritual deposition, in O. Davis, N. Sharples, K. Waddington (ed.) *Changing Perspectives on the First Millennium BC: Proceedings of the Iron Age*: 83-98. Oxbow Books.

Müller, H.-H. 1964. *Die Haustiere der mitteldeutschen Bandkeramiker*. Vol. 17. Akademie-Verlag.

Naze, G. 2014. Crécy-sur-Serre, la Croix Saint-Jacques et le Bois de Sort (Aisne, France). L'enceinte Michelsberg, une fosse de la Culture Villeneuve-Saint-Germain et les ensembles résiduels post-Rössen, Cerny tardif et de la période mésolithique. *Revue Archéologique de Picardie*: 43-136.

Neiss, R., F. Berthelot, J.-M. Doyen & P. Rollet. 2015. Les principales étapes de la formation du site urbain antique de Reims/Durocortorum, cité des Rèmes, in M. Reddé & W. Van Andringa (ed.) *Gallia* 72: 161-76.

Nespoulous, L. 2013. Les fosses pièges du site de Tama New Town au Japon: une approche du "Jomon des forêts", in N. Achard-Corompt, V. Riquier (ed.) *Chasse, culte ou artisanat? Les fosses "à profil en Y-V-W". Structures énigmatiques et récurrentes du Néolithique aux âges des Métaux en France et alentour*. Supplément 33: 283-94. Revue Archéologique de l'Est.

Nieszery, N. 1995. Linearbandkeramische Gräberfelder in Bayern. *Internationale Archäologie* 16.

Nuviala, P. 2015. La révolution zootechnique romaine et la diffusion des grands bœufs et des grands chevaux dans l'Est de la Gaule (IVème siècle avant. J.-C.-Vème siècle. après J.-C.). Dijon: Thèse de Doctorat, Université de Dijon.

–. 2016. La production des "grands bœufs" dans l'Est de la Gaule: entre évolutions gauloises et influences romaines, in G. Blancquaert, G., Malrain, F. (ed.) *Evolution des sociétés gauloises du second âge du Fer, entre mutations internes et influences externes*: 595-608. Numéro spécial 30. Revue Archéologique de Picardie.

Olsen, J. 2013. Hunting using permanent trapping systems in the northern section of the mountains of Southern Norway, in N. Achard-Corompt, V. Riquier (ed.) *Chasse, culte ou artisanat? Les fosses "à profil en Y-V-W". Structures énigmatiques et récurrentes du Néolithique aux âges des Métaux en France et alentour.* Supplément 33: 261-82. Revue Archéologique de l'Est.

Ours Debout 2014. *Souvenirs d'un chef Sioux.* Petite bibliothèque Payot. Payot et Rivages.

Paresys, C., C. Moreau, M. Saurel & G. Auxiette. 2009. La nécropole de Saint-Etienne-au-Temple. *Bulletin de la Société Archéologique Champenoise* 102: 153-92.

Paris, P.-E. 2016. *Au fil de l'os. Économie et société chez les Rèmes et les Suessions par le prisme de l'archéozoologie.* Leiden: Sidestone Press.

Parker Pearson, M. 2003. Food, identity and culture: an introduction and overview, in *Food, culture and identity in the neolithic and early Bronze age*, 1117: 1-30. British Archaeological Reports International Series.

Parkinson, W.A. & P.R. Duffy. 2007. Fortifications and Enclosures in European Prehistory: A Cross-Cultural Perspective. *Journal of Archaeological Research* 15: 97-141. https://doi.org/10.1007/s10814-007-9010-2.

Parmenter, P.C., E.V. Johnson & A.K. Outram. 2015. Inventing the Neolithic? Putting evidence-based interpretation back into the study of faunal remains from causewayed enclosures. *World Archaeology* 47. Taylor & Francis: 819-33.

Pastoureau, M. 2007. *L'ours. Histoire d'un roi déchu.* Editions du Seuil.

Pastre, J.F., C. Leroyer, N. Limondin-Lozouet, C. Chaussé, M. Fontugne, A. Gebhardt, C. Hatté & V. Krier. 2000. Le Tardiglaciaire des fonds de vallée du Bassin parisien (France). [The Late-Glacial from the Paris basin floodplains (France)]. *Quaternaire* 11. Association Française pour l'Etude du Quaternaire: 107-22.

Pavlù, I. 2000. *Life on a Neolithic site.* Praha: Institute of Archaeology, Czech Academy of Sciences.

Pavlù, I. 2010. *Činnosti na neolitickém sídlišti Bylany. (Activities on a Neolithic Site of Bylany).* Vydal Archeologicky ústav CR. ARUP. Praha.

–. 2016. Linear Pottery Houses and Their Inhabitants. *Open Archaeology*, 382-97. https://doi.org/10.1515/opar-2016-0027.

Payne, S. 1973. Kill-off Patterns in Sheep and Goats: the Mandibles from Aşvan Kale. *Anatolian Studies* 23: 281-303.

Peake, R. & V. Delattre. 2007. *Marolles-sur-Seine, la Croix Saint- Jacques (Seine-et-Marne), Nécropole et structures domestiques de l'étape initiale du Bronze final.* Rapport final d'opération, Inrap CIF.

Peake, R., V. Delattre & P. Pihuit. 1999. La nécropole de l'âge du Bronze de la Croix de la Mission à Marolles-sur-Seine (Seine-et-Marne). *Bulletin de la Société Préhistorique Française* 96: 581-605.

Peake, R. G. Allenet, G. Auxiette 2009, F. Boisseau, C. Chaussé, S. Coubray, C. Leroyer, C. Pautret-Homerville, J. Perrière, F. Toulemonde. Villiers-sur-Seine, le Gros Buisson. Un habitat aristocratique de la fin de l'âge du Bronze et du début du premier âge du Fer, in M.-J. Lambert, A. Daubigney, P.-Y. Milcent, M. Talon, J. Vital (ed.) *De l'âge du Bronze à l'âge du Fer en France et en Europe occidentale (Xè-VIIè siècle av. J.-C.).* Supplément 27: 559-64. Revue Archéologique de l'Est.

Peake, R., R. Issenmann & V. Delattre. 2011. Examples of Social Modelling in the Seine Valley during the Late Bronze Age and Early Iron Age, in T.Moore, X. Armada (ed.) *Atlantic Europe in the First Millennium BC. Crossing the Divide*: 319-35. Oxford: Oxford University Press.

Peake, R. (ed.). 2020. *Villiers-sur-Seine, un habitat aristocratique du IXe siècle avant notre ère*: (Recherches Archéologiques 18). Paris: co-édition INRAP/CNRS éditions.

Peltier, V. & S. Fournand. 2015. Pont-sur-Seine, le Haut de Launoy-Phase 3 (Aube). Une palissade au Néolithique récent et son système d'entrée, in F. Bostyn & L. Hachem (ed.) *Hommages à Mariannick Le Bolloch*, 3-4: 49-60. Revue Archéologique de Picardie.

Pennors, F. 2004. *Analyse fonctionnelle et pondérale des dépôts et trouvailles isolées du Bronze en France*. Paris: Thèse de Doctorat, Université de Paris 1 Panthéon Sorbonne.

Perlès, C. 2012. Le statut des échanges au Néolithique. *Rubricatum: revista del Museu de Gavà (en linia)*: 539-46.

Pétrequin, A.-M. & P. Pétrequin. 2006. *Objets de pouvoir en Nouvelle-Guinée*. Éditions de la Réunion des Musées Nationaux. Paris.

Pétrequin, P. (ed.). 1997. *Les Sites littoraux néolithiques de Clairvaux-les-Lacs et de Chalain (Jura): 3200-2900 av. J.-C.* Les Editions de la Maison des Sciences de l'Homme.

Pétrequin, P., R.-M. Arbogast, A.-M. Pétrequin, S. Van Willigen & M. Bailly. 2006. *Premiers chariots, premiers araires: la diffusion de la traction animale en Europe pendant les IVe et IIIe millénaires avant notre ère*. CNRS éditions, Paris.

Petrequin, P., E. Gauthier & A.-M. Pétrequin. 2017. *Jade 2. Objets-signes et interprétations sociales des jades alpins dans l'Europe néolithique*. Vol. 17, Dynamiques territoriales, 9 (Cahiers de la Maison des Sciences de l'Homme Ledoux). Besançon: Presses universitaires de Franche-Comté et Centre de recherche archéologique de la vallée de l'Ain. https://hal.archives-ouvertes.fr/hal-01516594.

Piliougine, C. 2015. Quelles fonctions pour l'enceinte de Bazoches-sur-Vesles, le Bois de Muisemont? Apport de l'analyse fonctionnelle des outils en matières dures animales (Néolithique Moyen II, Aisne, France). Paris: Mémoire de Master 2, Université de Paris 1 Panthéon Sorbonne.

Pinard, E. 2016. Que dire des traces de coup et/ou de découpe observées sur les restes humains gaulois (du IIIe au Ier s. avant notre ère) en contextes domestiques picards? *Revue Archéologique de Picardie*: 93-105.

Pinard, E. & S. Desenne (ed.). 2009. *Les gestuelles funéraires au second âge du Fer*, 3-4. Revue Archéologique de Picardie.

Pinard, E., S. Desenne, S. Gaudefroy & F. Gransar. 2010. Les gestuelles funéraires au second âge du Fer en Picardie, in P. Barral, B. Dedet, F. Delrieu, P. Giraud, I. Le Goff, S. Marion, A. Villard-Le Tiec (ed.) *Gestes funéraires au second âge du Fer*, 2: 37-50 (Environnement, Sociétés et Archéologie). Besançon: Presses universitaires de Franche-Comté.

Pinot Duclos, C. 1810. *Morceaux choisis*.

Pion, P., G. Auxiette, M. Boureux, P. Brun, J.-P. Demoule, C. Pommepuy & B. Robert. 1990. De la chefferie à l'Etat? Territoires et organisation sociale dans la vallée de l'Aisne aux âges des Métaux (2200-20 Av. J.C.), in *Archéologie et Espaces*: 183-260. Antibes: APDCA.

Pion, P., F. Gransar & G. Auxiette. 1996. Les établissements ruraux dans la vallée de l'Aisne, de la fin du second âge du Fer au début du haut-Empire romain (IIe siècle av. J.-C. / Ier siècle ap. J.-C): bilan provisoire des données et esquisse de synthèse. Numéro spécial. Revue Archéologique de Picardie, 55-107.

Pion, P., C. Pommepuy, G. Auxiette, B. Hénon & F. Gransar. 1997. L'oppidum de Condé-Sur-Suippe/Variscourt (Aisne) (fin IIe – début Ier siècle av. J.-C.): approche préliminaire de l'organisation fonctionnelle d'un quartier artisanal, in G. Auxiette, L. Hachem, B. Robert (ed.) *Espaces physiques, espaces sociaux dans l'analyse interne des sites du Néolithique à l'âge du Fer*: 276-309. Paris: Editions du Comité des Travaux Historiques et Scientifiques.

Plateaux, M. 1990. Approche régionale et différentes échelles d'observation pour l'étude du Néolithique et du Chalcolithique du Nord de la France. Exemple de la vallée de l'Aisne, in *Archéologie et espaces*, 10: 157-82. Antibes: APDCA.

Podborsky, V. 2002. *Dve pohrebiste neolitickeho lidu s linearni keramikou ve Vedrovicich na Morave*. Brno: Filozofická Fakulta Masarykovy Univerzity.

Polloni, A. 2008. Parures individuelles et sépultures collectives à la fin du Néolithique en Bassin parisien. *Préhistoires Méditerranéennes*. Association pour la promotion de la Préhistoire et de l'Anthropologie: 75-89.

Pommepuy, C., G. Auxiette & S. Desenne. 1998. Ruptures et continuités dans les pratiques funéraires de La Tène ancienne et moyenne/finale à Bucy-le-Long (Aisne), in J.-L. Brunaux, G. Leman-Delerive, C. Pommepuy (ed.) *Les rites de la mort en Gaule du Nord à l'âge du Fer*, 1-2: 85-98. Revue Archéologique de Picardie.

Pommepuy, C., F. Gransar. 1998. Bazoches-sur-Vesle, les Chantraines, in *Fouilles Protohistoriques de la Vallée de l'Aisne.* Document final de synthèse, Afan Centre-Nord.

Pommepuy, C. 1999. Le matériel de mouture de la vallée de l'Aisne de l'âge du Bronze à La Tène finale: formes et matériaux. *Revue Archéologique de Picardie* 3: 115-41.

Pommepuy, C., G. Auxiette, S. Desenne, F. Gransar & B. Henon. 2000. Des enclos à l'âge du Fer dans la vallée de l'Aisne, in *Le monde des vivants et le monde des morts*, 1-2: 197-216. Revue Archéologique de Picardie.

Poplin, F. 1993. Que l'homme cultive aussi bien le sauvage que le domestique, in *Exploitation des animaux sauvages à travers le temps*: 527-39. Antibes: APDCA.

Poulain, T. 1986. La faune de la fosse danubienne de Norrois (Marne), lieu-dit la Raie des Lignes. *Préhistoire et Protohistoire en Champagne Ardenne*: 41-43.

Poux, M. 2000. Espaces votifs-espaces festifs. Banquets et rites de libation en contexte de sanctuaires et d'enclos, in *Les enclos celtiques*, 1-2: 217-31. Revue Archéologique de Picardie.

–. 2004. *L'âge du vin. Rites et boissons, festins et libations en Gaule indépendante.* Montagnac: Editions Monique Mergoil.

Praud, I. 2012. L'architecture des bâtiments du Néolithique récent final. *Archéopages: Archéologie et société*: 110-13. https://doi.org/10.4000/archeopages.694.

Praud, I. 2015. Escalles: A Neolithic Causewayed Enclosure on the Pas-de-Calais Coast. *PAST*: 14-16.

Praud, I. (ed.) 2015. *Le Néolithique final dans la vallée de la Deûle: le site d'Houplin-Ancoisne, le Marais de Santes (Nord).* Recherches Archéologiques 9. Paris: INRAP/ CNRS Editions. https://hal-inrap.archives-ouvertes.fr/hal-01489255.

Praud, I. & E. Panloups. 2015. Escalles, Mont d'Hubert (Pas-de-Calais): premiers résultats de la fouille de l'enceinte du Néolithique Moyen II implantée sur le littoral de la Mer du Nord, in C. Laurelut & J. Vanmoerkerke (ed.) *Occupations et exploitations néolithiques, et si l'on parlait des plateaux? 107-4*: 189-204. Bulletin de la Société d'Archéologie Champenoise.

Quenez, J.-P. (ed.). 2020. *Le site de Villiers-sur-Seine, le Défendable.* Rapport de fouilles. Inrap CIF.

Quilliec, B. 2007. *L'épée atlantique: échanges et prestige au Bronze final.* Mémoire 42. Société Préhistorique Française.

Rast-Eicher, A. 2014. Bronze and Iron Age Wools in Europe, in C. Breniquet & C. Michel (ed.) *Wool Economy in the Ancient Near East and the Aegean. From the Begginings of Sheep Husbandry to Institutional Textile industry*: 224-51 (Ancient Textiles Series 17). Oxbow Books.

Rasteiro, R. & L. Chikhi. 2013. Female and male perspectives on the Neolithic transition in Europe: clues from ancient and modern genetic data. *PLoS one* 8. Public Library of Science.

Reichstein, H. 1977. Bemerkungen zu einigen Tierknochen aus frühneolithischen Siedlungsgruben von Rosdorf, Kr. Gottingen. *Nachrichten aus Niedersachsens Urgeschichte Hildesheim* 46: 1-26.

Revolon, S., M. Bailly & P. Lemonnier (ed.). 2012. *Objets irremplaçables.* Vol. 58 (Techniques & Culture). Paris: Editions de la Maison des Sciences de l'Homme.

Riche, C. 2004. Le vase zoomorphe d'Aubevoye (Eure): une découverte inédite dans un contexte Villeneuve-Saint-Germain en Haute-Normandie. *Bulletin de la Société Préhistorique Française* 101. JSTOR: 877-80.

Rigaud, S. 2014. Pratiques ornementales des premières communautés agro-pastorales de Bavière (Allemagne): Intégration? Acculturation? Convergence? Nouveaux apports de la nécropole de Essenbach-Ammerbreite. *Anthropologie* 52: 207-27.

Riquier, V. (ed.). 2017. *La Plaine de Troyes: évolution d'un territoire rural des premiers agriculteurs au premier réseau villageois.* Projet Collectif de Recherche, Rapport d'activité 2013-2016. Inrap GEN.

Riquier, V. & K. Meunier. 2014. L'apport des fouilles du Parc Logistique de l'Aube et de la chronologie céramique à la connaissance des occupations du Néolithique ancien et moyen I dans la plaine de Troyes, in C. Louboutin & C. Verjux (ed.) *Zones de production et organisation des territoires au Néolithique. Espaces exploités, occupés, parcourus.* Supplément 51: 355-67. Revue Archéologique du Centre de la France.

Riquier, V., G. Auxiette, K. Fechner, J. Grisard, S. Loicq, N. Théophane, E. Sehier, F. Toulemonde & K. Zipper. 2012. Oscillations et évolutions de l'habitat et des systèmes agraires en Champagne méridionale (2200-450 av. J.-C.): le terroir du Parc Logistique de l'Aube, in M. Honegger et C. Mordant (ed.) *L'Homme au bord de l'eau. Archéologie des zones littorales du Néolithique à la Protohistoire*: 65-88. Lausanne: Cahiers d'archéologie romande, 132 et Documents préhistoriques, 30.

Riquier, V., C. Paresys & K. Meunier. 2015. Buchères, Parc Logistique de l'Aube: nouvelles données sur l'occupation du sol du Néolithique ancien au Néolithique final, in C. Laurelut & J. Vanmoerkerke (ed.) *Occupations et exploitations néolithiques; et si on parlait des plateaux...*, 107, 4: 169-88. Bulletin de la société Archéologique Champenoise.

Robert, B. & O. Weller. 1995. Le commerce du sel au La Tène Final: une problématique enfin relancée. *Revue Archéologique de Picardie*: 87-96. https://doi.org/10.3406/pica.1995.2153.

Roffet-Salque, M., P. Gerbault & R. Gillis. 2017. L'exploitation laitière: approches génétique, archéozoologique et biomoléculaire, in M. Balasse & P. Dillmann (ed.) *Regards croisés: quand les sciences archéologiques rencontrent l'innovation*: 1-23 (Collections Sciences Archéologiques). Editions des Archives Contemporaines.

Roscio, M. 2011. Nouvelles approches des nécropoles de l'étape ancienne du Bronze final (Bz D-Ha A1) du Bassin parisien au Jura Souabe. Dijon: Thèse de Doctorat, Université de Bourgogne.

Rottier, S., J. Piette & C. Mordant. 2012. *Archéologie funéraire du Bronze final dans les vallées de l'Yonne et de la haute Seine: les nécropoles de Barbey, Barbuise et La Saulsotte*. Dijon: Éditions Universitaires de Dijon.

Ruby, P. & G. Auxiette. 2010. 1977-2007: trente années de recherches sur les «fossés en croix» de l'oppidum de Villeneuve-Saint-Germain (Aisne). *Revue Archéologique de Picardie*: 39-94.

Rück, O. 2013. From yard to house row: the Bandkeramik village-layouts in rows and feature-free areas provide a new view on settlement structure, in C. Hamon, P. Allard & M. Ilett (ed.) *The Domestic Space in LBK Settlements*: 201-30 (Internationale Archäologie). Rahden, Westf: Verlag Marie Leidorf GmbH.

Sahlins, M. 1965. On the sociology of primitive exchange, in M. Banton (ed.)*: The Relevance of Models for Social Anthropology.* London: Tavistock Publications.

Sahlins, M. 1968. *Tribesmen (Foundations of Modern Anthropology)*. Prentice Hall.

–. 1976. *Âge de pierre, âge d'abondance*. Bibliothèque des Sciences humaines. Gallimard.

Samzun, A., P. Pétrequin & E. Gauthier. 2012. Une imitation de hache alpine type Bégude à Buthiers-Boulancourt (Seine-et-Marne) au début du Ve millénaire, in É. Thirault & P.-A. Labriffe (de) (ed.) *Produire des haches au Néolithique, de la matière première à l'abandon*: 219-34. Société Préhistorique Française. https://halshs.archives-ouvertes.fr/halshs-00824727.

Schmitzberger, M. 2010. Die linearbandkeramische Fauna von Mold bei Horn, Niederösterreich, in E. Lenneis (ed.) *Die bandkeramische Siedlung von Mold bei Horn in Niederösterreich. 1. Naturwissenschaftliche Beiträge und Einzelanalyse.*

Schneider, M. 1980. Découverte néolithique à Dachschtein. *Cahiers Alsaciens d'Archéologie, d'Art et d'Histoire* 23: 47-58.

Séguier, J.-M. (ed.). 1996. *Varennes-sur-Seine, le Marais du Pont (Seine-et-Marne): occupations du Paléolithique supérieur et du Néolithique, nécropole de l'âge du Bronze et habitat groupé de La Tène finale*. Document Final de Synthèse, Afan IdF.

Séguier, J.-M. & G. Auxiette (ed.). 2006. *Souppes-sur-Loing, À l'Est de Beaumoulin (Seine-et-Marne)*. Rapport de fouille, Inrap CIF.

Séguier, J.-M., G. Auxiette, S. Coubray, C. Dunikowski, B. Lecomte-Schmitt & V. Zech-Matterne. 2008. Une ferme du début du IIIe s. av. J.-C. au Marais du Colombier, Varennes-sur-Seine (Seine-et-Marne): analyse archéologique et environnementale. *Revue Archéologique du Centre de la France*. https://journals.openedition.org/racf/1105.

Sergent, B. 1999. Le porc indo-européen, d'Ouest en Est, in P. Walter (ed.) *Mythologie du Porc*: 9-39. Editions Jérôme Millon.

Serjeantson, D. 2007. Intensification of animal husbandry in the Late Bronze Age? The contribution of sheep and pigs, in C. Haselgrove & R. Pope (ed.) *The Earlier Iron Age in Britain and the Near Continent*, Oxbow Books: 80-93.

Silar, B. 2012. Les Ilas et conopas des foyers andins. Des objets irremplaçables, un patrimoine vivant, in S. Revolon, M. Bailly & P. Lemonnier (ed.) *Objets irremplaçables* (Techniques et Cultures 58). Maison des Sciences de l'Homme.

Silver, A. 1969. The ageing of domestic animal, in R. Brothwell & E.S. Higgs (ed.) *Science in Archaeology*: 283-302. London: Thames and Hudson.

Sobocinski, M. 1978. Zwierzece szczatki kostne z Osady Neolitycznej w Gniechowicach. *Roczniki Akademii Rolniczej w Poznaniu*, 3-35.

Sohn, M. 2007. Du collectif à l'individuel: évolution des dépôts mobiliers dans les sépultures collectives d'Europe occidentale de la fin du IVe à la fin du IIIe millénaire av. J-C. *Bulletin de la Société Préhistorique Française*: 381-86.

–. 2008. Entre signe et symbole. Les fonctions du mobilier dans les sépultures collectives d'Europe occidentale à la fin du Néolithique. *Préhistoires Méditerranéennes*: 53-71.

Sonko, L. 1986. Les modes d'appropriation, de gestion et de conduite des animaux au sein d'un village diola (Boulandor). Contribution à l'étude du fonctionnement des systèmes agraires de Basse-Casamance, Sénégal. *Les Cahiers de la recherche-développement*, 9-10.

Soudský, B. 1962. The neolithic site of Bylany. *Antiquity* 36. Cambridge University Press: 190-200.

–. 1969. Etude de la maison néolithique. *Slovenská Archeologia*: 5-96.

Soudský, B. & I. Pavlù. 1972. The linear pottery culture settlement patterns of Central Europe, in P. Ucko, R. Tringham & G.W. Dimbleby (ed.) *Man Settlement and Urbanism*: 317-328. London: Duckworth.

Spatz, H. 1997. La nécropole du Néolithique moyen (Hinkelstein, Grossgartach) de Trebur (Gross-Gerau, hesse), in C. Jeunesse (ed.) *Le Néolithique danubien et ses marges entre Rhin et Seine,*. Supplément 3: 157-70. Strasbourg: Cahiers de l'Association pour la Promotion de la Recherche Archéologique en Alsace.

Spatz, H. & A. Von Den Driesch. 1999. *Das mittelneolithische Gräberfeld von Trebur,* *Kreis Groß-Gerau. Vol. 1*. Landesamt für Denkmalpflege Hessen. Vol. Band 19.

–. 2001. Zu den tierischen Beigaben aus dem Hinkelsteiner und Gross- gartacher Gräberfeld von Trebur, Kr. Gross-Gerau, in R.-M. Arbogast, C. Jeunesse & J. Schibler (ed.) *Rôle et statut de la chasse dans le Néolithique ancien danubien (5500-4900 av. J.-C)*. Rahden/Westf.: Marie Leidorf Verlag.

Stäuble, H. 1997. Häuser, Gruben und Fundverteilung, in J. Lüning (ed.) *Ein Siedlungsplatz der Ältesten Bandkeramik in Bruchenbrücken, Stadt Friedberg/ Hessen*, 39: 17-150. Bonn: Universitätsforschungen zur prähistorischen Archäologie.

Stehli, P. 1989. Merzbachtal. Umwelt und Geschichte einer bandkeramischen Siedlungskammer. *Germania* 67: 51-76.

Stépanoff, C. 2009. Devenir animal pour rester humain. Logiques mythiques et pratiques de la métamorphose en Sibérie méridionale. *Images Re-vues. Histoire, Anthropologie et Théorie de l'Art*. http://journals.openedition.org/imagesrevues/388.

Stephan, E. 2008. Die Tierknochenfunde aus den Michelsberger Erdwerken von Neckarsulm-Obereisesheim, Hetzenberg, und Heilbronn-Klingenberg, Schloßberg, in *Michelsberger Erdwerke im Raum Heilbronn: Neckarsulm-Obereisesheim, Hetzenberg, Ilsfeld, Ebene, Landkreis Heilbronn und Heilbronn-Klingenberg, Schlossberg, Stadtkreis Heilbronn*, Materialhefte zur Archäologie in Baden-Württemberg, 3: 131-242.

Steppan, K. 2003. *Taphonomie, Zoologie, Chronologie, Technologie, Ökonomie : die Säugetierreste aus den jungsteinzeitlichen Grabenwerken in Bruchsal, Landkreis Karlsruhe*. Vol. 66. Stuttgart: Materialhefte zur Archäologie in Baden-Württemberg.

Stordeur, D., L. Gourichon & D. Helmer. 2004. À l'aube de la domestication animale. Imaginaire et symbolisme animal dans les premières sociétés néolithiques du nord du Proche-Orient. *Anthropozoologica* 39: 143-63.

Strien, H.-C. 2005. Familientraditionen in der bandkeramischen Siedlung bei Vaihingen/Enz, in J. Lüning, C. Fridrich & A. Zimmerman (ed.)

Die Bandkeramik im 21. Jahrhundert, 21: 189-97. Rahden: Marie Leidorf.

Tappret, E. & A. Villes. 1996. Contribution de la Champagne à l'étude du Néolithique ancien, in P. Duhamel (ed.) *La Bourgogne entre les bassins rhénan, rhodanien et parisien: carrefour ou frontière*. Supplément 14: 175-256. Revue Archéologique de l'Est.

Tegel, W., J. Van Moerkerke, D. Hakelberg & U. Büntgen. 2016. Des cernes de bois à l'histoire de la conjoncture de la construction et à l'évolution de la pluviométrie en Gaule du Nord entre 500 BC et 500 AD, in G. Blancquaert, G., Malrain, F. (ed.) *Evolution des sociétés gauloises du second âge du Fer, entre mutations internes et influences externes*: 639-53. Numéro spécial 30. Revue Archéologique de Picardie.

Testart, A. 2014. *L'amazone et la cuisinière: anthropologie de la division sexuelle du travail*. Gallimard.

Thevenet. 2010. *Des faits aux gestes... des gestes aux sens? Pratiques funéraires et société durant le Néolithique ancien en Bassin parisien*. Paris: Thèse de Doctorat, Université de Paris I Panthéon-Sorbonne.

Thevenet, C. 2008. Les sépultures isolées du Néolithique moyen II, in C. Colas (ed.) *Beaurieux (Aisne), la Plaine, zone sud, tranche 2*. Rapport de fouilles: 52-69. Inrap NP.

– (ed.). 2013. *L'enceinte néolithique de Menneville, Derrière le Village (Aisne). Campagne de fouilles 2013*. Rapport de fouilles programmées. Picardie: Service Régional de l'Archéologie.

–. 2016a. L'enceinte rubanée de Menneville, Derrière le Village, et les structures associées (Aisne, France): de la diversité du traitement des défunts à la cohérence d'un système. *Gallia Préhistoire*: 29-92.

– (ed.). 2016b. *Menneville, Derrière le Village. Une enceinte du Néolithique ancien. Campagne de fouilles 2014*. Rapport de fouilles programmées. Picardie: Service Régional de l'Archéologie.

–. 2017. The final Linear Pottery Culture Enclosure at Menneville, Dep. Aisne, France: a complex ceremonial site, in *Salzmünde. Regel oder Ausnahme?*, Band 16: 561-74. Tagungen des Landesmuseums für Vorgeschichte Halle.

– (ed.). 2018. *Menneville, Derrière le Village. Une enceinte du Néolithique ancien (Aisne, Hauts-de-France). Campagne de fouilles 2017*. Rapport de fouilles programmées. Picardie: Service Régional de l'Archéologie.

–. 2019. La tombe à inhumation Michelsberg, in N. Vandamme & C. Colas (ed.) *Beaurieux, Cuiry-lès-Chaudardes, les Gravelines, Aisne, secteurs 2 et 3*, Rapport de fouilles: 83-91. Inrap NP.

– (ed.). 2020. *Menneville, Derrière le Village. Une enceinte du Néolithique ancien (Aisne, Hauts-de-France). Campagne de fouilles 2018*. Rapport de fouilles programmées. Picardie: Service Régional de l'Archéologie.

Thevenet, C., L. Hachem, M. Ilett, C. Hamon & P. Allard. forthcoming. Menneville, Derrière le Village. Nouvelles explorations, nouvelles observations sur l'enceinte rubanée, in *Le phénomène des enceintes dans le Néolithique du nord-ouest de l'Europe*.

Thiébault, S. 2005. L'apport du fourrage d'arbre dans l'élevage depuis le Néolithique. *Anthropozoologica* 40: 95-108.

Thomas, A. & F. Chenal. 2014. Le sexe et le genre dans la structuration funéraire: asymétrie ou dualité sociale des entités Cerny et Grossgartach du Néolithique moyen d'Europe occidentale, in 1839e réunion scientifique de la Société d'Anthropologie de Paris, 26: 39.

Thomas, A., F. Bostyn, P. Chambon, C. Hamon, Y. Maigrot & K. Meunier. 2018. Défunts et mobilier, in F. Bostyn, Y. Lanchon & P. Chambon (ed.) *Habitat du Néolithique ancien et nécropoles du Néolithique moyen I et II à Vignely, la Porte aux Bergers, Seine-et-Marne*: 375-85. Mémoire 64. Société Préhistorique Française.

Thomas, L.-V. 1975. *Anthropologie de la mort* (Bibliothèque Scientifique). Paris: Payot.

Toulemonde, F. 2010. L'alimentation végétale durant la Protohistoire ancienne en Île-de-France. Étude carpologique des sites de Gif-sur-Yvette, Rond-point de Corbeville (Essonne) et Villiers-sur-Seine, le Gros Buisson (Seine-et-Marne). *Revue Archéologique d'Ile de France*: 63-83.

–. 2013. Économie végétale et pratiques agricoles au Bronze final et au premier âge du Fer, de la côte de l'Île-de-France à la côte de Champagne.

Paris: Thèse de Doctorat, Université de Paris 1 Panthéon Sorbonne.

Tresset, A. 1996. Le rôle des relations homme/animal dans l'évolution économique et culturelle des sociétés des Vème et VIème millénaires en Bassin parisien. Approche éthno-zootechnique fondée sur les ossements animaux. Paris: Thèse de Doctorat, Université de Paris I – Panthéon-Sorbonne.

–. 1997. L'approvisionnement carné Cerny dans le contexte néolithique du Bassin parisien. *Mémoires du Musée de Préhistoire d'Ile-de-France*: 299-314.

–. 2005. Elevage, chasse et alimentation carnée, in F. Giligny (ed.) *Louvier, la Villette (Eure), un site néolithique moyen en zone humide.* Documents Archéologiques de l'Ouest: 249-62.

Twiss, K.C. 2008. Transformations in an early agricultural society: feasting in the southern Levantine Pre-Pottery Neolithic. *Journal of Anthropological Archaeology* 27: 418-42.

Uerpmann, M. 2001. Animaux sauvages et domestiques du Rubané "le plus ancien"(LBK 1) en Allemagne, in R.-M. Arbogast, C. Jeunesse & J. Schibler (ed.) *Rôle et statut de la chasse dans le Néolithique ancien danubien (5500-4900 av. J.-C), Premières rencontres danubiennes*: 57-75 (International Archäologie). Rahden/Westf.: Verlag Marie Leidorf GmbH.

Uerpmann, M. & H.-P. Uerpmann. 1997. Remarks on the faunal remains of some early farming communities in Central Europe. *Anthropozoologica* 25: 571-78.

Valéro, C., dir. 2008. Grez-sur-Loing, l'Epine (Seine-et-Marne). Vestiges du Néolithique, habitats de l'âge du Bronze final. Rapport de fouille, Inrap CIF.

Van de Velde, P. & L. Amkreutz. 2018. A world ends: the demise of the northwestern Bandkeramik. *Praehistorica Leidensia*: 19-36.

Van Gennep, A. 1981. *Les rites de passages*. Paris: Picard.

Viand, A., G. Auxiette & D. Bardel. 2008. L'habitat hallstattien de Milly-la-Forêt, le Bois Rond (Essonne). *Revue Archéologique d'Ile-de-France* 1: 133-68.

Vigne, J.-D. 1988. *Les mammifères post-glaciaires de Corse. Etude archéozoologique*. Paris: Gallia Préhistoire, CNRS, XXVIè supplément.

Villes, A. 1986. Une hypothèse: les sépultures de relégation dans les fosses d'habitat protohistorique en France septentrionale. *Anthropologie physique et Archéologie*, 167-74.

Von Den Driesch, A. 1992. Die Rolle der Tiere im Grablrult der Kulturgruppen Hinkelstein und Grossgartach, in *Der Tod im der Steinzeit*: 42-49. Hessisches Landesmuseum Darmstadt.

Weinstock, J. & K. Pasda. 2000. *Die tier-und menschenknochen aus dem erdwerk Calden.* Universitätsforschungen zur prähistorischen Archäologie.

Whittle, A. 2009. The people who lived in longhouses: what's the big idea, in *Creating communities: new advances in central European Neolithic research*: 249-63.

Whittle, A.W.R., F.M.A. Healy & A. Bayliss. 2011. *Gathering time: dating the early Neolithic enclosures of southern Britain and Ireland.* Oxbow Books.

Woimant, G.-P. 1990. Beauvais, les Aulnes du Canada: Viereckschanze ou l'enceinte quadrangulaire. *Revue Archéologique de Picardie*: 27-93. https://doi.org/10.3406/pica.1990.1611.

Yvinec, J.-H. 1987. Découpe, pelleterie et consommation des chiens gaulois à Villeneuve-Saint-Germain, in *La découpe et le partage du corps à travers le temps et l'espace*: 83-90. Numéro spécial 1. Anthropozoologica.

Zech-Matterne, V., G. Auxiette & F. Malrain. 2013. Essai d'approche des systèmes agricoles laténiens en France septentrionale. Données carpologiques, archéozoologiques et archéologiques, in S. Krausz, A. Colin, K. Gruel, I. Ralston, T. Dechezleprêtre (ed.) *L'âge du Fer en Europe. Mélanges offerts à Olivier Buchschenschutz*, Mémoire 32: 397-404. Bordeaux: Ausonius Editions.

Zimmerman, A. 2012. Das Hofplatzmodell – Entwicklung, Probleme, Perspektiven, in *Siedlungsstrukturen und Kulturwandel in der Bandkeramik. Neue Fragen zur Bandkeramik oder Alles beim Alten?* 25: 11-19. Dresden: Arbeits- und Forschungsberichte zur sächsischen Bodendenkmalpflege.

APPENDICES

Appendix 1: Aisne, Oise (Key: H:Hachem, AH: Auxiette and Hachem, HR: Hachem, and Robin M: Méniel)

Site	Missy-sur-Aisne, House 40	Missy-sur-Aisne, House 60	Missy-sur-Aisne, House 75	Missy-sur-Aisne, House 80	Missy-sur-Aisne, House 60-80	Missy-sur-Aisne, feature 14	Missy-sur-Aisne, feature 15	Missy-sur-Aisne, feature 55	Bucy-le-Long, House 10	Bucy-la-Fosselle, House 20	Bucy-le-Long, House 30	Bucy-le-Long, House 35	Bucy-le-Long, House 40	Bucy-le-Long, House 45	Bucy-le-Long, House 50	Bucy-le-Long, House 90	Bucy-le-Long, House 75	Bucy-le-Long, House 130	Berry-au-Bac, House 370	Berry-au-Bac, House 585	Berry-au-Bac, House 590	Berry-au-Bac, House 620
Field	Le Culot	Le Culot	Le Culot	Le Culot	Le Culot	Le Culot	Le Culot	Le Culot	La Fosselle	La Fosselle	La Fosselle	La Fosselle	La Fosselle	La Fosselle	La Fosselle	La Fosselle	La Fosselle	La Fosselle	Le Vieux Tordoir	Le Vieux Tordoir	Le Vieux Tordoir	Le Vieux Tordoir
Department	Aisne	Aisne	Aisne	Aisne	Aisne	Aisne	Aisne	Aisne	Aisne	Aisne	Aisne	Aisne	Aisne	Aisne	Aisne	Aisne	Aisne	Aisne	Aisne	Aisne	Aisne	Aisne
Cultural attribution	LBK	LBK	LBK	LBK	LBK	LBK	LBK	LBK	LBK	LBK	LBK	LBK	LBK	LBK	LBK	LBK	LBK	LBK	LBK	LBK	LBK	LBK
Ceramic stage	Aisne 3	Aisne 3	Aisne 3	Aisne 3?	Aisne 3?				Aisne 2	Aisne 3	Aisne 2	Aisne 2	Aisne 3	Aisne 2	Aisne 2?	Aisne 2?			Aisne 1	Aisne 1	Aisne 3	Aisne 1
Author	H	H	H	H	H	H	H	H	H	H	H	H	H	H	H	H	H	H	H	H	H	H
Cattle (*Bos taurus*)	12	402	17	90	41	27	2	24	204	882	608	11	54	71	32	149	8	5	25	51	74	93
Caprinae (*Ovis aries / Capra hircus*)	3	338	4	105	40	2			60	267	168	3	33	39	23	19			6	45	89	17
Pig (*Sus scrofa domesticus*)	5	67	8	24	10	3	1	2	34	111	82	4	16	23	5	10			1	13	15	13
Dog (*Canis familiaris*)											5											
Red-deer (*Cervus elaphus*)	3	21	2	12	3	5		2	4	16	33	1	1	10		6			1	5	39	12
Wild-boar (*Sus scrofa scrofa*)		11	2	0		1			3	3	24	1	3	5		1			1		10	6
Roe-deer (*Capreolus capreolus*)		7		1		1			1	9	11		1	5	1	4				2	20	
Aurochs (*Bos primigenius*)	2	16							2	12	13			5		1			2	3		20
Horse (*Equus sp.*)										1												
Wolf (*Canis lupus*)																						
Bear (*Ursus arctos*)																						
Beaver (*Castor fiber*)		46		9							3											
Badger (*Meles meles*)		3				2	1		1													
Hare (*Lepus europaeus*)																1						
Fox (*Vulpes vulpes*)									1	2						2				1		
Wild cat (*Felis sylvestris*)																						
Marten (*M. martes / M. foina*)									2							1						
Squirrel (*Sciurius vulgaris*)																1						
Hedgehog (*Erinaceus europaeus*)																						
Polecat (*Mustella putorius*)																						
Tortoise (*Emys orbicularis*)										1										1		
Mole (*Talpa europaea*)																						
Birds (*Aves*)		4		1						1				1								
Fish (*Pisces*)		5								1	2											
Amphibian (*Rana/Bufo sp.*)									1	5					3	1						
Microfauna																						
Cattle or Aurochs (*Bos sp.*)	1													1							1	5
Pig or Wild boar (*Sus sp.*)				1																		
Red-deer antlers						2			5					2	1	2			1	1		1
Roe-deer antlers	1	2							2	1	1	1		1	1							

Berry-au-Bac, House 630	Berry-au-Bac, pit 372	Berry-au-Bac, pit 627	Menneville, House 10	Menneville, House 35	Menneville, House 130	Menneville, House 140	Menneville, House 185	Menneville, House 200	Menneville eclosure, layer 1 (bottom), same level as the complex burials	Menneville eclosure, intermediate layer 2, same level as the isolated human remains	Menneville eclosure, upper layer 3 same level as the isolated human remains	Berry-au-Bac, House 195	Berry-au-Bac, House 200	Berry-au-Bac, House 300	Cuiry-lès-Chaudardes, Houses 45-90-112-126-390-640 (for details per house see Hachem 2011)	Cuiry-lès-Chaudardes, Houses 11-89-330-380-400-410-425-440-570-580 (for details per house see Hachem 2011)	Cuiry-lès-Chaudardes, Houses 225-245-280-360-420-500-520-530-690	Orainville	Vénizel, Features 134 and 136	Nogentel, pit 8	Vénizel, House 140	Orainville, House pits 37 and 32	Rivecourt	Saint-Ouen
Le Vieux Tordoir	Le Vieux Tordoir	Le Vieux Tordoir	Derrière le Village	Derrière le Village	Derrière le Village	Derrière le Village	Derrière le Village	Derrière le Village	Derrière le Village	Derrière le Village	Derrière le Village	Le Chemin de la Pêcherie	Le Chemin de la Pêcherie	Le Chemin de la Pêcherie	Les Fontinettes	Les Fontinettes	Les Fontinettes	La Bouguignotte	Le Creulet	Le Pré des Aulnaies	Le Creulet	Le Jardinet	La Saule Ferrée	L'Aumône
Aisne	Aisne	Aisne	Aisne	Aisne	Aisne	Aisne	Aisne	Aisne	Aisne	Aisne	Aisne	Aisne	Aisne	Aisne	Aisne	Aisne	Aisne	Aisne	Aisne	Aisne	Aisne	Aisne	Oise	Oise
LBK	LBK	LBK	LBK	LBK	LBK	LBK	LBK	LBK	LBK	LBK	LBK	LBK	LBK	LBK	LBK	LBK	LBK	LBK	LBK	LBK	LBK	BVSG	BVSG	BVSG
Aisne 3			Aisne 2	Aisne 3	Aisne 2?	Aisne 2	Aisne 3	Aisne 3				Aisne 1	Aisne 1	Aisne 1	Aisne 1	Aisne 2	Aisne 3	Aisne 3	Aisne 3				Final	
H	H	H	H	H	H	H	H	H	H	H	H	M	M	AH	H	H	H	H	H	H	H	H	HR	H
144	3	9	225	52	59	223	341	608	77	54	86	343	558	275	1414	3275	2877	66	7	20	47	24	86	6
142		8	40	13	13	61	140	231	45	15	14	130	190	51	471	749	1741	18	12	2	4	5	12	7
27	3		38	10	24	47	93	65		4	9	75	101	79	402	726	883	9		2		6	20	1
1			1	1	1	5		1	3						3	7	4							
36		2	16	5	2	23	19	16				6	13	11	235	303	198	34		1		2	4	3
22		1	17	5	4	5	3	4		1		6			245	326	204	9		1		3		
18			8	1	7	16	33	5				6	2	5	130	123	192	2				2	3	1
4			5	2		2	2					1	7	3	30	113	78	14		1		2	1	
																2	2						1	
																5	2							
							1									1								
					4	41	22				2				78	134	129							
		1		2	2					1				3	10	43	69							
					1						1				2	15	6	2						
															4	3	12							
								2							5	8	4						1	
1								3				2			2		3							
								1							2	10								
						1		1							2									
																1								
4						8	8	1		4	1				7	3	26	5						
					2	4										1	10							
3					1	7	7										11	2						
															9	2	1							
		1						1									8							
	1			1	1	1		3									3							
2				5		1	3	3	2	5		9			107	39	52	7					4	
2				1	2	1	3				1				9	12	6							1

Appendix 1: Aisne, Oise (Key: H:Hachem, AH: Auxiette and Hachem, HR: Hachem, and Robin M: Méniel)

Site	Missy-sur-Aisne, House 40	Missy-sur-Aisne, House 60	Missy-sur-Aisne, House 75	Missy-sur-Aisne, House 80	Missy-sur-Aisne, House 60-80	Missy-sur-Aisne, feature 14	Missy-sur-Aisne, feature 15	Missy-sur-Aisne, feature 55	Bucy-le-Long, House 10	Bucy-la-Fosselle, House 20	Bucy-le-Long, House 30	Bucy-le-Long, House 35	Bucy-le-Long, House 40	Bucy-le-Long, House 45	Bucy-le-Long, House 50	Bucy-le-Long, House 90	Bucy-le-Long, House 75	Bucy-le-Long, House 130	Berry-au-Bac, House 370	Berry-au-Bac, House 585	Berry-au-Bac, House 590	Berry-au-Bac, House 620
Total number of identified specimens	27	922	33	243	94	43	3	29	317	1315	949	21	112	164	62	198	8	5	35	122	250	167
Total of indeterminate fragments	81	1261	66	449		110	14	74	759	2455	1542	49	354	384	282	299	15	0	63	192	588	228
TOTAL	108	2183	99	692	94	153	17	103	1076	3770	2491	70	466	548	344	497	23	5	98	314	838	395
Total identified eaten	25	911	33	241	94	41	3	29	311	1307	945	20	108	158	61	195	8	5	34	120	250	161
Domestic mammals	20	807	29	219	91	32	3	26	298	1260	863	18	103	133	60	178	8	5	32	109	178	123
Wild mammals	5	104	4	22	3	9	0	3	13	47	81	2	5	25	1	17	0	0	2	10	72	38
% Domestic mammals	80	88.6	87.9	90.9	96.8	78		89.6	95.8	96.4	91.4	90	95.4	84.2	98.4	91.3				94.1	90.8	71.2	76.4
% Wild mammals	20	11.4	12.1	9.1	3.2	22		10.4	4.2	3.6	8.6	10	4.6	15.8	1.6	8.7				5.9	9.2	28.8	23.6

*Total number of identified remains without antlers, birds, fish, amphibians, microfauna, has been used to calcultate the % of domestic and wild animals

Appendix 2: Champagne-Ardenne

Site	Bréviandes, House 12	Bréviandes, House 13	Bréviandes, House 21	Bréviandes, House 22	Bréviandes, House 23	Buchères, House	Buchères, House (pits 91 and 108)	Buchères, isolated pit 123
Field	Zac Saint-Martin	Zac Saint-Martin	Zac Saint-Martin	Zac Saint-Martin	Zac Saint-Martin	PLA, décap. 39	PLA, décap 39	Les Bordes, décap. 11-12
Department	Aube	Aube	Aube	Aube	Aube	Aube	Aube	Aube
Cultural attribution	LBK	LBK	LBK	LBK	LBK	LBK	LBK	LBK
Ceramic stage								
Author	Hachem	Hachem	Hachem	Hachem	Hachem	Hachem	Hachem	Hachem
Cattle (*Bos taurus*)	244	846	374	957	44	35	221	103
Caprinae (*Ovis aries / Capra hircus*)	50	356	254	557	10	7	136	134
Pig (*Sus scrofa domesticus*)	48	212	87	83	8	7	18	67
Dog (*Canis familiaris*)		2	1				1	
Red-deer (*Cervus elaphus*)	52	13	22	21	5	5		21
Wild-boar (*Sus scrofa scrofa*)	3	3	2	1	1	3		6
Roe-deer (*Capreolus capreolus*)		13		5	1	1		3
Aurochs (*Bos primigenius*)	4	17	2	11	2	3	12	
Horse (*Equus sp.*)								
Wolf (*Canis lupus*)	1		1					
Bear (*Ursus arctos*)								
Beaver (*Castor fiber*)		1						2
Badger (*Meles meles*)								7
Hare (*Lepus europaeus*)		1						
Fox (*Vulpes vulpes*)				2				1

Berry-au-Bac, House 630	Berry-au-Bac, pit 372	Berry-au-Bac, pit 627	Menneville, House 10	Menneville, House 35	Menneville, House 130	Menneville, House 140	Menneville, House 185	Menneville, House 200	Menneville eclosure, layer 1 (bottom), same level as the complex burials	Menneville eclosure, intermediate layer 2, same level as the isolated human remains	Menneville eclosure, upper layer 3 same level as the isolated human remains	Berry-au-Bac, House 195	Berry-au-Bac, House 200	Berry-au-Bac, House 300	Cuiry-lès-Chaudardes, Houses 45-90-112-126-390-640 (for details per house see Hachem 2011)	Cuiry-lès-Chaudardes, Houses 11-89-330-380-400-410-425-440-570-580 (for details per house see Hachem 2011)	Cuiry-lès-Chaudardes, Houses 225-245-280-360-420-500-520-530-690	Orainville	Vénizel, Features 134 and 136	Nogentel, pit 8	Vénizel, House 140	Orainville, House pits 37 and 32	Rivecourt	Saint-Ouen
406	6	20	356	91	117	443	685	950	130	75	126	570	871	427	3154	5901	6522	180	19	27	51	44	133	
692	10	38	431	110	139	914	1527	1673	1	6	68	957	1539	704	6620	11448	15085	510	9	9	4	42	531	
1098	**16**	**58**	**787**	**201**	**256**	**1357**	**2212**	**2623**	**131**	**81**	**194**	**1527**	**2410**	**1131**	**9774**	**17349**	**21607**	**690**	**28**	**36**	**55**	**86**	**664**	
395	6	20	350	90	114	428	662	930	125	75	112	569	871	427	3031	5837	6415	154	19	27	51	44	128	
314	6	17	304	76	97	336	574	905	125	73	109	548	849	405	2290	4757	5505	93	19	24	51	35	118	
81	0	3	46	14	17	92	88	25	0	2	3	21	22	22	741	1080	910	61	0	3	0	9	10	
79.5			86.9	84.4	85.1	78.5	86.7	97.3	100	97.3	97.3	97.5	94.8	75.5	81.5	85.8	51.7					79.5	92.2	
20.5			13.1	15.6	14.9	21.5	13.3	2.7	0	2.7	2.7	2.5	5.2	24.4	18.5	14.2	38.3					20.5	7.8	

Buchères, isolated pits	Buchères, pits	Buchères	Buchères, House 1	Buchères, features 1563 and 1540	Buchères, House 1	Lesmont, House 1	Lesmont, House 2	Pont-sur-Seine, House 1	Pont-sur-Seine, House 2	Pont-sur-Seine, 7 isolated pits
Les Bordes, décap. 11-12	Les Bordes, décap. 11-12	Le Clos II	PLA, décap. 43-44	PLA, décap. 43-44	Rue de la Mairie	Les Graveries	Les Graveries	Marnay	Marnay	Marnay
Aube	Aube	Aube	Aube	Aube	Aube	Aube	Aube	Aube	Aube	Aube
LBK	BVSG	LBK	LBK	BVSG	LBK	LBK	LBK	LBK	LBK	LBK
					Final					
Hachem	Hachem	Djilali and Hachem	Hachem	Hachem	Hachem	Hachem	Hachem	Hachem	Hachem	Hachem
49	38	8	23		260	224	2	149	38	100
16	0	47	3	1	230	138	3	169	7	51
11	7	16	2		82	100		46	3	14
3	2		6	35	17	6		9		5
1	6			54	3	3		2		
		1		9	1	1		2		3
1				2	5	27		6		2
										1
										1
					1					

Appendix 2: Champagne-Ardenne

Site	Bréviandes, House 12	Bréviandes, House 13	Bréviandes, House 21	Bréviandes, House 22	Bréviandes, House 23	Buchères, House	Buchères, House (pits 91 and 108)	Buchères, isolated pit 123
Field	Zac Saint-Martin	Zac Saint-Martin	Zac Saint-Martin	Zac Saint-Martin	Zac Saint-Martin	PLA, décap. 39	PLA, décap 39	Les Bordes, décap. 11-12
Department	Aube	Aube	Aube	Aube	Aube	Aube	Aube	Aube
Cultural attribution	LBK	LBK	LBK	LBK	LBK	LBK	LBK	LBK
Ceramic stage								
Author	Hachem	Hachem	Hachem	Hachem	Hachem	Hachem	Hachem	Hachem
Wild cat (*Felis sylvestris*)								
Marten (*M. martes / M. foina*)		1						
Squirrel (*Sciurius vulgaris*)								1
Hedgehog (*Erinaceus europaeus*)		1	1	1				1
Polecat (*Mustella putorius*)								
Tortoise (*Emys orbicularis*)								
Mole (*Talpa europaea*)				1				
Birds (*Aves*)		2	1	2				5
Fish (*Pisces*)		1						2
Amphibian (*Rana/Bufo sp.*)								3
Microfauna								
Cattle or Aurochs (*Bos sp.*)	5				3		1	
Pig or Wild boar (*Sus sp.*)	2		1		1			
Red-deer antlers	5	5	6	4		1	2	
Roe-deer antlers	2	5		3		1		
Total number of identified specimens	416	1478	753	1648	75	63	391	356
Total of indeterminate fragments	1092	2423	1244	6347	76	41	171	562
TOTAL	1508	3901	1997	7995	151	104	562	918
Total identified eaten	402	1465	745	1638	71	61	388	346
Domestic mammals	342	1416	716	1597	62	49	376	304
Wild mammals	60	49	29	41	9	12	12	42
% Domestic mammals	82.2	96.7	96.1	97.3	82.7	80.3	96.9	87.9
% Wild mammals	14.4	3.3	3.9	2.7	12	19.7	3.1	12.1

*Total number of identified remains without antlers, birds, fish, amphibians, microfauna, has been used to calcultate the % of domestic and wild animals

Buchères, isolated pits	Buchères, pits	Buchères	Buchères, House 1	Buchères, features 1563 and 1540	Buchères, House 1	Lesmont, House 1	Lesmont, House 2	Pont-sur-Seine, House 1	Pont-sur-Seine, House 2	Pont-sur-Seine, 7 isolated pits
Les Bordes, décap. 11-12	Les Bordes, décap. 11-12	Le Clos II	PLA, décap. 43-44	PLA, décap. 43-44	Rue de la Mairie	Les Graveries	Les Graveries	Marnay	Marnay	Marnay
Aube	Aube	Aube	Aube	Aube	Aube	Aube	Aube	Aube	Aube	Aube
LBK	BVSG	LBK	LBK	BVSG	LBK	LBK	LBK	LBK	LBK	LBK
					Final					
Hachem	Hachem	Djilali and Hachem	Hachem	Hachem	Hachem	Hachem	Hachem	Hachem	Hachem	Hachem
		2				1				
					2					
										1
		1								
				1	2	6		2		1
					3			1		
2			1		4			6		1
	4							1		
83	57	75	35	102	609	507	5	393	48	179
42	40	56	26	51	335	83	0	364	20	165
125	97	131	61	153	944	590	5	757	68	344
81	53	74	34	101	600	501	5	383	48	177
76	45	71	28	1	572	462	5	364	48	165
5	8	3	6	100	28	39	0	19	0	12
93.8	84.9	95.9		1	95.4	92.2		95		93.2
6.2	15	4.1		99	4.6	7.8		5		6.8

Appendix 3: Ile-de-France

Site	Changis-sur-Marne, House 100	Changis-sur-Marne, House 115	Changis-sur-Marne	Marolles-sur-Seine	Balloy, House 1	Balloy, House 3	Balloy, House 4	Limay, pits
Field	Les Pétraux	Les Pétraux	Les Pétraux	Le Calvaire	La Haute Borne	La Haute Borne	La Haute Borne	Rue Nationale
Department	Seine-et-Marne	Seine-et-Marne	Seine-et-Marne	Seine-et-Marne	Seine-et-Marne	Seine-et-Marne	Seine-et-Marne	Yvelines
Cultural attribution	LBK	LBK	LBK	BVSG	BVSG	BVSG	BVSG	BVSG
Ceramic stage	Final	Final	Final	Final	Early	Early	Early	
Author	Hachem	Hachem	Hachem	Hachem	Hachem	Hachem	Hachem	Hachem
Cattle (*Bos taurus*)	180	116	71	32	109	7	3	9
Caprinae (*Ovis aries / Capra hircus*)	125	132	48	13	135	12		1
Pig (*Sus scrofa domesticus*)	66	38	21	25	49	2		1
Dog (*Canis familiaris*)	1							
Red-deer (*Cervus elaphus*)	7	10	1		9	2		
Wild-boar (*Sus scrofa scrofa*)	1	0	3		13	1		
Roe-deer (*Capreolus capreolus*)	1	1		1	5			
Aurochs (*Bos primigenius*)	3	2			6			
Horse (*Equus sp.*)								
Wolf (*Canis lupus*)								
Bear (*Ursus arctos*)								
Beaver (*Castor fiber*)				1				
Badger (*Meles meles*)	7	4						
Hare (*Lepus europaeus*)			2		2			
Fox (*Vulpes vulpes*)								
Wild cat (*Felis sylvestris*)								
Marten (*M. martes / M. foina*)								
Squirrel (*Sciurius vulgaris*)								
Hedgehog (*Erinaceus europaeus*)								
Polecat (*Mustella putorius*)								
Tortoise (*Emys orbicularis*)								
Mole (*Talpa europaea*)								
Birds (*Aves*)	2	2	1					
Fish (*Pisces*)					16			
Amphibian (*Rana/Bufo sp.*)					2			
Microfauna								
Cattle or Aurochs (*Bos sp.*)	11	4	1		2			
Pig or Wild boar (*Sus sp.*)	3	4					1	
Red-deer antlers	3	1	3					
Roe-deer antlers			2					
Total of indeterminate fragments	360	325	202	24	785	16	0	22
TOTAL	**770**	**639**	**355**	**96**	**1133**	**41**	**3**	**33**
Total identified eaten	391	303	146	72	328	24	3	11
Domestic mammals	372	286	140	70	293	21	3	11
Wild mammals	19	17	6	2	35	3	0	0
% Domestic mammals	95.2	94.4	95.9	97.2	89.3			
% Wild mammals	4.8	5.6	4.1	2.8	10.7			

*Total number of identified remains without antlers, birds, fish, amphibians, microfauna, has been used to calcultate the % of domestic and wild animals

Appendix 4: Aisne, Oise

Site	Osly-Courtil (enclosure)	Saint-Quentin (enclosure)	Beaurieux (cultural layer monumental builing)	Cuiry-lès-Chaudardes (pits)	Crécy-sur Serre (enclosure)	Maizy (enclosure)	Bazoches-sur-Vesle (enclosure)	Pontavert (enclosure)	Choisy-au-Bac (cultural layer)	Passel (enclosure)
Field	La Terre St Mard	Le Chemin de Harly	La Plaine	Les Fontinettes	La Croix St Jacques	Les Grands Aisements	Le Muisemont	Chemin de Beaurieux	Le Confluent	Le Vivier
Department	Aisne	Aisne	Aisne	Aisne	Aisne	Aisne	Aisne	Aisne	Oise	Oise
Cultural attribution	Middle Neolithic 1	Middle Neolithic 1	Middle Neolithic 1	Middle Neolithic 2	Middle Neolithic 2	Middle Neolithic 2	Middle Neolithic 2	Middle Neolithic 2	Middle Neolithic 2	Middle Neolithic 2
Ceramic stage	Post-Rössen	Post-Rössen	Cerny	Michelsberg	Michelsberg	Michelsberg	Michelsberg	Michelsberg	Chasséen	Chasséen
Author	Hachem	Hachem	Hachem	Hachem	Hachem	Hachem	Hachem	Hachem	Hachem	Hachem, Bedault, Leduc
Cattle (Bos taurus)	728	89	69	81	257	766	493	50	138	5587
Caprinae (Ovis aries / Capra hircus)	132	36	1	40	68	184	100	13	8	310
Pig (Sus scrofa domesticus)	383	24	9	77	103	457	215	20	15	2090
Dog (Canis familiaris)	3			0	1	2	5		0	246
Red-deer (Cervus elaphus)	13	1	6	43	69	102	20	1	27	36
Wild-boar (Sus scrofa scrofa)	9	2	1	18	12	13	2		1	84
Roe-deer (Capreolus capreolus)	1	1		3	3	45	7		1	8
Aurochs (Bos primigenius)	5		6	1	21	41	3	5	4	61
Horse (Equus sp.)		1			1	2	1			5
Wolf (Canis lupus)				1					1	2
Bear (Ursus arctos)							1			3
Beaver (Castor fiber)	1				4	4	1		1	
Badger (Meles meles)	2					7		1		
Hare (Lepus europaeus)	2	2		9	8	2			0	8
Fox (Vulpes vulpes)						9				
Wild cat (Felis sylvestris)						13				
Marten (M. martes / M. foina)						4				
Squirrel (Sciurius vulgaris)										
Hedgehog (Erinaceus europaeus)						1				
Polecat (Mustella putorius)						1				1
Tortoise (Emys orbicularis)				2						
Mole (Talpa europaea)		1								
Birds (Aves)	1	1		6		17		1		3
Fish (Pisces)				1	5	3		7		
Amphibian (Rana/Bufo sp.)		5		1		16		6		
Reptiles (Reptilis sp.)										1
Microfauna				1				2		
Cattle or Aurochs (Bos sp.)	3		1	2	8				18	92
Caprinae or Roe-deer (Ovis or Capreolus)	3									
Pig or Wild boar (Sus sp.)	6		1	9	12				1	20
Red-deer antlers	3	3	8		17		28	1	7	13
Roe-deer antlers					1		2			1
Total number of identified specimens	1295	166	102	295	590	1689	878	107	222	8571
Total of indeterminate fragments	1647	106	247	272	644	2482	3085	223	not counted	2091
TOTAL	**2942**	**272**	**349**	**567**	**1234**	**4171**	**3963**	**330**	**222**	**10662**
*Total identified eaten	1279	156	92	273	547	1653	848	90	196	8441
Domestic mammals	1246	149	79	198	429	1409	813	83	161	8233
*Wild mammals	33	7	13	75	118	244	35	7	35	208
% Domestic mammals	97.4	95.5	85.9	72.5	78.4	85.2	95.9	92.2	18.2	97.5
% Wild mammals	2.6	4.5	14.1	27.5	21.6	14.8	4.1	7.8	17.8	2.5

*Total number of identified remains without antlers, birds, fish, tortoise, amphibians, microfauna, has been used to calcultate the % of domestic and wild animals

Appendix 5: Ile-de-France

Site	Maisons-Alfort (enclosure)	Vitry-sur-Seine (pit)	Vignely	Maisons-Alfort (enclosure)	Limay (pits)
Field	ZAC d'Alfort, ALF III	Rue du Génie	La Noue Fenard	ZAC d'Alfort, ALF III	Rue Nationale
Department	Val-de-Marne	Val-de-Marne	Seine-et-Marne	Val-de-Marne	Yvelines
Cultural attribution	Middle Neolithic 1	Middle Neolithic 1	Middle Neolithic 2	Middle Neolithic 2	Middle Neolithic 2
Ceramic stage	Cerny	Cerny	Michelsberg	Chasséen	Chasséen
Author	Hachem	Hachem	Claudet, Hachem	Hachem	Hachem
Cattle (Bos taurus)	18		352	406	9
Caprinae (Ovis aries / Capra hircus)	2	3	34	76	3
Pig (Sus scrofa domesticus)	3		79	96	
Dog (Canis familiaris)			7	3	
Red-deer (Cervus elaphus)	5	10	39	53	1
Wild-boar (Sus scrofa scrofa)				43	
Roe-deer (Capreolus capreolus)			1	5	
Aurochs (Bos primigenius)			20	7	
Horse (Equus sp.)			3	1	
Wolf (Canis lupus)				2	
Bear (Ursus arctos)					
Beaver (Castor fiber)				4	
Badger (Meles meles)			1	4	
Hare (Lepus europaeus)					
Fox (Vulpes vulpes)			1		
Wild cat (Felis sylvestris)					
Marten (M. martes / M. foina)					
Squirrel (Sciurius vulgaris)					
Hedgehog (Erinaceus europaeus)					
Polecat (Mustella putorius)					
Tortoise (Emys orbicularis)					
Mole (Talpa europaea)					
Birds (Aves)			2		
Fish (Pisces)					
Amphibian (Rana/Bufo sp.)					
Microfauna					
Cattle or Aurochs (Bos sp.)				9	
Caprinae or Roe-deer (Ovis or Capreolus)				2	
Pig or Wild boar (Sus sp.)				8	
Red-deer antlers	1		22	7	
Roe-deer antlers		3	2	2	
Total number of identified specimens	29	16	563	728	13
Total of indeterminate fragments	50	4	904	1333	2
TOTAL	**79**	**20**	**1467**	**2061**	**15**
*Total identified eaten	28	13	537	700	13
Domestic mammals	23	3	472	581	12
*Wild mammals	5	10	65	119	1
% Domestic mammals			87.9	83	
% Wild mammals			12.1	17	

*Total number of identified remains without antlers, birds, fish, amphibians, microfauna, has been used to calcultate the % of domestic and wild animals

Appendix 6: Champagne-Ardenne

Site	Pont-sur-Seine, enclosure	La Saulsotte, enclosure	Buchères, monument STP
Field	Ferme de l'Ile	Le Vieux Bouchy	Parc Logistique, D39
Department	Aube	Aube	Aube
Cultural attribution	Middle Neolithic 2	Middle Neolithic 1	Middle Neolithic 1
Ceramic stage	finale	Cerny	Cerny
Author	Hachem	Hachem	Hachem
Cattle (*Bos taurus*)	294	12	27
Caprinae (*Ovis aries / Capra hircus*)	39	7	2
Pig (*Sus scrofa domesticus*)	129	6	10
Dog (*Canis familiaris*)	1		
Red-deer (*Cervus elaphus*)	167	5	
Wild-boar (*Sus scrofa scrofa*)	48	4	
Roe-deer (*Capreolus capreolus*)	8		
Aurochs (*Bos primigenius*)	55	2	
Horse (*Equus sp.*)	4		
Wolf (*Canis lupus*)			
Bear (*Ursus arctos*)			
Beaver (*Castor fiber*)	5		
Badger (*Meles meles*)	2		
Hare (*Lepus europaeus*)			
Fox (*Vulpes vulpes*)			
Wild cat (*Felis sylvestris*)			
Marten (*M. martes / M. foina*)			
Squirrel (*Sciurius vulgaris*)			
Hedgehog (*Erinaceus europaeus*)			
Polecat (*Mustella putorius*)			
Tortoise (*Emys orbicularis*)			
Mole (*Talpa europaea*)			
Birds (*Aves*)			
Fish (*Pisces*)			
Amphibian (*Rana/Bufo sp.*)			
Microfauna			
Cattle or Aurochs (*Bos sp.*)	11		
Pig or Wild boar (*Sus sp.*)	12		
Red-deer antlers	5		
Roe-deer antlers	1		
Total number of identified specimens	781	36	39
Total of indeterminate fragments	898	15	3
TOTAL	**1679**	**51**	**42**
Total identified eaten	752	36	39
Domestic mammals	463	25	39
Wild mammals	289	11	0
% *Domestic mammals*	61.6		
% *Wild mammals*	38.4		

*Total number of identified remains without antlers, birds, fish, amphibians, microfauna, has been used to calcultate the % of domestic and wild animals

Appendix 7: Somme, Pas-de-Calais

Site	Conty (pits)	Villers-Carbonnel (enclosure)	Carvin (enclosure)	Escalles (enclosure)
Field	Zac Dunant	La Sole d'Applincourt	La Gare d'Eau	Le Mont d'Hubert
Department	Somme	Somme	Pas-de-Calais	Pas-de-Calais
Cultural attribution	Middle Neolithic 1	Middle Neolithic 2	Middle Neolithic 2	Middle Neolithic 2
Ceramic stage	Cerny	Chasséen	Spiere	Spiere
Author	Hachem	Hachem	Hachem	Hachem, Chombart
Cattle (*Bos taurus*)	187	1680	147	4974
Caprinae (*Ovis aries / Capra hircus*)	21	395	60	2940
Pig (*Sus scrofa domesticus*)	146	591	37	1461
Dog (*Canis familiaris*)	4	7	27	24
Red-deer (*Cervus elaphus*)	9	21	7	6
Wild-boar (*Sus scrofa scrofa*)	10	16		8
Roe-deer (*Capreolus capreolus*)	3	7		7
Aurochs (*Bos primigenius*)	3	5	5	9
Horse (*Equus sp.*)				1
Wolf (*Canis lupus*)				
Bear (*Ursus arctos*)				
Beaver (*Castor fiber*)				
Badger (*Meles meles*)	7			
Hare (*Lepus europaeus*)	2		1	1
Fox (*Vulpes vulpes*)	1		4	1
Wild cat (*Felis sylvestris*)		1		1
Marten (*M. martes / M. foina*)				
Squirrel (*Sciurius vulgaris*)				
Hedgehog (*Erinaceus europaeus*)	—	1		
Polecat (*Mustella putorius*)				
Tortoise (*Emys orbicularis*)				
Mole (*Talpa europaea*)				
Birds (*Aves*)	3	1		28
Fish (*Pisces*)				
Amphibian (*Rana/Bufo sp.*)		2		4
Microfauna				
Cattle or Aurochs (*Bos sp.*)	1			3
Pig or Wild boar (*Sus sp.*)		5		
Red-deer antlers	2		10	4
Roe-deer antlers	2	10		
Total number of identified specimens	401	4	298	9472
Total of indeterminate fragments	784	1	377	3373
TOTAL	**1185**	**2747**	**675**	**12845**
Total identified eaten	393	1000	288	9433
Domestic mammals	358	**3747**	271	9399
Wild mammals	35	2724	17	34
% Domestic mammals	91	2673	94	99.6
% Wild mammals	9	51	6	0.4

Total number of identified remains without antlers, birds, fish, tortoise, amphibians, microfauna, has been used to calculate the % of domestic and wild animals

	98.1
	1.9

Appendix 8: Ile-de-France

Site	Cuiry-lès-Chaudardes (pits)	Mareuil-lès-Meaux (cultural layer)	Varenne-sur-Seine (pits)
Field	Les Fontinettes	les Lignères	Volstin
Department	Aisne	Seine-et-Marne	Seine-et-Marne
Cultural attribution	Late Neolithic	Late Neolithic	Late Neolithic
Ceramic stage		Middle	
Author	Hachem	Hachem	Hachem
Cattle (*Bos taurus*)	1	28	46
Caprinae (*Ovis aries / Capra hircus*)	2	6	8
Pig (*Sus scrofa domesticus*)	112	26	14
Dog (*Canis familiaris*)		1	
Red-deer (*Cervus elaphus*)	1	1	7
Wild-boar (*Sus scrofa scrofa*)			
Roe-deer (*Capreolus capreolus*)			
Aurochs (*Bos primigenius*)			
Horse (*Equus sp.*)			
Wolf (*Canis lupus*)			
Bear (*Ursus arctos*)			
Beaver (*Castor fiber*)			
Badger (*Meles meles*)			
Hare (*Lepus europaeus*)			
Fox (*Vulpes vulpes*)			
Wild cat (*Felis sylvestris*)			
Marten (*M. martes / M. foina*)			
Squirrel (*Sciurius vulgaris*)			
Hedgehog (*Erinaceus europaeus*)			
Polecat (*Mustella putorius*)			
Tortoise (*Emys orbicularis*)			
Mole (*Talpa europaea*)			
Birds (*Aves*)			
Fish (*Pisces*)			
Amphibian (*Rana/Bufo sp.*)			
Reptiles (*Reptilis sp.*)			
Microfauna			1
Cattle or Aurochs (*Bos sp.*)			
Caprinae or Roe-deer (*Ovis or Capreolus*)			
Pig or Wild boar (*Sus sp.*)		1	
Red-deer antlers	2		
Roe-deer antlers		63	76
Total number of identified specimens	118	47	31
Total of indeterminate fragments	57	110	107
TOTAL	**175**	**62**	**75**
Total identified eaten	116	62	75
Domestic mammals	115	61	68
Wild mammals	1	1	7
% Domestic mammals	99.1	98.4	90.6
% Wild mammals	0.9	1.6	9.3

*Total number of identified remains without antlers, birds, fish, tortoise, amphibians, microfauna, has been used to calcultate the % of domestic and wild animals

Appendix 9: Champagne-Ardenne

Site	Pont-sur-Seine, palissade	Pont-sur-Seine, cultural layer	Pont-sur-Seine, enclosure
Field	les Hauts de Launoy	les Hauts de Launoy	les Hauts de Launoy
Department	Aube	Aube	Aube
Cultural attribution	Late Neolithic	Late Neolithic	Late Neolithic
Ceramic stage			
Author	Hachem	Hachem	Hachem
Cattle (*Bos taurus*)	23	41	
Caprinae (*Ovis aries / Capra hircus*)			
Pig (*Sus scrofa domesticus*)		2	
Dog (*Canis familiaris*)			
Red-deer (*Cervus elaphus*)			
Wild-boar (*Sus scrofa scrofa*)			
Roe-deer (*Capreolus capreolus*)			
Aurochs (*Bos primigenius*)			
Horse (*Equus sp.*)			5
Wolf (*Canis lupus*)			
Bear (*Ursus arctos*)			
Beaver (*Castor fiber*)			
Badger (*Meles meles*)			
Hare (*Lepus europaeus*)			
Fox (*Vulpes vulpes*)			
Wild cat (*Felis sylvestris*)			
Marten (*M. martes / M. foina*)			
Squirrel (*Sciurius vulgaris*)			
Hedgehog (*Erinaceus europaeus*)			
Polecat (*Mustella putorius*)			
Tortoise (*Emys orbicularis*)			
Mole (*Talpa europaea*)			
Birds (*Aves*)			
Fish (*Pisces*)			
Amphibian (*Rana/Bufo sp.*)			
Microfauna			
Cattle or Aurochs (*Bos sp.*)			
Pig or Wild boar (*Sus sp.*)			
Red-deer antlers			
Roe-deer antlers			
Total number of identified specimens	23	43	5
Total of indeterminate fragments	17	17	0
TOTAL	**40**	**60**	**5**
Total identified eaten	23	43	5
Domestic mammals	23	43	0
Wild mammals	0	0	5
% Domestic mammals			
% Wild mammals			

*Total number of identified remains without antlers, birds, fish, amphibians, microfauna, has been used to calcultate the % of domestic and wild animals

Appendix 10: Somme

Site	Glisy, House 1	Glisy, House 2	Glisy, House 3
Field	Zac Jules Verne	Zac Jules Verne	Zac Jules Verne
Department	Somme	Somme	Somme
Cultural attribution	Final Neolithic	Final Neolithic	Final Neolithic
Ceramic stage			
Author	Hachem	Hachem	Hachem
Cattle (*Bos taurus*)	18		2
Caprinae (*Ovis aries / Capra hircus*)	33	1	6
Pig (*Sus scrofa domesticus*)	19	2	5
Dog (*Canis familiaris*)			1
Red-deer (*Cervus elaphus*)	1		
Wild-boar (*Sus scrofa scrofa*)			
Roe-deer (*Capreolus capreolus*)			
Aurochs (*Bos primigenius*)			
Horse (*Equus sp.*)	1		
Wolf (*Canis lupus*)			
Bear (*Ursus arctos*)			
Beaver (*Castor fiber*)			
Badger (*Meles meles*)			
Hare (*Lepus europaeus*)			
Fox (*Vulpes vulpes*)			
Wild cat (*Felis sylvestris*)			
Marten (*M. martes / M. foina*)			
Squirrel (*Sciurius vulgaris*)			
Hedgehog (*Erinaceus europaeus*)			1
Polecat (*Mustella putorius*)			
Tortoise *(Emys orbicularis)*			
Mole (*Talpa europaea*)			
Birds (*Aves*)			
Fish (*Pisces*)			
Amphibian *(Rana/Bufo sp.)*			
Reptiles (*Reptilis sp.*)			
Microfauna			
Cattle or Aurochs (*Bos sp.*)			
Caprinae or Roe-deer (*Ovis or Capreolus*)			1
Pig or Wild boar (*Sus sp.*)			
Red-deer antlers			
Roe-deer antlers			
Total number of identified specimens	72	3	16
Total of indeterminate fragments	70	4	22
TOTAL	142	7	38
Total identified eaten	72	3	15
Domestic mammals	70	3	14
**Wild mammals*	2	0	1
% Domestic mammals	97.2	100	93.3
% Wild mammals	2.8	0	6.7

*Total number of identified remains without antlers, birds, fish, tortoise, amphibians, microfauna, has been used to calcultate the % of domestic and wild animals

Appendix 11: Aisne

Site	Bazoches-sur-Vesle	Beaurieux	Beaurieux	Berry-au-Bac	Bucy-le-Long	Ciry-Salsogne	Cuiry-lès-Chaudardes	Limé
Field	La Foulerie	Les Grèves	Les Grèves	Le Chemin de la Pêcherie	Le Grand Marais	la Bouche à Vesle	Les Fontinettes	Le Gros Buisson
Department	Aisne	Aisne	Aisne	Aisne	Aisne	Aisne	Aisne	Aisne
Cultural attribution	Bronze	Bronze	Bronze	Bronze	Bronze	Bronze	Bronze	Bronze
Phase/Etape	BzeIIIbHallC	BzeIIbIIIa	BzeIIIb	BzefIIbIIIa	BzeIIIb	BzeIII	BzeIIIbHallB	BzeIIIb
Author	Auxiette	Auxiette	Auxiette	Auxiette	Auxiette	Auxiette	Auxiette	Auxiette
Cattle (*Bos taurus*)	27	10	111	14	12	4	39	38
Caprinae (*Ovis aries / Capra hircus*)		8	45	25	2	9	36	11
Pig (*Sus scrofa domesticus*)	6	40	38	26	10	23	80	10
Horse (*Equus caballus*)	1		3	1	1		1	3
Dog (*Canis familiaris*)			11	3			13	
Red-deer (*Cervus elaphus*)	3		21	8	3	3	7	2
Wild-boar (*Sus scrofa scrofa*)							4	1
Roe-deer (*Capreolus capreolus*)				2			1	
Aurochs (*Bos primigenius*)			7					
Wolf (*Canis lupus*)								
Bear (*Ursus arctos*)								
Beaver (*Castor fiber*)			1					
Badger (*Meles meles*)			5					
Hare (*Lepus europaeus*)			1					
Fox (*Vulpes vulpes*)								
Wild cat (*Felis sylvestris*)								
Marten (*M. martes / M. foina*)								
Hedgehog (*Erinaceus europaeus*)								
Polecat (*Mustella putorius*)								
Mole (*Talpa europaea*)								
Tortoise (*Emys orbicularis*)								
Birds (*Aves*)			3	4		8	1	1
Fish (*Pisces*)			8				1	
Total number of identified specimens	37	58	254	83	28	47	183	66
Total of indeterminate fragments	29	63	243	83	0	55	156	38
TOTAL	**66**	**121**	**497**	**166**	**28**	**102**	**339**	**104**
Domestic mammals	34	58	208	69	25	36	169	62
Wild mammals	3	0	35	10	3	3	12	3
% Domestic mammals	91.9	100	81.9	83.1	89.3	76.6	92.3	93.9
% Wild mammals	8.1	0	13.8	12.0	10.7	6.4	6.6	4.5
% birds	0	0	1.2	4.8	0.0	17.0	0.5	1.5
% fish	0	0	3.1	0.0	0.0	0.0	0.5	0.0

Limé	Limé	Menneville	Menneville	Menneville	Osly-Courtil	Pasly	Pasly	Soupir	Vasseny
Le Gros Buisson	Les Fussis	La Bourguignotte	La Bourguignotte	Derrière le Village	La Terre Saint Mard	Les Côteaux	Derrière Longpont	le Parc	le dessous des Groins
Aisne	Aisne	Aisne	Aisne	Aisne	Aisne	Aisne	Aisne	Aisne	Aisne
Bronze	Bronze	Bronze	Bronze	Bronze	Bronze	Bronze	Bronze	Bronze	Bronze
BzeII	BzeIIIb	BzeIII	BzeIIIaIIIb	BzeIIIbHallB	BzeIIIb	BzeIIIa	BzeIIIbHallC	Bronze final	Bronze final
Auxiette	Auxiette	Auxiette	Auxiette	Auxiette	Auxiette	Auxiette	Auxiette	Auxiette	Auxiette
38	15	36	54	95	132	48	24	22	46
11	1	155	167	112	259	30	8	13	15
10	24	73	76	160	247	78	29	6	63
3		4	4	5	5	1	1		11
			3	2	3			7	
2		4		3	41	7	3	4	2
1	1	1			61	4	2		
	2				7	4	9		
		8			12				1
				1	4	17	1		
						1	2		
		1							
						1			
1				1	1				
							1		
66	43	282	304	379	772	191	80	52	138
24	88	63		557	449	58	50	6	51
90	**131**	**345**	**304**	**936**	**1221**	**249**	**130**	**58**	**189**
62	40	408	304	374	646	157	62	48	135
3	3	236	0	4	125	33	17	4	3
93.9	93.0	10.0	100.0	98.7	83.7	82.2	77.5	92.3	97.8
4.5	7.0	374.6	0.0	1.1	16.2	17.3	21.3	7.7	2.2
1.5	0.0	15.9	0.0	0.3	0.1	0.0	0.0	0.0	0.0
0.0	0.0	0.0	0.0	0.0	0.0	0.0	1.3	0.0	0.0

Appendix 12: Ile-de-France

Site	Choisy-au-Bac	Choisy-au-Bac	Changis-sur-Marne	Changis-sur-Marne	Larchant	Lieusaint	Lieusaint	Noyen-sur-Seine
Field	Canal Seine-Nord	Canal Seine-Nord	Les Pétreaux	Les Pétreaux	Les Groues	les Perpignans-Zac du Levant	les Perpignans-Zac du Levant	Nord du Bois du Chêne
Department	Oise	Oise	Seine-et-Marne	Seine-et-Marne	Seine-et-Marne	Seine-et-Marne	Seine-et-Marne	Seine-et-Marne
Cultural attribution	Bronze	Hallstatt ancien	Bronze	Bze/Hall ancien	Bronze	Bronze	Bronze	Bronze
Phase/Etape	BzeIIIb	HallC	RSFO	BzeIIIbHallC	BzeIIIb	BzeIIb	BzeIIIaIIIb	BzeIIb
Author	Auxiette & Bedault	Auxiette & Bedault	Auxiette	Auxiette	Auxiette	Auxiette	Auxiette	Auxiette
Cattle (*Bos taurus*)	2079	331	11	1084	3	33	8	32
Caprinae (*Ovis aries / Capra hircus*)	762	34	22	1115	2	55	3	40
Pig (*Sus scrofa domesticus*)	2511	540	14	616	12	131	3	20
Horse (*Equus caballus*)	38	5		129		2	3	2
Dog (*Canis familiaris*)	43	1	4	13	1	6		2
Red-deer (*Cervus elaphus*)	121	17	2	73		1		6
Wild-boar (*Sus scrofa scrofa*)	20	3		79				1
Roe-deer (*Capreolus capreolus*)	2			3				
Aurochs (*Bos primigenius*)				13				
Wolf (*Canis lupus*)								
Bear (*Ursus arctos*)		1		1				
Beaver (*Castor fiber*)	2	1		5				1
Badger (*Meles meles*)	3							1
Hare (*Lepus europaeus*)				5				
Fox (*Vulpes vulpes*)				1				1
Wild cat (*Felis sylvestris*)								
Marten (*M. martes / M. foina*)								
Otter (*Lutra lutra*)								
Hedgehog (*Erinaceus europaeus*)								
Polecat (*Mustella putorius*)								
Mole (*Talpa europaea*)		1						
Tortoise (*Emys orbicularis*)								1
Birds (*Aves*)	6			24				5
Fish (*Pisces*)	1			1				
Total number of identified specimens	5588	934	53	3162	18	228	17	112
Total of indeterminate fragments	4233	970	83	4419	4	169	6	37
TOTAL	**9821**	**1904**	**136**	**7581**	**22**	**397**	**23**	**149**
Domestic mammals	5433	911	51	2957	18	227	17	96
Wild mammals	148	22	2	180	0	1	0	10
% Domestic mammals	97.2	97.5	96.2	93.5	100.0	99.6	100.0	85.7
% Wild mammals	2.6	2.4	3.8	5.7	0.0	0.4	0.0	8.9
% birds	0.1	0.0	0.0	0.8	0.0	0.0	0.0	4.5
% fish	0.0	0.0	0.0	0.0	0.0	0.0	0.0	0.0

Noyen-sur-Seine	Varennes-sur-Seine	Varennes-sur-Seine	Varennes-sur-Seine	Varennes-sur-Seine	Varennes-sur-Seine	Ville-Saint-Jacques	Ville-Saint-Jacques	Villiers-sur-Seine	Maisse	Wissous
Nord du Bois du Chêne	La Justice	La Justice	Le Marais du Colombier	Beauchamps	La Justice	Le Bois d'Echalat	Le Bois d'Echalat	le Gros Buisson	La Plaine-Saint-Eloi	Zone Sud-Ouest aéroport Orly
Seine-et-Marne	Seine-et-Marne	Seine-et-Marne	Seine-et-Marne	Seine-et-Marne	Seine-et-Marne	Seine-et-Marne	Seine-et-Marne	Seine-et-Marne	Essonne	Essonne
Bronze	Bronze	Bronze	Bronze	Bronze	Bronze	Bronze	Bronze	Bronze	Bronze	Bronze
BzeIIbIIIa	BzeI-IIa	BronzeIIIb	BzeI	RSFO	BzeIIIb	BzeI-IIa	BzeIIIbHallB	BzeIIIb	Bronze final	BzeIIIa
Auxiette	Auxiette	Auxiette	Auxiette	Auxiette	Auxiette	Auxiette	Auxiette	Auxiette	Auxiette	Auxiette
94	15	2	24	3	2	2	48	1598	24	179
73	4	5	25	4	5	4	15	1695	6	46
51	1	5	6	2	5	8	81	10497	26	70
5							1	48	2	3
7			1	1		1	2	38		11
1				1		1		1862	1	18
								578	1	
							1	76		2
								27		
								9		
								7		
								40		7
								10		
				1				3		
								2		
1								4		
								1		
				1				7		
								41		2
								4		
232	20	12	56	13	12	17	147	16547	60	338
329	2	14	87	10	14	12	76	7341	47	333
561	**22**	**26**	**143**	**23**	**26**	**29**	**223**	**23888**	**107**	**671**
230	20	12	56	10	12	16	146	13876	35	309
2	0	0	0	2	0	1	1	2619	1	27
99.1	100.0	100.0	100.0	76.9	100.0	94.1	99.3	83.9	58.3	91.4
0.9	0.0	0.0	0.0	15.4	0.0	5.9	0.7	15.8	1.7	8.0
0.0	0.0	0.0	0.0	0.0	0.0	0.0	0.0	0.2	0.0	0.6
0.0	0.0	0.0	0.0	0.0	0.0	0.0	0.0	0.0	0.0	0.0

Appendix 13: Champagne-Ardenne, Hauts-de-France, Normandie

Site	Saint-Léger-Près-Troyes/Buchères	Saint-Léger-Près-Troyes/Buchères	Saint-Léger-Près-Troyes/Buchères	Saint-Léger-Près-Troyes/Buchères	Saint-Léger-Près-Troyes/Buchères	Bezanne
Field	Parc logistique de l'Aube-Décapage 31	Parc logistique de l'Aube-Décapage 19	Parc logistique de l'Aube-Décapage 19	Parc logistique de l'Aube-Décapage 19	Parc logistique de l'Aube-Décapage 39	la Bergerie
Department	Aube	Aube	Aube	Aube	Aube	Marne
Cultural attribution	Bronze	Bronze	Bronze	Bronze	Bronze	Bronze
Phase/Etape	BzeIIa	BzeIIa	BzeIIbIIIa	BzeIIIbHallC	BzeIIIb	BzeIIIb
Author	Auxiette & Bandelli	Auxiette	Auxiette	Auxiette	Auxiette	Auxiette
Cattle (*Bos taurus*)	138	176	124	550	61	52
Caprinae (*Ovis aries / Capra hircus*)	8	240	30	643	33	24
Pig (*Sus scrofa domesticus*)	2	204	26	641	68	14
Horse (*Equus caballus*)	4	11		51	6	7
Dog (*Canis familiaris*)		33	2	62	1	
Red-deer (*Cervus elaphus*)		26	5	88	1	33
Wild-boar (*Sus scrofa scrofa*)		2		7	2	
Roe-deer (*Capreolus capreolus*)		2		4		
Aurochs (*Bos primigenius*)					1	
Wolf (*Canis lupus*)					1	
Bear (*Ursus arctos*)				1		
Beaver (*Castor fiber*)		1		1		
Badger (*Meles meles*)						
Hare (*Lepus europaeus*)		4	4	8		
Fox (*Vulpes vulpes*)				4		2
Wild cat (*Felis sylvestris*)	1					
Marten (*M. martes / M. foina*)				1		
Otter (*Lutra lutra*)						
Hedgehog (*Erinaceus europaeus*)						
Polecat (*Mustella putorius*)				1		
Mole (*Talpa europaea*)						
Tortoise (*Emys orbicularis*)						
Birds (*Aves*)			2	1		3
Fish (*Pisces*)						
Cetaceans						
Total number of identified specimens	153	699	193	2063	174	135
Total of indeterminate fragments	45	1158	323	3619	142	36
TOTAL	**198**	**1857**	**516**	**5682**	**316**	**171**
Domestic mammals	152	664	182	1947	169	97
Wild mammals	1	35	9	115	5	35
% Domestic mammals	99.3	95.0	94.3	94.4	97.1	71.9
% Wild mammals	0.7	5.0	4.7	5.6	2.9	25.9
% birds	0.0	0.0	1.0	0.0	0.0	2.2
% fish	0	0	0	0	0	0

Sézanne	Hérouvilette	Ifs	Etaples-sur-Mer	Guines	Maroeuil	Abbeville
La Maladrerie	Les Pérelles	Object'Ifs Sud	Le Chemin des Prés	Jardins du Couvent 2	Rue Curie	Mont à Cailloux
Marne	Calvados	Calvados	Nord	Nord	Pas-de-Calais	Somme
Bronze	Bronze	Bronze	Bronze	Bronze	Bronze	Bronze
BzeIIIb	BzeA1A2	BzeIIIbHallC	Bze moyen II	BzeIIIbHallB	Bze	BzeB1C2
Auxiette	Auxiette	Auxiette	Auxiette	Auxiette	Auxiette	Auxiette
21	17	183	119	97	6	332
8	1	306	32	16	14	30
5	3	58	7	26	9	31
1		9	1	11		
1		1	1	29		1
1	1	1	1		4	1
		1				
			1			
37	22	559	162	179	33	395
34	3	329	38	11	71	58
71	**25**	**888**	**200**	**190**	**104**	**453**
36	21	557	160	179	29	394
1	1	2	1	0	4	1
97.3	95.5	99.6	98.8	100.0	87.9	99.7
2.7	4.5	0.4	0.6	0.0	12.1	0.3
0.0	0.0	0.0	0.0	0.0	0.0	0.0
0	0	0	0	0	0	0

Appendix 14: Aisne

Site	Beaurieux	Beaurieux	Beaurieux	Beaurieux	Barenton	Bazoches-sur-Vesle	Bazoches-sur-Vesle	Braine	Berry-au-Bac	Bucy-le-Long
Field	les Grèves	les Grèves	les Grèves	les Grèves	Site N-Zac du Griffon	les Chantraines	la Foulerie	la Grange des Moines	le Chemin de la Pêcherie	le Grand Marais
Department	Aisne	Aisne	Aisne	Aisne	Aisne	Aisne	Aisne	Aisne	Aisne	Aisne
Cultural attribution	Hallstatt	Hallstatt	Hallstatt	Hallstatt	Hallstatt	Hallstatt	Hallstatt	Hallstatt	Hallstatt	Hallstatt
Phase/Etape	HallD1	HallD	HallD3LTA	HallD3LTA	HallDLTA	HallDLTA	HallC	HallD3	HallDLTA	HallDLTA
Author	Auxiette	Auxiette	Auxiette	Auxiette	Auxiette	Auxiette	Auxiette	Auxiette	Auxiette	Auxiette
Cattle (*Bos taurus*)	3	14	6	9	15	17	27	21	7	519
Caprinae (*Ovis aries / Capra hircus*)	26	19	49	104	24	23	0	63	0	444
Pig (*Sus scrofa domesticus*)	2	12	22	281	23	6	6	15	4	309
Horse (*Equus caballus*)		3	1	1		1	1	4	2	27
Dog (*Canis familiaris*)		1			6	4		4	1	18
Red-deer (*Cervus elaphus*)			2	2	1	1	3		2	15
Wild-boar (*Sus scrofa scrofa*)						1				4
Roe-deer (*Capreolus capreolus*)			1							
Aurochs (*Bos primigenius*)										
Wolf (*Canis lupus*)										
Bear (*Ursus arctos*)										
Beaver (*Castor fiber*)										
Badger (*Meles meles*)										
Hare (*Lepus europaeus*)										
Fox (*Vulpes vulpes*)										
Wild cat (*Felis sylvestris*)										
Marten (*M. martes / M. foina*)										
Hedgehog (*Erinaceus europaeus*)										
Polecat (*Mustella putorius*)										
Mole (*Talpa europaea*)										
Tortoise (*Emys orbicularis*)										
Birds (*Aves*)					5	1	2	1		4
Fish (*Pisces*)					1					
Total number of identified specimens	31	49	81	398	74	54	39	108	16	1340
Total of indeterminate fragments	39	54	60	223	65	52	29	42	6	2133
TOTAL	**70**	**103**	**141**	**621**	**139**	**106**	**68**	**150**	**22**	**3473**
Domestic mammals	31	49	78	395	68	51	34	107	14	1317
Wild mammals	0	0	3	2	1	2	3	0	2	19
% Domestic mammals	100	100.0	96.3	99.2	91.9	94.4	87.2	99.1	87.5	98.3
% Wild mammals	0	0.0	3.7	0.5	1.4	3.7	7.7	0.0	12.5	1.4
% birds	0	0.0	0.0	0.0	6.8	1.9	5.1	0.9	0.0	0.3
% fish	0	0.0	0.0	0.3	0.0	0.0	0.0	0.0	0.0	0.0

Bucy-le-Long	Charly-sur-Marne	Ciry-Salsogne	Condé-sur-Suippe	Limé	Limé	Limé	Menneville	Menneville	Saint-Quentin	Sermoise	Tergnier	Villeneuve-Saint-Germain	Villers-en-Prayeres
le Fond du Petit Marais	Rue Pierre Legivre	la Bouche à Vesle	le Déprofundis	la Prairie	la Fosse aux Chevaux	Le Gros Buisson	Derrière le Village	Derrière le Village	le Dessus du Champ Bossu III	les Prés du Bout de la Ville	les Hauts Riez	les Etomelles	les Mauchamps
Aisne	Aisne	Aisne	Aisne	Aisne	Aisne	Aisne	Aisne	Aisne	Aisne	Aisne	Aisne	Aisne	Aisne
Hallstatt	Hallstatt	Hallstatt	Hallstatt	Hallstatt	Hallstatt	Hallstatt	Hallstatt	Hallstatt	Hallstatt	Hallstatt	Hallstatt	Hallstatt	Hallstatt
HallDLTA	HallCD1	HallDLTA	HallDLTA	HallDLTA	HallDLTA	HallDLTA	HallC	HallDLTA	HallCD	HallDLTA	HallDLTA	HallDLTA	HallDLTA
Auxiette	Auxiette	Auxiette	Villa	Auxiette	Auxiette	Auxiette	Auxiette	Auxiette	Auxiette	Auxiette	Auxiette	Auxiette	Auxiette
157	54	38	154	104	80	4	95	209	51	2	83	28	336
272	38	18	72	76	124	81	112	277	203	15	18	43	277
184	35	36	51	61	37	3	160	213	37	2	12	55	194
22	2	5	6	20	10		5	12	6	5	25	2	34
9	4	8	17	6	2		2	9			5	2	26
2	10	9	2		1	1	3	7	1		2		3
													5
3		20	1					3			2		1
							1						
											2		
								4					
4			1	1	2		1	14					2
653	143	134	304	268	256	89	379	748	298	24	147	130	880
542	76	162	1037	678	129	11	557	998	46	28	522	125	694
1195	**219**	**296**	**1341**	**946**	**385**	**100**	**936**	**1746**	**344**	**52**	**669**	**255**	**1574**
644	133	105	300	267	253	88	374	720	297	24	143	130	867
5	10	29	3	0	1	1	4	14	1	0	4	0	11
98.6	93.0	78.4	98.7	99.6	98.8	98.9	98.7	96.3	99.7	100.0	97.3	100.0	98.5
0.8	7.0	21.6	1.0	0.0	0.4	1.1	1.1	1.9	0.3	0.0	2.7	0.0	1.3
0.6	0.0	0.0	0.3	0.4	0.8	0.0	0.3	1.9	0.0	0.0	0.0	0.0	0.2
0.0	0.0	0.0	0.0	0.0	0.0	0.0	0.0	0.0	0.0	0.0	0.0	0.0	0.0

Appendix 15: Ile-de-France

Site	Tremblay	Gonesse	Saint-Denis	Mantes-la-Jolie	Changis-sur-Marne	Bussy-Saint-Georges	Lieusaint
Field	Zac sud Charles de Gaulle	Parc des Tulipes	Nozal Chadron	Zac des Bords de Seine	Les Pétreaux	Coudraies/ Pigeonneaux	Jadins de la Méridienne
Department	Seine-Saint-Denis	Val d'Oise	Val d'Oise	Yvelines	Seine-et-Marne	Seine-et-Marne	Seine-et-Marne
Cultural attribution	Hallstatt	Hallstatt	Hallstatt	Hallstatt	Hall.anc./moy.	Hallstatt	Hallstatt
Phase/Etape	HallD3LTA	HallLDTA	HallLDTA	HallD	HallC2D1	HallD	HallD
Author	Auxiette	Auxiette	Auxiette	Auxiette	Auxiette	Auxiette	Auxiette
Cattle (Bos taurus)	15	144	17	19	49	8	143
Caprinae (Ovis aries / Capra hircus)	3	97	17	79	32	18	58
Pig (Sus scrofa domesticus)	3	192	17	12	49	6	52
Horse (Equus caballus)	2	1	12	1	5	2	20
Dog (Canis familiaris)		9		3			7
Red-deer (Cervus elaphus)		26		1	4		3
Wild-boar (Sus scrofa scrofa)					10		
Roe-deer (Capreolus capreolus)							
Aurochs (Bos primigenius)				6			
Wolf (Canis lupus)							
Bear (Ursus arctos)				1	1		
Beaver (Castor fiber)							
Badger (Meles meles)		1			2		
Hare (Lepus europaeus)		1					
Fox (Vulpes vulpes)		2					
Wild cat (Felis sylvestris)					1		
Marten (M. martes / M. foina)						1	
Otter (Lutra lutra)							
Hedgehog (Erinaceus europaeus)							
Polecat (Mustella putorius)							
Mole (Talpa europaea)							
Tortoise (Emys orbicularis)							
Birds (Aves)					1	13	
Fish (Pisces)					5		
Total number of identified specimens	23	473	63	122	159	48	283
Total of indeterminate fragments	16	417	89	27	169	71	313
TOTAL	**39**	**890**	**152**	**149**	**328**	**119**	**596**
Domestic mammals	23	443	63	114	135	34	280
Wild mammals	0	30	0	8	18	1	3
% Domestic mammals	100	93.7	100.0	93.4	84.9	70.8	98.9
% Wild mammals	0	6.3	0.0	6.6	11.3	2.1	1.1
% birds	0	0.0	0.0	0.0	0.6	27.1	0.0
% fish	0	0.0	0.0	0.0	3.1	0.0	0.0

Lieusaint	Lieusaint	Ville-Saint-Jacques	Changis-sur-Marne	Ecuelles	Egligny	Varennes-sur-Seine	Wissous	Wissous	Milly-la-Forêt
les Perpignans-Zac du Levant	?	Le Bois d'Echalat	Les Pétreaux	Malassis et Charmoy	le Bois de la Pêcherie	Beauchamps	Zone Sud-Ouest aéroport Orly	Le Pérou	Le Bois Rond
Seine-et-Marne	Seine-et-Marne	Seine-et-Marne	Seine-et-Marne	Seine-et-Marne	Seine-et-Marne	Seine-et-Marne	Essonne	Essonne	Essonne
Hallstatt	Hallstatt	Hallstatt	Hallstatt	Hallstatt	Hallstatt	Hallstatt	Hallstatt	Hallstatt	Hallstatt
HallD2D3	HallDLTA	HallD3	HallDLTA	HallDLTA	HallDLTA	HallDLTA	HallC	HallDLTA	HallDLTA
Auxiette	Auxiette	Auxiette	Auxiette	Auxiette	Auxiette	Auxiette	Auxiette	Auxiette	Auxiette
9	41	14	159	142	66	24	8	326	226
7	42	97	317	488	32	35	2	320	1014
6	37	63	204	302	45	41	8	344	283
6	6	1	1	11	4	1		4	5
1	2	1	12	152	1	7	7	1	157
	4	3	7		2		1	12	8
			11	1				5	
								2	
			2						1
			6	2			1		6
								1	12
		6							2
				1	1				1
			7	2	3			5	9
									1
29	132	185	726	1101	154	109	26	1020	1725
9	126	160	308	614	67	63	2	292	569
38	**258**	**345**	**1034**	**1715**	**221**	**172**	**28**	**1312**	**2294**
29	128	176	693	1095	148	108	25	681	1685
0	4	9	26	3	2	1	1	8	29
100.0	97.0	95.1	95.5	99.5	96.1	99.1	96.2	66.8	97.7
0.0	3.0	4.9	3.6	0.3	1.3	0.9	3.8	0.8	1.7
0.0	0.0	0.0	1.0	0.2	1.9	0.0	0.0	0.5	0.5
0.0	0.0	0.0	0.0	0.0	0.0	0.0	0.0	0.0	0.1

Appendix 16: Champagne-Ardenne, Normandie

Site	Saint-André-les-Vergers	Saint-Léger-Près-Troyes/Buchères	Caurel	Reims	Reims	Reims
Field	Echevilly	Parc logistique de l'Aube-Décapage 31	Le Puisard	Zac Dauphinot	Croix Blandin	Croix Blandin
Department	Aube	Aube	Marne	Marne	Marne	Marne
Cultural attribution	Hallstatt	Hallstatt	Hallstatt	Hallstatt	Hallstatt	Hallstatt
Phase/Etape	HallC	HallD2D3	HallC	HallD	HallD1D2	HallD2D3
Author	Auxiette	Auxiette&Bandelli	Auxiette	Auxiette	Auxiette	Auxiette
Cattle (*Bos taurus*)	30	22	9	103	74	40
Caprinae (*Ovis aries / Capra hircus*)	20	9	3	224	1	59
Pig (*Sus scrofa domesticus*)	11	3	5	85	42	12
Horse (*Equus caballus*)	4	5		8	6	3
Dog (*Canis familiaris*)		2		8		4
Red-deer (*Cervus elaphus*)	1		1	4		2
Wild-boar (*Sus scrofa scrofa*)						
Roe-deer (*Capreolus capreolus*)						
Aurochs (*Bos primigenius*)			1			
Wolf (*Canis lupus*)						
Bear (*Ursus arctos*)						
Beaver (*Castor fiber*)						
Badger (*Meles meles*)						
Hare (*Lepus europaeus*)	2			16		2
Fox (*Vulpes vulpes*)						
Wild cat (*Felis sylvestris*)						
Marten (*M. martes / M. foina*)						
Otter (*Lutra lutra*)						
Hedgehog (*Erinaceus europaeus*)						
Polecat (*Mustella putorius*)						
Mole (*Talpa europaea*)						
Tortoise (*Emys orbicularis*)						
Birds (*Aves*)				2		
Fish (*Pisces*)						
Cétacés						
Total number of identified specimens	68	41	19	450	123	122
Total of indeterminate fragments	50	21	11	292	17	59
TOTAL	**118**	**62**	**30**	**742**	**140**	**181**
Domestic mammals	65	41	17	428	123	118
Wild mammals	3	0	2	20	0	4
% Domestic mammals	95.6	100.0	89.5	95.1	100.0	96.7
% Wild mammals	4.4	0.0	10.5	4.4	0.0	3.3
% birds	0.0	0.0	0.0	0.4	0.0	0.0
% fish	0.0	0.0	0.0	0.0	0.0	0.0

Reims	Romain	Gueux	Ifs	Ifs	Hérouvilette	Ifs	Ifs
Ormes et Thillois-site 14	Cense Sauvage	les Batailles	Object'Ifs Sud	Object'Ifs Sud	Les Pérelles	AR67	AR67
Marne	Marne	Marne	Calvados	Calvados	Calvados	Calvados	Calvados
Hallstatt	Hallstatt	Hallstatt	Hallstatt	Hallstatt	Hallstatt	Hallstatt	Hallstatt
HallD2D3	HallD3LTAB	HallDLTA	HallD1D2	HallD3	HallD3LTA	HallD3LTA	HallCD1
Auxiette	Auxiette	Auxiette	Auxiette	Auxiette	Auxiette	Auxiette	Auxiette
135	10	3	51	6	165	382	140
41	13	7	97	19	127	261	150
52	24	64	41	5	43	133	59
30	3	5	1	2	9	63	13
5	1		2		11	16	
2	5	1	2		4	4	3
			1		1		1
	3					1	
							1
	1		7				
					1		
	1		1				1
					3		
265	61	80	203	32	364	860	368
86	116	42	157	24	148	371	121
351	**177**	**122**	**360**	**56**	**512**	**1231**	**489**
263	51	79	192	32	355	855	362
2	9	1	10	0	6	5	5
99.2	83.6	98.8	94.6	100.0	97.5	99.4	98.4
0.8	14.8	1.3	4.9	0.0	1.6	0.6	1.4
0.0	1.6	0.0	0.5	0.0	0.0	0.0	0.3
0.0	0.0	0.0	0.0	0.0	0.0	0.0	0.0

Appendix 17: Aisne, Ile-de-France

Site	Villeneuve-Saint-Germain	Limé	Lieusaint	Poincy	Changis-sur-Marne	Saint-Mard	Varennes-sur-Seine	Larchant	Ville-Saint-Jacques	Champagne-sur-Oise	Bailly
Field	les Grèves	le Gros Buisson	Zac de la Pyramide	Près le Pont de Trilport	Les Pétreaux	Zac de la Fontaine aux Bergers	Beauchamps	Les Groues	Le Bois d'Echalat	Les Basses Coutures	Les Merisiers
Department	Aisne	Aisne	Seine et Marne	Seine et Marne	Seine et Marne	Seine et Marne	Seine et Marne	Seine et Marne	Seine et Marne	Val d'Oise	Yvelines
Cultural attribution	La Tène	La Tène	La Tène	La Tène	La Tène	La Tène	La Tène	La Tène	La Tène	La Tène	La Tène
Phase/Etape	LTA	LTA	LTA	LTA	LTAB	LTAB	LTAB	LTB1	LTB	LTA	LT A
Author	Auxiette	Auxiette	Auxiette	Auxiette	Auxiette	Auxiette	Auxiette	Auxiette	Auxiette	Auxiette	Auxiette
Cattle (*Bos taurus*)	109	4	76	38	178	85	14	192	40	33	407
Caprinae (*Ovis aries / Capra hircus*)	351	81	71	84	282	196	89	225	42	121	429
Pig (*Sus scrofa domesticus*)	341	3	51	87	120	47	118	266	26	27	687
Horse (*Equus caballus*)	18		9	2	4	5	1	11	2	3	32
Dog (*Canis familiaris*)	12		6	9	25	36	25	23	5	25	113
Red-deer (*Cervus elaphus*)	2	1	4		17	4				1	55
Wild-boar (*Sus scrofa scrofa*)	1				6						7
Roe-deer (*Capreolus capreolus*)	1				1	1					13
Aurochs (*Bos primigenius*)											
Wolf (*Canis lupus*)											
Bear (*Ursus arctos*)											
Beaver (*Castor fiber*)											
Badger (*Meles meles*)											
Hare (*Lepus europaeus*)	2				22						66
Fox (*Vulpes vulpes*)											4
Wild cat (*Felis sylvestris*)											
Marten (*M. martes / M. foina*)											
Hedgehog (*Erinaceus europaeus*)											
Polecat (*Mustella putorius*)											
Mole (*Talpa europaea*)											
Tortoise (*Emys orbicularis*)											
Birds (*Aves*)	9			2	21	6	13	17	4	2	7
Fish (*Pisces*)	1				1		1				
Total number of identified specimens	847	89	217	222	677	380	261	734	119	212	1820
Total of indeterminate fragments	1090	11	126	88	421	80	122	343	106	48	1280
TOTAL	1937	100	343	310	1098	460	383	1077	225	260	3100
Domestic mammals	831	88	213	220	609	369	247	717	115	209	1668
Wild mammals	6	1	4	0	46	5	0	0	0	1	145
% Domestic mammals	98.1	98.9	98.2	99.1	90.0	97.1	94.6	97.7	96.6	98.6	91.6
% Wild mammals	0.7	1.1	1.8	0.0	6.8	1.3	0.0	0.0	0.0	0.5	8.0
% birds	1.1	0.0	0.0	0.9	3.1	1.6	5.0	2.3	3.4	0.9	0.4
% fish	0.1	0.0	0.0	0.0	0.1	0.0	0.4	0.0	0.0	0.0	0.0

Appendix 18: Champagne-Ardenne, Hauts-de-France, Normandie

Site	Buchères	Cernay-lès-Reims	Cernay-lès-Reims	Cernay-lès-Reims	Reims	Caurel	Dourges	Ifs	Ifs	Ifs	Ifs
Field	Les Vignes Neuves-D41	La Borne Saint Laid	Les Champs Virés	Le Puisard	les Hauts des Nervas	Le Puisard	le Marais de Dourges	ObjectTfs sud	ObjectTfs sud	ObjectTfs sud	ObjectTfs sud
Department	Aube	Marne	Marne	Marne	Marne	Marne	Nord	Calvados	Calvados	Calvados	Calvados
Cultural attribution	La Tène	La Tène	La Tène	La Tène	La Tène	La Tène	La Tène	La Tène	La Tène	La Tène	La Tène
Phase/Etape	LTA2B1	LTA	LTA	LTA	LTA	LTAB	LTA	LTAB	LTA	LTB	LTBC1
Author	Auxiette	Auxiette	Auxiette	Auxiette	Auxiette	Auxiette	Auxiette	Auxiette	Auxiette	Auxiette	Auxiette
Cattle (*Bos taurus*)	66	90	4	29	232	25	54	265	45	31	40
Caprinae (*Ovis aries / Capra hircus*)	256	103	66	110	87	12	19	136	9	17	4
Pig (*Sus scrofa domesticus*)	61	37	4	27	41	11	14	64	0	5	1
Horse (*Equus caballus*)	24	34	11	8	10	6	21	17	6	30	1
Dog (*Canis familiaris*)	4	2		2	1	1	7	8		1	
Red-deer (*Cervus elaphus*)	1				6						
Wild-boar (*Sus scrofa scrofa*)											
Roe-deer (*Capreolus capreolus*)											
Aurochs (*Bos primigenius*)											
Wolf (*Canis lupus*)											
Bear (*Ursus arctos*)											
Beaver (*Castor fiber*)											
Badger (*Meles meles*)											
Hare (*Lepus europaeus*)		3							1		
Fox (*Vulpes vulpes*)	5				1						
Wild cat (*Felis sylvestris*)											
Marten (*M. martes / M. foina*)											
Hedgehog (*Erinaceus europaeus*)											
Polecat (*Mustella putorius*)											
Mole (*Talpa europaea*)											
Tortoise (*Emys orbicularis*)											
Birds (*Aves*)	4				1		1	1			
Fish (*Pisces*)	3										
Total number of identified specimens	424	269	85	176	379	55	116	492	60	84	46
Total of indeterminate fragments	330	260	6	214	166	33	90	127	6	22	5
TOTAL	754	529	91	390	545	88	206	619	66	106	51
Domestic mammals	411	266	85	176	371	55	115	490	60	84	46
Wild mammals	6	3	0	0	7	0	0	1	0	0	0
% Domestic mammals	96.9	98.9	100.0	100.0	97.9	100.0	99.1	99.6	100.0	100.0	100.0
% Wild mammals	1.4	1.1	0.0	0.0	1.8	0.0	0.0	0.2	0.0	0.0	0.0
% birds	0.9	0.0	0.0	0.0	0.3	0.0	0.9	0.2	0.0	0.0	0.0
% fish	0.7	0.0	0.0	0.0	0.0	0.0	0.0	0.0	0.0	0.0	0.0

Appendix 19: Aisne

Site	Limé total	Chambry	Beaurieux	Bazoches-sur-Vesle	Urvillers	Barenton	Barenton	Ciry-Salsogne
Field	Le Gros Buisson	Zac du Griffon	Les Grèves	les Chantraines	Zac de l'Epinette	Site L-Zac du Griffon	Site M-Zac du Griffon	Le Bruy
Department	Aisne	Aisne	Aisne	Aisne	Aisne	Aisne	Aisne	Aisne
Cultural attribution	La Tène	La Tène	La Tène	La Tène	La Tène	La Tène	La Tène	La Tène
Phase/Etape	LTB2C1	LTC1	LTC1C2	LTC2	LTC2	LTC2D1	LTC2D1	LTCD
Author	Auxiette	Auxiette	Auxiette	Auxiette	Auxiette	Auxiette	Auxiette	Auxiette
Cattle (*Bos taurus*)	106	166	1	20	32	89	28	128
Caprinae (*Ovis aries / Capra hircus*)	91	584	2	4	126	513	36	81
Pig (*Sus scrofa domesticus*)	151	231	9	10	50	41	28	44
Horse (*Equus caballus*)	10	8		22	7	21	3	26
Ane (*Equus asinus*)								
Dog (*Canis familiaris*)	17	63	1		4	30	1	2
Red-deer (*Cervus elaphus*)	8	1				3		
Wild-boar (*Sus scrofa scrofa*)			1					
Roe-deer (*Capreolus capreolus*)	1	1						
Aurochs (*Bos primigenius*)								
Wolf (*Canis lupus*)								
Bear (*Ursus arctos*)								
Beaver (*Castor fiber*)								
Badger (*Meles meles*)								
Hare (*Lepus europaeus*)								
Fox (*Vulpes vulpes*)								
Wild cat (*Felis sylvestris*)								
Marten (*M. martes / M. foina*)						2		
Hedgehog (*Erinaceus europaeus*)								
Polecat (*Mustella putorius*)								
Mole (*Talpa europaea*)						7		
Tortoise (*Emys orbicularis*)								
Birds (*Aves*)		19	1			15		
Fish (*Pisces*)				1				
Total number of identified specimens	384	1073	15	57	219	721	96	281
Total of indeterminate fragments	73	109	54	47	37	467	27	202
TOTAL	457	1182	69	104	256	1188	123	483
Domestic mammals	375	1052	13	56	219	694	96	281
Wild mammals	9	2	1	0	0	5	0	0
% Domestic mammals	97.7	98.0	86.7	98.2	100.0	96.3	100.0	100.0
% Wild mammals	2.3	0.2	6.7	0.0	0.0	0.7	0.0	0.0
% birds	0.0	1.8	6.7	0.0	0.0	2.1	0.0	0.0
% fish	0.0	0.0	0.0	1.8	0.0	0.0	0.0	0.0

Bucy le Long	Vasseny	Limé	Villeneuve-Saint-Germain	Bazoches-sur-Vesle	Berry-au-Bac	Braine	Juvincourt et Damary	Condé-sur-Suippe
le Fond du Petit Marais	le dessous des Groins	les Sables sud	les Etomelles	les Chantraines	le Chemin de la Pêcherie	La Grange des Moines	le Ruisseau de Fayau	la Sucrerie
Aisne	Aisne	Aisne	Aisne	Aisne	Aisne	Aisne	Aisne	Aisne
La Tène	La Tène	La Tène	La Tène	La Tène	La Tène	La Tène	La Tène	La Tène
LTD1D2	LTD	LTD	LTD1a	LTD1a	LTD1a	LTD1a	LTD1a	LTD1a
Auxiette	Auxiette	Auxiette	Auxiette	Auxiette	Auxiette	Auxiette	Auxiette	Auxiette&Paris&Méniel
99	18	74	459	602	65	41	793	6885
145	24	31	308	415	148	4	648	6847
98	9	38	743	471	58	13	785	19943
48	1	32	57	194	8	9	220	1174
		8						6
10	1		48	48	3		105	1000
	1		1	3	5			17
			2				1	29
			1		1			3
1				4				111
								1
								2
				6				
2			8	24		1	28	101
								7
403	54	183	1627	1767	288	68	2580	36126
707	30	98	190	1118	126	169	2855	33264
1110	84	281	1817	2885	414	237	5435	69390
400	53	183	1615	1730	282	67	2551	35855
1	1	0	4	13	6	0	1	163
99.3	98.1	100.0	99.3	97.9	97.9	98.5	98.9	99.2
0.2	1.9	0.0	0.2	0.7	2.1	0.0	0.0	0.5
0.5	0.0	0.0	0.5	1.4	0.0	1.5	1.1	0.3
0.0	0.0	0.0	0.0	0.0	0.0	0.0	0.0	0.0

Appendix 20: Ile-de-France, Oise

Site	Passel	Nanteuil-le-Haudouin	Nanteuil-le-Haudouin	Allones	Nanteuil-le-Haudouin	Wissous	Wissous	Nanterre	Nanterre	Nanterre	Changis-sur-Marne	Jossigny	Varennes-sur-Seine	Varennes-sur-Seine
Field	Le Vivier	ZBF	ZBF	Extension ZAC du Ther	ZBF	Zone Sud-Ouest aéroport Orly	Zone Sud-Ouest aéroport Orly	les Guignons	Avenue Jules Quentin	Passage du Quignons	Les Pêtreaux	Pré aux Chênes	Le Marais du Colombier	La Justice
Department	Oise	Oise	Oise	Oise	Oise	Essonne	Essonne	Haut-de-Seine	Hauts-de-Seine	Hauts-de-Seine	Seine-et-Marne	Seine-et-Marne	Seine-et-Marne	Seine-et-Marne
Cultural attribution	La Tène	La Tène	La Tène	La Tène	La Tène	La Tène	La Tène	La Tène	La Tène	La Tène	La Tène	La Tène	La Tène	
Phase/Etape	LTB2C1	LTC2	LTD1aD1b	LTD1	LTD1a	LTC2	LTD2	LTD2	LTD2	LTD2	LTB2C1	LTB2C1	LTBC1	LTC
Author	Auxiette	Auxiette	Auxiette	Auxiette	Auxiette	Auxiette	Auxiette	Auxiette	Auxiette	Auxiette	Auxiette	Auxiette	Auxiette	Auxiette
Cattle (*Bos taurus*)	76	0	303	7	35	389	1834	5115	1354	403	14	110	75	146
Caprinae (*Ovis aries / Capra hircus*)	259	16	93	13	4	189	602	2009	917	196	154	122	175	86
Pig (*Sus scrofa domesticus*)	56	33	149	7	7	225	962	2422	622	267	18	105	245	76
Horse (*Equus caballus*)	4	0	34	5		20	250	586	214	31	0	27	3	49
Dog (*Canis familiaris*)	1		21		4	33	32	236	155	27	7	21	14	7
Red-deer (*Cervus elaphus*)	1		1	5	1	11	10	3	2	2		2	1	3
Wild-boar (*Sus scrofa scrofa*)		2	2			1	2		1					
Roe-deer (*Capreolus capreolus*)		1						1	1		2		1	
Aurochs (*Bos primigenius*)														
Wolf (*Canis lupus*)														
Bear (*Ursus arctos*)														
Beaver (*Castor fiber*)														1
Badger (*Meles meles*)														
Hare (*Lepus europaeus*)		1		1			2	7	8	5	11			1
Fox (*Vulpes vulpes*)						1	1		2				1	
Wild cat (*Felis sylvestris*)										4				
Marten (*M. martes / M. foina*)														
Hedgehog (*Erinaceus europaeus*)														
Polecat (*Mustella putorius*)														
Mole (*Talpa europaea*)														
Tortoise (*Emys orbicularis*)						1			1					
Birds (*Aves*)		9	4			4	44	73	30	9		1	20	
Fish (*Pisces*)						1			6	3	1			
Total number of identified specimens	397	62	607	38	51	875	3740	10458	3309	945	206	388	535	369
Total of indeterminate fragments	83	7	94	2	6	72	444	6077	1515	188	31	209	376	207
TOTAL	480	69	701	40	57	947	4184	16535	4824	1133	237	597	911	576
Domestic mammals	396	49	600	32	50	856	3680	10368	3262	924	193	385	512	364
Wild mammals	1	4	3	6	1	13	16	11	13	11	13	2	3	5
% Domestic mammals	99.7	79.0	98.8	84.2	98.0	97.8	98.4	99.1	98.6	97.8	93.7	99.2	95.7	98.6
% Wild mammals	0.3	6.5	0.5	15.8	2.0	1.5	0.4	0.1	0.4	1.2	6.3	0.5	0.6	1.4
% birds	0.0	14.5	0.7	0.0	0.0	0.5	1.2	0.7	0.9	1.0	0.0	0.3	3.7	0.0
% fish	0.0	0.0	0.0	0.0	0.0	0.1	0.0	0.1	0.1	0.1	0.0	0.0	0.0	0.0

Villiers-sur-Seine	Changis-sur-Marne	Lieusaint	Egligny	Gonesse	Varennes-sur-Seine	Varennes-sur-Seine	Poincy	Varennes-sur-Seine	Souppes-sur-Loing	Meaux	Tremblay	Tremblay	Roissy	Champagne-sur-Oise	Champagne-sur-Oise	Roissy	Louvres	Ivry-sur-Seine	Soindres
le Gros Buisson	La Pelle à Four	Jardins de la Méridienne	Le Bois de la Pêcherie	Zac des Tulipes Nord-site 3	Beauchamps	La Justice	Près le Pont de Trilport	La Justice	A l'Est de Beaumoulin	Jean Rose	Zac sud Charles de Gaulle	Zac sud Charles de Gaulle	Zac de la Demi-Lune	Les Basses Coutures	Les Basses Coutures	Zac de la Demi-Lune	Le Vieux Moulin	Saint-Just	le Bordel
Seine-et-Marne	Seine-et-Marne	Seine-et-Marne	Seine-et-Marne	Seine-et-Marne	Seine-et-Marne	Seine-et-Marne	Seine-et-Marne	Seine-et-Marne	Seine-et-Marne	Seine-et-Marne	Seine-Saint-Denis	Seine-Saint-Denis	Val d'Oise	Val d'Oise	Val d'Oise	Val d'Oise	Val d'Oise	Val-de-Marne	Yvelines
LTC2	LTC2D1	LTC2D1	LTD	LTD	LTD	LTD	LTD1aD1b	LTD1bD2a	LTD2	LTD2aD2b	LTC2	LTD	LTC1C2	LTC2	LTC2D1	LTCD	LTD1	LTC2	LTD
Auxiette	Auxiette	Auxiette	Auxiette	Auxiette	Auxiette	Auxiette	Auxiette	Auxiette	Auxiette	Auxiette	Auxiette	Auxiette	Auxiette	Auxiette	Auxiette	Auxiette	Auxiette	Auxiette	Auxiette
2116	241	238	95	100	39	346	631	376	4292	187	176	96	12	130	93	26	44	213	114
755	285	154	45	46	11	287	535	157	1161	106	62	29	15	29	24	61	39	179	61
1777	141	84	58	12	7	628	659	205	1213	911	49	70	28	54	24	4	9	125	39
404	15	33	6	31	9	17	138	89	758	8	73	15	1	23	13	5	12	39	12
132	11	11	2	7	3	7	31	33	149	4	9	8	2	3	3	3		11	6
13	1		2	2		9		8	133	10		3	6			1	1		2
6			1			1			15										
1	1					1			59		2		2						
		1																	
						1													
34	1						1												
1						2									1				
2	5						1		7	1									
4						4		2	13			1							
			1																
5			1				1		9						1				
15	9		3	7		24	6	4	27	47	2	12	1		3			8	
4						4	4		7	6	1								
5269	710	521	212	207	69	1331	2007	874	7843	1280	374	234	67	239	162	100	105	575	234
2587	327	253	101	100	30	652	450	557	1469	173	99	80	13	21	33	18	26	230	106
7856	1037	774	313	307	99	1983	2457	1431	9312	1453	473	314	80	260	195	118	131	805	340
5184	693	520	206	98	69	1285	1994	860	7573	1216	369	218	58	239	157	99	104	567	232
61	8	1	2	2	0	18	2	10	227	11	2	4	8	0	1	1	1	0	2
98.4	97.6	99.8	97.2	47.3	100.0	96.5	99.4	98.4	96.6	95.0	98.7	93.2	86.6	100.0	96.9	99.0	99.0	98.6	99.1
1.2	1.1	0.2	0.9	1.0	0.0	1.4	0.1	1.1	2.9	0.9	0.5	1.7	11.9	0.0	0.6	1.0	1.0	0.0	0.9
0.3	1.3	0.0	1.4	3.4	0.0	1.8	0.3	0.5	0.3	3.7	0.5	5.1	1.5	0.0	1.9	0.0	0.0	1.4	0.0
0.1	0.0	0.0	0.0	0.0	0.0	0.3	0.2	0.0	0.1	0.5	0.3	0.0	0.0	0.0	0.0	0.0	0.0	0.0	0.0

Appendix 21: Champagne-Ardenne, Hauts-de-France

Site	Buchères	Sézanne	Betheny	Reims	Reims	Reims	Reims	Reims	Reims	Buchères	Glisy
Field	Parc logistique de l'Aube	La Maladrerie	Les Equiernolles	Le Haut des Nervas	Rue d'Anjou	Villa des Capuccins	Chanzy villa du théâtre	Rue Carnot	Rue Rockfeller	Parc logistique de l'Aube	Zac Jules Verne-site C
Department	Aube	Marne	Marne	Marne	Marne	Marne	Marne	Marne	Marne	Aube	Somme
Cultural attribution	La Tène	La Tène	La Tène	La Tène	La Tène	La Tène	La Tène	La Tène	La Tène	La Tène	La Tène
Phase/Etape	LTD1D2a	LTB2C1	LTD1	LTD1	LTD2b	LTD2b	LTD2b	LTD2bAug	LTD2bAug	LTD1D2a	LTB2C1
Author	Auxiette	Auxiette	Auxiette	Auxiette	Auxiette	Auxiette	Auxiette	Auxiette	Auxiette	Auxiette	Auxiette
Cattle (*Bos taurus*)	393	81	128	110	279	92	130	482	475	393	147
Caprinae (*Ovis aries / Capra hircus*)	222	115	186	66	264	104	106	56	248	222	179
Pig (*Sus scrofa domesticus*)	214	33	165	45	425	192	358	172	580	214	106
Horse (*Equus caballus*)	60	28	43	55	38	23	13	33	48	60	95
Dog (*Canis familiaris*)	8	8	24	19	29	3	9	8	35	8	16
Red-deer (*Cervus elaphus*)	10	2	1	2						10	
Wild-boar (*Sus scrofa scrofa*)	4							1		4	
Roe-deer (*Capreolus capreolus*)		1									1
Aurochs (*Bos primigenius*)											
Wolf (*Canis lupus*)											
Bear (*Ursus arctos*)											
Beaver (*Castor fiber*)											
Badger (*Meles meles*)											
Hare (*Lepus europaeus*)	3		18						6	3	
Fox (*Vulpes vulpes*)			1								
Wild cat (*Felis sylvestris*)											
Marten (*M. martes / M. foina*)											
Hedgehog (*Erinaceus europaeus*)											
Polecat (*Mustella putorius*)											
Mole (*Talpa europaea*)											
Tortoise (*Emys orbicularis*)											
Birds (*Aves*)	5	2	5					2	11	5	14
Fish (*Pisces*)				3	3	1			1		
Total number of identified specimens	919	270	571	300	1038	415	616	754	1404	919	558
Total of indeterminate fragments	569	82	283	193	368	187	134	160	683	569	562
TOTAL	1488	352	854	493	1406	602	750	914	2087	1488	1120
Domestic mammals	897	265	546	295	1035	414	616	751	1386	897	543
Wild mammals	17	3	20	2	0	0	0	1	6	17	1
% *Domestic mammals*	97.6	98.1	95.6	98.3	99.7	99.8	100.0	99.6	98.7	97.6	97.3
% *Wild mammals*	1.8	1.1	3.5	0.7	0.0	0.0	0.0	0.1	0.4	1.8	0.2
% birds	0.5	0.7	0.9	0.0	0.0	0.0	0.0	0.3	0.8	0.5	2.5
% fish	0.0	0.0	0.0	1.0	0.3	0.2	0.0	0.0	0.1	0.0	0.0

Méaulte	Glisy	Méaulte	Amiens-Renancourt	Glisy	Pont-de-Metz	Amiens-Renancourt	Amiens-Renancourt	Amiens-Renancourt site 5	Etaples	Dourges	Carvin
Zac du Coquelicot	Les Champs Tortus	Zac du Coquelicot	Site 5-Zac de Renancourt-site 5	Zac Jules Verne-site H	Nouvel Hôpital	Site 2-Zac de Renancourt-site 2	Site 5-Zac de Renancourt-site 5	Site 5-Zac de Renancourt-site 5	Pièce à Liards	Le Marais de Dourges	Gare d'Eau
Somme	Somme	Somme	Somme	Somme	Somme	Somme	Somme	Somme	Nord	Nord	Pas-de-Calais
La Tène	La Tène	La Tène	La Tène	La Tène	La Tène	La Tène	La Tène	La Tène	La Tène	La Tène	La Tène
LTB2C1	LTC2	LTC2	LTCD1	LTD	LTD1D2	LTD2b	LTC2D1	LTD2bHtEmp	LTB2	LTCD	LTD2
Auxiette	Auxiette	Auxiette	Auxiette	Auxiette	Auxiette	Auxiette	Auxiette	Auxiette&Dréano	Auxiette	Auxiette	Auxiette
131	956	78	323	45	77	546	328	838	421	121	540
112	541	70	118	56	20	104	126	667	177	84	216
83	1066	173	183	29	26	413	199	2003	116	56	180
35	169	17	100	5	32	121	101	12	1	14	153
18	117	2	30	3	11	15	31	63	180	11	13
2			2	1		2	2	44	2	1	3
					1						
	2	1									1
											3
	9										
1	12		1	1		2	6	192			
	1								6	3	
							4	4			
	1										
	1										
6	53	6	5	1		9	41	796	6	4	1
							2	132	4		
388	2928	347	762	141	166	1213	840	4751	913	294	1110
119	301	50	87	82	4	204	125	1937	54	198	264
507	3229	397	849	223	170	1417	965	6688	967	492	1374
379	2849	340	754	138	166	1199	785	3583	895	286	1102
3	26	1	3	2	0	5	12	240	8	4	7
97.7	97.3	98.0	99.0	97.9	100.0	98.8	93.5	75.4	98.0	97.3	99.3
0.8	0.9	0.3	0.4	1.4	0.0	0.4	1.4	5.1	0.9	1.4	0.6
1.5	1.8	1.7	0.7	0.7	0.0	0.7	4.9	16.8	0.7	1.4	0.1
0.0	0.0	0.0	0.0	0.0	0.0	0.0	0.2	2.8	0.4	0.0	0.0

Appendix 22: Normandie

Site	Ifs	Ifs	Mondeville	Loucelles	Ifs	Mondeville	Bretteville-l'Orgueilleuse	Mondeville
Field	Object'Ifs sud	Object'Ifs sud	l'Etoile	Sainte Croix et Grande Tonne	Object'Ifs sud	l'Etoile	Le Bas des Prés et Résidence du Parc	l'Etoile
Department	Calvados	Calvados	Calvados	Calvados	Calvados	Calvados	Calvados	Calvados
Cultural attribution	La Tène	La Tène	La Tène	La Tène	La Tène	La Tène	La Tène	La Tène
Phase/Etape	LTB2C1	LTBC1	LTC	LTC1	LTC2	LTC2	LTC2D1	LTC2D1
Author	Auxiette	Auxiette	Auxiette	Auxiette	Auxiette	Auxiette	Auxiette	Auxiette
Cattle (Bos taurus)	346	40	234	70	687	446	65	440
Caprinae (Ovis aries / Capra hircus)	140	4	182	8	411	344	72	204
Pig (Sus scrofa domesticus)	71	1	143	8	260	153	33	117
Horse (Equus caballus)	16	1	22	16	20	95	2	139
Dog (Canis familiaris)	32		13	9	64	83	29	16
Red-deer (Cervus elaphus)								
Wild-boar (Sus scrofa scrofa)								1
Roe-deer (Capreolus capreolus)								
Aurochs (Bos primigenius)								
Wolf (Canis lupus)						1	1	
Bear (Ursus arctos)								
Beaver (Castor fiber)								
Badger (Meles meles)								
Hare (Lepus europaeus)						1		
Fox (Vulpes vulpes)								9
Wild cat (Felis sylvestris)								
Marten (M. martes / M. foina)								
Hedgehog (Erinaceus europaeus)								
Polecat (Mustella putorius)								
Mole (Talpa europaea)								
Tortoise (Emys orbicularis)								
Birds (Aves)	1				16	2	1	8
Fish (Pisces)								
Total number of identified specimens	606	46	594	111	1458	1125	203	934
Total of indeterminate fragments	81	5	153	36	298	257	82	140
TOTAL	687	51	747	147	1756	1382	285	1074
Domestic mammals	605	46	594	111	1442	1121	201	916
Wild mammals	0	0	0	0	0	2	1	10
% Domestic mammals	99.8	100.0	100.0	100.0	98.9	99.6	99.0	98.1
% Wild mammals	0.0	0.0	0.0	0.0	0.0	0.2	0.5	1.1
% birds	0.2	0.0	0.0	0.0	1.1	0.2	0.5	0.9
% fish	0.0	0.0	0.0	0.0	0.0	0.0	0.0	0.0

Ifs	Ifs	Ifs	Mondeville	Ifs	Bretteville-l'Orgueilleuse	Ifs	Ifs	Bretteville-l'Orgueilleuse
Object'Ifs sud	Object'Ifs sud	AR67	l'Etoile	Object'Ifs sud	Le Bas des Prés et Résidence du Parc	Object'Ifs sud	Object'Ifs sud	Le Bas des Prés et Résidence du Parc
Calvados	Calvados	Calvados	Calvados	Calvados	Calvados	Calvados	Calvados	Calvados
La Tène	La Tène	La Tène	La Tène	La Tène	La Tène	La Tène	La Tène	La Tène
LTC2	LTD1	LTD	LTD	LTD1ab	LTD1bD2a	LTD2a	LTD2b	LTD2bAug
Auxiette	Auxiette	Auxiette	Auxiette	Auxiette	Auxiette	Auxiette	Auxiette	Auxiette
43	437	204	64	747	59	126	284	5
8	273	188	22	543	64	35	118	4
15	113	81	6	243	17	34	63	6
1	13	13	12	53	7	10	43	4
5	5	38	4	68	10	18	25	
		1	1		1		2	
	3							
	1			1				
		1						
		16						
	5	3		4		1	2	
72	850	545	109	1659	158	224	537	19
6	295	191	38	393	17	47	130	1
78	1145	736	147	2052	175	271	667	20
72	841	524	108	1654	157	223	533	19
0	4	18	1	1	1	0	2	0
100.0	98.9	96.1	99.1	99.7	99.4	99.6	99.3	100.0
0.0	0.5	3.3	0.9	0.1	0.6	0.0	0.4	0.0
0.0	0.6	0.6	0.0	0.2	0.0	0.4	0.4	0.0
0.0	0.0	0.0	0.0	0.0	0.0	0.0	0.0	0.0

Appendix 23:

Cattle Estimated meat weight by age range	Abbeville Early/Middle Bronze Age	Etaples Early/Middle Bronze Age	Villiers/S. Late Bronze Age	Villiers/S. Late La Tène (LTC2)	Braine Late La Tène (LTD1b)	Souppes/L. Late La Tène (LTD2)
1/6 mois			40	0	0	0
6/12 mois	96	192	0	0	96	144
12/18 mois	84		84	168	168	588
18/24 mois			300	0	0	0
24/30 mois	408		0	1904	952	544
> 36 mois	960	800	1280	1760	3840	14560
Total	1548	992	1704	3832	5056	15836

Estimated meat weight for cattle, according to Minimum Number of Individual by age range

Pig Estimated meat weight by age range	Abbeville Early/Middle Bronze Age	Etaples Early/Middle Bronze Age.	Villiers/S. Late Bronze Age	Villiers/S. Late La Tène (LTC2)	Braine Late La Tène (LTD1b)	Souppes/L. Late La Tène (LTD2)
1/6 mois			2046	198	132	528
6/12 mois	32	32	4896	192	128	640
12/24 mois	51		4896	3111	3723	1887
>24 mois			1472	1024	1408	640
Total	83	32	13310	4525	5391	3695

Estimated meat weight for pig, according to Minimum Number of Individual by age range

Sheep(Goat) Estimated meat weight by age range	Abbeville Early/Middle Bronze Age	Etaples Early/Middle Bronze Age	Villiers/S. Late Bronze Age	Villiers/S. Late La Tène (LTC2)	Braine Late La Tène (LTD1b)	Souppes/L. Late La Tène (LTD2)
1/6 mois			40	4	0	8
6/12 mois	16	8	80	48	48	128
12/18 mois			0	80	30	20
18/24 mois			99	77	165	77
> 24 mois		12	168	216	120	528
Total	16	20	387	425	363	761

Estimated meat weight for Sheep, according to Minimum Number of Individual by age range

BIOGRAPHIES

Ginette Auxiette has a Ph.D. in archaeology from the Université de Paris I, Panthéon-Sorbonne and is currently a researcher specialized in archaeozoology with the French National Institute for Preventive Archaeological Research (INRAP). Her research carried out with the team Trajectoires « De la sédentarisation à l'Etat» (UMR 8215 of CNRS) deals with different aspects of material culture in the Metal Ages in the northern half of France, where she has participated in archaeological excavations for more than two decades. Her work is based on multiscalar approaches to societies from the Bronze Age to the end of the Iron Age, economic and societal approaches to animal resources within settlements, evolutionary, qualitative and quantitative approaches to rites and cults in the domestic context, evolutionary approaches and specificities of rites in the funerary context.

Lamys Hachem is currently a researcher in zooarchaeology and pre-history at the French National Institute for Preventive Archaeological Research (INRAP). She graduated with a Ph.D. in archaeology from the Université de Paris I, Panthéon-Sorbonne where she also has been teaching for several years. Within the team Trajectoires « De la sédentarisation à l'Etat» (UMR 8215 of CNRS) she has focused her research on the societies of the Early, Middle and Final Neolithic period, particulary in the northern half of France where she has participated in archaeological excavations for more than two decades. She has performed research leading to understanding the customs and evolution of animal herding and hunting, the chronological evolution of feeding, the internal analysis of habitats such as houses, villages or cultural layers, and the characterization of animal deposits in the funerary context and enclosures.